Applied Mathematical Sciences
Volume 163

Applied Mathematical Sciences

1. *John:* Partial Differential Equations, 4th ed.
2. *Sirovich:* Techniques of Asymptotic Analysis.
3. *Hale:* Theory of Functional Differential Equations, 2nd ed.
4. *Percus:* Combinatorial Methods.
5. *von Mises/Friedrichs:* Fluid Dynamics.
6. *Freiberger/Grenander:* A Short Course in Computational Probability and Statistics.
7. *Pipkin:* Lectures on Viscoelasticity Theory.
8. *Giacaglia:* Perturbation Methods in Non-linear Systems.
9. *Friedrichs:* Spectral Theory of Operators in Hilbert Space.
10. *Stroud:* Numerical Quadrature and Solution of Ordinary Differential Equations.
11. *Wolovich:* Linear Multivariable Systems.
12. *Berkovitz:* Optimal Control Theory.
13. *Bluman/Cole:* Similarity Methods for Differential Equations.
14. *Yoshizawa:* Stability Theory and the Existence of Periodic Solution and Almost Periodic Solutions.
15. *Braun:* Differential Equations and Their Applications, 3rd ed.
16. *Lefschetz:* Applications of Algebraic Topology.
17. *Collatz/Wetterling:* Optimization Problems 4th ed.
18. *Grenander:* Pattern Synthesis: Lectures in Pattern Theory, Vol. I.
19. *Marsden/McCracken:* Hopf Bifurcation and Its Applications.
20. *Driver:* Ordinary and Delay Differential Equations.
21. *Courant/Friedrichs:* Supersonic Flow and Shock Waves.
22. *Rouche/Habets/Laloy:* Stability Theory by Liapunov's Direct Method.
23. *Lamperti:* Stochastic Processes: A Survey of the Mathematical Theory.
24. *Grenander:* Pattern Analysis: Lectures in Pattern Theory, Vol. II.
25. *Davies:* Integral Transforms and Their Applications, 2nd ed.
26. *Kushner/Clark:* Stochastic Approximation Methods for Constrained and Unconstrained Systems.
27. *de Boor:* A Practical Guide to Splines: Revised Edition.
28. *Keilson:* Markov Chain Models–Rarity and Exponentiality.
29. *de Veubeke:* A Course in Elasticity.
30. *Sniatycki:* Geometric Quantization and Quantum Mechanics.
31. *Reid:* Sturmian Theory for Ordinary Differential Equations.
32. *Meis/Markowitz:* Numerical Solution of Partial Differential Equations.
33. *Grenander:* Regular Structures: Lectures in Pattern Theory, Vol. III.
34. *Kevorkian/Cole:* Perturbation Methods in Applied Mathematics.
35. *Carr:* Applications of Centre Manifold Theory
36. *Bengtsson/Ghil/Källén:* Dynamic Meteorology: Data Assimilation Methods.
37. *Saperstone:* Semidynamical Systems in Infinite Dimensional Spaces.
38. *Lichtenberg/Lieberman:* Regular and Chaotic Dynamics, 2nd ed.
39. *Piccini/Stampacchia/Vidossich:* Ordinary Differential Equations in R^n.
40. *Naylor/Sell:* Linear Operator Theory in Engineering and Science.
41. *Sparrow:* The Lorenz Equations: Bifurcations, Chaos, and Strange Attractors.
42. *Guckenheimer/Holmes:* Nonlinear Oscillations, Dynamical Systems, and Bifurcations of Vector Fields.
43. *Ockendon/Taylor:* Inviscid Fluid Flows.
44. *Pazy:* Semigroups of Linear Operators and Applications to Partial Differential Equations.
45. *Glashoff/Gustafson:* Linear Operations and Approximation: An Introduction to the Theoretical Analysis and Numerical Treatment of Semi-Infinite Programs.
46. *Wilcox:* Scattering Theory for Diffraction Gratings.
47. *Hale:* Dynamics in Infinite Dimensions/Magalhães/Oliva, 2nd ed.
48. *Murray:* Asymptotic Analysis.
49. *Ladyzhenskaya:* The Boundary-Value Problems of Mathematical Physics.
50. *Wilcox:* Sound Propagation in Stratified Fluids.
51. *Golubitsky/Schaeffer:* Bifurcation and Groups in Bifurcation Theory, Vol. I.
52. *Chipot:* Variational Inequalities and Flow in Porous Media.
53. *Majda:* Compressible Fluid Flow and Systems of Conservation Laws in Several Space Variables.
54. *Wasow:* Linear Turning Point Theory.
55. *Yosida:* Operational Calculus: A Theory of Hyperfunctions.
56. *Chang/Howes:* Nonlinear Singular Perturbation Phenomena: Theory and Applications.
57. *Reinhardt:* Analysis of Approximation Methods for Differential and Integral Equations.
58. *Dwoyer/Hussaini/Voigt (eds):* Theoretical Approaches to Turbulence.
59. *Sanders/Verhulst:* Averaging Methods in Nonlinear Dynamical Systems.
60. *Ghil/Childress:* Topics in Geophysical Dynamics: Atmospheric Dynamics, Dynamo Theory and Climate Dynamics.

(continued after index)

M. di Bernardo C.J. Budd A.R. Champneys
P. Kowalczyk

Piecewise-smooth Dynamical Systems

Theory and Applications

Mario di Bernardo, Laurea; PhD
Department of Engineering Mathematics
University of Bristol, Bristol BS8 1TR, UK
and
Department of Systems and Computer Science
University of Naples Federico II, Naples, Italy

Christopher J. Budd, MA; DPhil
Department of Mathematical Sciences
University of Bath, Bath BA2 7AY, UK

Alan R. Champneys, BSc; DPhil
Department of Engineering Mathematics
University of Bristol, Bristol BS8 1TR, UK

Piotr Kowalczyk, MSc; PhD
School of Engineering
Computer Science and Mathematics,
University of Exeter, Exeter EX4 4QF, UK
and
Department of Engineering Mathematics
University of Bristol, Bristol BS8 1TR, UK

The authors gratefully acknowledge the help of Dr Petri Piiroinen of the National University of Ireland at Galway, who used his novel numerical routines for piecewise-smooth dynamical systems to create the cover figure.

British Library Cataloguing in Publication Data
A catalogue record for this book is available from the British Library

ISBN: 978-1-84628-039-9 e-ISBN: 978-1-84628-708-4

Printed on acid-free paper

Library of Congress Control Number: 2007937295

Mathematics Subject Classification (2000): 34A36; 70K50; 37D45; 93C65; 70E55

9 8 7 6 5 4 3 2 1

Springer Science + Business Media
springer.com

Preface

Many dynamical systems that occur naturally in the description of physical processes are piecewise-smooth. That is, their motion is characterized by periods of smooth evolutions interrupted by instantaneous events. Traditional analysis of dynamical systems has restricted its attention to smooth problems, thus preventing the investigation of non-smooth processes such as impact, switching, sliding and other discrete state transitions. These phenomena arise, for example, in any application involving friction, collision, intermittently constrained systems or processes with switching components.

Literature that draws attention to piecewise-smooth systems includes the comprehensive work of Brogliato [38, 39], the detailed analysis of Kunze [165], the books on bifurcations in discontinuous systems [193, 177] and various related edited volumes [268, 35]. These books contain many examples largely drawn from mechanics and control. Also there is a significant literature in the control and electronics communities; see for example the book [193], which has many beautiful examples of chaotic dynamics induced by non-smooth phenomena. Earlier studies of non-smooth dynamics appeared in the Eastern European literature; for instance the pioneering work of Andronov *et al.* on non-smooth equilibrium bifurcations [5], Feigin [98, 80] on C-bifurcations, Peterka [216] and Babitskii [19] on impact oscillators, and Filippov [100] on sliding motion. Delving into this and other literature, one finds that piecewise-smooth systems can feature rich and complex dynamics.

In one sense, jumps and switches in a system's state represent the grossest form of nonlinearity. On the other hand, many examples appear benign at first glance since they are composed of pieces of purely linear systems, which are solvable closed form. However, this solvability is in general an illusion since one does not know *a priori* the times at which the switches occur. Nevertheless, the analysis of such dynamics is not intractable, and indeed, many tools of traditional bifurcation theory may be applied. However, it has become increasingly clear that there are distinctive phenomena unique to discontinuous systems, which can be analyzed mathematically but fall outside the usual methodology for smooth dynamical systems.

Indeed, for smooth systems, governed by ordinary differential equations, there is now a well established qualitative, topological theory of dynamical systems that was pioneered by Poincaré, Andronov and Kolmogorov among others. This theory has led to a mature understanding of bifurcations and routes to chaos—see, for example the books by Kuznetsov [168], Wiggins [273], Arrowsmith & Place [9], Guckenheimer & Holmes [124] and Seydel [232]. The key step in the analysis is to use topological equivalence, Poincaré maps, center manifolds and normal forms to reduce all possible transitions under parameter variation to a number of previously analyzed cases. These ideas have also informed modern techniques for the numerical analysis of dynamical systems and, via time-series analysis, techniques for the analysis of experimental data from nonlinear systems. The bifurcation theory methodology has shown remarkable success in describing dynamics observed in many areas of application including, via center-manifold and other reduction techniques, spatially extended systems. However, most of these successes are predicated on the dynamical system being smooth.

The purpose of this book then is to introduce a similar qualitative theory for non-smooth systems. In particular we shall propose general techniques for analyzing the bifurcations that are unique to non-smooth dynamical systems, so-called *discontinuity-induced* bifurcations (DIBs for short). This we propose as a general term for all transitions in dynamics specifically brought about through interaction of invariant sets of the system ('attractors') with a boundary in phase space across which the system has some kind of discontinuity. First and foremost, we shall give a consistent classification of all known DIBs for piecewise-smooth continuous-time dynamical systems (flows), including such diverse phenomena as sliding, chattering, grazing and corner collision. We will then describe a unified analytical framework for reducing the analysis of each such bifurcation involving a periodic orbit to that of an appropriately defined Poincaré map. This process is based on the construction of so-called *discontinuity mappings* [198, 64], which are analytical corrections made to account for crossing or tangency with discontinuity boundaries. We introduce the notion of the *degree of smoothness* depending on whether the state, the vector field or one of its derivatives has a jump across a discontinuity boundary. We show how standard examples such as impact oscillators, friction systems and relay controllers can be put into this framework, and show how to construct discontinuity mappings for tangency of each kind of system with a discontinuity boundary.

The analysis is completed by a classification of the dynamics of the Poincaré maps so-obtained. Thus we provide a link between the theory of bifurcations in piecewise-smooth flows and that associated with discontinuity crossings of fixed points of piecewise-smooth maps—so-called *border-collision* bifurcations [207, 21], which are just particular examples of a DIB. The presentation is structured in such a manner to make it possible for a reader to follow a series of steps to take a non-smooth dynamical system arising in an application from an outline description to a consistent mathematical characterization.

Throughout, the account will be motivated and illustrated by copious examples drawn from several areas of applied science, medicine and engineering; from mechanical impact and friction oscillators, through power electronic and control systems with switches, to neuronal and cardiac and models. In each case, the theory is compared with the results of a numerical analysis or, in some cases, with data from laboratory experiments. More general issues concerning the numerical and experimental investigation of piecewise-smooth systems are also discussed.

The manner of discourse will rely heavily on geometric intuition through the use of sketch figures. Nevertheless, care will be taken to single out as theorems those results that do have a rigorous proof, and where the proof is not presented, a reference will be given to the appropriate literature.

The level of mathematics assumed will be kept to a minimum: nothing more advanced than multivariable calculus, differential equations and linear algebra traditionally taught at undergraduate level on mathematics, engineering or applied science degree programs. A familiarity with the basic concepts of nonlinear dynamics would also be useful. Thus, although the book is aimed primarily at postgraduates and researchers in any discipline that impinges on nonlinear science, it should also be accessible to many final-year undergraduates.

We now give a brief outline each chapters.

Chapter 1. Introduction. This serves as a non-technical motivation for the rest of the book. It can in fact be read in isolation and is intended as a primer for the non-specialist. After a brief motivation of why piecewise-smooth systems are worthy of study, the main thrust of the chapter is to immerse the reader in the kind of dynamics that are unique to piecewise-smooth systems via a series of case studies. The first case study is the single-degree-of-freedom impact oscillator. The notion of grazing bifurcation is introduced along with the dynamical complexity that can result from this seemingly innocuous event. Agreement is shown among theory, numerics and physical experiment. After brief consideration of bi-linear oscillators, we then consider two mathematically related systems that can exhibit recurrent sliding motion: a relay controller and a stick-slip friction system. The next case study concerns a well-used electronic circuit with a switch, the so-called DC–DC converter. Finally, we consider one-dimensional maps that arise through the study of these flows, including a simple model of heart attack prediction. Here we introduce the ubiquitous *period-adding cascade* that is unique to non-smooth systems.

Chapter 2. Qualitative theory of non-smooth dynamical systems. The aim here is to set out concisely the mathematical and notational framework of the book. We present a brief introduction to the qualitative theory of dynamical systems for smooth systems, including a brief review of standard bifurcations, stressing which of these also makes sense for piecewise-smooth systems. The formalism of piecewise-smooth systems is introduced, although no specific attempt is made to develop an existence and uniqueness theory. However, a brief introduction is given to the extensive literature on other more rigorous mathematical formulations for non-smooth dynamics, such as differential inclusions, complementarity systems and hybrid dynamical systems. A working definition of discontinuity-induced bifurcation is given from a topological point of view, which motivates a brief list of the kinds of discontinuity-induced bifurcations that are likely to occur as a single parameter is varied. The notion of discontinuity mapping is introduced, and such a map is carefully derived in the case of transverse crossing of a discontinuity boundary. The chapter ends with a discussion on numerical techniques, both direct and indirect, that will be used throughout the rest of the book for investigating the dynamics of example systems and calculating the appropriate bifurcation diagrams.

Chapter 3. Border collision in piecewise-smooth continuous maps. This chapter contains results on the dynamics of discrete-time continuous maps that are locally composed of two linear pieces. First border-collision bifurcations are analyzed whereby a simple fixed point passes through the boundary between the two map pieces. General criteria are established for the existence and stability of simple period-one and -two fixed points created or destroyed in such transitions, by using information only on the characteristic polynomial of the matrix representation of the two sections of the map. Analogs of simple fold and period-doubling bifurcations are shown to occur, albeit where the bifurcating branch has a non-smooth rather than quadratic character. The cases of one and two dimensions are considered in detail. Here, more precise information can be established such as conditions for the existence of period-adding, and cascades of such as another parameter (representing the slope of one of the linear pieces) is varied. Finally, we consider maps that are noninvertible in one part of their domain. For such maps, conditions can be found for the creation of robust chaos, which has no embedded periodic windows.

Chapter 4. Bifurcations in general piecewise-smooth maps. Here the analysis of the previous chapter is generalized to deal with maps that crop up as normal forms of the grazing and other non-smooth bifurcations analyzed in subsequent chapters, and which change their form across a discontinuity boundary. First, we treat maps that are piecewise-linear but discontinuous. We then proceed to study continuous maps that are a combination of a linear and a square-root map, and finally maps that combine a linear map with an $\mathcal{O}(3/2)$ or a quadratic map. In each case we study the existence of both periodic and chaotic behavior and look at the transitions between these states. Of particular interest will be the identification of *period-adding* behavior in which, under the variation of a parameter, the period of a periodic state increases in arithmetic progression, accumulating onto a chaotic solution.

Chapter 5. Boundary equilibrium bifurcations in flows. This chapter collects and reviews various results on the global consequences of an equilibrium point encountering the boundary between two smooth regions of phase space in a piecewise-smooth flow. Cases are treated where the vector field is continuous across the boundary and where it is not (and indeed where the boundary may itself be attracting—the Filippov case). In two dimensions, a more or less complete theory is possible since the most complex attractor is a limit cycle, which may be born in a non-smooth analog of a Hopf bifurcation. In the Filippov case, so-called pseudo-equilibria that lie inside the sliding region can be created or destroyed on the boundary, as they can for impacting systems.

Chapter 6. Limit cycle bifurcations in impacting systems. We return to the one-degree-of-freedom impact oscillator from the Introduction, stressing a more geometrical approach to understanding the broad features of its dynamics. Within this approach, grazing events are thought of as leading to singularities in the phase space of certain Poincaré maps. These singularities are shown to organize the shape of strange attractors and also the basins of attraction of competing attractors. An attempt is made to generalize such geometrical considerations to general n-dimensional hybrid systems of a certain class. The narrative then switches to dealing with grazing bifurcations of limit cycles within this general class. The discontinuity mapping idea is used to derive normal form maps that have a square-root singularity. The technique is shown to work on several example systems. The chapter also includes a treatment of chattering (a countably infinite sequence of impacts in a finite time) and multiple impacts, including a simple example of a triple collision.

Chapter 7. Limit cycle bifurcations in piecewise-smooth flows. This chapter treats the general case of non-Filippov flows and two specific kinds of bifurcation event where a periodic orbit grazes with a discontinuity surface. In the first kind the periodic orbit becomes tangent to a smooth surface. In the second kind the periodic orbit passes through a non-smooth junction between two surfaces. For both kinds, discontinuity mappings are calculated and normal form mappings derived that can be analyzed using the techniques of the earlier chapters. Examples of the theory are given including general bilinear oscillators, a certain stick-slip system and the DC–DC convertor introduced in Chapter 1.

Chapter 8. Sliding bifurcations in Filippov systems. The technique of discontinuity mappings is now applied to the situations where flows can slide along the attracting portion of a discontinuity set in the case where the vector fields are discontinuous. Four non-generic ways that periodic orbits can undergo sliding are identified that lead to four bifurcation events. Each event involves the fundamental orbit involved in the bifurcation gaining or losing a sliding portion. The mappings derived at these events typically have the property of being non-invertible due to the loss of initial condition information inherent in sliding. So a new version of the theory of Chapters 3 and 4 has to be derived, dealing with this added complication. Examples of relay controllers and friction oscillators introduced in Chapter 1 are given further treatment in the light of this analysis.

Chapter 9. Further applications and extensions. This chapter contains a series of additional case study applications that serve to illustrate further bifurcations and dynamical features, a detailed analysis of which would be beyond the scope of this book. Each application arises from trying to understand or model some experimental or in service engineered or naturally occurring system. The further issues covered include the notion of parameter fitting to experimental data, grazing bifurcations of invariant tori and examples of codimension-two bifurcations.

Acknowledgments

There are numerous people we should like to thank for their part in the writing of this book.

Arne Nordmark has been a constant inspiration. His unique insight into the behavior of piecewise-smooth systems and how to analyze them has been a revelation to us, and we only hope that we have done justice to his ideas. ARC and MDiB would also like to thank him and Harry Dankowicz for a pivotal conversation at Snowbird in 1999 in which they first came to realize that grazing in piecewise-linear flows does not lead to piecewise-linear maps. An earlier motivation for this book was the research literature on impact oscillators. CJB would like to thank Steve Bishop, David Chillingworth, Markus Kunze, Tassilo Küpper, Tom Mullin, Steve Shaw and Michael Thompson for sharing their ideas with him, and also give special thanks to George Whiston who introduced him to impacting systems in general and the significance of grazing in particular. We all miss Karl Popp whose untimely death robbed us of a marvelous colleague and the dynamical systems community of a truly gifted experimentalist. We are pleased to be able to use this book to present the results of some of his very careful experiments. John Hogan has also been instrumental, not least by inculcating an atmosphere of interdisciplinary collaboration at Bristol, but also for encouraging a research focus on piecewise-smooth dynamics. His frank criticisms of a very much earlier draft of Chapters 3 and 4 proved especially constructive at a key stage of writing this book. Others who have provided comments on earlier drafts include Bernard Brogliato, Giles Hunt, Xiaopeng Zhao, Michael Shantz, Victor Avrutin, Yuri Kuznetsov and the anonymous referees appointed by Springer-Verlag. All the comments have been helpful. Where errors remain, the authors take full responsibility. We should also like to thank Phil Holmes who encouraged us to write a coherent single text rather than bolting together several published papers. Also, his book with John Guckenheimer started all of us off on our interest in dynamical systems, and the bouncing ball problem described there remains a seminal example of a piecewise-smooth dynamical system.

Petri Piiroinen deserves a special mention. He is in many ways the implicit fifth author of this book. Not only has he contributed many of the numerical algorithms, and performed computations that have led to presented figures, but many of his scientific ideas have made it into the book. We should also like to thank Gerard Olivar, Yuri Kuznetsov and Arne Nordmark for allowing us to use some of their unpublished material in Chapter 5. Thanks too go to other PhD students, postdocs and collaborators whose results have formed an integral part of the book. We mention in particular Felix Dux, Harbir Lamba, Gordon Lee, Richard Eyres, Chris Halse, Sally-Anne Myles, Stephen Pring, Gustavo Osorio, Ricardo Alzate, Alessandro Colombo, Nick Lieven and Laurie Virgin. PK would also especially like to thank Jan Sieber who has served as an unofficial checker of mathematical rigor. Others with whom we have had insightful conversations have included Remco Liene, Maurice Heemels, Nathan Van der Wouw, Henk Nijmejer, Hans Schumacher, Michele Shatzman, Jim Yorke, Christoph Glocker, Enric Fossas, Marian Wiercigroch, Mark Feigin and David Wagg.

The study of piecewise-smooth dynamical systems as a subject in its own right is relatively new. We have tried to use a consistent terminology and notation, which we accept may not be to everyone's taste. We have been helped in this task by the unofficial Non-smooth Standardization Agency that has provided a sounding board for agreeing on certain basic nomenclature. Its members include Harry Dankowicz, John Hogan, Yuri Kuznetsov, Arne Nordmark, Petri Piiroinen and Gerard Olivar.

We say a big thank you to Steph Harding, Helen Desmond and Karen Borthwick, our Editors at Springer, for the professional manner in which they have handled our ever-too-optimistic estimates of when the book would be finished. The fact that the period of writing has outlasted several editors, says it all! Thanks are also due to the production team at Springer for their care and attention to detail. Also to Philip Naylor who created a file exchange system that enabled us to work collaboratively even when in far flung parts. The UK Engineering and Physical Sciences Research Council (EPSRC), the European Union under the FP5 grant SICONOS and the British Council also deserve acknowledgment for funding much of the travel and basic research necessary to put this book together.

Finally, we should like to thank those close to us—family, friends and colleagues—for their love, support and understanding as we have attempted to write this book while also having busy day jobs. Those who have kept us sane during this time have included Sue, Bryony and Jeremy Budd; Renata, Diego and Daria di Bernardo; Valerio Rinaldi; Sharon, Max, Dominic and Emma Champneys; Janina, Bolesław and Paweł Kowalczyk; Giles Hunt and members of the Bath Centre for Nonlinear Mechanics; all at the Bristol Centre for Applied Nonlinear Mathematics and at the SINCRO research group of the University of Naples Federico II.

Bath, Bristol, Exeter, Naples August 2007

Contents

Glossary

$:=$	Is equal to by definition.
α	The free parameter in Filipov's method, $\alpha \in [0,1]$; see, (2.30).
β	Utkin's equivalent control for a Filippov system, $\beta \in [-1,1] = 2\alpha - 1$; see (2.29).
$\beta(x,y)$	Vector multiplied by $y = \sqrt{-H_{\min}}$ in the leading-order expression for the discontinuity mapping at grazing.
δ, δ_i	Unknown small time(s) in construction of discontinuity mapping.
ε	Small perturbation in state variable x.
λ, λ_i	Eigenvalue of a matrix, eigenvalue of an equilibrium point of a flow, multiplier of a fixed point of a map. Unknown parameter in Complementarity system. Unknown parameter in sticking flow, see (2.43).
μ	Parameter; $\mu \in \mathbb{R}^p$ or $\mu \in \mathbb{R}$, so that we write $F(x,\mu)$ etc. where the dependence on a parameter is imporant.
ϕ	Abstract dynamical system $\phi^t : X \to X$, $x_t = \phi^t(x_0)$.
Ψ	Notation for the flow operator of a hybrid system.
Φ, Φ_i	Flow operator corresponding to ODE $\dot{x} = F_i(x)$.
Σ_{ij}, Σ	Discontinuity manifold, sometimes called impacting set, switching set, or discontinity boundary.
$\widehat{\Sigma}_{ij}$	Sliding region of a switching set Σ_{ij} in a Filippov system.
$\partial\widehat{\Sigma}_{ij}^+$, $\partial\widehat{\Sigma}^+$	Boundary of sliding region corresponding to $\beta = +1$.
$\partial\widehat{\Sigma}_{ij}^-$, $\partial\widehat{\Sigma}^-$	Boundary of sliding region corresponding to $\beta = -1$.
a	Symbol to represent iterate of two-zone nonsmooth map in region S_1. Acceleration in a single zone impacting system $a(x) = (H_x F)_x F = \mathcal{L}_F^2 H(x)$.
A	Symbol representing iterate in S_1 of part of a stable orbit of a two-zone nonsmooth map.

b	Symbol to represent iterate of two-zone nonsmooth map in region S_2.
B	Symbol representing iterate in S_2 of part of a stable orbit of a two-zone nonsmooth map.
BEB	Boundary equilibrium bifurcation; see Chapter 5.
C^r	The space of continuously r times differentiable functions.
C^T	Row vector multiplying linearisation with respect to x of smooth function $H(x) = C^T x + D\mu$ representing discontinuity surface Σ.
D	Linearisation of discontinuity surface H with respect to parameter μ.
$\mathcal{D}$	Domain of definition of a piecewise-linear system, $x \in \mathcal{D} \subset \mathbb{R}^n$.
DAE	Differential algebraic equation.
DIB	Discontinuity-induced bifurcation.
DM	Discontinuity mapping (for transversal or non-transversal crossing).
DoF	Degree of freedom.
E	Rank-one matrix representing difference between linearisations $N_1 - N_2$ in piecewise-linear map $x \to N_1 x$, if $C^T x > 0$, $x \to N_2 x$ if $C^T x < 0$. Vector E multiplying the scalar $y = \sqrt{-H(x)}$ for $x < 0$ in square root map written in simpest form.
f	General expression for a (smooth or nonsmooth) vector field $\dot{x} = f(x)$, or map $x \mapsto f(x)$.
F_i	Smooth vector field or map applying in region S_i; $\dot{x} = F(x_i)$.
F_{ij}, F_s	Sliding vector field for $x \in \widehat{\Sigma}_{ij}$.
G	Grazing region of a discontinuity boundary Σ. If $\Sigma = \{x \in \mathcal{D} \| H(x) = 0\}$, then $G := \{x \in \Sigma \| H_x F(x) = 0\}$.
$\mathcal{G}$	Grazing manifold. Image of G under the flow.
G_Π	Intersection of grazing manifold with Poincaré section Π.
H_{ij}, H	Smooth function defining discontiny manifold $\Sigma_{ij} := \{x \in \mathcal{D} \subset \mathbb{R}^n \| H(x) = 0\}$.
$\mathcal{L}_f h(x)$	Lie derivative $\mathcal{L}_f h(x) = \frac{\partial h}{\partial x} f(x)$.
n	The dimension of phase space, $x \in \mathbb{R}^p$.
N, N_i	Linearisation with respect to state of (piecewise) linear map. $x \mapsto N_i x + M_i \mu$ if $x \in S_i$.
M, M_i	Linearisation with respect to parameter of (piecewise) linear map.
O	"Big O". $O(\varepsilon^n)$ means of the same size as ε^n as $\varepsilon \to 0$. That is, $O(\varepsilon^n)/\varepsilon^n$ tends to a finite limit as $\varepsilon \to 0$. Also $O(n)$ means $O(\varepsilon^n)$ for an implied small quantity ε.
o	"Little o". $o(\varepsilon^n)$ means asymptotically smaller than ε^n. That is, $o(\varepsilon^n)/\varepsilon^n \to 0$ as $\varepsilon \to 0$.

ODE	Ordinary differential equations.
P	General notation for a Poincaré map, $x \to P(x)$.
p	The dimension of parameter space, $\mu \in \mathbb{R}^p$.
PDM	Poincaré-section discontinuity mapping; see Definition 2.34.
PWS	Piecewise smooth.
Q	Discontinuity mapping $x \to Q(x)$.
R or R_{ij}	Reset (or restitution) map in a hybrid system, sometimes called the impact map $x \to R(x)$ or $x^+ = R(x^-)$.
S_i	Open region of phase space in which dynamics is governed by $\dot{x} = F_i(x)$ or $x \to F_i(x)$.
$\overline{S}_i$	The closed region S_i plus its boundary.
S^+	Physical region of phase space for which $H(x) > 0$ in an impacting hybrid system with a single impact boundary $\Sigma = \{x \| H(x) > 0\}$.
S^-	Unphysical region for which $H(x) < 0$ in an impacting hybrid system with a single impact boundary.
s	Phase variable $s = \omega t/2\pi$ where ω is the angular frequency of a forcing function $w(t)$.
t	Time.
u	Co-ordinate of a single-degree-freedom system.
v	Velocity. In an impacting hybrid system we have $v(x) = H_x F(x) = \mathcal{L}_F H(x)$.
W	Smooth vector representing simplest form of impact law $x + W(x)H_x F$.
w	Forcing function $w(t)$ in a single-degree-of-freedom system.
x	State variable $x \in \mathbb{R}^n$.
x_*	Grazing point of a trajectory. Sometimes also used for equilibrium point of flow or fixed point of map.
x^*	Equilibrium point of flow, or fixed point of map.
y	Scalar variable $\sqrt{-H(x)}$ for $H(x) < 0$.
$\mathcal{Z}$	Sticking region of impacting system.
ZDM	Zero-time discontinuity mapping; see Definition 2.35.

1

Introduction

1.1 Why piecewise smooth?

Dynamical systems theory has proved a powerful tool to analyze and understand the behavior of a diverse range of problems. There is now a well-developed qualitative, or geometric, approach to dynamical systems that typically relies on the system evolution being defined by a smooth function of its arguments. This approach has proved extremely effective in helping to understand the behavior of many important physical phenomena such as fluid flows, elastic deformation, nonlinear optical and biological systems. However, this theory excludes many significant systems that arise in practice. These are dynamical system containing terms that are *non-smooth* functions of their arguments. Problems of this nature arise everywhere! Important examples are electrical circuits that have switches, mechanical devices in which components impact with each other (such as gear assemblies) or have freeplay, problems with friction, sliding or squealing, many control systems (including their implementation via adaptive numerical methods) and models in the social and financial sciences where continuous change can trigger discrete actions. Such problems are all characterized by functions that are piecewise-smooth but are *event driven* in the sense that smoothness is lost at instantaneous events, for example, upon application of a switch. They have fascinating dynamics with significant practical application and a rich underlying mathematical structure. It a serious omission that their behavior is not easily described in terms of the modern qualitative theory of dynamical systems.

A commonly expressed reason for this omission is that there is strictly speaking no such thing as a piecewise-smooth dynamical system and that in reality all physical systems are smooth (at least at all length scales greater than the molecular). However, this statement is misleading. The timescales over which transitions such as an impact or a control-law switch occur in an engineering system can be remarkably small compared with that of the overall dynamics, and thus, the correct global model is certainly discontinuous on a macroscopic timescale. Furthermore, relatively simple phenomena when considered from the point of view of piecewise-smooth systems often turn out to be natural limits of far more complex scenarios observed in smoother systems. For example, it is quite natural for a piecewise-smooth system to undergo a sudden jump from strongly stable periodic motion to full scale chaotic motion under variation of a parameter. In a smooth system, such a scenario would typically require an infinite sequence of bifurcations to occur, such as the famous Feigenbaum cascade of period-doubling bifurcations, leading to chaos.

A second reason for the exclusion of piecewise-smooth systems from the established literature is that they challenge many of our assumptions about dynamics. For example, how can we define concepts such as structural stability, bifurcation and qualitative measures of chaos in such systems? By making careful assumptions about the problems we investigate, which are not inconsistent with the physical problems leading to them, it will become apparent that many of the concepts once thought to be the domain of smooth systems only, naturally extend to piecewise-smooth ones as well. But, and this is the main thrust of this book, there are also dynamical phenomena that are unique to piecewise-smooth systems that are, nevertheless, straightforward to analyze.

The purpose of this introductory chapter is to be a self-contained and non-technical guide to piecewise-smooth dynamical systems, which will outline the more detailed treatment given in the later chapters; but can be read independently from them. We will establish the basic foundations for discussion of non-smooth dynamics in an informal, non-technical and applications-oriented setting, through the description of case study examples arising from physical models. We will also show how bifurcations in piecewise-smooth *flows* (systems of ordinary differential equations) naturally generate piecewise-smooth *mappings*, or *maps* (discrete-time iteration processes), which is a connection that lies at the heart of this book. The chapter is essentially designed to be read like an extended essay. *Italicized terms* are used to introduce mathematical concepts that will be defined more accurately later on in the book. Also, the application-oriented nature of the essay is aimed at answering the question of why piecewise-smooth systems are worth studying.

As a first motivating example of a piecewise-smooth system, consider the operation of a domestic central heating system that is trying to achieve a desired temperature θ. If this temperature is exceeded, a thermostat causes a switch to turn off the power supply to a boiler. The system then evolves smoothly with the heating *off*, until the temperature falls below θ. At this point the system dynamics changes, as the boiler is turned *on* and a different set of evolution rules apply. Thus, if we view the switching process as taking an infinitesimally short time compared with the heating and cooling phases, we can view the dynamics of the temperature $T(t)$ as being that of a continuous piecewise-smooth flow. Two different smooth flow regimes describe the off and on states, with switching occurring when the dynamics crosses the boundary $T(t) = \theta$ between them.

Let us suspend belief for a moment and imagine an instantly responsive heating system. The natural dynamics would then be a state known as *sliding* in which $T(t)$ is permanently set to the threshold value θ, with the thermostat poised between the on and off positions. As the temperature rises above threshold, the boiler is switched off, which instantaneously causes the temperature to fall below threshold. Thus, the boiler is reignited, causing the temperature to rise above threshold, and so on. We shall see shortly that sliding corresponds to a natural state of so-called *relay* controllers and also to the stick phase of systems with dry friction that can exhibit stick-slip motion.

Returning to the more realistic situation where changes in temperature lag behind the turning on or off of the boiler, we can consider the dynamics of this example as being driven by *events*. The events are the times t at which $T(t) = \theta$ and switching occurs. The system evolves smoothly between events such that we can easily define a discrete-time *event map* that expresses the system state at one switching as a function of the state at the previous switch. This map, which may be smooth or non-smooth, effectively has a lower-dimensional state space since we know that the temperature is at threshold. Suppose now that the heating system has a timer device that switches on and off the heating at fixed times each day. In this case we could consider sampling the temperature at 24-hour intervals, producing a *stroboscopic* map that expresses the system state at a fixed time each day as a function of the state at the same time the previous day. This map is unlikely to be smooth, because the dynamics of a system that starts with a temperature above θ is likely to be different from one that starts below.

This simple example demonstrates that any discussion of piecewise-smooth systems should naturally include both flows and maps. A third naturally arising kind of piecewise-smooth system is a combination of a flow and a map, and we shall call these *hybrid* systems. Such systems arise when the effect of the flow reaching the switching threshold is to cause an instantaneous jump in the flow (which in effect becomes discontinuous). In the heating system, this might occur if the result of the temperature dropping to θ is to instantaneously turn on an electric fire that heats the house very much more rapidly than the boiler, so that (on a 24-hour timescale) we see an effectively instantaneous temperature rise. We begin our more detailed discussion of case studies with a class of hybrid systems that have played a key role in the historical development of the theory of piecewise-smooth systems.

1.2 Impact oscillators

Consider the motion of an elastic ball bouncing vertically on a rigid surface such as a table. In unconstrained motion the ball falls smoothly under gravity between impacts and has an 'instantaneous' reversal of its velocity at each impact. Suppose that a simple Newtonian restitution law applies such that reversed velocity is a coefficient $0 \leq r \leq 1$ times the incoming velocity. Typical

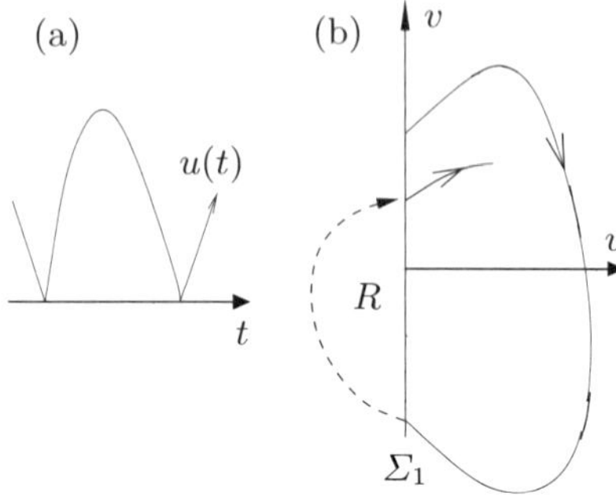

Fig. 1.1. Sketch figure of both (a) the position $u(t)$ of a bouncing ball against time and in (b), (u, v)-*phase space*, where $v(t)$ is the velocity of the ball. Here R is the map that takes v to $-rv$.

motion of the ball is represented in Fig. 1.1. Note that if $r < 1$ then a state where the ball is at rest (*stuck*) on the table can be reached by simply releasing the ball. After an infinite number of impacts (an accumulation of a *chattering sequence*), but a *finite* time, the ball comes to rest. If we were to allow the possibility of an oscillating rigid table (like a tennis player bouncing the ball on his racket between rallies), then the dynamics can be incredibly rich [124, Ch. 2], as we are about to see in a related model.

A bouncing ball is just a simple example of what are termed *impact oscillators*, which are low-degree-of-freedom mechanical systems with hard constraints that feature *impacts* (like the bounce of the ball on the table). impact oscillators have played an important role in the historical development of piecewise-smooth systems. Their dynamics has been studied in the Czech and Russian literature since the 1950s (see, e.g., [98, 19] and references therein, especially [216, 217]), much or which was essentially rediscovered in the Western literature in the 1980s and 1990s [237, 236, 251, 264, 197, 43, 102, 18, 67].

Impacting behavior is found in a large number of mechanical systems ranging from gear assemblies [146, 149, 229, 249], impact print hammers [128, 256], walking robots [138], boiler tube dynamics [212, 122], metal cutters [267], car suspensions [29], vibration absorbers [234], [20], percussive drilling and moling [269, 163] and many-body particle dynamics [228]; see also Fig. 1.1. The effect of the rigid collisions is to make these systems highly nonlinear, and chaotic behavior becomes the rule rather than the exception. Collisions also lead to associated wear on the components of the system. If these components are, for example, the tubes in a boiler [122] or gear teeth [146], then it is crucial to estimate the average wear that might occur in certain operating conditions.

We will not consider the detailed physics of the impacting process in this book. Such processes can be highly subtle, especially when involving the impact of rough bodies, which may also involve friction. It is well covered in the many texts on impact mechanics and tribology; see for example [243, 279]. Instead, like in the bouncing ball example, we shall take simple coefficient of restitution impact laws, which, despite their simplicity, we will see can give a close match to experimental observations.

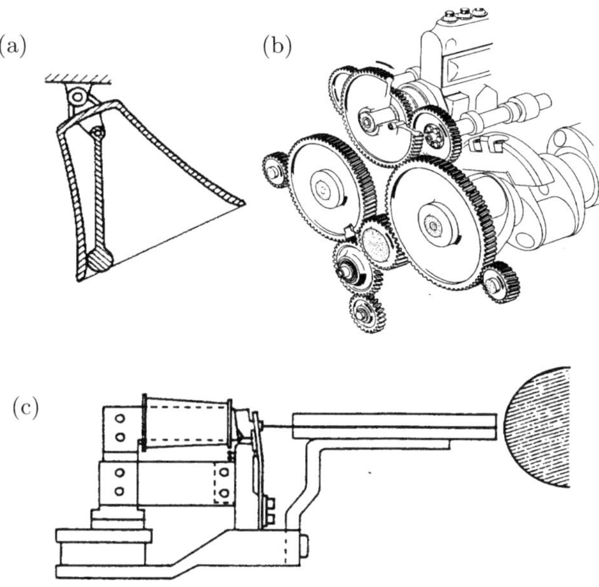

Fig. 1.2. Some examples of vibro-impact systems taken from [208], (a) a bell, (b) a gear assembly and (c) an impact print hammer. (Reprinted from [208] with permission from ASME).

We shall look at two case study examples of one-degree-of-freedom impacting systems. First, we consider a simple model that contains an instantaneous impact, where we find analytical, numerical and experimental evidence for complex dynamics and the period-adding route to chaos. Second, in Sec. 1.2.4 we consider how this dynamics might arise via taking the limit of a sequence of, possibly more realistic, continuous models that feature compliant impact. Chapters 4, 6 and 7 will complete these studies by first presenting a general theory of non-smooth maps, then of hybrid systems of arbitrary dimensions (which includes impact oscillators as a special case), and finally of continuous flows. The presentation of these case studies will draw heavily on work by Peterka [216, 217], Nordmark [197], Whiston [264, 263, 265], Chillingworth [53], Shaw & Holmes [237, 236], Thompson & Bishop [251, 103, 30, 31, 260], Budd [171, 42, 43, 44, 45] and their co-workers.

1.2.1 Case study I: A one-degree-of-freedom impact oscillator

Consider the motion of a body in one spatial dimension, which is completely described by the position $u(t)$ and velocity $v(t) = \frac{du}{dt}$ of its center of mass. Thus we think of this body as a single particle in space. When in free motion, we suppose that there is a linear spring and dashpot that attach this particle to a datum point so that its position satisfies the dimensionless differential equation

$$\frac{d^2u}{dt^2} + 2\zeta\frac{du}{dt} + u = w(t), \quad \text{if} \quad u > \sigma. \tag{1.1}$$

Here, the mass and stiffness have been scaled to unity, 2ζ measures the viscous damping coefficient, and $w(t)$ is an applied external force. We assume that motion is free to move in the region $u > \sigma$, until some time t_0 at which $u = \sigma$ where there is an impact with a rigid obstacle. Then, at $t = t_0$, we assume that $(u(t_0), v(t_0)) := (u_-, v_-)$ is mapped in zero time to (u^+, v^+) via an *impact law*

$$u^+ = u^- \quad \text{and} \quad v^+ = -rv^-, \tag{1.2}$$

where $0 < r < 1$ is Newton's coefficient of restitution. An idealized mechanical model of this system is given in Fig. 1.3.

The simplest form of forcing function $w(t)$ can arise from an excitation of the lower part of the oscillator. An equivalent problem is to set $w(t) = 0$ in (1.1) but to introduce an excitation on the whole system by moving the obstacle (so that σ becomes a function of time) and using a collision law that takes into account the relative velocity between the particle and the moving obstacle so that

$$v^+ - d\sigma/dt = -r\left(v^- - d\sigma/dt\right).$$

A simple translation in space, setting $\widehat{u}(t) = u(t) - \sigma(t)$, and $\widehat{v}(t) = v(t) - d\sigma/dt$, and dropping the hats recovers (1.1).

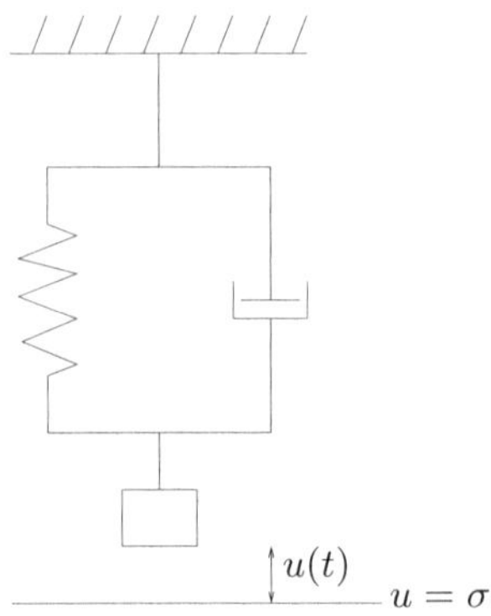

Fig. 1.3. A simple impact oscillator.

Note that r^2 measures the percentage of the kinetic energy that is absorbed in the impact. The case $r = 1$ gives an *elastic* collision [170] (often assumed in simulations of granular media, for example, [228]) and $r = 0$ is a completely dissipative collision [238] (modeling, for example, the behavior of a clapper inside a church bell [33]). In experiments, e.g. [209], a value of $r = 0.95$ is found to be reasonable to model the case of a steel bar impacting with a rigid point, whereas in [260], a different value of r was found to provide the best fit for an impacting cantilever beam; see also [91]. This indicates that the value r depends not only on the material properties of the impacting components, but also on their geometry. This is because the restitution law represents the overall effect of a much more rapid process of energy dissipation through the propagation of shock waves (those of which in the audible range we hear as the crack or bang associated with impact).

There have been many analytical and experimental investigations of the forced impact oscillator with different types of forcing function $w(t)$; see [30] for a survey. In this case study we concentrate on periodic sinusoidal forcing:

$$w(t) = \cos(\omega t), \qquad \text{with period} \quad T = 2\pi/\omega. \tag{1.3}$$

However, the literature also includes discussions of forcing caused by an external flow such as vortices shed from a boiler tube [57] or from an ocean wave [172], quasi-periodic forcing [215, 214], stochastic forcing [276, 45] and problems where $w(t)$ is the solution of another problem, for example, a further impact oscillator. The latter case arises quite commonly when energy is transmitted via impacts in a loosely fitting mechanical structure, of which the executive toy 'Newton's cradle' is a simple example.

It is difficult, in practice, to realize such a system exactly in an experiment. There is no such thing as a perfect, instantaneous impact, as the action of the impact excites higher oscillatory modes in almost any vibrating system. This difficulty can be reduced (although not entirely eliminated) by using a highly massive moving object. Such an experimental impact oscillator used by Popp and co-authors [209, 132] is depicted in Fig. 1.4. Here, a massive beam is mounted on an almost frictionless air bearing and is allowed to move freely under the restoring force of two springs that are carefully engineered to behave elastically. The beam is excited by an electromagnetic field and repeatedly comes into contact with a rigid stop. The velocity of the beam is measured at discrete time intervals by using a laser-Doppler device, and this measurement converted to a position measurement by integration. Results from this experiment will be referred to several times in what follows and will be compared with the results of numerical simulation of (1.1)–(1.3).

In the absence of impacts, the system (1.1) is *linear* and is therefore easy to analyze. Its solutions comprise exponentially decaying free oscillations converging to driven periodic motions at frequency ω. The form of these periodic solutions is unique, up to phase, independent of initial conditions, and does not change a great deal under parameter variation, provided that we avoid

Fig. 1.4. An experimental impact oscillator, after [208]. In this figure (a) shows (1) the beam (2) the restoring springs and (3) the frictionless air bearing. Panel (b) shows the electromagnetic excitation and (c) the impact with the rigid obstacle. (Reprinted from [208] with permission from ASME).

natural resonances $\omega = n$ for any integer n. This state of affairs changes completely when impacts occur, introducing a strong *nonlinearity* into the system. Then we observe a multitude of different possible recurrent behaviors, which include *periodic motions* of both higher and lower frequency than ω, and much more irregular *chaotic* motions in which the *orbit* $u(t)$ is highly irregular and is acutely sensitive to its initial conditions The number and nature of these different types of behavior now depend sensitively on the different parameters in the system.

We can easily look at the dynamics of different types of such orbit by plotting the solution *trajectories* of the solution in the *phase plane* (u, v). Note that the phase space of this system is actually three-dimensional because for a complete description of the dynamics we must include the phase variable

$$s = t \bmod 2\pi/\omega.$$

Examples of three qualitatively different solutions of the idealized simple impact oscillator are given in Fig. 1.5 for three differing, nearby input frequencies. Here we see (a) periodic motion with two impacts per period, (b) more complicated periodic motion and (c) chaotic motion.

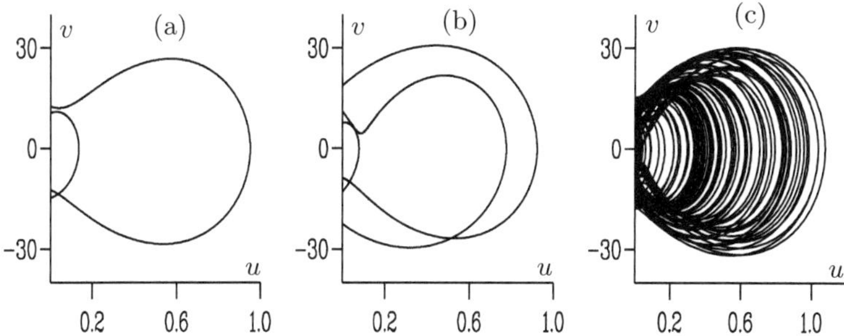

Fig. 1.5. Solutions of the idealized impact oscillator (1.1)–(1.3) in which $\sigma = 0$, $r = 0.95$, $\zeta = 0$ and (a) $\omega = 3$, (b) $\omega = 2.76$, (c) $\omega = 2.9$. (Reprinted from [208] with permission from ASME).

It is valuable to compare the solutions of this simple model with those seen in an experiment. For the experimental set up illustrated above, at the same parameter values as in the simulation, we have the phase plane plots seen in Fig. 1.6. The quantitative and qualitative agreement with the simulations is striking. The main difference between model and experiment is the excitation of a rapidly decaying higher mode of oscillation immediately after the impact. However, this does not seem to have significant effect upon the global dynamics. Note that the chaotic motion is entirely the result of the impacting behavior and is quite different from solutions to a linear differential equation, even though the motion between impacts is completely described by a linear model.

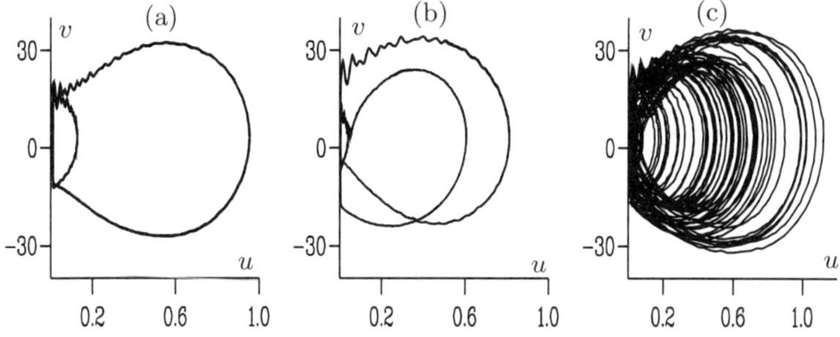

Fig. 1.6. The dynamics of the experimental impact oscillator at the corresponding parameter values to those in Fig. 1.5. (Reprinted from [208] with permission from ASME).

Consider now the general motion of an impact oscillator. It is simplest to start the analysis by assuming that the particle described by the oscillator starts at the obstacle with an initial velocity of $v_0 > 0$, at a time t_0 and a corresponding phase s_0. The motion of the particle is then described by the linear system (1.1) with initial data $u(t_0) = \sigma$ and $v(t_0) = v_0$. Provided that $v_0 > 0$, this motion will initially lie in the region $u > \sigma$, and in general (certainly if $\zeta = 0$), the trajectory will strike the obstacle at a later time t_1 with velocity $-v_1/r < 0$. After the impact, the velocity is v_1. Setting $v = v_1$ and $t = t_1$ the motion then continues as above. The overall dynamics is thus a series of smooth flows, interrupted by discontinuous changes in velocity.

Things are rather different if $v_0 = 0$ at the point of release at $t = t_0$. If $d^2u/dt^2 = f(t_0) - \sigma < 0$, then the particle cannot move and remains *stuck* to the obstacle until it has a positive acceleration. (A simple example of this being the motion of any particle under gravity, which, when placed on an obstacle with zero velocity will simply stay stuck to that obstacle.) The region over which sticking occurs is called the *sticking region* $\mathcal{Z} = \{(u, v, t) = (\sigma, 0, t) | w(t) - \sigma < 0\}$. If $0 < r < 1$, then the particle generically enters a sticking phase via an *infinite* sequence of impacts, a *chattering sequence.* (If $r = 0$, a particle impacting with $f(t_0) - \sigma < 0$ will stick immediately.)

Returning to the case with $v_0 > 0$, let us try to construct solutions analytically. It is easiest to look at the case of no viscous damping $\zeta = 0$ (which we shall henceforth assume unless otherwise stated), which is with little loss of generality if $r < 1$, because the restitution law provides the largest source of damping on the system. The differential equation (1.1) is linear and so can be solved using elementary methods. Taking the initial condition $u(0) = s_0$, $\frac{du}{dt}(0) = v(0) = v_0$, we get

$$u(t; v_0, s_0) = (\sigma - \gamma C_0)\cos(t - s_0) + (v_0 + \omega\gamma S_0)\sin(t - s_0) + \gamma C(t), \quad (1.4)$$

where

$$\gamma = \frac{1}{1-\omega^2}, \quad C(t) = \cos(\omega t), \quad S(t) = \sin(\omega t), \quad C_0 = C(s_0), \quad S_0 = S(s_0). \quad (1.5)$$

Now suppose that the orbit described by the flow (1.4) impacts with the obstacle at a later time t_1 so that

$$u(t_1; v_0, s_0) = \sigma, \quad (1.6)$$

with a velocity $-v_1/r$ before impact, and velocity v_1 after impact [Fig. 1.7(a)]. Such trajectories implicitly define an *impact map* P_I relating the time (phase) and velocity of one impact to that of the next,

$$P_I(s_0, v_0) = (s_1, v_1). \quad (1.7)$$

We can continue this analysis further to look at subsequent impacts at times t_i with velocities $v_i > 0$ immediately after impact, so that $(t_{i+1}, v_{i+1}) =$

$P_I(t_i, v_i)$. As we are considering a system that is periodically forced with period T, we can also define an alternative *stroboscopic* Poincaré map:

$$P_S(u(t), v(t)) = (u(t+T), v(t+T)), \tag{1.8}$$

which we use a lot in the later analysis of the impacting system. Note that in computing P_S we must determine all impacts in the interval $(t, t+T)$. Even for the simple linear system described in (1.1) the computation of the impact time t_1 from (1.6) using (1.4) involves solving a (nonlinear) transcendental equation. Hence, even though the system is piecewise-linear, we should regard the system as fully nonlinear, since its evolution requires knowledge of t_1. Indeed, the general impossibility of solving such equations as (1.6) in closed form renders the distinction between piecewise-linear and piecewise-smooth systems essentially meaningless. For both, the grossest nonlinearity is usually that introduced by interaction with a discontinuity surface. Fortunately, efficient numerical methods exist to compute the smooth flows, to determine the impact times and to follow these as the solution parameters vary.

A key feature of all the analysis in this book is a study of how solutions close to certain distinguished trajectories of piecewise-linear systems behave. Let us consider such analysis in the context of the impact oscillator, for the case of a trajectory that impacts. To begin with consider the case in Fig. 1.7(a) where the velocity v_1 is not small and the trajectory τ impacts with the obstacle at times t_1, t_2, etc. In this case if we look at a trajectory that starts close to τ (so that it leaves the obstacle at a time close to t_0 with an initial velocity close to v_0), then it will remain close to τ at least up to the time t_2.

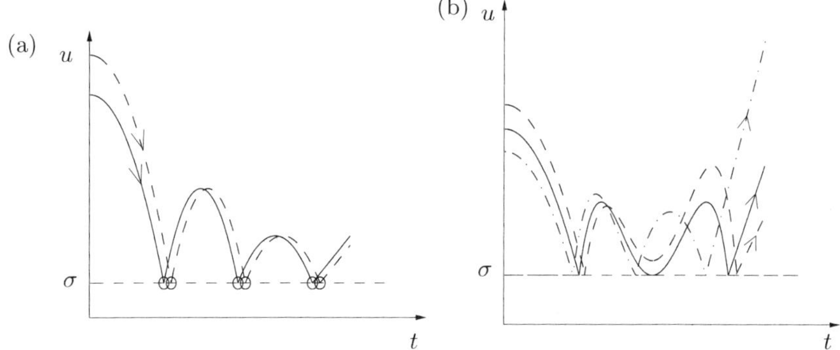

Fig. 1.7. (a) An impacting trajectory (solid) with a high-velocity impact and a nearby trajectory (dashed) projected onto the (t, u)-plane. (b) An impacting trajectory (solid) with a zero velocity impact at t_1 and two nearby trajectories, one (dashed) with no impact close to t_1 and one (dot-dashed) with a low velocity impact close to t_1.

In fact, even though the algebraic expression for P_I defined above by (1.7) using (1.5) and (1.6) is not easily written down in closed form, it is possible to linearize the impact map P_I about the point (t_0, v_0). To avoid interruption to the flow of the text, we do not give the details here; but see [237] and related calculations in Sec. 6.3 of Chapter 6. Specifically we find that the linearization DP_I of P_I is given by

$$DP_I = \begin{pmatrix} \frac{r}{v_1} & 0 \\ -\frac{r^2 A_1}{v_1} & -r \end{pmatrix} \begin{pmatrix} \cos(\Lambda) & \sin(\Lambda) \\ -\sin(\Lambda) & \cos(\Lambda) \end{pmatrix} \begin{pmatrix} -v_0 & 0 \\ -A_0 & 1 \end{pmatrix}. \tag{1.9}$$

Here the time interval Λ and the accelerations A_i at impact are given by

$$\Lambda = t_1 - t_0, \quad A_0 = \cos(\omega t_0) - \sigma \quad \text{and} \quad A_1 = \cos(\omega t_1) - \sigma.$$

The detailed form of DP_I is much less important than the fact that it exists at all. Although the trajectory τ has impacts, the dynamics of the map close to τ is just the same as a map derived from a smooth system and can be analyzed in the same way.

The situation changes completely when the impact velocity v_1 drops to zero. This case is illustrated in Fig. 1.7(b) in which the trajectory τ (represented by a solid line) has impact velocity $v_1 = 0$ and a subsequent higher-velocity impact at time t_2 with velocity v_2. This trajectory is compared to two nearby trajectories τ^- (dot-dash) and τ^+ (dash) where τ^- has a (low velocity) impact close to t_1 with a second impact close to t_2, where in contrast, τ^+ does not impact close to t_1 but does impact at a later time close to t_2.

Rather surprisingly the effect of 'losing' an impact close to t_1 has an enormous effect on the subsequent dynamics. It is possible to explain some of this by looking again at the linearized map DP_I. It follows from (1.9) that the determinant of the map P_I is given by

$$J = \det[DP_I] = r^2 v_0 / v_1.$$

This measures the contraction or expansion in phase space caused by the evolution of the dynamics from a point of time immediately after an impact, to a point immediately after the next. If $|J| > 1$, then nearby trajectories diverge, whereas they become closer together if $|J| < 1$. If $v_1 \to 0$, then $|J| \to \infty$. This gives a hint that something special occurs to trajectories close to those with a zero velocity impact when there is a tangential *grazing impact* between the particle and the obstacle. The effect of the grazing is a local (infinite) stretching of phase space, which in turn has a profound destabilizing effect on the dynamics. Indeed, if the initial velocity of the trajectory τ^- is v_0^- and that of τ^+ is v_0^+ and the velocity of the impacts close to t_2 are v_2^- and v_2^+ respectively, then the analysis presented in Chapter 6 will show that $|v_2^+ - v_2|$ is proportional to $|v_0^+ - v_0|$. In contrast, $|v_2^- - v_2|$ is proportional to $|v_0^- - v_0|^{1/2}$, which is asymptotically much more significant. In the rest of this section, we will present some simple numerical and analytical evidence for the possible behavior that can arise as a result of such grazing events.

1.2.2 Periodic motion

The simplest type of motion of the impact oscillator is a *periodic orbit*, which we now investigate in detail. However, we do urge caution. It is easy to get carried away in the study of periodic orbits and miss the much more subtle chaotic dynamics that is the hallmark of most impacting systems.

A periodic orbit of (1.1)–(1.3) is a trajectory that exactly repeats after a fixed number m of impacts and a fixed number n of forcing periods T. We shall call such a periodic orbit an (m, n) orbit. This orbit corresponds to a *fixed point* of m iterations of the map P_I and of n iterations of the map P_S. The *winding number* [216] z of such an (m, n) orbit is defined by $z = m/n$ and is the average number of impacts per period of the forcing. (This definition can be extended to chaotic orbits for which z may or may not be a rational number.) The numbers m and n are independent, and each can take an arbitrary value. Non-impacting orbits have $m = 0$ and chattering orbits have *finite* n but have $m = \infty$. Similarly, we can find $(1, n)$ orbits for all values of n. As is common with many nonlinear systems, there are often several co-existing (stable) periodic orbits. Peterka [216] refers to $(1, n)$ orbits as *beating* motions and $(m, 1)$ orbits (where we have several impacts per forcing period) as the *fundamental* orbits. In practice, we see all sorts of orbits existing simultaneously and the $(m, 1)$ orbits are often either unstable or unphysical. The number of impacts of an orbit gives a convenient classification of its complexity. The more impacts an orbit has, the more complex are the equations that need to be solved to study it and the more likely the orbit is to be unstable, or indeed non-physical, due to the effects of grazing. It is relatively easy to study single impact orbits analytically, and harder, but not impossible, to study double impact orbits.

For the special case of a single impact per period, $m = 1$, we can construct periodic orbits explicitly and analyze how their existence changes as we vary a parameter. This exercise is useful and instructive as it allows us to give explicit examples of various phenomena that will be covered by the more general theory developed later in this book. Let us start by allowing n to be any positive integer. Then, a $(1, n)$-orbit corresponds to a fixed point (s_n, v_n) of the map P_I with the corresponding trajectory taking time $T = 2n\pi/\omega$ between impacts. To find such a fixed point we must find a phase s_n and a velocity v_n. That is, if $u(t; v_n, s_n)$ is the trajectory given by (1.4), then after an elapsed time T this trajectory intersects the obstacle, so that it has a position $u(T + s_n; v_n, s_n) = \sigma$, and has an impact velocity $v = du/dt$ that is precisely $v = -v_n/r$. If these conditions are satisfied, then after the impact the trajectory exactly repeats. The necessary conditions for the existence of such a fixed point are thus that

$$u(T + s_n; s_n, v_n) = \sigma \quad \text{and} \quad \dot{u}(T + s_n; s_n, v_n) = -\frac{v_n}{r}.$$

Using the explicit expression (1.4) for u and v, we can write these conditions as

$$(\sigma - \gamma C)c_n + (v_n + \omega\gamma S)s_n + \gamma C = \sigma, \tag{1.10}$$

$$-(\sigma - \gamma C)s_n + (v_n + \omega\gamma S)c_n - \omega\gamma S = -\frac{v_n}{r}, \tag{1.11}$$

where we have used the shorthand

$$C = \cos(\omega s_n), S = \sin(\omega s_n), C_n = \cos(2\pi n/\omega), S_n = \sin(2\pi n/\omega).$$

To find the fixed point, we must solve this nonlinear system for s_n and for v_n. To do this we rearrange (1.10),(1.11) in the form

$$\begin{bmatrix} \gamma(1 - C_n) & \omega\gamma S_n \\ \gamma S_n & \omega\gamma(C_n - 1) \end{bmatrix} \begin{bmatrix} C \\ S \end{bmatrix} = \begin{bmatrix} \sigma(1 - C_n) - v_n S_n \\ \sigma S_n - S_n v_n - v_n/r \end{bmatrix}.$$

This is a linear equation for each of the terms C and S, which we can solve immediately to give

$$S = \sin(\omega s_n) = \frac{\beta}{2\gamma r} v_n, \quad \beta = \frac{1 - r}{\omega}. \tag{1.12}$$

Solving similarly for C in terms of v_n and σ, and expanding the expression $C^2 + S^2 = 1$, gives the following quadratic equation for v_n:

$$(\alpha^2 + \beta^2)\frac{v_n^2}{r^2} - \frac{4\alpha\sigma}{r} v_n + 4(\sigma^2 - \gamma^2) = 0, \tag{1.13}$$

where

$$\alpha = \frac{S_n(1 + r)}{(1 - C_n)}.$$

Suppose that $\gamma < sigma$. If we think of v_n as a function of σ, then the set of points (σ, v_n) satisfying (1.13) lie on an *ellipse*. Because the velocity v_n of the particle immediately after impact is *positive*, the only part of the ellipse that makes physical sense are those solutions for which $v_n \geq 0$.

Equation (1.13) has two solutions $v_n^{\pm}$ given by

$$v_n^{\pm} = r(2\alpha\sigma \pm \Delta)/(\alpha^2 + \beta^2) \quad \text{with} \quad \Delta^2 = 4\alpha^2\sigma^2 - 4(\sigma^2 - \gamma^2)(\alpha^2 + \beta^2). \tag{1.14}$$

Substitution of these values for v_n into (1.12) enables us to find the phase s_n. However, not all of the values of s_n and v_n obtained by such a procedure actually correspond to a $(1, n)$ periodic orbit. They must also satisfy conditions that guarantee that such an orbit is physically possible. From the previous discussion, we already know that one such restriction is that

$$v_n \geq 0. \tag{1.15}$$

However, there is a more subtle *global* condition that the trajectory $u(t)$ does not penetrate the obstacle for any time t between the two impacts, $s_n < t < s_n + T$. Such a condition is satisfied provided that

$$u(t) \geq \sigma \quad \text{for} \quad s_n < t < s_n + 2\pi n/\omega. \tag{1.16}$$

As parameters in the system vary, either (1.15) or (1.16) may be violated at certain isolated values, giving rise to *grazing bifurcation points* at which the qualitative properties of the solution changes, often in a dramatic manner.

Let us now consider an experiment where we allow the clearance σ to vary, and plot solutions $v_n^{\pm}$ to (1.14) as a function of σ. In such a way we will plot a *bifurcation diagram* for the existence of $(1, n)$ periodic orbits in the (σ, v_n)-plane given by the part of the ellipse that has $v_n > 0$; see Fig. 1.8. This ellipse intersects the line $v_n = 0$ at $\sigma = \pm|\gamma|$. The slope of the major axis of the ellipse is proportional to $s_n = \sin(2\pi n/\omega)$, and hence it rotates clockwise as ω is increased.

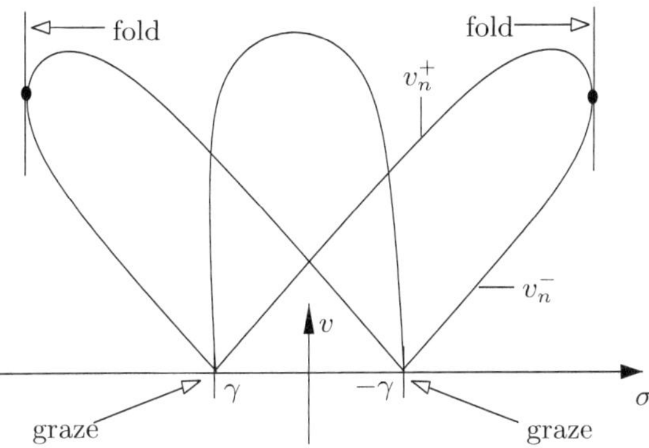

Fig. 1.8. Schematic diagram of the rotated ellipse for different ω, which indicates the two solution branches $v_n^{\pm}$ that are destroyed at the saddle-node bifurcations when $\sigma = \sigma_{SN}$ (see text for details). In this figure we see a sub-resonant ellipse on the left and a super-resonant ellipse on the right. Grazing bifurcations occur as indicated when $v_n = 0$ and $|\sigma| = |\gamma|$.

Consider first the case that ω is just a little bigger than $2n$, so that $s_n > 0$ (the so-called *super-resonant* case). Then the major axis of the ellipse has positive slope, and there are two solutions $v_n^{\pm}(\sigma)$ for $\sigma > |\gamma|$ up to a σ-value, σ_{SN}, at which $v_n^+ = v_n^-$. There are no solutions for $\sigma > \sigma_{SN}$ or for $\sigma < -|\gamma|$. The parameter value σ_{SN} is an example of a *saddle-node* or *fold* bifurcation point. It has the property that the number of fixed points changes from two to none as σ increases through σ_{SN}. Typically if σ is close to (and just less than) σ_{SN}, one of the solutions is stable (usually v_n^+) and the other is unstable. Physically, if we were to slowly increase σ, we would see a dramatic change in the behavior of the system at this value. saddle-node bifurcations are common to all dynamical systems, both smooth and non-smooth. If $s_n < 0$ (the *sub-resonant case*), a similar pattern occurs. For example, if $n < \omega < 2n$, then the major axis has a *negative slope* and σ_{SN} is less than $-\gamma$, so that there are no solutions for $\sigma < \sigma_{SN}$ or $\sigma > +\gamma$. As σ (or indeed as ω) is varied, the $(1, n)$ orbits can also lose stability through a *period-doubling* (or *flip*) bifurcation and typically we see a smooth evolution from a $(1, n)$ orbit to a $(2, 2n)$ orbit. The saddle-node bifurcation occurs when the Jacobian of either of the associated maps has an eigenvalue $+1$ and the period-doubling bifurcation when it has an eigenvalue -1. The locations of such bifurcation points determines part of the range of parameter values over which the $(1, n)$ orbits are likely to be observed.

If saddle-node and period-doubling bifurcations were the only ways in which the periodic orbits of the impact oscillator could change their qualitative behavior as the parameters governing the system vary, then the impact oscillator would behave in an almost identical manner to a smooth system. However, there are particular parameter values at which we see changes in behavior that are quite different from smooth systems. These are the values for which the periodic orbits have *grazing intersections* with the obstacle.

Grazing arises where $v_n = 0$, that is, at the foot of the semi-ellipse when $\sigma = \pm|\gamma|$. For a small change in the parameters, it is then possible to have a $(0, n)$ orbit with *no* impacts. For example, in the sub-resonant case, the orbit with no impacts co-exists with the $(1, n)$ periodic orbit if σ is just less than $-|\gamma|$.

Grazing also occurs when, under changes in the parameters, the $(1, n)$ orbit has an additional zero velocity impact *between* the existing (non-zero velocity) impacts. This is a harder condition to verify from analytical calculations as it involves the global failure of the global condition (1.16). One way of viewing finding such grazing points is to demand that the impact point of the periodic orbit should lie on a curve G in the (v_n, s_n)-plane that leads to a grazing impact. We shall see in Chapter 6 that the set G and its image under the dynamics can have a complex geometry. We illustrate the two kinds of grazing orbits via the solution trajectories in Fig. 1.9.

If grazing occurs at a particular value, say σ_g, of σ, then various things may occur as σ varies through σ_g. We may see a transition from one periodic orbit to another; for example, in the super-resonant case we see a transition from a $(0, n)$ orbit to a $(1, n)$ orbit as σ increases through $-|\gamma|$. Alternatively we may see a coalescence of two periodic orbits; for example the co-existence of the $(0, n)$ and $(1, n)$ orbits for $\sigma < -|\gamma|$ in the sub-resonant case. More dramatically, we may also see the instantaneous creation of an infinite number of periodic orbits together with a (robust) chaotic attractor, which we will give examples of presently. We call all such transitions examples of *grazing bifurcations* and will study these in depth in Chapters 4 and 6.

As a second experiment, consider what happens as we vary ω for fixed $\sigma = 0$. Thus we consider the points of intersection of the semi-ellipses in Fig. 1.8 with the σ axis. In general v_n is single valued, has a large peak when $\omega = 2n$ and decreases rapidly away from this value. If ω is changed slowly, we see the series of $(1, n)$ solutions shown in Fig. 1.10(a) for different n, with

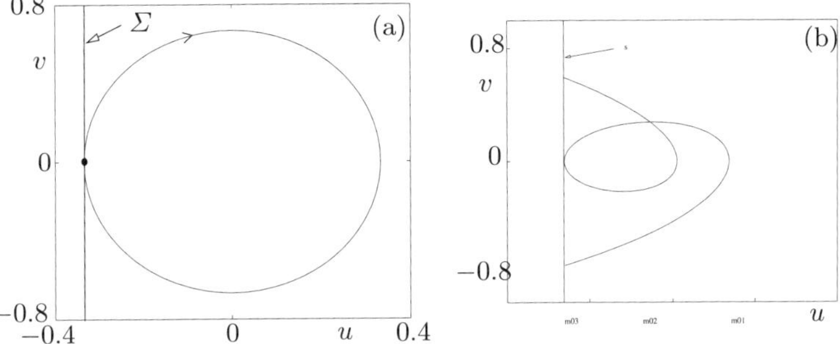

Fig. 1.9. Periodic orbits undergoing grazing bifurcations projected onto the (u, v)-plane for $\zeta = 0$, $r = 0.8$ and $\omega = 2$. (a) A periodic orbit with a single grazing impact, where $v_n = 0$ and $\varphi_n = 0$. This orbit occurs when $\sigma = -1/3$. (b) A periodic orbit containing both a high-velocity impact with $v_n = 0.5932$ and $\varphi_n = 1.6265$ and a grazing impact with zero velocity. This orbit occurs when $\sigma = 0.331265$.

resonant peaks at $\omega = 2n$, that is, *twice* the values of the usual resonant frequencies.

As a contrast, consider the equivalent picture when $\sigma = -5$ [Fig. 1.10(b)]. In this case the obstacle is a large distance from the mean position of the oscillator and the dynamics is more nearly that of a free oscillator without impact. The regions of existence of the periodic orbits are now much closer to the resonant values $\omega = n$.

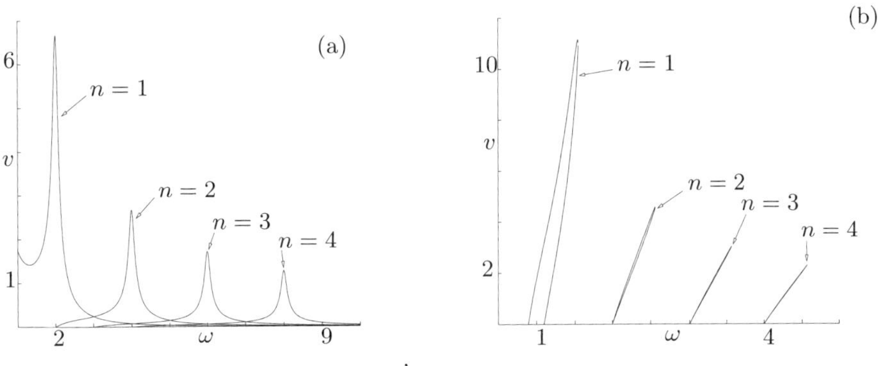

Fig. 1.10. Resonance curves upon which $(1, n)$ orbits exist, as defined in the text; (a) close to even values of ω for $\sigma = 0$ and (b) close to $\omega = n$ when $\sigma = -5$.

Although the $(1, n)$ orbits exist for a wide range of parameter values and are relatively easy to analyze, they are not the only form of periodic behavior observed. Indeed, we find examples of multiple impact orbits, orbits with chatter and chaotic orbits. For example, *double impact* $(2, n)$ periodic orbits are fixed points of the second-iterate map P_I^2 and can arise at period-doubling bifurcations of $(1, n)$ orbits or through more complex transitions. It is possible to derive algebraic equations for such orbits [216], which are considerably simplified when $\omega \approx 3$ and can be solved explicitly in that case [45]. Indeed, when $\sigma = 0$ a stable orbit with two low-velocity impacts can be shown to exist. This result has also been confirmed in experiments, see Fig. 1.13. Indeed, the double-impact orbit appears from the experimental evidence to exist, and to be stable, for an appreciable range of excitation frequencies. This conclusion is potentially useful for anyone planning to operate a system modeled by an impact oscillator, as the double-impact orbit is a stable, controllable, low-wear operating state.

In contrast, a high-wear state would be one that involves a large sequence of impacts per forcing period. In fact, we have already encountered motion that has an infinite number of impacts, in the example of a bouncing ball. The ball rebounds with a coefficient of restitution $r < 1$ and comes to rest after an *infinite* number of impacts, but in a *finite* time. A similar phenomenon is observed in our simple one-degree-of-freedom impact oscillator if forcing is slow (ω is small) and an impact occurs during a phase when the forcing acts in a direction that pushes the particle towards the obstacle. Then after a finite time, and an infinite sequence of impacts, the particle becomes *stuck* to the obstacle, until such time that the forcing reverses its direction and pushes the particle off. We call this a *chattering sequence*, a detailed analysis of which is given in [42, 203]. All orbits with initial conditions that eventually end up in the stuck phase, do so through a chattering sequence and exit it with the same values of u, v and s. If the forward trajectory from this exit point itself has a chattering sequence, then the result is a an (∞, n) *periodic orbit with chatter.* Such an orbit arises both experimentally and in simulations when $r = 0.87$, $\omega = 0.3$ and $\sigma = 0$ as illustrated in Fig. 1.11. Owing to the extreme contraction of phase space resulting from chattering, periodic chattering orbits are also always *super-stable.* That is, small perturbations return to the chattering orbit exactly after a *finite* time.

1.2.3 What do we actually see?

When studying any type of dynamical system we are often interested in the *ω-limit sets* of the motion, that is, the set of all possible long time behavior of the motion. Loosely speaking, these are the *attractors* of the system and represent what is observed physically after transients have decayed. If we look at impact oscillators with coefficient of restitution $r < 1$, then the overall behavior of trajectories that impact is dissipative; indeed there is an average contraction of the area of phase space by the factor r^2 at each impact. As a result, the ω-limit sets of the motion are simpler than the general dynamics and comprise periodic or chaotic motions that occupy only a small fraction of the total phase space. Such motion, when chaotic, is said to evolve on a *strange attractor* whose dimension is strictly less than that of the underlying

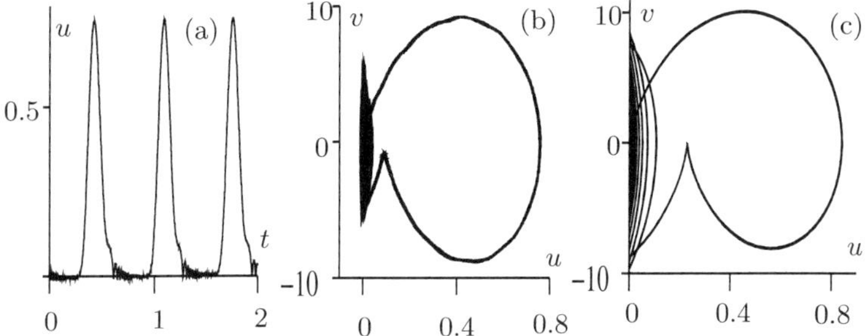

Fig. 1.11. Chattering periodic motion for $r = 0.87$, $\omega = 0.3$ and $\sigma = 0$. Panels (a) and (b) give an experimental periodic chattering orbit, with (c) containing the corresponding numerical simulation. (Reprinted from [208] with permission from ASME).

phase space (three in this case). The high stretching associated with grazing means that chaotic behavior is observed in impact oscillators for wide ranges of the parameters. Often, the strange attractors associated with grazing events have a fingered appearance when considered as a sequence of iterations of the maps P_I or P_S (see Fig. 1.19 below). Chaos can also arise at other points, with the chaotic attractors having the usual fractal form associated with smooth dynamical systems.

Thus, the analytically calculated periodic orbits represent the tip of the iceberg of the possible dynamical behavior. To extend our understanding we must carry out experimental investigation of the system, either numerical or physical *experiments*:

Numerical simulations. To determine stable behavior, direct numerical simulation methods can be used, in an *event-driven* manner. That is, for a given initial state with $u > 0$, the linear equation (1.1) is solved either exactly or via a high-order accurate time-stepping scheme such as the Runge–Kutta method. The *test function* $u - \sigma$ is monitored, and changes of sign are sought using the bisection method, or an in-built hit-crossing detector of the time-stepping algorithm. The next impact point is thus determined accurately. After applying the impact law, this point is then used as the starting point for the next calculation. Provided that care is taken with accumulation points, this process can be easily applied to calculate flows with many impacts. To determine the possible ω−limit sets for a particular parameter value a *random set* of initial data is chosen, and the orbits from each such point are calculated over many, say 1000, impacts without storing. The flow is then continued for another sequence of, say, 200 impacts and stored. To see how this limit set changes, a small parameter adjustment is made and the process is repeated. The solution data thus obtained can then be plotted in a numerical bifurcation diagram where some measure of the solution state is plotted against the parameter. Plots obtained in this way via direct numerical simulation are sometimes referred to as *Monte Carlo* bifurcation diagrams. Rather than plot the whole solution over the 200 impacts, it is usual to sample the data, e.g., by plotting v or s at every impact $u = \sigma$, or by plotting x or v at fixed values of the forcing phase $s = s_0$. This method has the advantage of capturing most, if not all, of the long time dynamics, but has the disadvantage of not being able to capture unstable behavior, which requires a path-following algorithm, see e.g. [233]. More details of numerical methods for piecewise-smooth problems are given in Chapter 2.

Experimental methods. Experimentally, a similar bifurcation diagram of the ω-limit sets can be produced by slowly increasing a parameter (typically the excitation frequency) over a period of several hours, with the system allowed to reach a stable state for each frequency value. It is necessary in experiments, where data are typically extracted at discretely sampled times, to plot the ω-limit set of the values of u at sample times kT (the iterates of the stroboscopic map P_S). Hence we adopt this convention for both the numerical and the analytical plots. More details of experiments on impact oscillators will be given in Chapter 9.

Let us now describe the results of two such experiments: first varying the forcing frequency ω for fixed clearance σ; and then varying σ for fixed ω.

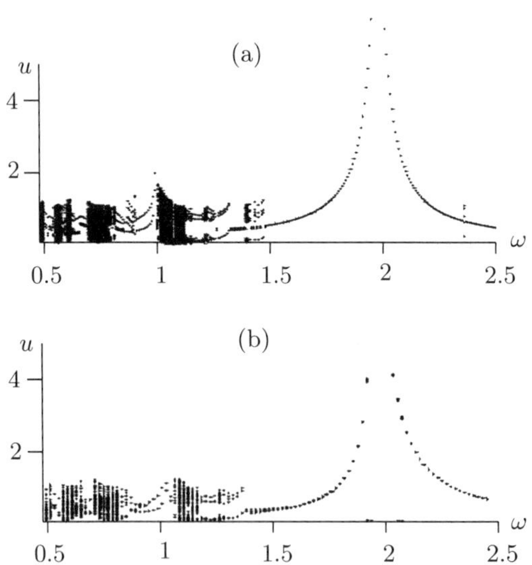

Fig. 1.12. The bifurcation diagram for increasing $\omega \in (0.5, 2.5)$ for $\sigma = 0$ and $r = 0.93$. (a) Analytical and (b) experimental results. (Reprinted from [208] with permission from ASME).

For a first investigation, we consider the effect of *increasing* ω from 0.5 to 4, taking $\sigma = 0$, $r = 0.93$. The resulting analytical and experimental bifurcation diagrams are presented in Figs. 1.12 and 1.13 for two different ranges of values of the forcing frequency. As one can see, there is close agreement between the numerically computed and the experimental bifurcation diagrams. As we would expect, the figures are not identical, but they are surprisingly close given the simple restitution law model of impact. In a neighborhood of $\omega = 2$, we observe a $(1, 2)$ periodic orbit in both numerical simulations and experiments. A similar resonance peak of stable $(1, n)$ periodic orbits is observed near $\omega = 2n$ for any integer n, as predicted by the analytical calculations of periodic orbits above. In contrast, close to the *odd integer* values of ω, e.g., $\omega = 1$ and $\omega = 3$, we either see stable multiple-impact periodic motions (visible as having a finite set of points in their ω-limit sets) or chaotic orbits (visible as bands in the diagram).

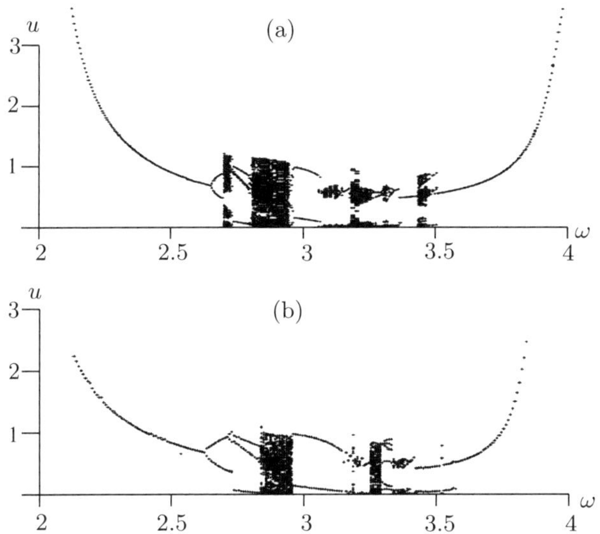

Fig. 1.13. Similar to Fig. 1.12, but for $\omega \in (2, 4)$. (Reprinted from [208] with permission from ASME).

Note that the $(1, 1)$ orbit that is stable for $\omega \approx 2$ loses stability upon increasing ω via a period-doubling bifurcation (see Fig. 1.13(a) at $\omega = 2.6528$, where the characteristic pitchfork shape of the period-doubled branch beyond the bifurcation can be seen). In a smooth system we might then expect to see a (Feigenbaum) cascade of period-doubling bifurcations to $(2^k, 2^k)$ orbits, for $k = 1, 2, \ldots, \infty$ leading to chaos [62]. Such behavior is not observed, however, in the impact oscillator, because the resulting orbits tend to lose stability through grazing bifurcations. Thus, as ω is increased, within the range $(2.5, 3.5)$ we see a dramatic increase in complexity of the observed dynamics. We can also find bi-stability between competing attractors for the same ω-value. For example, in numerical results for the lower value $r = 0.8$, the $(1, 1)$ periodic orbit is found to coexist with a $(6, 6)$ orbit when ω is close to 2.6.

The two orbits $(1,1)$ and $(6,6)$ orbits for $\omega = 2.6$ have complex *domains of attraction* (sets of initial conditions that end up at particular ω-limit sets) in which the effects of grazing make themselves apparent. Several authors have considered the question of computing such domains by using cell mapping and related methods; see e.g. [196, 30]. The domains of attraction in this case are given in Fig. 1.14(a). As can be seen, the two domains are beautifully interwoven and lead to acute sensitivity to the initial data.

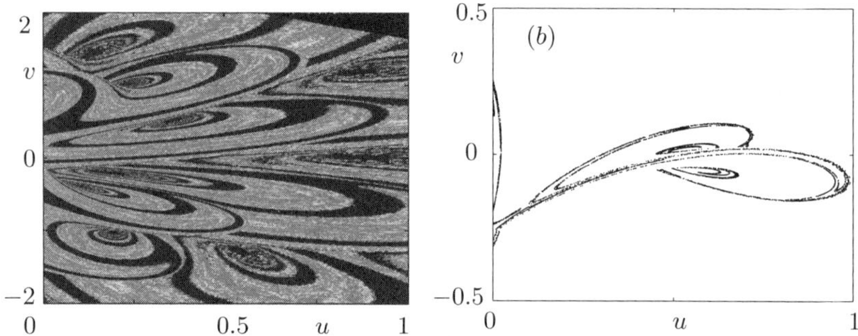

Fig. 1.14. (a) Domains of attraction within P_S for $\sigma = 0, r = 0.8$ and $\omega = 2.6$. In this figure, the dark regions are attracted to a period-one periodic orbit and the light regions to a period-six orbit. (b) A strange attractor plotted via P_S for the same parameter values but with $\omega = 2.8$.

For larger values of ω, in particular over an interval containing $\omega = 2.8$, we see the chaotic orbit that corresponds to the strange attractor plotted as a set of iterations of the map P_S in Fig. 1.14(b). This attractor takes a familiar *fractal* form, often encountered in chaotic systems, demonstrating both stretching and folding in the map P_S. If you look at this figure, it seems to echo some of the structure seen in portions of the domain of attraction plotted in Fig 1.14(a). This is no coincidence, as both reflect the complicated geometry of the grazing set G, as we shall see in Chapter 6. For values of ω close to 3, the $(2,2)$ orbit created in the period-doubling bifurcation at $\omega = 2.6528$ restabilizes, leading to a parameter interval (a *window*) of stable $(2,2)$ periodic motion. This is followed by more intervals of chaos and periodic windows, before, around $\omega = 3.5$, a stable $(1,2)$ orbit is born.

As a second (this time purely numerical) experiment, consider the effect of changing σ, with ω fixed close to a resonance value, $\omega = 2$. Specifically we shall probe the effects of grazing bifurcations. We consider the *grazing of a non-impacting orbit.* If $\sigma = -\infty$, a non-impacting $(0,1)$ periodic orbit exits. As σ is increased, there is a first value σ_g at which this orbit has a grazing impact with the obstacle. It turns out that this causes a precise coincidence with the point at which the $(1,1)$-periodic orbit has zero impact velocity.

Close to σ_g, we see complex behavior that is qualitatively different depending on whether the forcing is sub-resonant ($\omega < 2$), resonant ($\omega = 2$) or super-resonant ($\omega > 2$). Fixing $r = 0.8$, we present the results of one-parameter sweeps in σ passing through the value σ_g in Figs. 1.15–1.17 for $\omega = 1.8$, $\omega = 2$ and $\omega = 2.2$, respectively. We also extend the calculation to the case of non-zero damping, taking $\zeta = 0.01, 0.5, 1, 2$. For general ζ the grazing bifurcation occurs when

$$\sigma = \sigma_g := \frac{-1}{\sqrt{(\omega^2 - 1)^2 + 4\zeta^2}}.$$

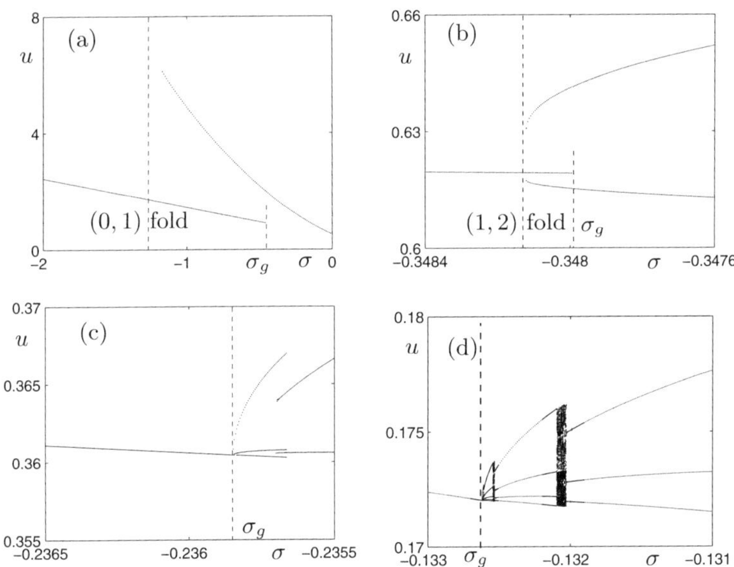

Fig. 1.15. Stroboscopic bifurcation diagram for $\omega = 1.8$, $r = 0.8$ and (a) $\zeta = 0.01$, (b) $\zeta = 0.5$, (c) $\zeta = 1$ and (d) $\zeta = 2$.

The case of low damping $\zeta = 0.01$ is almost identical to that of $\zeta = 0$. When $\omega = 1.8$ and $\sigma < \sigma_g$, the non-impacting orbit exists for $\sigma < \sigma_g$ and co-exists in this range of values of σ with an unstable $(1, 1)$ orbit that is created when $\sigma = \sigma_g$. This orbit restabilizes at the saddle-node bifurcation and continues to exist for $\sigma > \sigma_g$. When $\omega = 2$, the non-impacting orbit evolves smoothly into the impacting one as σ increases, and when $\omega = 2.2$, there is a sudden jump to chaos. For larger values of ζ, rather different behavior is observed. If $\zeta = 0.05$, for all values of ω, the non-impacting orbit co-exists with an unstable $(1, 2)$ orbit, which is also created when $\sigma = \sigma_g$ and which restabilizes at a saddle node bifurcation.

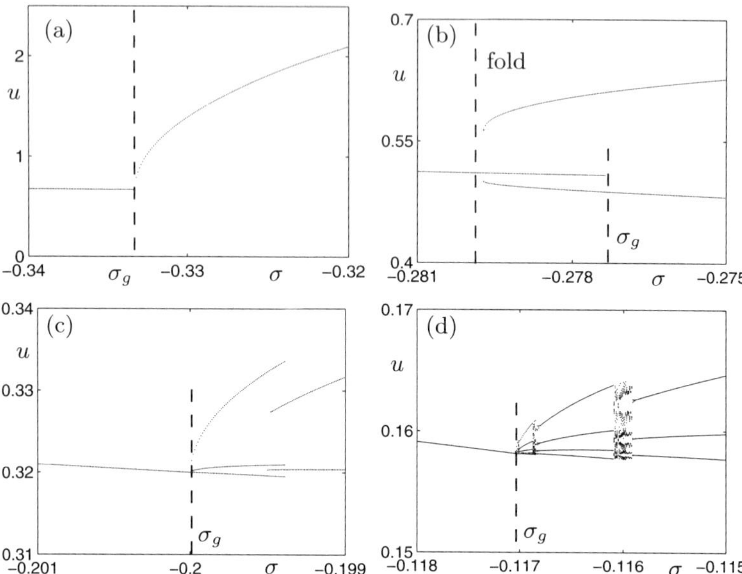

Fig. 1.16. Similar to Fig. 1.15 but for $\omega = 2$.

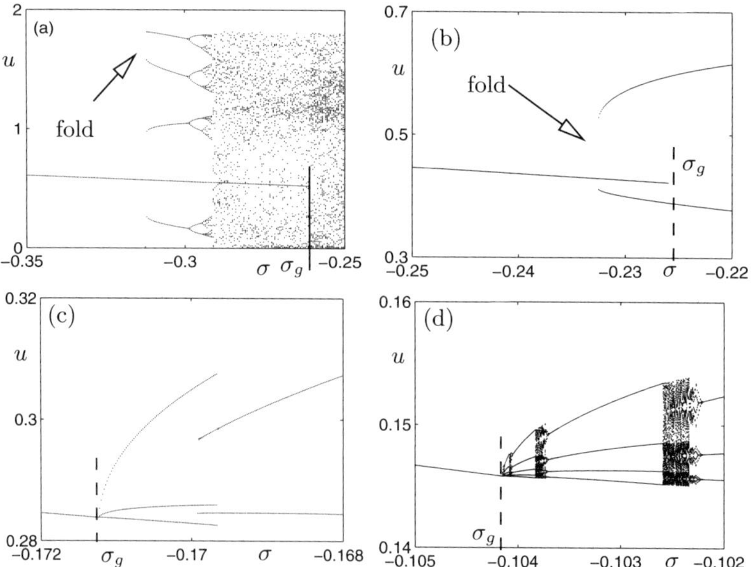

Fig. 1.17. Similar to Fig. 1.15 but for $\omega = 2.2$.

If $\zeta = 1$, for all values of ω, the non-impacting orbit evolves smoothly into an impacting $(1,3)$ orbit as σ increases through σ_g. This orbit is replaced by a $(1,2)$ orbit for larger values of σ. Finally, if $\zeta = 2$, we see in all cases the phenomenon of *period-adding*. Here, as σ is decreased towards σ_g there is the creation of a sequence of periodic orbits, the period of which increases in an arithmetic sequence. The range of values of σ over which periodic orbits of period n and $n+1$ are observed are separated by intervals of chaotic behavior. This is quite different from the *period-doubling cascade* associated with chaotic behavior in smooth dynamical systems. All of these phenomena will be explained in Chapters 4 and 6.

Let us consider finally the case of *grazing of an orbit that is already impacting*. Fixing $\omega = 2$, $r = 0.8$ and $\zeta = 0$, a resonant $(1,n)$ periodic orbit with high velocity impact exists when $\sigma = 0$. As σ is increased, this orbit has an additional zero velocity impact at $\sigma_g = 0.331265$. Fig. 1.18 shows what

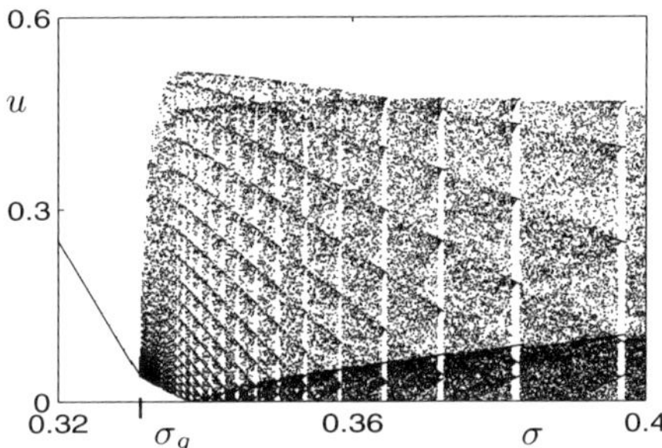

Fig. 1.18. Stroboscopic bifurcation diagram fixing $\omega = 2$ and varying σ for $r = 0.8$.

is observed when σ is close to this value. Here we see a further example of the complex dynamics that may result from a grazing bifurcation. As σ is increased, the $(1,1)$ periodic orbit immediately evolves into a robust chaotic orbit. The corresponding strange attractor is plotted in Fig. 1.19 just after the grazing bifurcation. Note that this has a very different form than that in Fig. 1.14. The characteristic *fingered* shape of this attractor, first reported in the work of Thompson & Ghaffari [251], arises because the grazing map stretches phase space strongly in one direction and compresses it in the other. Returning to Fig. 1.18, note how the amplitude of the attractor increases rapidly as σ is increased, just after the grazing bifurcation point. As σ is increased further, the chaotic orbit is interrupted by *windows* of σ-values at which there are high-period periodic orbits that have just one low-velocity impact per period. Initially at $\sigma = 0.336$ we see a stable 20 impact orbit that, as σ is increased further, evolves into stable periodic orbits with 19,18,17 16,... impacts per period. These periodic windows are separated by bands of chaotic behavior.

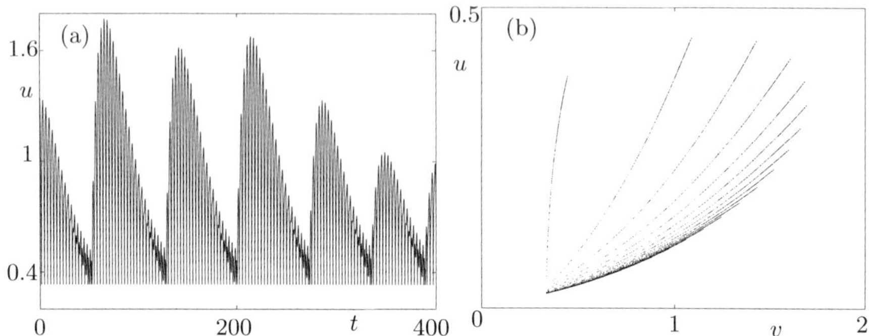

Fig. 1.19. The fingered strange attractor arising when $\omega = 2$, $\sigma = 0.3333$: (a) time history and (b) iterations of P_S.

We shall return in Chapter 6 to an explanation of the dynamics we have observed in this case study.

1.2.4 Case study II: A bilinear oscillator

A model for rigid impacts that has an instantaneous jump in velocity is unrealistic in practice as it would require an infinite force (even though, as we have seen, such a model gives good agreement with experiments). A natural generalization is to replace the rigid impact by a highly stiff, elastic deformation that takes a short but finite time, over which the velocity changes continuously by a large amount. The simplest compliant oscillator with sinusoidal forcing may be written in the bi-linear form

$$\frac{d^2u}{dt^2} + 2\zeta\frac{du}{dt} + k_1 u = \cos(\omega t), \qquad \text{for } u \geq 0 \tag{1.17}$$

and

$$\frac{d^2u}{dt^2} + 2\zeta\frac{du}{dt} + k_2 u = \cos(\omega t), \qquad \text{for } u < 0. \tag{1.18}$$

In this system, we assume that $K = k_2/k_1$ is large, so that the oscillator spends relatively short time intervals with $u < 0$ before returning to $u > 0$. It is natural to impose the conditions that u and du/dt are continuous across the boundary $\{u = 0\}$, from which it follows that d^2u/dt^2 is then also continuous, but d^3u/dt^3 is discontinuous. Such an oscillator was has been used to model the behavior of a moored ship [252, 253] and other offshore structures. See those references and [237, 196] for details of the dynamics of such a system. Other mechanical oscillators with non-smooth stiffness characteristics include models of rocking blocks [133], and walking robots [221]. In this case study, we restrict ourselves to showing how when K is large the dynamics closely resembles that of the impact oscillator. In particular, provided the velocity on crossing $u = 0$ is never small, one can derive an approximate restitution law of an equivalent rigid impact oscillator.

Suppose the particle moves through $u = 0$ at time t_0 with velocity $v_0 < 0$ that is not too small. Then its behavior for $u < 0$ closely approximates that of an unforced, high-frequency harmonic oscillator, with rapidly reversing velocity. If we suppose that ζ is small and that the time spent with $u < 0$ is much smaller than the period $2\pi/\omega$ of the forcing, then the particle for $u < 0$ moves to good approximation under the law

$$\frac{d^2u}{dt^2} + k_2 u = g, \quad \text{where } g \approx \cos(\omega t_0) < 0.$$

If the maximum penetration occurs at a later time $t = t_1$ where $du/dt = 0$, we have that

$$u(t) = A\cos(\sqrt{k_2}(t_1 - t)) + g/k_2, \tag{1.19}$$

for some A to be determined. As $u(t_0) = 0$ and $v(t_0) = -v_0$, it follows that

$$A\cos(\sqrt{k_2}(t_1 - t_0)) + g/k_2 = 0 \quad \text{and} \quad \sqrt{k_2}A\sin(\sqrt{k_2}(t_1 - t_0)) = -v_0,$$

so that $\tan(\sqrt{k_2}(t_1 - t_0)) = \sqrt{k_2}v_0/g$. Thus, if

$$k_2 \gg 1 \quad \text{and} \quad \sqrt{k_2}v_0 \gg 1, \tag{1.20}$$

we have $\sqrt{k_2}(t_1 - t_0) \approx \pi/2$, and $A \approx v_0/\sqrt{k_2}$. Hence, the total time Δ spent with $u < 0$ is given by

$$\Delta = 2(t_1 - t_0) \approx \frac{\pi}{\sqrt{k_2}},$$

and in this region, the particle moves as one half-wave of a sinusoid of amplitude $v_0/\sqrt{k_2}$. Using this approximation, we can make a more precise estimate of the motion. Suppose that the particle has $u < 0$ over the interval $t \in [t_0, t_2]$; then, multiplying (1.18) by du/dt and integrating, we have

$$\left[\frac{1}{2}\left(\frac{du}{dt}\right)^2\right]_{t_0}^{t_2} + 2\zeta\int_{t_0}^{t_2}\left(\frac{du}{dt}\right)^2 dt = 0. \tag{1.21}$$

Differentiation of (1.19) with respect to t and substitution into the second integrand in (1.21) using $t_2 = t + \Delta$, results in the expression

$$\left(\frac{du}{dt}(t_2)\right)^2 = v_0^2\left(1 - \frac{\zeta\pi}{\sqrt{k_2}}\right).$$

We conclude that, provided (1.20) is satisfied, the period of time spent in with $u < 0$ is short. Neverthcless, within this short period of time, the velocity of the particle is reversed. Indeed, this process can be well approximated by an instantaneous impact with coefficient of restitution r given by

$$r^2 = \left(1 - \frac{\zeta\pi}{\sqrt{k_2}}\right).$$

The second condition (1.20) for the compliant oscillator is equivalent to a non-grazing condition for the impact oscillator. Things are much more subtle when the impact velocity is small (i.e., when $v_0 = O(1/\sqrt{k_2})$), which we will consider in detail in Chapter 7 where we show it typically leads to $O(3/2)$ power-law behavior of the induced Poincaré map.

An elegant series of calculations showing the similarity of the behavior of the compliant oscillator with the impact oscillator is given by Nordmark [196], which we repeat in Fig. 1.20. By increasing the value of $K = k_2/k_1$ we can how see the complex dynamics of the impact oscillator, leading to a fingered strange attractor (produced at a grazing bifurcation of the impact oscillator considered earlier), arises through a series of bifurcations. Motion which is originally regular, becomes chaotic through a series of period-doubling bifurcations, the first of which occurs for $K = 3.8$. The figure shows the period-two orbit for $K = 18$, period-four for $K = 18.5$ and fully developed chaos at $K = 19$. As k is further increased, (e.g., for $K = 40$ and $K = 100$), the shape of the strange attractor evolves into the fingered structure characteristic of the impact oscillator.

1.3 Other examples of piecewise-smooth systems

Note that the above bilinear oscillator has the feature that the dynamics across the *discontinuity set* $\{u = 0\}$ is continuous and differentiable. That is, both $u(t)$ and du/dt are continuous functions of time. However, there is typically a jump in d^2u/dt^2. For such systems, the discontinuity set can never be simultaneously attracting from both sides; that is, we cannot have $(du/dt) < 0$ for $u > 0$ and $du/dt > 0$ for $u < 0$ for the same values of velocity and phase. We now look at three examples of piecewise-smooth systems that arise in engineering applications for which the first derivative of the switching state (effectively du/dt) undergoes a jump as we cross a discontinuity set. In these so-called *Filippov systems*, the jump in first derivative can cause the dynamics to evolve towards the discontinuity set from both sides and hence can cause the dynamics to evolve within the discontinuity set itself. This kind of evolution is termed *sliding* motion in the context of control theory. Such motion is also common in systems modeling the dynamics of dry friction, where just to confuse matters, it is the 'sticking' phase that corresponds to what we have called sliding motion, rather the 'sliding' or 'slipping' phase of free movement between two surfaces. However, we will continue to use the term *sliding* in the control-theory sense. Quite often, complex dynamics in these systems can be observed when a particular trajectory (such as a periodic orbit) undergoes a transition that changes its number of sliding segments.

1.3.1 Case study III: Relay control systems

The idea of using a switching action (or *relay*) has been widely employed in control engineering since the 1950s. Indeed, relay control is the main idea

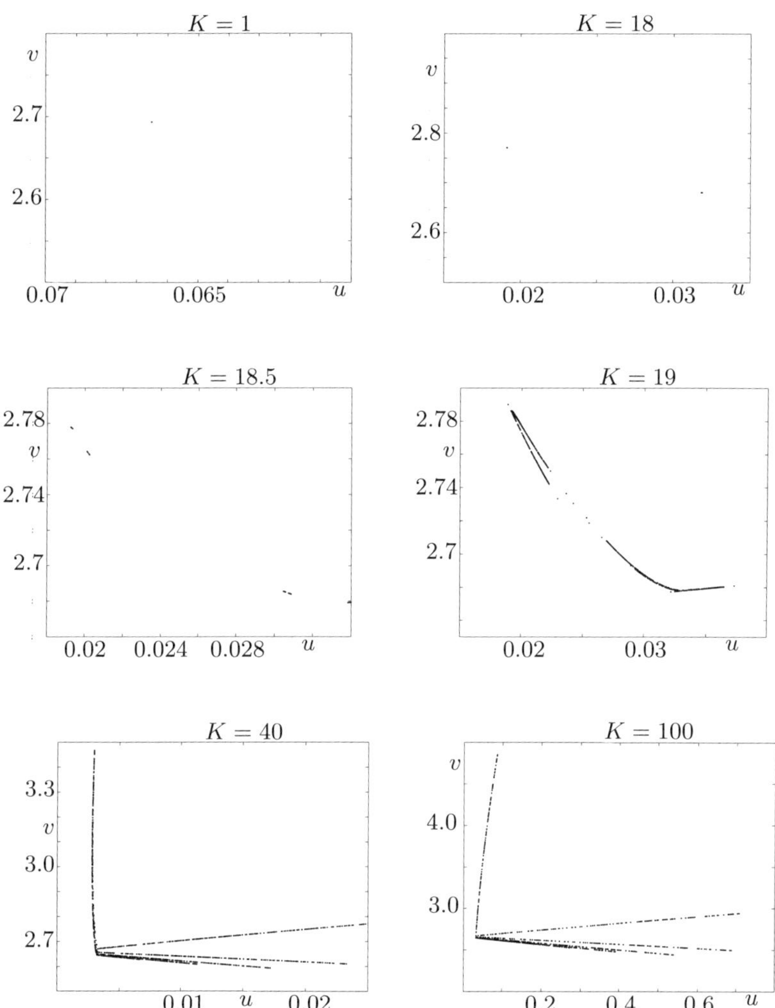

Fig. 1.20. The evolution of a fingered attractor of the bilinear oscillator (1.17),(1.18) as the stiffness increases. Here $\omega = 1$, $\zeta = 0.25$ and $k_1 = 1/100$ are fixed, whereas $K = k_2/k_1$ is varied in each successive panel.

behind the central heating example with which we started this chapter, and it has been used, for instance, in pulsed servomechanisms [28], tuning controllers in the process industry [13]. More generally, relay systems play an important role in the theory of variable structure controllers [257], of hybrid systems [71]. Although systems with a relay feedback have been studied for a long time (for example, in the work of Andronov *et al.* [5] and Flugge-Lotz [101] from the 1950s and 1960s), the dynamics of these systems is not fully understood. It is well known that relay systems have a tendency to self-oscillate if used outside their desired operating regime. To estimate the amplitude and frequency of such oscillations, specialized methods are typically used; see, for example, the book by Tsypkin [255]). These methods assume that a simple, non-sliding *limit cycle* (isolated periodic orbit) exists; but, as pointed out by Johansson [144], there is no general proof that these are the only kinds of periodic motion that can occur, or indeed that a given initial state actually converges to such a cycle.

It has also been shown that even low-order relay feedback systems can exhibit more complex self-oscillations (either periodic or chaotic), which include segments of sliding motion [84, 258, 144]. Other complicated solutions in relay systems include orbits with a multiple number of fast and slow switches per period, termed *higher-order sliding modes* [109, 144]. These solutions are akin to the chattering motion for impact oscillators discussed earlier. Examples of engineering control systems with relay elements featuring chaotic behavior as well as quasi-periodic solutions are discussed in [4, 58, 59, 116].

Here we consider a simple class of model problems corresponding to single-input–single-output, linear, time-invariant relay control systems with unit negative feedback of the output variable. Such problems can be written in the general form:

$$\dot{x} = Ax + Bu, \tag{1.22}$$

$$y = C^T x, \tag{1.23}$$

$$u = -\text{sgn}(y), \tag{1.24}$$

where the n-dimensional vector $x \in \mathbb{R}^n$ represents the system state, the scalar $y \in \mathbb{R}$ is a measure of the output of the system, and the discrete variable $u \in \{-1, 1\}$ is the control input. Also, $A \in \mathbb{R}^{n \times n}$, $B \in \mathbb{R}^{n \times 1}$ and $C^T \in \mathbb{R}^{1 \times n}$ are assumed to be constant matrices. The input u and output y of the linear part are scalar functions, whereas x, the state vector, has $n \geq 1$ components. Furthermore, it is assumed that the system matrices are given in *observer canonical form* [48]; i.e.

$$A = \begin{pmatrix} -a_1 & 1\,0 \cdots 0 \\ -a_2 & 0\,1 \cdots 0 \\ \vdots & \ddots \vdots \\ -a_{n-1} & 0\;0\;\;0\;\;1 \\ -a_n & 0\;0\;\;0\;\;0 \end{pmatrix}, \quad B = \begin{pmatrix} b_1 \\ b_2 \\ \vdots \\ b_4 \\ b_5 \end{pmatrix}, \quad C^T = \begin{pmatrix} 1 \\ 0 \\ \vdots \\ 0 \\ 0 \end{pmatrix}^T.$$

In Chapter 8, we shall explore in some detail the dynamics of a system of the form (1.22)–(1.24), finding several transitions between the number of sliding segments that a trajectory can undergo. For now, we merely report the results of numerical computation of trajectories of a simple three-dimensional relay system; see Fig. 1.21. The solution displayed in panel (a) is an attractor for all initial conditions of the system at the stated parameter values. It represents a symmetric (i.e., invariant under $x \to -x$) limit cycle that has a

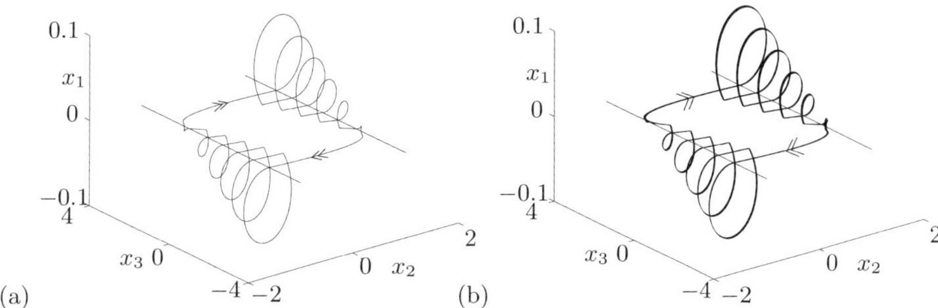

Fig. 1.21. Orbits of the three-dimensional relay system (1.22)–(1.24) with $b = (1,-2,1)^T$, $a_{31} = -5$ and (a) $a_{11} = 1.206$, $a_{21} = -99.9372$ and (b) $a_{11} = 1.35$, $a_{21} = -99.93$. The sliding region is the subset of $\{y = 0\}$ bounded by the two horizontal lines.

total of 12 separate pieces of sliding motion (the horizontal orbit segments). In Chapter 2 we shall explain how to define the dynamics on the switching set (or *switching manifold*) $\{y = 0\}$ in a consistent way as the limit of the dynamics above and below the set.

Now, panel (b) of Fig. 1.21 shows the attractor of the same system under a slight adjustment of its parameter values. The fact that this is chaotic attractor can be seen from the 'thickness' of the apparent single curve in the figure, which is in fact many different trajectories almost overlaid. This is not a transient; as one continues to compute the solution, so the thick region continues to be filled out. Not also that this attractor is asymmetric, which can be seen in the way the trajectories enter their final long pieces of sliding at the thin end of the two horn-like structures. On the right-hand side, there is a definite piece of non-sliding motion (the upwards blip) between the final piece of sliding in the horn and the long piece of sliding that causes the transition to the left-hand horn. At the end of the left-hand horn however, there is no distinguishable blip separating the two intervals of sliding motion.

One might imagine that a bifurcation happens as one continuously varies the parameters between those values used in panels (a) and (b) of Fig. 1.21. The bifurcation in question seems to involve interaction with the two horizontal lines in the figures. These represent the boundaries within the switching manifold $\{y = 0\}$ that delineate the region where sliding is possible. This is an example of a *discontinuity-induced bifurcation* (DIB), as were the grazing bifurcations that we encountered in the last section. The analysis of DIBs associated with sliding forms the subject of Chapter 8. In fact, a detailed explanation of the dynamical transitions involved in the relay system simulated in Fig. 1.21 forms the entire subject of Sec. 8.2.

1.3.2 Case study IV: A dry-friction oscillator

Another source of Filippov dynamics occurs in systems exhibiting dry friction. Friction plays an important role in engineering. It is the source of self-sustained stick-slip vibrations, which can cause undesired effects such as noisy machine operation, wear of components, squeaking doors and squealing railway wheels; see, e.g., [5, 70, 225]. Only in recent years, due to the introduction of new analytical techniques, have these systems been studied from the standpoint of bifurcation theory. For example, Popp and Stelter [224], introduced four different models for stick-slip motion, perhaps the simplest of which is a single degree-of-freedom oscillator with external forcing where chaotic behavior characterized by stick-slip was found. Several different routes to chaos were identified (e.g., period-doubling and intermittency). In later work, Popp and collaborators [131, 223] verified many of these theoretical results experimentally. Later, Galvanetto [110, 111, 113, 114] studied the yet more complex dynamics that can occur in a two block stick-slip system first envisaged by den Hartog [70]. Here a one-dimensional map is introduced for studying the discontinuity-induced bifurcations in the four-dimensional system that lead to transitions between pure slip and stick-slip motion. Similar bifurcations were also detected in a simplified version of the system [112].

Here, though, we shall focus on the forced vibrating system with dry-friction first studied by Yoshitake and Sueoka [277]. They consider a block that is free to move in a single horizontal axis, with mass M, and is attached to a fixed point by a spring of stiffness K. The block is subject to sinusoidal forcing and rests on a rough drive belt that moves with a constant velocity V (which without loss of generality can be scaled to be equal to 1) such that the interaction between the block and the belt is well approximated by a kinematic dry-friction law. Under non-dimensionalization, the mass, stiffness and drive-belt velocity can all be scaled to unity and the resulting equations of motion can then be expressed dimensionless form as

$$\ddot{u} + u = C(1 - \dot{u}) + A\cos(\nu t), \tag{1.25}$$

where

$$C(v) = \alpha_0 \text{sgn}(v) - \alpha_1 v + \alpha_2 v^3 \tag{1.26}$$

is the kinematic friction characteristic. Here $v = (1 - \dot{u})$ corresponds to the relative velocity between the driving belt and the moving block and α_i, $i = 0, 1, 2$ are positive constants that depend on the material characteristics of the block and belt, which can be fit to a particular set of experimental measurements. There is a large literature on the derivation of so-called Coulomb friction laws of the form (1.26) (see, for example, references in [131, 63]), which we shall not go into here. The parameter α_0 here represents the modulus of static friction times the assumed normal force onto the friction surface, and the motivation for the next two terms is that the size of dynamic friction (the tangential force that applies with $v \neq 0$) is typically lower than the static friction, at least for small relative velocities v.

The two remaining dimensionless parameters in the model (1.25) are the amplitude A and frequency ν of the sinusoidal forcing. Following [277] we shall take the values

$$\alpha_0 = \alpha_1 = 1.5, \quad \alpha_2 = 0.45, \quad A = 0.1$$

and allow the frequency ν to be a parameter that varies; that is, the *bifurcation parameter*.

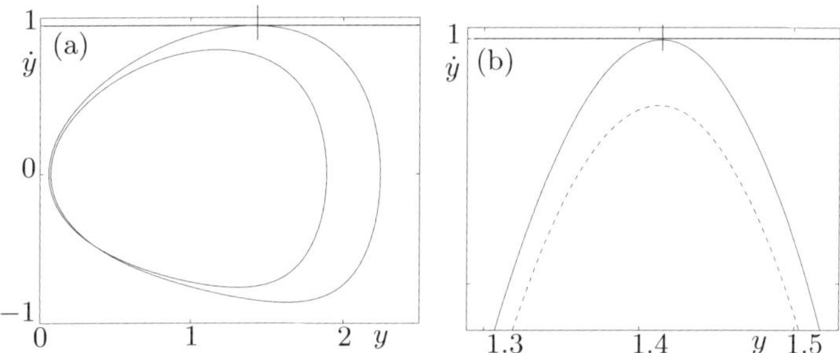

Fig. 1.22. (a) Orbit of (1.25) of period $4T$ $(8\pi/\nu)$ undergoing grazing-sliding bifurcation for $\nu = 1.7077997$. (b) Enlargement of the region where grazing-sliding occurs; the dot-dashed line corresponds to a stable periodic orbit for $\nu = 1.7082$ that clearly does not reach the switching manifold.

Let us focus in particular on the bifurcation scenario for $\nu \approx 1.7078$ where there is a transition from pure slip to stick-slip motion. Figure 1.22(a) shows an attracting $4T$-periodic orbit at precisely the transition (bifurcation) point. Here the orbit grazes the *switching manifold* (or discontinuity set) $\{\dot{u} = 1\}$. The point of tangency occurs at precisely the edge of the region where the switching manifold becomes attractive (that is the *sliding region*), which is indicated in the figure as the region to the left of the short vertical line. Figure 1.22(b) shows a zoom of this trajectory near the tangency and a non-sliding attracting period-$4T$ trajectory for slightly larger ν-values.

As the parameter ν is decreased, the Monte Carlo bifurcation diagram in Fig. 1.23(a) shows that the ω-limit set becomes a chaotic attractor. Details of the chaotic trajectory are shown in Fig. 1.23(b). Here we see that the chaotic behavior is composed of stick-slip motion; the trajectory repeatedly enters the stick phase (which corresponds to sliding motion of the equivalent relay control system) at different points. However, the trajectory always exits from stick with the same velocity and position, albeit with a different phase each time. We shall return to this example in Chapter 8, where the DIB in question is given the name *grazing-sliding bifurcation*.

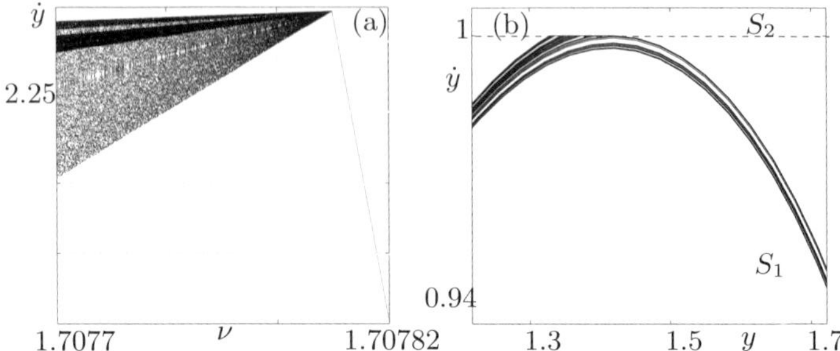

Fig. 1.23. (a) Bifurcation diagram obtained from the direct numerical simulation of (1.25), (1.26). (b) A portion of a chaotic trajectory zoomed into a neighborhood of the switching manifold for $\nu = 1.706$.

1.3.3 Case study V: A DC–DC converter

DC–DC converters are circuits that are used to change one DC voltage to another. In the past, this was done by converting the DC voltage to an AC one, passing this through a transformer, and then transforming the resulting AC voltage back to a DC one. This procedure results in significant energy loss and rather bulky devices. To convert between voltages with domestic electronic devices, such as laptop computers, something more compact and with less energy loss is needed. The DC–DC converters frequently employed use electronic switches to convert from one DC voltage to another, with negligible energy loss. Significantly, such mechanisms can be implemented using small solid state devices; see, e.g., [150]. The use of switches means that DC–DC converters represent inherently non-smooth dynamical systems, which when driven beyond their designed operating limits can give rise to complex dynamics of the form studied in this book. In fact, there is already a rich literature on the many possible forms of dynamics of DC–DC converters, including rapidly switching periodic and chaotic motions [104, 68, 82, 278, 75]. Our interest is, of course, in the special types of dynamics that arise due to the non-smooth nature of the switching process.

The simple DC–DC converter circuit illustrated in Figs. 1.24 and 1.25 aims to convert a constant input voltage E to a constant higher or lower voltage $\widehat{\gamma}$, by switching on and off the part of the circuit containing the input. If this switching process were governed by whether the output voltage $V(t)$ is greater or less than the constant desired value $\widehat{\gamma}$, then we would be in the situation of the central heating example. So, ignoring time-delays and latency in the circuit, the motion would slide along the surface $V = \widehat{\gamma}$. In practice this would lead to rapid switching, which would lead to undesirable effects like inefficiency, overheating and the excitation of overtones. Instead, a standard technique is to use a pulse-width modulated feedback control where the output $V(t)$ is compared with a reference voltage $V_r(t)$ taking the form of a low-amplitude periodic ramp function centered around $\widehat{\gamma}$ (see Fig. 1.25).

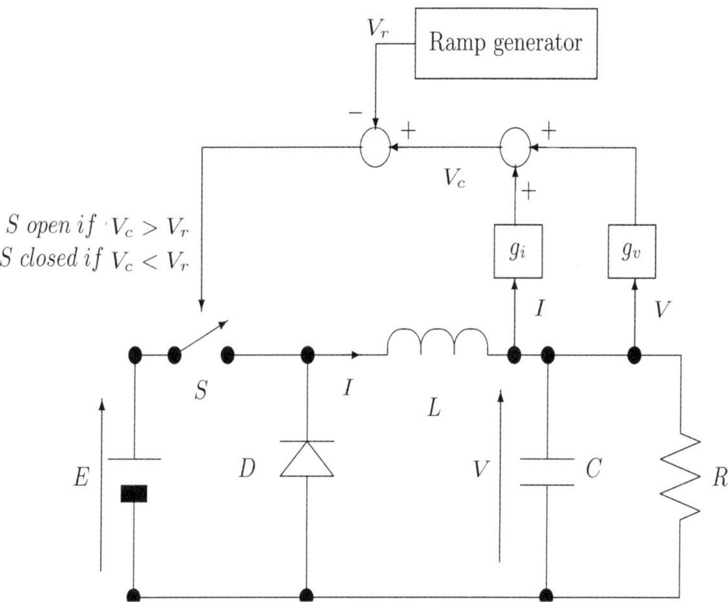

Fig. 1.24. Schematic diagram of a simple DC–DC buck converter circuit.

Thus, the current $I(t)$ and voltage $V(t)$ inside the circuit evolve in a smooth (in fact, to good approximation, linear) manner between switching events, which arise when $V(t) = V_r(t)$.

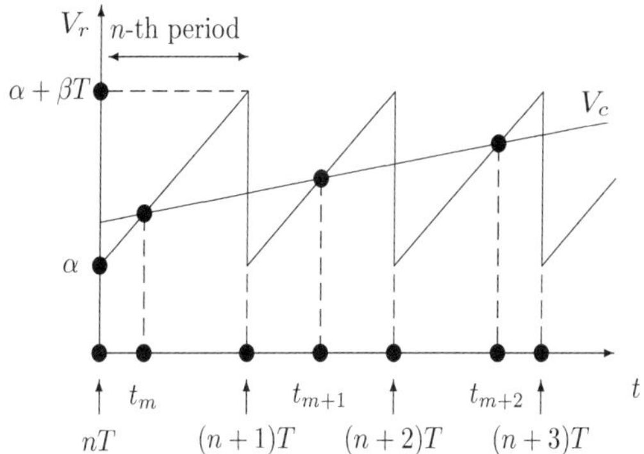

Fig. 1.25. The pulse-width modulated signal $V_r(t)$ and the normal operating condition, with one crossing of the ramp and one crossing of the discontinuity per period T. Here $\gamma = \widehat{\gamma} - \eta T/2$.

Straightforward linear circuit theory (Kirchhoff's laws) can be used to describe the dynamics of this circuit, leading to the equations [75]

$$\dot{V} = -\frac{1}{RC}V + \frac{I}{C}, \tag{1.27}$$

$$\dot{I} = -\frac{V}{L} + \begin{cases} 0 & \text{for } V \geq V_r(t), \\ E/L & \text{for } V < V_r(t) \end{cases} \tag{1.28}$$

for the output voltage $V(t)$ and corresponding current $I(t)$. Here C, E, L and R are positive constants representing the capacitance, battery voltage, inductance and resistance, respectively, of the components depicted in Fig. 1.24. The reference voltage V_r is a piecewise-linear but discontinuous 'ramp' signal

$$V_r(t) = \gamma + \eta(t \bmod T), \qquad \gamma,\, \eta,\, T > 0,$$

represented in Fig.1.25. The parameter values taken in this case study are those used in the experiments of Deane and Hamill [68];

$$\begin{aligned} &R = 22\,\Omega,\ C = 4.7\mu\,\mathrm{F},\ L = 20\,\mathrm{mH},\ T = 400\,\mu\mathrm{s}, \\ &\gamma = 11.75238\,\mathrm{V},\ \eta = 1309.524\,\mathrm{Vs}^{-1}. \end{aligned} \tag{1.29}$$

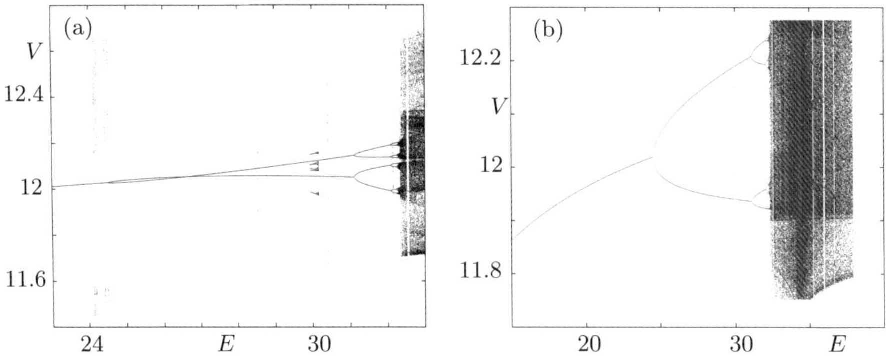

Fig. 1.26. DC–DC converter bifurcation diagram, obtained by direct numerical simulation. (a) Using a stroboscopic Poincaré map, sampling every time the ramp signal has its discontinuity; adopting a Monte Carlo approach to show competing attractors. (b) Using a 'crossing map' sampled every time the smooth part of the ramp is crossed, just showing the fine structure of the fundamental attractor for $E > 32.34$.

The Monte Carlo bifurcation diagrams in Fig. 1.26 summarize the dynamics observed upon varying the input voltage E as a bifurcation parameter. For sufficiently small E, the ω-limit set comprises a single stable periodic $(1, 1)$ orbit — that is, having one crossing of the ramp function per period

T. As the input voltage is increased through $E = 24.516$, this periodic orbit undergoes a period-doubling cascade [seen more clearly in the 'crossing map' in Fig. 1.26(b)]. However, the usual sequence leading to chaos is suddenly interrupted by an abrupt enlargement of the resulting chaotic attractor for $E \approx 32.342$. Note in addition, that there are ranges of E-values (e.g., around 24.5 and 30) where several periodic and chaotic attractors with small basins of attractions coexist with the fundamental dynamics (the additional streaks of points in Fig. 1.26, which are *not* printing errors!). These attractors are built around structures that repeat every 3T, 6T or 12T.

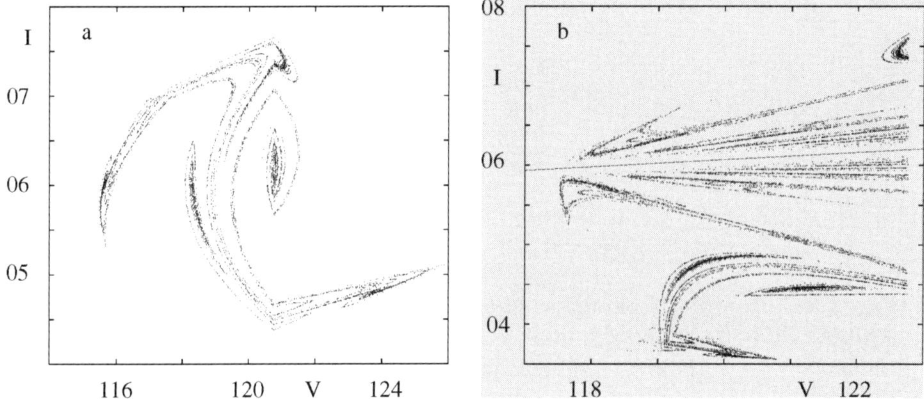

Fig. 1.27. Large scale chaotic attractor for $E = 35V$: (a) using a stroboscopic Poincaré map; (b) plotting every time the ramp signal is crossed. The dashed line corresponds to the sliding line corresponding to trajectories that slide along the ramp $V(t) \equiv V_r(t)$.

The broadband chaos that is observed for $E > 32.342$ is depicted in Fig. 1.27(a) in a *Poincaré section* where points are plotted every time they cross the ramp discontinuity in V_r (i.e., at fixed time intervals T). This attractor is built around structures that repeat every 5T. Note that there are five highly populated, dark regions in Fig. 1.27(a) interspersed by clouds of rarer points. One of these dark regions forms a spiral centered on $(V, I) = (12.28, 0.62)$ which appears to have sharp *corners* when $v \approx 12.28$. When the same attractor is viewed in a Poincaré section defined by each crossing of the ramp, Fig. 1.27(b), the spiral region corresponds to the fingered structure, not unlike that observed for the impact oscillator in Fig. 1.19(b). Plotting graphs of solutions, Fig. 1.28, reveals that this fingering is associated with multiple switchings occurring in one of the five periods owing to the trajectory being close to a *sliding orbit* lying entirely on the switching set $\{V = V_r\}$. Further intricate details of this chaotic attractor are described in [104]. Note that such chaos has also been observed experimentally, where when output audibly was described as "*a raucous whine, like frying bacon*"; see [68].

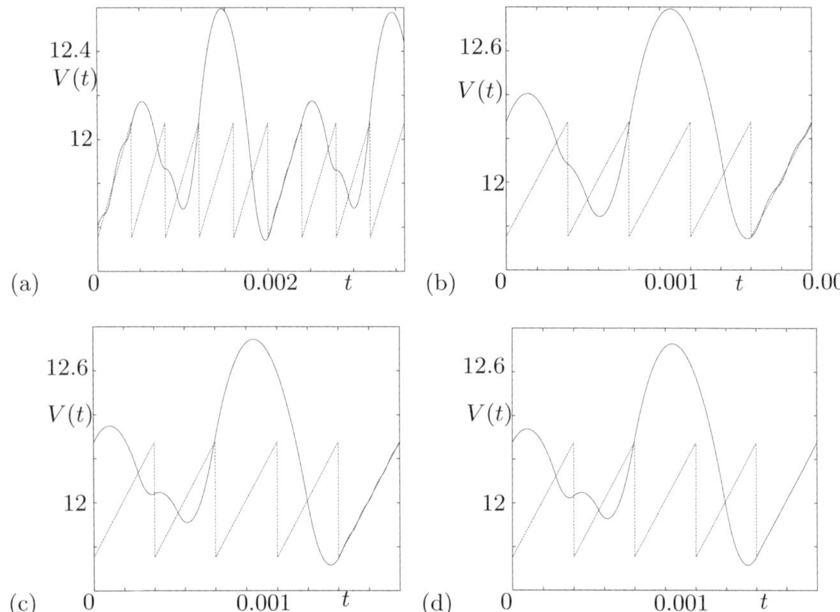

Fig. 1.28. Trajectories plotted as graphs of $V(t)$ for $E = 34.33998$: (a) a chaotic solution, (b) a $(5, 5)$ periodic orbit, (c) a $(10, 5)$ periodic orbit, and (d) a (near) sliding periodic trajectory.

Aside from the small-basin-of-attraction 3-, 6- and 12-periodic windows, the observed dynamics (including period-doubling and chaos) for $E < 32.342$ is all of type (n, n), that is, with one switching per ramp cycle. The sudden enlargement can best be characterized by the point at which the average number of switchings per period on the observed attractor first goes above 1. This quantity reaches a peak at around $E = 34.34$, which coincides with the parameter value at which we can compute a 5-periodic orbit that appears to slide, that is, to become tangent to the ramp [see Fig. 1.28(d)]. In a formal sense, this trajectory resembles a chattering orbit and may be labeled $(\infty, 5)$, and numerically we can find nearby (unstable) $(m, 5)$ orbits for m apparently arbitrarily large. Figure 1.28 depicts two such periodic solutions. Numerical calculations and analytical evidence reported in [75] suggest that these five-periodic orbits lie in an approximate double spiral accumulating on a such a sliding trajectory. The corners in the bifurcation diagram (and indeed the corresponding sharp corners in the spiral structure of Fig. 1.27) can be explained by the theory that we shall present in Chapter 7.

1.4 Non-smooth one-dimensional maps

A substantial part of this book will be devoted to a detailed study of non-smooth discrete-time dynamical systems. That is, iterated maps whose functional form is smooth in separate regions of their domain of definition, but that may have discontinuities across certain sets. These discontinuities may be in a derivative of the map, or, more severely, in its value. In Chapters 3 and 4, we will give a detailed account of the theory of piecewise-linear and related maps. Such maps are often intrinsic models of interest in their own right, or they may arise as Poincaré maps in the neighborhood of cyclic behavior of a continuous-time dynamical system. At the heart of this book, in Chapters 6, 7 and 8, is the derivation of approximate non-smooth maps in the neighborhood of DIB points in continuous-time dynamical systems.

In this introduction we shall introduce the topic by briefly presenting the dynamics of three case study one-dimensional maps, each with a different kind of discontinuity. Now, in the history of nonlinear dynamics research, the theory of one-dimensional maps has played a crucial role. Indeed it was in this context that period-doubling cascades and chaos were first described; see e.g., [73]. In the context of describing the chaotic dynamics of such maps, one-dimensional piecewise-linear maps with a single corner (so-called *tent maps*) are often considered as simple explicitly calculable examples. For ease of explanation, it is often assumed that no iterate of the map passes through the point of discontinuity. In this book however, we shall focus on the specific consequences of when a fixed or periodic point of the map passes through the discontinuity in a so-called *border-collision* bifurcation [206].

1.4.1 Case study VI: A simple model of irregular heartbeats

We begin with a simple conceptual model that attempts to use mathematical modeling to describe human physiology; see for example the book [151] for general background. Irregular heartbeats can have a variety of causes and, depending on the type, can range from being fatal to just mildly unpleasant. A particularly dangerous kind of problem occurs when re-entrant waves are set up that cause the heart tissue to no longer function as a pump [275]. A much less extreme irregularity occurs due to poor conduction in the atrioventricular (AV) node, a critical collection of cells on the surface of the heart that transfers the electrical signal spreading through the atria to the ventricles. Conduction is slow through the AV node, but then rapidly spreads out along approximately one-dimensional fibers (the so-called bundle of HIS) into the ventricular heart tissue. Thus the AV node is crucial in setting up the synchronous contraction of the ventricles that pumps blood around the body. Failure of the AV node to 'fire' is not in general life-threatening but can be distinctly unpleasant as it leads to temporal disruption to a regular heart rhythm, such as skipped beats.

Keener and Sneyd [151, Sec. 14.3] consider a simple model for such irregular heartbeats. The model supposes that the AV node is a collection of cells subjected to a periodic signal $\varphi(t)$ arriving from the atria. The cells are excitable, in that when the input reaches a threshold $\theta(t)$, they 'fire' electrical impulses into the ventricles. After firing (at time t_n), the threshold increases

$$\theta(t_n^+) = \theta(t_n^-) + \Delta = \varphi(t_n) + \Delta,$$

by some constant $\Delta > 0$, to allow the cells time to recover. Then, the threshold slowly relaxes according to a law

$$\theta(t) = \theta_0 + (\theta(t_n^+) - \theta_0)e^{-\gamma(t-t_n)}, \qquad t > t_n,$$

where γ and θ_0 are positive constants representing the decay rate and the base value of the threshold, respectively. Finally, firing occurs again when $\theta(t_{n+1}) = \varphi(t_n)$.

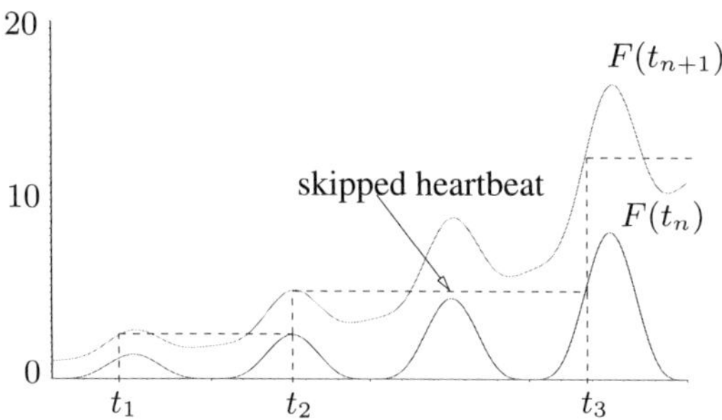

Fig. 1.29. Construction of the firing time map for the AV-node model according to the implicit equations (1.30) and (1.31) for the specific case $\varphi = \sin^4(\pi t)$, $\Delta = 1$, $\gamma = 0.55$.

Thus, we have a map of firing times $t_{n+1} = G(t_n)$ defined implicitly by

$$\varphi(t_{n+1}) = \theta_0 + [\varphi(t_n) + \Delta - \theta_0]e^{-\gamma(t_{n+1}-t_n)}.$$

However, this map can be simplified by setting

$$F(t) = (\varphi(t) - \theta_0)e^{\gamma t}, \tag{1.30}$$

to obtain

$$F(t_{n+1}) = F(t_n) + \Delta e^{\gamma t_n}, \tag{1.31}$$

which must be solved for the smallest $t_{n+1} > t_n$. Note that this map has a fundamental *discontinuity* corresponding to a t_n value at which $G(t_n)$ has a local extremum; see Fig. 1.29, which depicts the construction of the sequence of firing times t_1, t_2, etc. according to this implicit map.

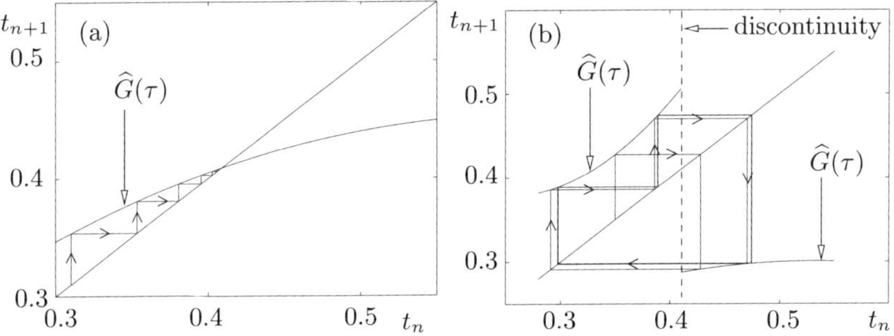

Fig. 1.30. The Keener and Sneyd map 1.30, 1.31 in the case (a) $\gamma = 0.55$ and (b) $\gamma = 0.8$, depicting the dynamics of the map via cobweb diagrams.

Despite being only implicitly defined, Keener and Sneyd show that the map can be represented graphically by using a re-scaled firing time variable $\tau_n = \frac{t_n - k_n T}{T}$, where k_n is the uniquely defined integer that puts $\tau \in [0, 1)$. Figure 1.30 depicts the dynamics of the map defined in this way for two values of the decay rate γ. Here we present the results in the form of a *cobweb diagram*. Starting from a value for τ_0, we compute $\widehat{G}(\tau_0)$ and reflect in the 45° line to produce a new value $\tau_1 = \widehat{G}(\tau_0)$. And so the process repeats. At the second iterate, we compute $\widehat{G}(\tau_1)$ and reflect in the 45° line to obtain τ_2, and so on. This is a good way of visualizing trajectories of one-dimensional discrete-time dynamical systems.

If γ is sufficiently small, as in Fig. 1.30(a), we see that the period-one fixed point is stable. All initial τ-values are eventually attracted to it via the cobweb process. This fixed point corresponds to a regular heartbeat where the AV node fires every time it receives a stimulus. However, transient effects may be important; for example, a large positive perturbation to t_n can cause a failure to fire (an iterate to the right of the discontinuity) for one beat. Systems that display this kind of dynamics, where small positive perturbations from a stable state can lead to large excursions, are often referred to as *excitable*, and examples abound in bio-medical systems and temperature-sensitive chemical reactions (see, e.g., [151, 231]). For $\gamma = 0.8$ [Fig. 1.30(b)], the simple fixed point has disappeared through interaction with the discontinuity. Instead we now see a pattern of iterates that involves several firings in a row interspersed by a skipped beat.

The theory of *discontinuous maps* in Chapter 4 indicates that the precise pattern of skipped beats is likely to be highly sensitive to changes in γ but will in general be periodic (with possibly very high period) for almost all γ-values. Changes to the pattern occur whenever one point on the attractor passes through the discontinuity point.

1.4.2 Case study VII: A square-root map

Piecewise square-root maps are continuous but have an unbounded slope on one side of a discontinuity boundary. Consider the simple continuous square-root map illustrated in Fig. 1.31 [197, 171, 198]:

$$x \mapsto f(x), \quad \text{where } f(x) = \begin{cases} \sqrt{\sigma - x} + \lambda\sigma, & \text{if} \quad x < \sigma, \\ \lambda x, & \text{if} \quad x \geq \sigma. \end{cases} \tag{1.32}$$

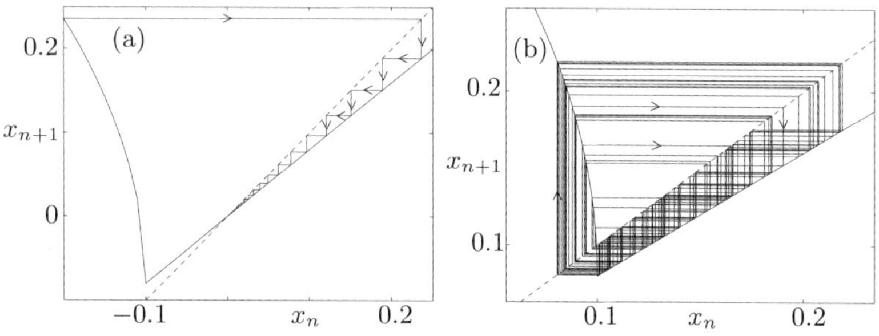

Fig. 1.31. Dynamics of the one-dimensional square root map (1.32) (a) for $\sigma = -0.1$ and (b) for $\sigma = 0.1$, depicting the dynamics of the map via cobweb diagram.

This map is designed to give insight into the dynamics observed in Figs. 1.15–1.17 close to a grazing bifurcation in an impact oscillator, as illustrated in Fig. 1.18. This connection is made more precise in Chapter 6. For example, for the single degree-of-freedom impact oscillator, the fixed point set of the associated Poincaré mapunder variation of σ, Fig. 1.15(a) has exactly the form of a linear piece and a piece with a square-root singularity. This is no accident. Although for such a system the true stroboscopic or impact map will be two-dimensional, we can understand a lot about the behavior of impacting systems by looking at a much simpler one-dimensional map with a square-root singularity. Consider

Here we take σ as the primary bifurcation parameter, which plays the role of the obstacle position in the impact oscillator. If $\sigma < 0$, then the map has a single, stable fixed point at $x = 0$. A *border-collision* of this fixed point then occurs as σ passes through zero. For $\sigma > 0$ we see more complicated behavior. The behavior at the grazing bifurcation depends crucially on λ, which is like $e^{-\zeta}$, where ζ is the damping coefficient of the impact oscillator.

Three different cases can be identified:

Low damping, $2/3 < \lambda < 1$: As σ passes through zero, there is an immediate creation of an *intermittent* chaotic motion characterized by a regular sequence of iterates on the linear side $x > \sigma$ interspersed by a single iterate on the square-root side $x < \sigma$. There is an interval $0 < \sigma < \sigma'$ for which this motion is the only stable behavior. The x-interval spanned by the chaotic motion has size proportional to $\sqrt{\sigma}$.

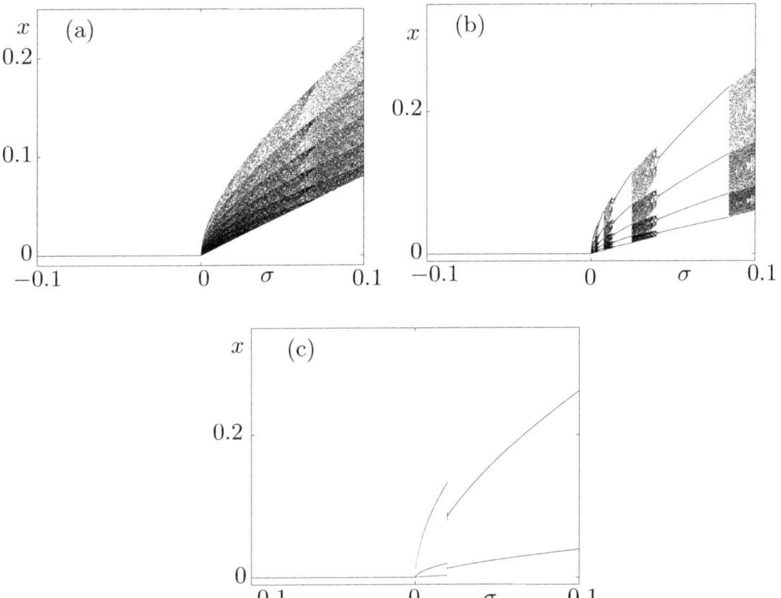

Fig. 1.32. Bifurcation diagram of the one-dimensional square-root map (1.32) in the cases (a) $\lambda = 0.8$ (fully chaotic), (b) $\lambda = 0.6$ (period-adding with alternating chaos) and (c) $\lambda = 0.15$ (period-adding).

Intermediate damping, $1/4 < \lambda < 2/3$: Here there is an infinite sequence of windows of stable periodic motion alternating with bands of chaos. Each periodic window contains a unique stable period-n orbit (repeats after n iterates of the map) with all but one iterate lying on the linear side. The windows are arranged in a *period-adding* cascade so that a window containing an n-periodic orbit is preceded (upon increasing σ) by an $(n-1)$-periodic window. The width and the location of the windows decrease geometrically as $\sigma \to 0$.

High damping, $0 < \lambda < 1/4$: In this case no chaotic motion occurs, but the period-adding windows continue to exist. The windows overlap, for small intervals of σ, thus giving rise to multiple periodic attractors at the same parameter values.

Figure 1.32 plots the ω-limit sets of the iterations of the map F as functions of σ for the cases of $\lambda = 0.8, \lambda = 0.6$ and $\lambda = 0.15$. The figure illustrates each of the three cases above: chaotic, period-adding with chaos and pure period-adding behavior. Note from each figure that there is clear evidence of scaling laws governing both the size of the orbits and the width of the windows. An explanation of how the these scalings arise will be given in Chapter 4, along with a discussion of the dynamics of multi-dimensional square-root maps.

That maps of this form arise naturally from grazing bifurcations in impact oscillators will be the main subject of Chapter 6.

1.4.3 Case study VIII: A continuous piecewise-linear map

Even simpler than square-root maps are those that are completely linear in each of two halves of their domain and yet are continuous across the region where these two domains join. As we shall see in Chapters 7 and 8, maps of this form can be used to explain the dynamics observed in the friction oscillator and DC–DC converter case studies. Partly owing to the relative ease of analyzing such systems, there is already a considerable literature on the border-collision bifurcations of piecewise-linear maps; e.g., [204, 21, 80, 99]. However, care should be taken in interpreting these results as having meaning for understanding the dynamics of piecewise-smooth flows. For example, in Chapter 7 we shall see that a grazing in a piecewise-smooth ordinary differential equation does *not* lead to a locally piecewise-linear Poincaré map. Chapter 3 is devoted to a detailed analysis of border-collisions in general continuous locally piecewise-smooth maps.

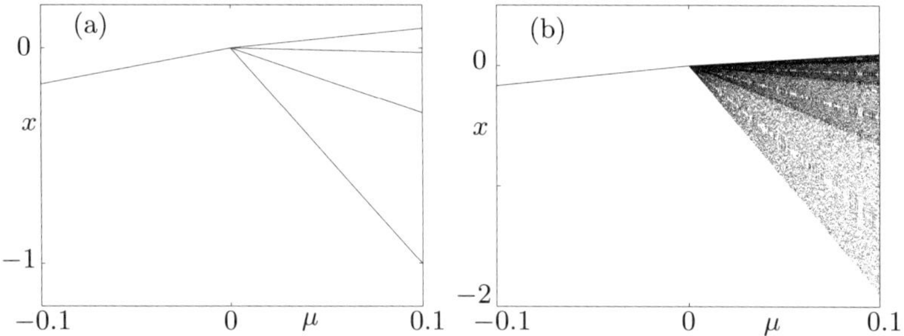

Fig. 1.33. Monte Carlo bifurcation diagrams of the piecewise-linear map (1.33) for $\alpha = 0.4$ and (a) $\beta = -12$, (b) $\beta = -20$.

Let us focus here on the particular case of one-dimensional maps that, without loss of generality, can be written in the form

$$x \mapsto f(x), \quad i = 1, 2 \qquad \text{where} \quad f = \begin{cases} F_1 = \alpha x + \mu, \text{ if } x \leq 0, \\ F_2 = \beta x + \mu, \text{ if } x > 0, \end{cases} \tag{1.33}$$

depending on three real parameters μ, α and β. The most interesting dynamics occurs for $\alpha > 0$ and $\beta < 0$. Note that by introducing the rescaling $\tilde{x} = x/|\mu|$, we can assume without loss of generality that $\mu = \pm 1$. However, our primary interest is the border-collision bifurcation that occurs as μ varies through zero, for which parameter value there is a trivial fixed point at $x = 0$. Thus, treating μ as the bifurcation parameter, we see that the dynamics is scale invariant;

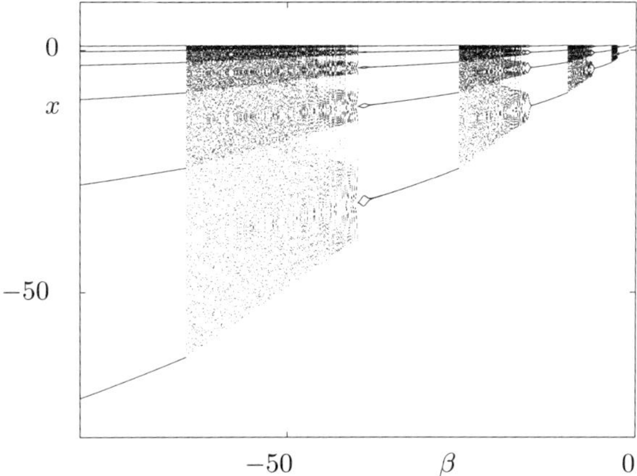

Fig. 1.34. Monte Carlo bifurcation diagram of (1.33) for $\mu = 1$, $\alpha = 0.4$ and $\beta \in (-80, 0)$.

that is, all dynamics for μ of a certain sign can be mapped trivially into the dynamics for $|\mu| = 1$.

Suppose, for example, that $0 < \alpha < 1$. Then for $\mu < 0$ it is easy to see that there is a unique stable fixed point at $x = \mu/(1-\alpha) > 0$. Figure 1.33 shows what can happen for $\mu > 0$. Depending on the values of α and β, the fixed point can spontaneously bifurcate into either a higher period orbit or to a chaotic attractor. It is perhaps more interesting to see what happens as we vary α or β for $\mu < 1$. Figure 1.34 shows results for variation of β for fixed α. Note that we see a period-adding cascade very much like for the square-root map in the previous case study. However, the scaling of the size of the periodic windows are rather different, and the limit of a period-∞ orbit is reached as $\beta \to -\infty$ rather than $\sigma \to 0$ as it was for the square-root map. An explanation of this scaling, in fact precise values for the boundaries of the periodic windows, will be given in Chapter 3. In between the periodic windows one sees chaotic motion. Here we see perhaps the simplest example of a system that generates chaos. Moreover the chaos is *robust* [23] in that inside one of these chaotic parameter intervals there is no *small* adjustment one can make to the parameters of the system that collapses the system onto a periodic rather than a chaotic attractor. For this reason, this simple chaotic map would be a good candidate for a chaotic signal generator, for example, for use in chaotic communications [139]

One way of motivating the rest of this book is as a means of completing an analysis of the above case study examples, and indeed we shall return to each of them again. Before we do so, to enable us to generalize such analyses to other example system, we will need to set up a mathematical framework with which to describe different classes of piecewise-smooth systems and their discontinuity-induced bifurcations. That is the purpose of the next chapter.

2

Qualitative theory of non-smooth dynamical systems

In this chapter, we give an overview of the basic theory of both smooth and non-smooth dynamical systems, to be expanded upon in later chapters. In particular we shall define what we meant by each of the *italicized terms* encountered in Chapter 1. We start with the definition of a dynamical system and review the essential concepts from the theory of smooth dynamical systems that can also apply to non-smooth systems. This material is available in the now many textbooks on nonlinear dynamics and chaos, and so only the briefest of details are given, with appropriate references. Next, in Sec. 2.2, we define carefully what we mean by the different classes of piecewise-smooth dynamical systems that we treat. In Sec. 2.3, we point out the relation to some of the other mathematical formalisms that exist for defining non-smooth systems. Section 2.4 considers notions of stability and bifurcation in non-smooth systems and introduces the key concept of the book, that of *discontinuity-induced bifurcation* (DIB), where an invariant set changes its topology with respect to the set of discontinuity surfaces. This is naturally followed by Sec. 2.5, which explains the idea of a *discontinuity mapping* (DM) which is the main analytical tool to be used in Chapters 6–8. The chapter ends with a brief discussion in Sec. 2.6 of numerical methods for simulation, parameter continuation and bifurcation detection in non-smooth systems.

2.1 Smooth dynamical systems

The qualitative theory of differential equations [7, 124, 273, 168] begins with a quite general definition of a dynamical system. This is written in terms of an n-dimensional state space (or phase space) $X \subset \mathbb{R}^n$ (with the usual topology) and an evolution operator ϕ that takes elements x_0 of the phase space and evolves them through a 'time' t to a state x_t

$$\phi^t : X \to X, \qquad x_t = \phi^t(x_0).$$

The time t takes values in an index set T, which we usually consider to be either discrete (the integers $\mathbb{Z}$) or continuous (the real numbers $\mathbb{R}$). Note that ϕ may not be uniquely defined for all $t \in T$. For example, so called *noninvertible* dynamical systems may not be defined for $t < 0$; or in certain systems, some initial states x_0 may diverge to infinity in a finite time. Formally, in these cases we need to define a space-dependent subset $T^*(X) \subset T$ such that $\phi^t(x)$ is uniquely defined for $x \in X$ provided $t \in T^*(X)$. We shall, however, ignore such technicalities, other than to state that only positive time should be taken for noninvertible systems.

Definition 2.1. *A state space X, index set T and evolution operator ϕ^t are said to define a* **dynamical system** *if*

$$\phi^0(x) = x, \quad \textit{for all} \quad x \in X, \tag{2.1}$$
$$\phi^{t+s}(x) = \phi^s(\phi^t(x)) \quad \textit{for all} \quad x \in X, \quad t, s \in T. \tag{2.2}$$

The set of all points $\phi^t(x)$ for all $t \in T$ is called the **trajectory** *or* **orbit** *through the point x.*

The **phase potrait** *of the dynamical system is the partitioning of the state space into orbits.*

Remarks

1. Properties (2.1) and (2.2) define ϕ^t to be a *semi-group*.
2. When the dynamical system is invertible (uniquely defined for $t < 0$ as well as for $t > 0$), then we have the additional property that is a consequence of (2.1) and (2.2)
$$\phi^t \phi^{-t} = \text{id}.$$

Definition 2.2. *A dynamical system satisfying (2.1) and (2.2) is said to be* **smooth of index** *r, or C^r, if the first r derivatives of ϕ with respect to x exist and are continuous at every point $x \in X$.*

We shall often be interested in dynamics that is, in some sense, recurrent or repeatable. Specifically, we will gain an understanding of the phase space structure by from specific sets that remain invariant under the system dynamics.

Definition 2.3. *An* **invariant set** *of a dynamical system (2.1), (2.2) is a subset $\Lambda \subset X$ such that $x_0 \in \Lambda$ implies $\phi^t(x_0) \in \Lambda$ for all $t \in T$. An invariant set that is closed (contains its own boundary) and bounded is called an* **attractor** *if*

1. *for any sufficiently small neighborhood $U \subset X$ of Λ, there exists a neighborhood V of Λ such that $\phi^t(x) \in U$ for all $x \in V$ and all $t > 0$, and*
2. *for all $x \in U$, $\phi^t(x) \to \Lambda$ as $t \to \infty$.*

The set of all attractors of a given system typically describes the long-term observable dynamics. A given dynamical system may have many competing attractors, with their relative importance being indicated by the size of the set of initial conditions that they attract; that is, their *domain of attraction*

Definition 2.4. *The* **domain of attraction** *(also known as the basin of attraction) of an attractor Λ is the maximal set U for which $x \in U$ implies $\phi^t(x) \to \Lambda$ as $t \to \infty$.*

We already saw in Fig. 1.14(a) that domains of attraction in non-smooth systems can have remarkably complicated structures, which can be true in smooth systems too.

Another useful notion is to define points in phase space that are eventually approached infinitely often in the future, or were approached infinitely often in the past.

Definition 2.5. *A point p is an* **ω-limit point** *of a trajectory $\phi^t(x_0)$ if there exists a sequence of times $t_1 < t_2 < \ldots$ with $t_i \to \infty$ as $i \to \infty$ such that $\phi^{t_i}(x_0) \to p$ as $t_i \to \infty$. If instead there exists a sequence of times with $t_1 > t_2 > \ldots$ and $t_i \to -\infty$ and $\phi^{t_i}(x_0) \to p$, then we say that p is an* **α-limit point** *of x_0. The ω- (α-) limit set of x_0 is the set of all possible ω- (α-) limit points. The set of all such ω-limit points (or α-limit points) for all $x_0 \in X$ is called the* **ω-limit set** *(or* **α-limit set***) of the system. This set is closed and invariant.*

An ω-limit point is sometimes called a *recurrent point* of the dynamical system.

There is only so much that can be gained from this abstract definition of a dynamical system. Its usefulness is that it defines properties like attractors, and domains of attraction for quite general classes of system such as partial differential equations, systems with time delays and discrete-valued systems. However, when dealing with smooth systems, we shall largely only be interested in cases where the state space X is (possibly some subset of) Euclidean space $\mathbb{R}^n$ and the evolution is either described by a discrete-time map or a continuous-time flow. We now take each in turn.

2.1.1 Ordinary differential equations (flows)

Given a system of *ordinary differential equations* (ODEs)

$$\dot{x} = f(x), \quad x \in \mathcal{D} \subset \mathbb{R}^n, \tag{2.3}$$

where $\mathcal{D}$ is a domain, then $\{X, T, \phi^t\}$ defines a dynamical system if we set $X = \mathcal{D}$ and $T = \mathbb{R}$ and let $\phi^t(x) := \Phi(x,t)$ be the solution operator or *flow* that takes initial conditions x up to their solution at time t:

$$\frac{\partial}{\partial t}\Phi(x,t) = f(\Phi(x,t)), \qquad \Phi(x,0) = x\,. \tag{2.4}$$

Remarks

1. If we suppose the vector field f is C^{r-1} for some $r > 2$, then (2.4) implies the flow $\Phi(x,t)$ is one index smoother; that is, the dynamical system is C^r, since f is a derivative of Φ.
2. Note that we have not included in the above the possibility that the vector field f depends explicitly on time t. However such systems can be treated within the general framework by allowing time to be an additional dynamical state. For example, taking the $(n+1)$st state $x_{n+1} = t$, implies $\dot{x}_{n+1} = 1$ so that the $(n+1)$st component of f is unity. In many examples time appears periodically, and then it can be helpful to consider the phase space to be cylindrical:

Example 2.1 (A periodically forced system). Consider the forced system

$$\ddot{u} + 2\zeta\dot{u} + ku = a\cos(\omega t). \tag{2.5}$$

If we set $X = \mathbb{R}^2 \times S^1 \subset \mathbb{R}^3$, with $x_3 = t \bmod(2\pi/\omega)$, we obtain

$$\begin{aligned}
\dot{x}_1 &= x_2, \\
\dot{x}_2 &= -kx_1 - 2\zeta x_2 + ax_3, \\
\dot{x}_3 &= 1,
\end{aligned}$$

with corresponding phase potrait depicted in Fig. 2.1.

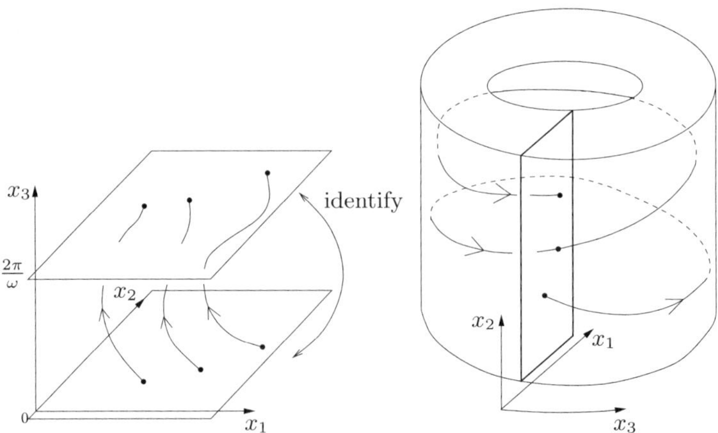

Fig. 2.1. Schematic description of the cylindrical phase space associated with the periodically forced system (2.5).

We shall be concerned with systems that depend on parameters. So we shall often write

$$\dot{x} = f(x, \mu),$$

where $\mu \in \mathbb{R}^p$ is a set of parameters. If we say that f is smooth, we mean that the dependence on μ is as smooth as it is on x. Unless it is crucial, we shall often drop the explicit parameter dependence of f and return to the more compact notation $f(x)$.

Systems of ODEs can exhibit the following kinds of invariant sets, see Fig. 2.2.

Equilibria. The simplest form of an invariant set of an ODE is an *equilibrium* solution x^* which satisfies $f(x^*) = 0$. These are also sometimes called *stationary points* of the flow since $\Phi(x^*, t) = \Phi(x^*, 0)$ for all t.

limit cycles. The next most complex kind of invariant set would be a *periodic orbit*, which is determined by an initial condition x_p and a period T. Here T is defined as the smallest time $T > 0$ for which $\Phi(x_p, T) = x_p$. periodic orbits form closed curves in phase space (topologically they are circular). A periodic orbit that is isolated (does not have any other periodic orbits in its neighborhood) is termed a *limit cycle*.

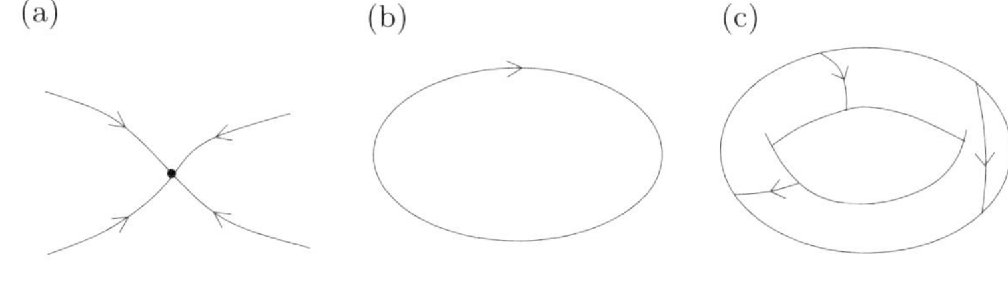

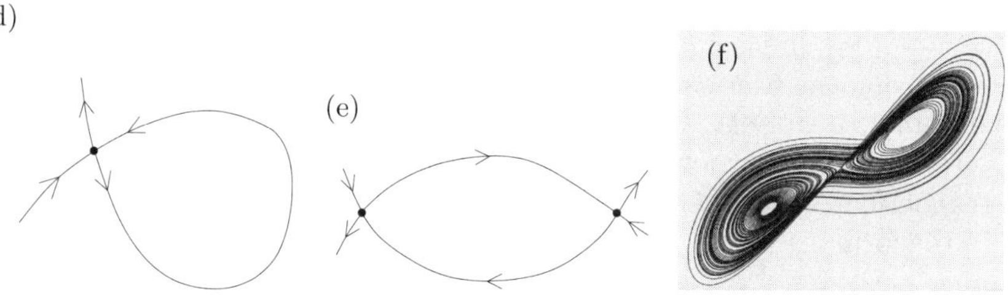

Fig. 2.2. Phase potrait representation of invariant sets of smooth flows: (a) equilibrium, (b) limit cycle, (c) invariant torus, (d) homoclinic orbit, (e) heteroclinic orbit, (f) chaotic attractor.

Invariant tori. Invariant tori are the nonlinear equivalent of two-frequency motion (see Fig. 2.3). Flow on a torus may be genuinely quasi-periodic in that it contains no periodic orbits, or it may be *phase locked* into containing a stable and an unstable periodic orbit, which wind a given number of times around the torus.

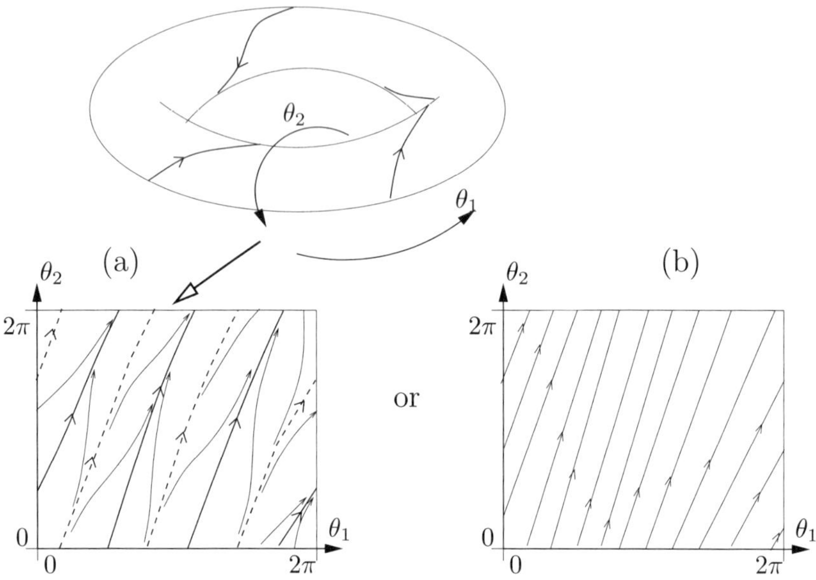

Fig. 2.3. Possible motion on an invariant torus, (a) phase locked, and (b) quasi-periodic.

Homoclinic and heteroclinic orbits. Another important class of invariant sets are *connecting orbits*, which tend to other invariant sets as time asymptotes to $+\infty$ and to $-\infty$. Consider, for example, orbits that connect equilibria. A *homoclinic orbit* is a trajectory $x(t)$ that connects an equilibrium x^* to itself; $x(t) \to x^*$ as $t \to \pm\infty$. A *heteroclinic orbit* connects two different equilibria x_1^* and x_2^*; $x(t) \to x_1^*$ as $t \to -\infty$ and $x(t) \to x_2^*$ as $t \to +\infty$. Homoclinic and heteroclinic orbits play an important role in separating the basins of attraction of other invariant sets.

Other invariant sets. It is quite possible for dynamical systems to contain certain simple geometric subsets of phase space where trajectories must remain for all time once they enter. For example, an ODE system written in the form

$$\begin{aligned}\dot{x}_1 &= f_1(x_1, x_2, x_3)\\ \dot{x}_2 &= x_1 f_2(x_1, x_2, x_3)\\ \dot{x}_3 &= x_1 f_3(x_1, x_2, x_3)\end{aligned}$$

for smooth functions f_i, $i = 1, \ldots, 3$ has as an invariant set the plane $\{(x_1, x_2, x_3) \in \mathbb{R}^3 : x_1 = 0\}$. The dynamics on this invariant plane could contain equilibria, periodic orbits and other attractors. Similarly, in addition to invariant tori, flows can contain invariant spheres, cylinders, etc. invariant sets that are everywhere locally smoothly described by an m-dimensional set of co-ordinates are called *invariant manifolds*, important examples of which are stable and unstable manifolds of saddle points, which we shall encounter shortly.

Chaos. More complex invariant sets are *chaotic*, a term that might be defined in a number of different ways, but we suppose:

Definition 2.6. *A closed and bounded invariant set Λ is called* **chaotic** *if it satisfies the two additional conditions:*

1. *It has* **sensitive dependence on initial conditions***; i.e.:*

   There exists an $\varepsilon > 0$ such that, for any $x \in \Lambda$, and any neighborhood $U \subset \Lambda$ of x, there exists $y \in U$ and $t > 0$ such that $|\phi^t(x) - \phi^t(y)| > \varepsilon$
2. *There exists a* **dense trajectory** *that eventually visits arbitrarily close to every point of the attractor, i.e.:*

   There exists an $x \in \Omega$ such that for each point $y \in \Omega$ and each $\varepsilon > 0$ there exists a time t (which may be positive or negative) such that $|\phi^t(x) - y| < \varepsilon$.

The first property says that initial conditions in the invariant set diverge from each other locally. The second property says that there is at least one trajectory in the invariant set such that not only eventually comes back arbitrarily close to itself, but to *every* point of the invariant set. This property ensures that we are talking about an attractor composed of a single piece, not two separate ones. This property is also known as *topological transitivity*.

We saw several examples of chaotic attractors of non-smooth systems in Chapter 1. For flows (smooth or non-smooth), it can be shown that the dimension of phase space must be at least three in order for a flow to exhibit chaos. Various techniques for analyzing and quantifying chaotic motion exist, such as Lyapunov exponents, time series analysis, invariant measures, fractal dimension, etc. For a more thorough treatment of the statistical properties of chaos see for example the book by Sprott [242]. Some of these notions have counterparts in non-smooth systems, see for example the work of Kunze [165].

Flows naturally lead to maps through the process of taking a (Poincaré) section through the flow and considering the map of that section to itself induced by the flow; see Fig. 2.5. We will make this important concept precise in Sec. 2.1.5 below.

2.1.2 Iterated maps

Given a **map** defined by the rule

$$x \mapsto f(x), \quad x \in \mathcal{D} \subset \mathbb{R}^n, \tag{2.6}$$

then $T = \mathbb{Z}$; that is, 'time' is integer-valued, and the operator ϕ is just f. Evolving through time $m > 0$ involves taking the mth iterate of the map;

$$\phi^m(x_0) = x_m = f(x_{m-1}) = f(f(x_{m-2})) = \ldots := f^{(m)}(x_0),$$

where a superscript (m) means m-fold composition

$$f^{(m)}(x_0) = \overbrace{f \circ f \circ \ldots \circ f(x_0)}^{m \text{ times}} .$$

Again we shall write $f(x, \mu)$ for systems that depend on parameters $\mu \in \mathbb{R}^p$.

A useful way of studying one-dimensional maps is via *cobweb diagrams* that plot x_{n+1} against x_n by reflecting in the main diagonal

Example 2.2 (logistic map). An example of a cobweb diagram for the *logistic map* given by

$$x \to \mu x(1 - x), \quad x \in [0, 1], \quad 0 < \mu \leq 4 \tag{2.7}$$

is given in Fig. 2.4.

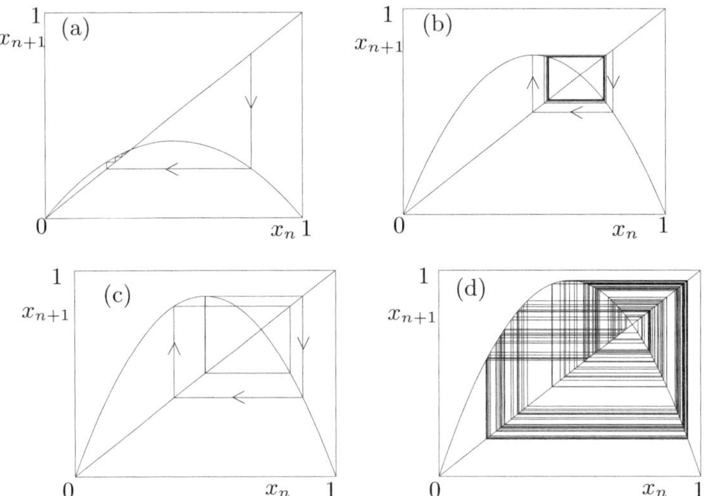

Fig. 2.4. Cobweb diagrams for the logistic map (2.7) starting with $x_0 = 0.8$ showing: (a) convergence to a stable fixed point for $\mu = 1.5$; (b) convergence to a period-two attractor for $\mu = 3.1$; (c) period-four attractor (note that here the initial condition is set to $x = 0.5$), and (d) chaotic behavior for $\mu = 4$.

Definition 2.7. *A mapping (2.6) is said to be* **invertible** *for $x \in \mathcal{D} \subset \mathbb{R}^n$ if given any $x_1 \in \mathcal{D}$ there is a* unique *$x_0 \in \mathcal{D}$ such that $x_1 = f(x_0)$. In such a case we define the inverse mapping $f^{(-1)}$ by $x_0 = f^{(-1)}(x_1)$ for all points x_1 in $f(\mathcal{D})$.*

Note that the smoothness of the dynamical system in the case of maps is given simply by the smoothness of the function f. Smooth (that is, at least

C^1) invertible maps, with smooth inverses are referred to as *diffeomorphisms*. We will now list some important types of invariant sets of maps.

fixed points. The simplest kind of invariant set of a map is a *fixed point*, which is a point x^* such that $f(x^*) = x^*$. fixed points of maps have a close connection to periodic orbits of flows, through the induced (Poincaré) map; see Fig. 2.5

periodic points. Next in order of complexity come periodic points, which satisfy $f^{(m)}(x^*) = x^*$ for some (least value of) $m > 0$. We refer to such a point as a *period-m point* of the map and its orbit as a period-m orbit. Clearly each point $f^{(i)}(x^*)$, $i \leq m-1$ of a period-m orbit is also a period-m point. These again are the close analogs of periodic orbits of flows (of a higher period), implying more intersections with a Poincaré section; see Fig. 2.5(b).

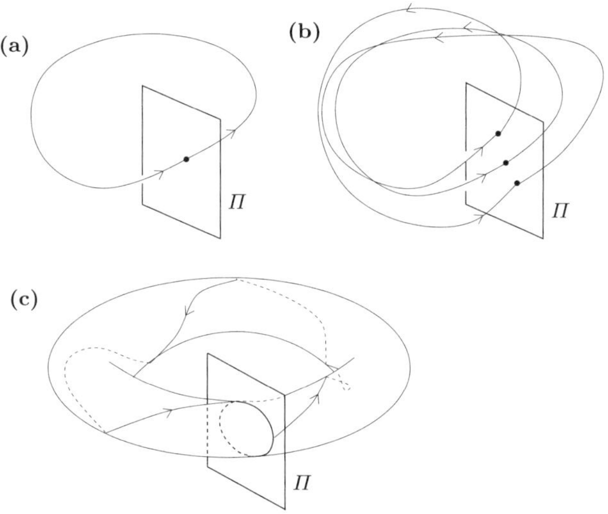

Fig. 2.5. Depicting the relation between maps and flows obtained by taking a Poincaré section Π through the phase space of the flow and considering the induced map from $\Pi \to \Pi$. (a) The correspondence between fixed points and period-T limit cycles; (b) between period-m points and higher-period limit cycles ($m = 3$ in this case); and (c) between invariant circles and invariant tori.

Invariant circles. Analogous to invariant tori of flows are *invariant closed curves* of a map, which again may be defined by taking a Poincaré section of a torus; see Fig. 2.5(c). Such closed curves are topologically circles, and we can reduce the dynamics *on* an invariant closed curve to that of a map of the unit circle to itself, a so-called *circle map*. The dynamics *off* or transverse to an invariant closed curve can also be complex. Typically, as parameters vary, such curves lose their smoothness and eventually fail to exist as continuous invariant sets; see for example [9, 168, 8] for the kind of dynamics one expects under such bifurcation sequences for smooth systems. Recently Zhusubaliyev & Mosekilde [281, 282, 280] and Dankowicz, Piiroinen & Nordmark [65] found yet more complex bifurcation sequences can occur near an invariant circle of certain piecewise-smooth systems; we shall return to non-smooth circle maps in Chapter 4 and non-smooth torus bifurcations in Chapter 9.4.3. For comparison with the non-smooth case, we will recall here just a few standard results for the dynamics *on* smooth invariant circles. For more details, see for example the book by Arrowsmith and Place [9].

Consider a map $f : S^{(1)} \to S^{(1)}$, where $S^{(1)}$ is the unit circle.

Example 2.3 (Arnol'd circle map). A canonical example of a circle map is the Arnol'd circle map (or standard map)

$$\theta \to f(\theta) = \theta + \alpha + \varepsilon \sin(\theta) \pmod{2\pi}, \tag{2.8}$$

where $0 \le \varepsilon < 1$. When $\varepsilon = 0$, clearly the map describes a rigid rotation through an angle α. If $\alpha = p/q$ is rational, then all points are periodic with period q. If α is irrational, then motion never repeats and all initial conditions θ_0 are quasi-periodic and $\cup_{n=1}^{\infty} f^{(n)}(\theta_0)$ fills out the entire circle. However, the dynamics is not chaotic since nearby initial conditions remain close.

For $\varepsilon > 0$, then one can use the notion of *rotation number* to define the equivalent of these two behaviors.

Definition 2.8. *Consider a circle map $f : S^{(1)} \to S^{(1)}$, which can be written in functional form as $\overline{f}(\theta \bmod 2\pi) \bmod 2\pi$, where $\overline{f} : \mathbb{R} \to \mathbb{R}$ is called a* lift *of f. We define the* **rotation number** *ρ of a point $x \in [0, 2\pi)$ by*

$$\rho(f, x) = \left(\lim_{n \to \infty} \frac{f^{(n)}(x) - x}{n} \right) \quad (mod\ 2\pi). \tag{2.9}$$

Now, we have the standard result; see for example [73, 151]:

Theorem 2.1. *Suppose a circle map f is continuous and has a continuous inverse; then the rotation number is independent of initial condition x; that is, $\rho(f, x) = \rho(f)$.*

If the rotation number is irrational, it can be shown that (under the additional assumption that both the map f and its inverse are differentiable) the dynamics is topological equivalent to a rigid rotation through angle ρ; thus, the dynamics is non-chaotic and the forward iterate of any initial condition eventually fills the whole circle. In contrast, if $\rho = p/q$ is rational, then the dynamics is said to be *mode locked* and there is at least one orbit of period q. Typically there will be two such orbits, with one stable and one unstable. Given a family of circle maps parameterized by α, then the rotation number will generically be rational over intervals of α-values. Both irrational and rational rotation numbers occur for sets of α-values that have positive measure. We will return to the study of circle maps in Chapter 4, where we show that they are closely linked to maps that are discontinuous on an interval.

Chaos. Definition 2.6 of chaotic invariant sets also applies to maps. In contrast to flows where the phase space must be at least three-dimensional, in the noninvertible case, maps of dimension one can exhibit chaos. We have already seen this for the square-root map in case study VII in Chapter 1. Smooth one-dimensional maps can be chaotic too, as the following well-known example shows:

Example 2.4 (logistic map continued). Consider again the logistic map (2.7). For $\mu > 1$, there are two fixed points at $x = 0$ and $x = (\mu - 1)/\mu$. For $1 < \mu < 3$, the non-trivial one is the unique attractor of the system. For $\mu > 3$, there are also two period-two points

$$x = \frac{1 + \mu \pm \sqrt{\mu^2 - 2\mu - 3}}{2\mu}.$$

As μ is further increased, a chaotic attractor is born via a so-called *period-doubling* cascade; see Fig. 2.6. Note that in the 'chaotic' range of μ-values, the attractor actually alternates between parameter intervals of chaos and intervals of periodic orbits (so-called *periodic windows*) appearing in the bifurcation diagram.

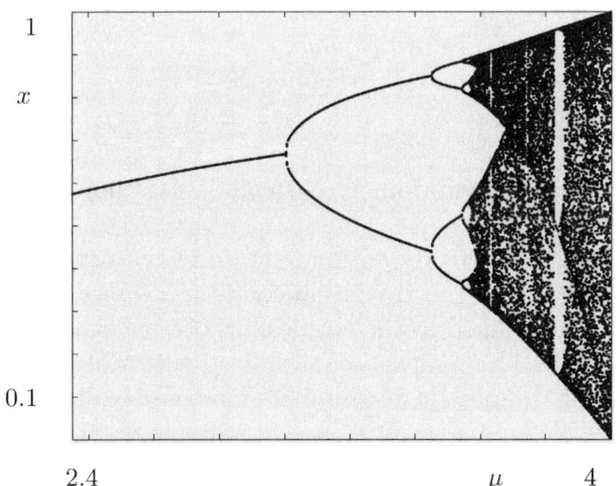

Fig. 2.6. The bifurcation diagram of the logistic map (2.7) showing the period-doubling cascade to chaos as the parameter μ is increased and the presence of periodic windows within the chaos.

For invertible maps, at least two dimensions are required in order for there to be chaotic invariant sets.

2.1.3 Asymptotic stability

When considering dynamical systems with physical application, we are usually only interested in stable behavior. Important notions of stability in dynamical systems include that of either *Lyapunov* or *asymptotic* stability of an invariant set. In general, the former means stability in the weak sense that trajectories starting nearby to the invariant set remain nearby for all time, whereas the latter is more or less synonymous with the concept of an attractor (Definition 2.3). Both refer to stability of invariant sets with respect to perturbations of initial conditions, at fixed parameter values. There are other less restrictive versions of this kind of stability, such as input–output stability, orbital stability, and controllability of arbitrary trajectories (not just invariant sets), but these will not concern us. For simplicity we shall define stability only of equilibria of flows. Similar definitions can be given for fixed points of discrete-time systems, or for other invariant sets of either continuous-time or discrete-time systems.

To formally define Lyapunov stability, consider a generic nonlinear system of the form (2.3) and assume that it has an equilibrium point that, without loss of generality, is at the origin; that is, $f(0) = 0$:

Definition 2.9. *The equilibrium state at the origin is said to be* **(Lyapunov) stable** *if for any* $\varepsilon > 0$*, there exists a* $\delta > 0$ *such that*

$$\|x_0\| < \delta \Rightarrow \|\Phi(x_0, t)\| < \varepsilon, \forall t > 0.$$

Definition 2.10. *The equilibrium state at the origin is said to be* **asymptotically stable** *(in the sense of Lyapunov) if*

1. *it is stable;*
2. $\lim_{t\to\infty} \Phi(x_0, t) = 0$.

We will say that an equilibrium is *unstable* if it is not stable according to Definition 2.9.

Thus, stability refers to the ultimate state of the dynamics not being altered under small changes to the initial conditions. For equilibria, the notion of stability is closely linked to the eigenvalues of the corresponding linearized ODEs, with a sufficient condition for stability being that all eigenvalues lie in the left half of the complex plane. Similar sufficient conditions for asymptotic stability exist for other invariant sets. For example, for fixed points of maps, stability is guaranteed if all eigenvalues (often called multipliers) of the linearization of the map lie inside the unit circle. Proving stability can be more tricky in the case that eigenvalues lie on the imaginary axis. One technique is to construct so-called *Lyapunov functions* that act like the energy of a perturbation from the invariant set in question, and then prove that this energy decreases with time as one follows the dynamics. As we shall see in Sec. 2.4, for non-smooth systems, proving asymptotic stability even in the case of equilibria whose eigenvalues in the left half-plane can be highly tricky. There, the Lyapunov function technique can be extremely useful.

We next deal with a quite different notion of stability, that of invariance of the dynamics under perturbation to the system itself rather than to initial conditions.

2.1.4 Structural stability

Dynamical systems theory aims to classify dynamics qualitatively. structurally stable systems are ones for which all 'nearby' systems have qualitatively 'equivalent' dynamics. Thus we need a precise notion of *nearby* and also of *equivalence.*

'Nearby' refers to any possible perturbation of the system itself [the function $f(x)$], including variation of the system's parameters. We want to call two systems 'equivalent' if their phase spaces have the same dimension and there phase potraits contain the same number and type of invariant sets, which in the same general position with respect to each other. To achieve such a definition, we use topology, which is the mathematics of 'rubber sheet geometry.' Mathematically we want to say that two phase potraits are the same if there is a smooth transformation that stretches, squashes, rotates, but not folds one phase potrait into the other. Such transformations are called *homeomorphisms*, which are continuous functions defined over the entire phase space whose inverses are also continuous.

Definition 2.11. *Two dynamical systems* $\{X, T, \phi^t\}$ *and* $\{X, T, \psi^t\}$ *are* **topological equivalent** *if there is a homeomorphism* h *that maps the orbits of the first system onto orbits of the second one, preserving the direction of time.*

For discrete time systems, two topological equivalent maps f and g that satisfy

$$f(x) = h^{-1}(g(h(x))), \quad \text{implying} \quad h(f(x)) = g(h(x)),$$

for some homeomorphism h, are said to be topologically *conjugate*, and we can write more simply

$$f = h^{-1} \circ g \circ h. \tag{2.10}$$

For ODEs, the homeomorphism should apply at the level of the flow.

Definition 2.12. *Two flows* $\Phi(x, t)$ *and* $\Psi(h(x), t))$ *that correspond, respectively, to ODEs* $\dot{x} = f(x)$ *and* $\dot{y} = g(y)$ *are said to be* **topologically conjugate** *if there exists a homeomorphism* h *such that*

$$\Phi(x, t) = h^{-1}(\Psi(h(x), t)). \tag{2.11}$$

Actually, for topological equivalence of flows, the conjugacy does not need to apply at each time t. Rather, we require the weaker condition that there is an invertible, continuous mapping of time $t \mapsto s(t)$ so that we can write

$$\Phi(x,t) = h^{-1}(\Psi(h(x), s(t)))\,. \tag{2.12}$$

Note, though, that conditions (2.11) and (2.12) are hard to check in practice, because one must solve the ODE exactly in order to construct an explicit expression for the flow operator Φ. A more restrictive condition, which is easier to check in practice, is that two ODEs be *smoothly topologically conjugate*; that is, the homeomorphism h in (2.11) is differentiable, with differentiable inverse (a *diffeomorphism*). Then we can write

$$f(x) = \left(\frac{dh(x)}{dx}\right)^{-1} f(h(x)).$$

Having defined what we mean by topological equivalence, we can now define structural stability.

Definition 2.13. *A flow (or discrete-time map) is* **structurally stable** *if there is an $\varepsilon > 0$ such that all C^1 perturbations of maximum size ε to the vector field (map) f lead to topological equivalent phase potraits.*

One key application of topological equivalence is to show that 'normally' dynamical systems in the neighborhood of an invariant set are topological equivalent to the linearization of the system about that set. We consider this in the two specific contexts of equilibria of flows and fixed points of maps. As we shall see in the next subsection, the result for maps implies an analogous result for periodic orbits of flows.

Consider first an equilibrium x^* of $\dot{x} = f(x)$. Now, for small $y = x - x^*$, we can expand f as a Taylor series about x^* to write

$$\dot{y} = f_x(x^*)y + O(y^2),$$

and drop the $O(y^2)$-term. (Here $f_x(x^*)$ given by $(f_x)_{i,j} = \partial f_i / \partial x_j$ is the Jacobian derivative of the vector field evaluated at x^*.) The general solution to the linear system is

$$y(t) = \exp(f_x(x^*)t)y(0).$$

Usually this can be expressed in terms of the eigenvalues and eigenvectors of $f_x(x^*)$. For example, in the case that Jacobian has a full set of n independent eigenvalues $\{\lambda_i : i = 1, 2, \ldots n\}$, then we can write

$$y(t) = V^{-1}\mathrm{diag}\{e^{-\lambda_1 t}, e^{-\lambda_2 t}, \ldots e^{-\lambda_n t}\}Vy(0),$$

where the ith column of V contains the eigenvectors of f_x corresponding to eigenvalue λ_i. (Here diag$\{\cdot\}$ means the diagonal matrix whose entries on the main diagonal are those stated.) So if the spectrum (set of eigenvalues) of $f_x(x^*)$ is in the left half-plane, then the solution of the linear system tends to zero as $t \to \infty$ and the equilibrium of the linear system is stable.

Definition 2.14. *We shall refer to the* **eigenvalues** *of an equilibrium* x^* *of an ODE* $\dot{x} = f(x)$ *to mean the eigenvalues of the associated Jacobian matrix* $f_x(x^*)$. *An equilibrium is said to be* **hyperbolic** *if none of its eigenvalues lie on the imaginary axis.*

It can be proved [168, Ch. 2] that the flow local to any two hyperbolic equilibria of n-dimensional systems that have the same number of eigenvalues with negative real part are topologically equivalent to each other. In particular, we have

Theorem 2.2 (Hartman–Grobman). *The dynamics close to a hyperbolic equilibrium point are topologically equivalent to that of the system linearized about that point.*

An equilibrium x^* with $n_s > 0$ eigenvalues of negative real part and $n_u > 0$ eigenvalues of positive real part is called a saddle point. Close to x^* we can define the stable [unstable] manifold $W^s(x^*)$ [$W^u(x^*)$], which is an invariant manifold of the flow that is composed of all trajectories that tend to x^* as $t \to \infty$ ($t \to -\infty$). $W^s(x^*)$ is of dimension n_s and is tangent at x^* to the stable eigenspace of $f_x(x^*)$; similarly $W^s(x^*)$ is of dimension n_u and is tangent at x^* to the unstable eigenspace of $f_x(x^*)$. See Fig. 2.7.

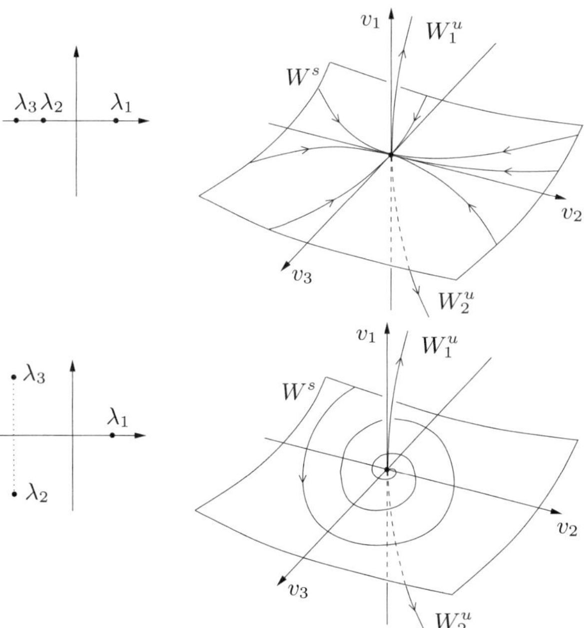

Fig. 2.7. Stable and unstable manifolds near 3 dimensional saddle equilibria with (a) purely real eigenvalues and (b) complex stable eigenvalues (*a saddle focus*). The vectors v_k are the eigenvectors corresponding to the eigenvalues λ_k.

Similarly, consider a fixed point x^* of a map $x \mapsto f(x)$ (period-m points can be treated as well, since they are fixed points of $f^{(m)}$). Linearizing about this fixed point, we get

$$y \mapsto f_x(x^*)y, \quad \text{with solution} \quad y_n = [f_x(x^*)]^n y_0.$$

Hence $y_i \to 0$ as $i \to \infty$, satisfying the second of the conditions for asymptotic stability of the linearized system, if all eigenvalues μ_i of $f_x(x^*)$ lie inside the unit circle.

Definition 2.15. *We shall refer to the* **multipliers** λ_i *of a fixed point* x^* *of a map* $x \to f(x)$ *to mean the eigenvalues of the associated linearization* $f_x(x^*)$. *A fixed point is said to be* **hyperbolic** *if none of the multipliers lie on the unit circle.*

For a general map in n-dimensions, one can define the *orientability* of the map close to a fixed point as the sign of the product of all its multipliers $\prod_{i=1}^n \lambda_i$. If this product is positive, the map is locally orientable; if negative, the map is non-orientable. If the product is zero, then the map is noninvertible. Note that any map that arises as the Poincaré mapof a smooth flow must be orientable [271]. Figure 2.8 shows the two possible types of orientable saddle point in two-dimensional maps. That with negative multipliers [Fig. 2.8(b)] is sometimes referred to as a *flip saddle.* One can also define stable and unstable

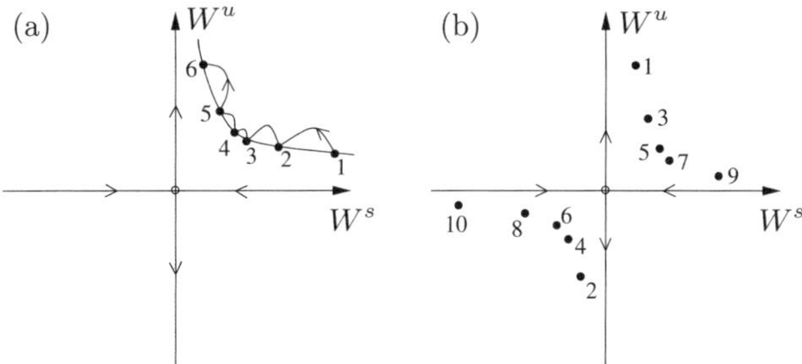

Fig. 2.8. The dynamics near a saddle fixed point of a map in the cases of (a) two positive multipliers, and (b) two negative multipliers. Numbers denote subsequent iterations of the map starting from the point labeled '1'.

manifolds at saddle points analogously as for equilibria of flows. Note, though, a distinction with the case of flows. In a flow, a one-dimensional manifold is composed of a single trajectory. In a map, a one-dimensional manifold contains many orbits; see Fig. 2.9. Hence stable and unstable manifolds in maps can intersect transversally (at a non-zero angle), whereas if a stable and unstable manifold intersect in a flow, they must do so along a line; that is, they must share a common trajectory.

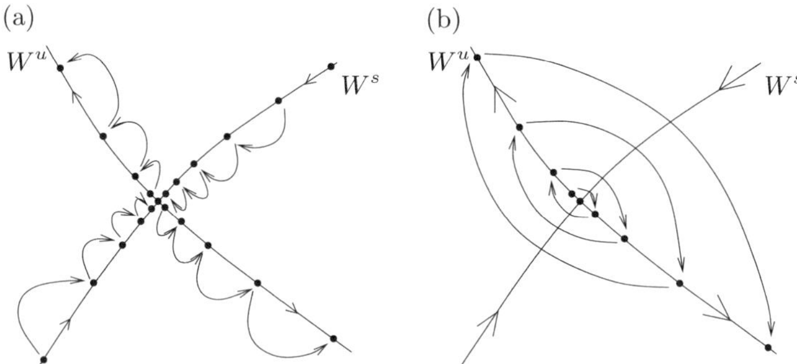

Fig. 2.9. (a) The stable and unstable manifolds close to a saddle fixed point in two dimensions with positive multipliers. (b) Similar figure in the case of negative multipliers, where, for clarity, only the dynamics along the unstable manifold is depicted. Similar behavior is observed on the stable manifold with the difference that the direction of 'hoppings' is reversed.

There are similar notions of hyperbolicity for other invariant sets. Loosely speaking, an invariant set is hyperbolic (sometimes called *normally hyperbolic*) if the dynamics in directions transverse to the set is exponentially attracting or repelling at rates that are faster than the dynamics in the invariant set. See, for example, [272]. Generally speaking, hyperbolic dynamics are structurally stable.

Many dynamical systems that arise in applications are not structurally stable. For example, systems can have persistent non-hyperbolic equilibria (center points) if they preserve a first integral such as energy. An important such class is that of Hamiltonian systems, which have very different dynamics than the systems in question here; see, for example, the reprint collection by Miess and MacKay [184]. Alternatively, the system may be invariant under the action of a symmetry, which again leads to certain structurally unstable things happening generically. The dynamics of systems with symmetry is a large subject in its own right, and one that we do not deal with here; see, for example, the book by Golubitsky, Schaeffer & Stewart [120]. Largely speaking, we shall avoid Hamiltonian or symmetric systems in what follows.

2.1.5 Periodic orbits and Poincaré maps

We have already hinted at the important connection between flows and maps. We now make this connection more precise. One of the main building blocks of the dynamics of a set of ODEs are its periodic solutions, and these provide a natural way to transform between flows and maps. Consider a limit cycle solution $x(t) = p(t)$ to (2.3) of period $T > 0$; that is, $p(t+T) = p(t)$. To study the dynamics near a such a cycle, we construct a *Poincaré section*, which is an $(n-1)$-dimensional surface Π that contains a point $x_p = p(t^*)$ on the limit cycle and which is *transverse* to the flow at x_p. Let us introduce a notation

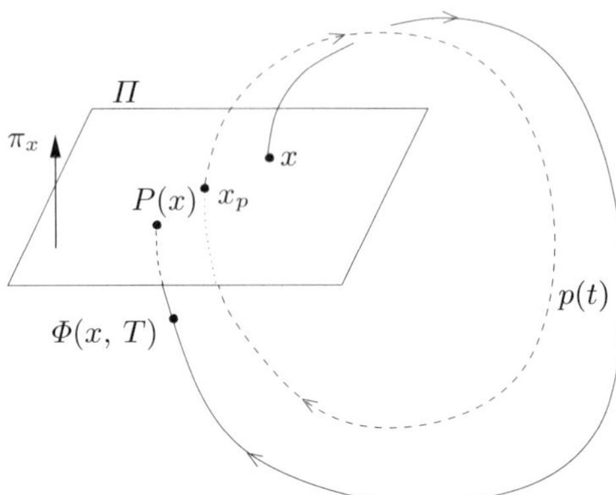

Fig. 2.10. The construction of a Poincaré mapclose to a periodic orbit $p(t)$.

that

$$\Pi = \{x \in \mathbb{R}^n : \pi(x) = 0\}, \tag{2.13}$$

for some smooth scalar function π. Then the transversality condition is that the normal vector $\pi_x(x_p)$ to Π at x_p has a non-zero component in the direction of the $\Phi_t(x_p, 0) = f(x_p)$. (Here a subscript x or t means differentiation with respect to that variable). That is, we require

$$\pi_x(x_p)f(x_p) \neq 0. \tag{2.14}$$

where a subscript x or t means partial differentiation with respect to that variable, so that $\pi_x(x_p)$ is the normal vector to Π at $x = x_p$.

Now, we can use the flow Φ to define a map P from Π to Π, called the *Poincaré map*, which is defined for x sufficiently close to x_p via

$$P(x) = \Phi(x, \tau(x)),$$

where $\tau(x)$ is defined implicitly as the time closest to T for which

$$\pi(\Phi(x, \tau(x))) = 0. \tag{2.15}$$

By the Implicit Function Theorem (see Theorem 2.4 below), the transversality (2.14) guarantees that there is a locally unique solution for $\tau(x)$. Note that we can then define the Poincaré mapas a *smooth projection* S of the time-T map $\Phi(\cdot, T)$ for $x \in \Pi$

$$P(x) = S(\Phi(x, T), x), \quad \text{where} \quad S(y, x) = \Phi(y, \tau(x) - T); \tag{2.16}$$

see Fig. 2.11. Thus, x_p becomes a fixed point of the map P.

We can study the stability (and possible bifurcations of) the periodic solution by studying the linearization P_x of the Poincaré map at x_p. It will be important for us to be able to compute this linearization when we consider grazing bifurcations of periodic orbits in Chapters 6–8. Computing the total derivative with respect to x, we have

$$P_x(x_p) = \Phi_x(x_p, T) + \Phi_t(x_p, T)\tau_x(x_p),$$

and, from implicit differentiation of (2.15),

$$\tau_x(x_p) = -\frac{\pi_x(x_p)\Phi_x(x_p, T)}{\pi_x(x_p)\Phi_t(x_p, T)}.$$

Hence

$$\begin{aligned} P_x(x_p) &= \left(I - \frac{\Phi_t(x_p, T)\pi_x(x_p)}{\pi_x(x_p)\Phi_t(x_p, T)}\right)\Phi_x(x_p, T) \\ &= \left(I - \frac{f(x_p)\pi_x(x_p)}{\pi_x(x_p)f(x_p)}\right)\Phi_x(x_p, T). \end{aligned} \tag{2.17}$$

Note that (2.17) is a *rank-one update* of the time-T map $\Phi_x(x_p, T)$ around $p(t)$ (the multiplying factor is the linearization of S defined in (2.16)). The $n \times n$ matrix $\Phi_x(x_p, T)$ is referred to as the *Monodromy matrix* and corresponds to the fundamental solution matrix up to time T of the linear variational equations

$$\dot{y} = f_x(p(t))y, \tag{2.18}$$

around the periodic orbit $p(t)$. The direction of the flow $\Phi_t(x_p, t) = f(x_p)$ can easily be shown to solve (2.18) and, hence, $f(x_p)$ is an eigenvector of $\Phi_x(x_p, T)$ corresponding to the multiplier 1. Letting the expression (2.17) act on $f(x_p)$, we see that this corresponds to an eigenvalue 1 of the linearized Poincaré mapP_x. However, since this eigenvector does not lie in the linear approximation to Π we will never see its effect when computing the Poincaré maptaking only points $x \in \Pi$.

Other than this trivial eigenvalue, the eigenvalues of the Monodromy matrix are precisely the multipliers λ_i of the Poincaré map. This can be argued as follows (see for example [168, Thm. 1.6] for a more careful proof). The non-trivial eigenvectors of the Monodromy matrix form an $(n-1)$-dimensional invariant subspace $\tilde{\Pi}$, say, that does not contain the direction $f(x_p)$. Hence $\tilde{\Pi}$ can be chosen to be a Poincaré section, as it satisfies the transversality condition (2.14). Now all we need to show is that the multipliers of two Poincaré maps P and $\tilde{P}$ defined via two different sections Π and $\tilde{\Pi}$ are the same. Let Π be given by (2.13), (2.15), and let $\tilde{\Pi} := \{x : \tilde{\pi}(x, \tilde{\tau}(x)) = 0\}$. The equivalence of these two maps arises because we can write

$$\tilde{P} = \tilde{S}^{-1} \circ P \circ \tilde{S}, \quad \text{where} \quad \tilde{S}(x) = \Phi(x, \tau(x) - \tilde{\tau}(x)), \tag{2.19}$$

where $\tilde{S}$ is the smooth mapping that takes points in $\tilde{\Pi}$ to Π using the flow; see Fig. 2.11(b). Linearizing (2.19) we obtain that

$$\tilde{P}_x = \tilde{S}_x^{-1} P_x \tilde{S}_x.$$

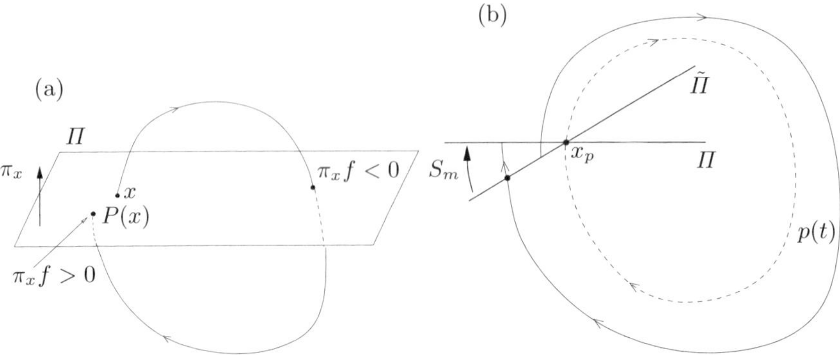

Fig. 2.11. (a) Illustrating a Poincaré mapdefined by the first intersection with a surface $\Pi : \{\pi(x) = 0\}$ in the direction of increasing $\pi(x)$. (b) A representation in two dimensions of the smooth projection S_m along the flow lines between Poincaré sections Π and $\tilde{\Pi}$.

Hence $\tilde{P}_x$ and P_x are *similar* matrices that must have the same eigenvalues. In fact, the expression (2.19) applies for any two Poincaré sections along the orbit $p(t)$. It also shows that the two Poincaré maps satisfy the condition (2.10) to be topological equivalent. Summarizing, we have

Theorem 2.3. *All Poincaré maps defined with respect to any Poincaré section that is transverse to the flow around a periodic orbit* $p(t)$ *of a smooth ODE (2.3) are locally topological equivalent. Moreover, they have the same non-zero multipliers* $\lambda_1, \ldots \lambda_{n-1}$. *The linearization of the corresponding time-*T *map around* $p(t)$ *is related by the formula (2.17) and has eigenvalues* $1, \lambda_1, \ldots \lambda_{n-1}$.

Now, we say that a *hyperbolic* periodic orbit $p(t)$ is one whose Poincaré maphas multipliers λ_i, $i = 1, \ldots n-1$ that are all off the unit circle. The Hartman–Grobman theorem for maps then tells us that flow around the orbit is locally topologically equivalent to the linearization. An obvious consequence of this is that hyperbolic periodic orbits are necessarily isolated in phase space.

Poincaré maps do not necessarily require a periodic orbit in order to be defined. A Poincaré section Π can be taken anywhere in the phase space, provided the flow is everywhere transverse to it, as for example in Fig. 2.5(c) where Π is chosen transverse to the flow on an invariant torus. For transversality, we require that a condition equivalent to (2.14) applies throughout Π. So if we define Π as before to be the zero-set of a smooth function (2.13), then we are only interested in defining a Poincaré mapfor points x for which

$$\pi(x) = 0 \quad \text{and} \quad \pi_x(x)f(x) \neq 0.$$

The map is defined by the first intersection with Π *in the same sense*. That is, $P(x) = \Phi(x, \tau(x))$, where $\tau(x)$ is the first time $t > 0$ such that $\pi(\Phi(x,t)) = 0$ and $\pi_x f(\Phi(x,0)) \cdot \pi_x f(\Phi(x,t)) > 0$; see Fig. 2.11(a). Note that the map P

may not be defined for the whole Poincaré section, since not all points need to return.

One of the benefits of studying Poincaré maps rather than flows is that they drop by one the dimension of the sets we need to consider. Thus, limit cycles of flows correspond to isolated fixed points of Poincaré maps; invariant tori correspond to closed curves of the map; and chaotic invariant sets decrease their fractal dimension by one.

2.1.6 Bifurcations of smooth systems

Broadly speaking, there two notions of 'bifurcation', one analytical and the other topological. From the first point of view, bifurcations are branching points of parameterized sets of solutions $x(\mu)$ to nonlinear operators $G(x, \mu) = 0$. Simply put, a 'bifurcation' is a point at which the Implicit Function Theorem fails; see, for example, [141, 55, 47, 119]).

Theorem 2.4 (Implicit Function Theorem). *Suppose that for some $\mu = \mu_0$ there exists a solution $x = x_0$ to a smooth nonlinear equation $G(x, \mu) = 0$, where $G : \mathbb{R}^n \times \mathbb{R} \to \mathbb{R}^n$; then, provided $G_x(x_0, \mu_0)$ is nonsingular, a smooth path of solutions $x(\mu)$ can be continued locally, with $x(\mu_0) = x_0$.*

This analytic point of view does not adapt naturally to the study of nonsmooth systems, and so the notion adopted in this book is that a bifurcation is a change in the topology of the phase portraits of a dynamical system as a parameter is varied. Of particular importance are changes to the number and nature of the attractors of the system. A rich theory now exists for smooth systems, which we shall briefly review here. Many more details can be found in the books by Guckenheimer & Holmes [124], Kuznetsov [168] and Wiggins [273]; hence, we give only a quick introduction to bifurcation theory applied to parameterized systems either in the form of a smooth vector field or map

$$\dot{x} = f(x, \mu), \quad \text{or} \quad x \mapsto f(x, \mu) \tag{2.20}$$

for $x \in \mathbb{R}^n$, $\mu \in \mathbb{R}^p$.

We define a bifurcation simply in terms of loss of structural stability upon varying a parameter.

Definition 2.16. *A* **bifurcation** *occurs at a parameter value μ_0 if the dynamical system $\{X, T, \phi^t\}$ is not structurally stable.*

An **unfolding** *(or* versal unfolding*) of a bifurcation is a simplified system that for small $\mu - \mu_0$ contains all possible structurally stable phase potraits that arise under small perturbations of the system at the bifurcation point.*

The **codimension** *of a bifurcation is the dimension of parameter space required to unfold the bifurcation.*

A **bifurcation diagram** *is a plot of (some measure of) the invariant sets of a dynamical system against a single bifurcation parameter μ, which indicates stability.*

We can distinguish between two kinds of bifurcation:

Definition 2.17. *A* **local bifurcation** *arises due to the loss of hyperbolicity of an invariant set upon varying a parameter. All other bifurcations are called* **global bifurcations**.

Many kinds of local bifurcations of smooth systems have been studied and classified; see for example Kuznetsov [165, Chs. 2–5]. Figure 2.12 summarizes the main types of codimension-one *local bifurcations* of smooth vector fields and an associated representative bifurcation diagram. In each case, under appropriate defining and non-degeneracy conditions, one can calculate an appropriate *normal form* that can be obtained by smooth co-ordinate transformations from any system that undergoes the bifurcation in question.

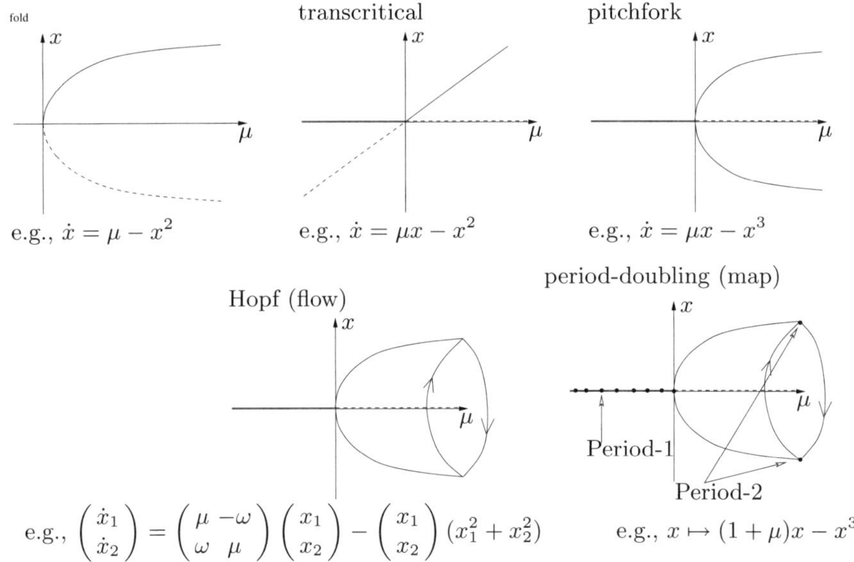

Fig. 2.12. Main codimension-one local bifurcations in smooth dynamical systems.

Note that steady bifurcations of equilibria of flows— fold (or saddle-node) bifurcations and the associated pitchfork or transcritical bifurcations for systems with symmetry or invariance— have a direct analogy for limit cycle bifurcations; that is, bifurcations of fixed points of maps. The defining condition for the former is that there is an eigenvalue at zero and for the latter that there is a multiplier at 1. The case of the Hopf bifurcation is more subtle. The direct analog for maps is when a complex pair of eigenvalues crosses the unit circle. This *torus* or *Neimark–Sacker* bifurcation causes the birth of invariant circles of the map, with all inherent complications associated with the dynamics of circle maps that we outlined earlier. There are also special cases when the multipliers concerned are low-order roots of unity. Finally, for maps there is the case of the *period-doubling* or *flip* bifurcation, which has no analog in equilibrium bifurcations. Many of these bifurcations come in *super-* or *subcritical* subcases, depending on whether a stable nontrivial invariant set is created as the trivial equilibrium (or fixed point) becomes unstable, or vice versa.

In contrast, global bifurcations typically occur because of a change in the topology of stable and unstable manifolds of invariant sets (see, for example, [168, Chs. 6 and 7]). A typical example is a *homoclinic bifurcation* when the stable and unstable manifold of the same invariant set form an intersection or tangency at a fixed parameter value. See Fig. 2.13 for two examples. Also, stable and unstable manifolds of other invariant sets can form an intersection in a *heteroclinic connection* that can cause the sudden appearance or disappearance of a chaotic attractor in a *boundary crisis* bifurcation [168].

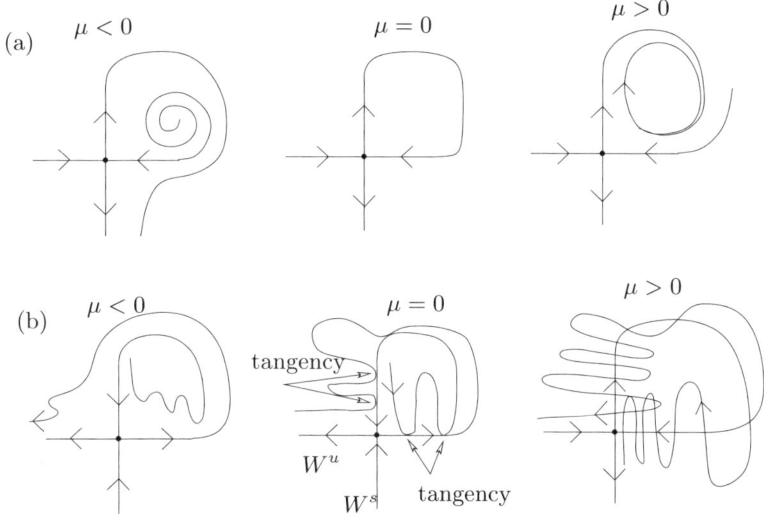

Fig. 2.13. Two global bifurcations. (a) A homoclinic bifurcation to a saddle equilibrium creating a single stable limit cycle. (b) A homoclinic tangency to a saddle point in a two-dimensional map creating a homoclinic tangle, which implies the existence of a chaotic invariant set through the Smale-Birkhoff homoclinic theorem [124, Thm. 5.3.5].

An interesting feature of smooth dynamical systems is that they can exhibit a *cascade* of local bifurcations under parameter variation. A well-known example is the period-doubling cascade. Here, a supercritical period-doubling at a parameter value μ_1 creates a stable period-2 orbit, followed by a further period-doubling of the period-2 orbit at $\mu = \mu_2$, creating a stable period-4 orbit, and so on, as we saw in Fig. 2.6. Remarkably, we observe a universal scaling law, established by Feigenbaum [97], that

$$\lim_{k\to\infty} \frac{\mu_k - \mu_{k-1}}{\mu_{k+1} - \mu_i} = \delta = 4.6692\ldots. \tag{2.21}$$

That is, the period-doubling sequence converges to a finite μ-value, and in the limit, the rate of convergence is the same for 'all' systems! For more precise details see, for example, [56, 73]. This universality of the period-doubling cascade has only been shown for a certain class of one-dimensional maps that have 'a single hump' like the logistic map, but it also applies to many ODEs because of the folded structure of their Poincaré maps. For one-dimensional maps, one can say much more, and we find cascades of periodic orbits described by the following theorem

Theorem 2.5 (Sharkovskii [235]). *Consider the following ordering of all positive integers:*

$$\begin{aligned}
&1 < 2 < 4 < \ldots 2^k < 2^k + 1 < \ldots \\
&\ldots \\
&< 2^{k+1} \cdot (2l+1) < 2^{k+1} \cdot (2l-1) < \ldots < 2^{k+1} \cdot 5 < 2^{k+1} \cdot 3 < \\
&\ldots < 2^k \cdot (2l+1) < 2^k \cdot (2l-1) < \ldots < 2^k \cdot 5 < 2^k \cdot 3 < \\
&\ldots \\
&\ldots < 2 \cdot (2l+1) < 2 \cdot (2l-1) < \ldots < 2 \cdot 5 < 2 \cdot 3 < \\
&\ldots < (2l+1) < (2l-1) < \ldots < 5 < 3.
\end{aligned}$$

If f is a continuous map of the interval $[-1, 1]$ to itself with a periodic point of period p, then, for any $q < p$ (where the inequality sign refers to the ordering above), f has a periodic point of period q.

Remarks

1. This result contains the statement 'period 3 implies chaos' that was the title of the paper by Li & Yorke [178] from which the word *chaos* was first used to describe bounded non-repeating motion.
2. Often in applications a period-q ($q \neq 2^k$) orbit first appears by a fold bifurcation upon increasing a parameter beyond the end of a period-doubling cascade. This leads to a *periodic window* of parameter values within which this orbit is stable, with the windows separated by chaotic regions. Thus the Sharkovskii ordering often gives the ordering of stable periodic windows that are observed in simulations of bifurcation diagrams 'inside' the chaotic regime after the end of period-doubling cascade (see, for example, Fig. 2.6).

Sharkovskii's Theorem relies heavily on the smoothness assumption for the map. An important feature of this book, and especially the results in Chapters 3 and 4, will be the identification of other types of cascades of stable periodic orbits close to a bifurcation point. We shall see that these cascades do not generally follow the Sharkovskii ordering, in that either the chaos is *robust* (i.e., has no periodic windows), or if windows exist they obey *period-adding* type orderings for which we see intervals of periodic motions of period n obeying the simple ordering $n < n+1 < n+2 < \ldots$. Indeed, as we shall see, period-adding is one of the unifying features of the behavior of non-smooth systems.

2.2 Piecewise-smooth dynamical systems

We now move onto the main theme of this chapter where we set the scene for a systematic study of the dynamics of non-smooth systems. Motivated by the case studies in Chapter 1, we shall introduce three classes of piecewise-smooth system: *maps*, *flows* and *hybrid systems*. Note that a complete existence and uniqueness theory does not exist, as far as we are aware, for these quite broad classes of system. Instead, in Sec. 2.3 below, we shall show the relation of these classes to other more precise formulations for the description of non-smooth dynamics for which such theory does exist. Nevertheless, our rather loose classification, while perhaps lacking mathematical rigor, shall prove highly useful in explaining the dynamics observed in example systems.

2.2.1 Piecewise-smooth maps

Definition 2.18. *A* **piecewise-smooth map** *is described by a finite set of smooth maps*

$$x \mapsto F_i(x,\mu), \quad \textit{for} \quad x \in S_i, \tag{2.22}$$

where $\cup_i S_i = \mathcal{D} \subset \mathbb{R}^n$ and each S_i has a non-empty interior. The intersection Σ_{ij} between the closure (set plus its boundary) of the sets S_i and S_j (that is, $\Sigma_{ij} := \bar{S}_i \cap \bar{S}_j$) is either an $\mathbb{R}^{(n-1)}$-dimensional manifold included in the boundaries ∂S_j and ∂S_i, or is the empty set. Each function F_i is smooth in both the state x and the parameter μ for any open subset U of S_i.

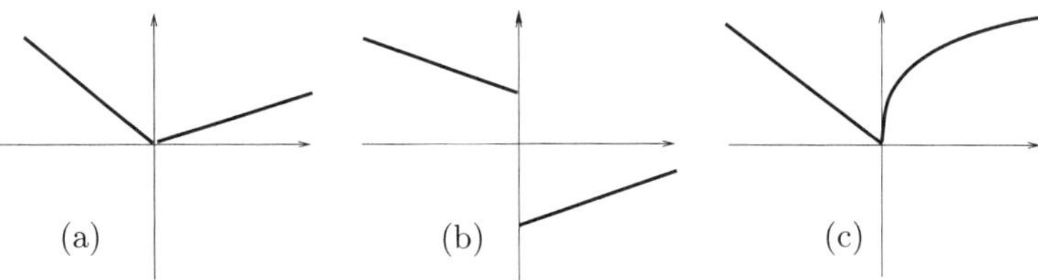

Fig. 2.14. Examples of piecewise-smooth one-dimensional maps: (a) piecewise-linear continuous map; (b) piecewise-linear discontinuous map; (c) square-root piecewise smooth map. In each case $S_1 = \{x < 0\}$, $S_2 = \{x > 0\}$ and $\Sigma_{12} = \{x = 0\}$.

A set Σ_{ij} for a piecewise-smooth map is usually termed a **border** or **discontinuity boundary** that separates regions of phase space where different smooth maps apply. Examples of piecewise-smooth one-dimensional maps are given in Fig. 2.14. Note that in the above definition we allow the possibility that one of the component maps F_i may itself be non-smooth (in the sense of having infinite or ill-defined derivatives) at the boundary Σ_{ij}. For example, the square-root map in Fig. 2.14(a) is such that the first derivative of $F_2(x)$ tends to ∞ as $x \to 0$. We also include the case that $F_i \neq F_j$ along Σ_{ij}, so that the map has a jump in state as in Fig. 2.14(b). Such maps are discontinuous piecewise-smooth maps. In this case, there are a number of choices that one can make about the value of the map for points in Σ_{ij}: for example, taking the average of F_i and F_j there; or allowing the map to be set valued at this point, taking all possible convex combinations $\{F_i + \lambda(F_j - F_i) : 0 \leq \lambda \leq 1\}$. In practice, such choices make little practical difference to the dynamics of the map, since they describe what happens to a set of points of zero measure.

Definition 2.19. *The* **order of singularity** *of a point $\widehat{x} \in \Sigma_{ij}$ of a continuous piecewise-smooth map is the order of the first non-zero term in the formal power-series expansion of $F_1(x) - F_2(x)$ about $x = \widehat{x}$.*

Remarks

1. This order is -1 times the usual definition of the order of a singularity in complex variable theory. That is, a complex $f(z)$ is said to have a pole with singularity of $O(n)$ if its Laurant series expansion starts with a term of order z^{-n}. Here we are saying that the map has singularity of order n if the Taylor series expansion of $F_1(x) - F_2(x)$ starts with a term of order x^n.
2. Note that we allow this order to be non-integer:

   Example 2.5 (square-root map). Consider the square-root map described in case study VII. According to the functional form (1.32), we have

$$S_1 = \{x < \sigma\}, \quad S_2 = \{x > \sigma\}, \quad \Sigma_{12} = \{x = \sigma\},$$

$$F_1 = \sqrt{\sigma - x} + r\sigma, \quad F_2 = rx,$$

   and hence

$$[F_1 - F_2](\sigma + \varepsilon) = \varepsilon^{1/2} + O(\varepsilon).$$

   In this case we say that this map has an $O(1/2)$ singularity.

Maps that are locally piecewise-linear and continuous such as Fig. 2.14(a) and case study VIII are said to have an $O(1)$ singularity. Clearly differentiation of these one-dimensional maps with respect to x leads to maps with singularities of one order lower. For this reason we shall say that a point of discontinuity of a map with a jump, as in Fig. 2.14(b) and the heart attack map, case study VI, has an $O(0)$ singularity at a point $x \in \Sigma_{ij}$ if $0 < \|F_1(x) - F_2(x)\| < \infty$.

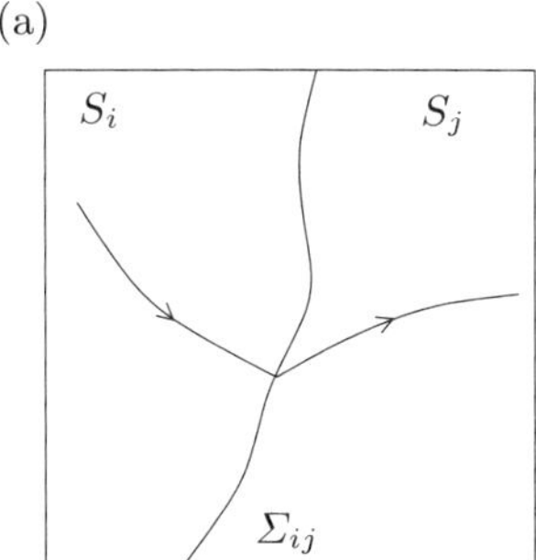

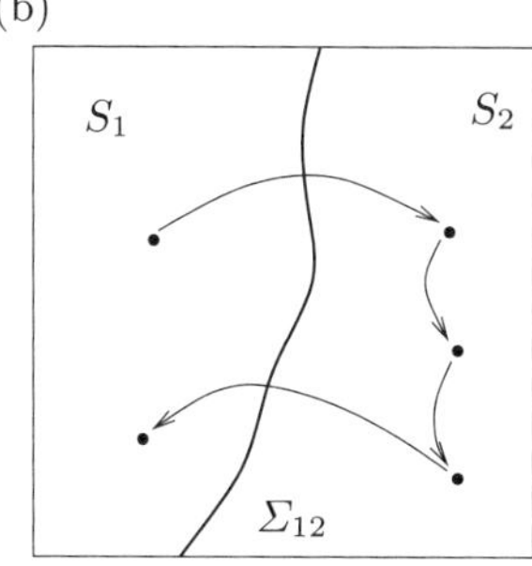

Fig. 2.15. Illustrating schematically trajectories of (a) a piecewise-smooth flow, and (b) a piecewise-smooth map.

2.2.2 Piecewise-smooth ODEs

Definition 2.20. *A* **piecewise-smooth flow** *is given by a finite set of ODEs*

$$\dot{x} = F_i(x, \mu), \quad \text{for} \quad x \in S_i, \tag{2.23}$$

where $\cup_i S_i = \mathcal{D} \subset \mathbb{R}^n$ *and each* S_i *has a non-empty interior. The intersection* $\Sigma_{ij} := \bar{S}_i \cap \bar{S}_j$ *is either an* $\mathbb{R}^{(n-1)}$*-dimensional manifold included in the boundaries* ∂S_j *and* ∂S_i*, or is the empty set. Each vector field* F_i *is smooth in both the state* x *and the parameter* μ*, and defines a smooth flow* $\Phi_i(x,t)$ *within any open set* $U \supset S_i$*. In particular, each flow* Φ_i *is well defined on both sides of the boundary* ∂S_j*.*

A non-empty border between two regions Σ_{ij} will be called a **discontinuity set**, **discontinuity boundary** or, sometimes, a **switching manifold**. We suppose that each piece of Σ_{ij} is of codimension-one, i.e., is an $(n-1)$-dimensional smooth *manifold* (something locally diffeomorphic to $\mathbb{R}^n$) embedded within the n-dimensional phase space. Moreover, we shall demand that each such Σ_{ij} is itself piecewise-smooth. That is, it is composed of finitely many pieces that are as smooth as the flow. See Fig. 2.15(a).

Note that Definition 2.20 does not uniquely specify a rule for the evolution of the dynamics within a discontinuity set. One possibility is to assign each Σ_{ij} as belonging to a single region $\bar{S}_i$ only. That is, F_i rather than F_j applies on Σ_{ij}. In fact, such notions make little difference except in the case where the flow becomes confined to the boundary (Filippov trajectories). Before we get to that case, let us first consider what might happen to the flow of the piecewise-smooth ODE as we cross a discontinuity boundary Σ_{ij}.

Definition 2.21. *The* **degree of smoothness** *at a point* x_0 *in a switching set* Σ_{ij} *of a piecewise-smooth ODE is the highest order* r *such the Taylor series expansions of* $\Phi_i(x_0,t)$ *and* $\Phi_j(x_0,t)$ *with respect to* t*, evaluated at* $t=0$*, agree up to terms of* $O(t^{r-1})$*. That is, the first non-zero partial derivative with respect to* t *of the difference* $[\Phi_i(x_0,t) - \Phi_j(x_0,t)]|_{t=0}$ *is of order* r*.*

Remarks

- This definition almost accords with the usual definition of smooth functions, thinking of the flow at a point as being a function of t. Thus, if we say that a piecewise-smooth flow has degree of smoothness r across a discontinuity boundary, then it is C^{r-1} but not C^r. The vector field is one degree less smooth (because it is by definition the time derivative of the flow). Thus for a flow with degree of smoothness r according to the definition, the vector field will be C^{r-2} but not C^{r-1}.
- Note the subtle distinction between this definition and the corresponding Definition 2.19 for the singularity of a map. Here we do not allow the possibility for the degree of smoothness to be non-integer. [Although there is a growing literature on differential equations with fractional order right-hand sides (see, for example, [155]) we shall not treat them here.]

Now, consider an ODE local to a single discontinuity set Σ_{12} that can be written

$$\dot{x} = \begin{cases} F_1(x, \mu), & \text{if } x \in S_1 \\ F_2(x, \mu), & \text{if } x \in S_2 \end{cases},$$

where F_1 generates a flow Φ_1, F_2 a flow Φ_2. We have

$$\left.\frac{\partial \Phi_i(x,t)}{\partial t}\right|_{t=0} = F_i(x),$$

$$\left.\frac{\partial^2 \Phi_i(x,t)}{\partial t^2}\right|_{t=0} = \frac{\partial F_i}{\partial t} = \frac{\partial F_i}{\partial \Phi_i}\frac{\partial \Phi_i}{\partial t} = F_{i,x}F_i(x),$$

where a second subscript x means partial differentiation with respect to x. Similarly

$$\left.\frac{\partial^3 \Phi_i(x,t)}{\partial t^3}\right|_{t=0} = F_{i,xx}F_i^2 + F_{i,x}^2 F_i,$$

etc. So, if F_1 and F_2 differ in an mth partial derivative with respect to the state x, we find that the flows Φ_1 and Φ_2 differ in their $(m+1)$st partial derivative with respect to t.

Therefore, if $F_1(x) \neq F_2(x)$ at a point $x \in \Sigma_{12}$, then we have degree of smoothness one there. Systems with degree one are said to be of *Filippov* type. Examples of Filippov systems from Chapter 1 include case studies III, IV and V; the relay controller, friction oscillator and DC–DC converter examples.

Alternatively if $F_1(x) = F_2(x)$ but there is a difference in the Jacobian derivatives $F_{1,x} \neq F_{2,x}$ at x, then the degree of smoothness is said to be 2. A difference in the second-derivative tensor $F_{1,xx} \neq F_{2,xx}$ gives smoothness of degree three, etc. Systems with smoothness of degree two or higher may be called *piecewise-smooth continuous systems*, typified by the next example

Example 2.6 (bi-linear oscillator). The bi-linear oscillator, case study II, can be written as the first-order system by setting $u = x_1$,$v = x_2$ and $t = x_3$ so that

$$\dot{x}_1 = x_2 \tag{2.24}$$
$$\dot{x}_2 = -2\zeta x_2 - k_i x_1 + \cos(x_3) \tag{2.25}$$
$$\dot{x}_3 = 1. \tag{2.26}$$

where the value of k_i depends on region S_i, with $S_1 = \{x_1 < 0\}$, $S_2 = \{x_1 > 0\}$. Clearly the flow here has degree of smoothness two at all points in $\Sigma = \{x_1 = 0\}$. If instead $k_1 = k_2$ and the coefficient ζ in (2.24)–(2.26) had been allowed to vary across Σ, then the degree of smoothness would be one, at all points in Σ except where $x_2 = 0$; in which case, the degree would be two. Thus we have cases where the degree of smoothness is the same at all points in Σ and cases where it is not. This distinction shall become crucial when we consider grazing bifurcations in Chapters 6 and 7.

Definition 2.22. *A discontinuity boundary Σ_{ij} is said to be* **uniformly discontinuous** *in some domain $\mathcal{D}$ if the degree of smoothness of the system is the same for all points $x \in \Sigma_{ij} \cap \mathcal{D}$. We say that the discontinuity is* **uniform with degree m** *if the first non-zero partial derivative of $F_i - F_j$ evaluated on Σ_{ij} is of order $m-1$. Furthermore, the degree of smoothness is one if $F_i(x) - F_j(x) \neq 0$ for $x \in \Sigma_{ij} \cap \mathcal{D}$.*

In fact, the assumption of uniform discontinuity imposes a great restriction on the form that $F_i - F_j$ can take. Consider a general piecewise-smooth continuous system with a single boundary Σ that can be written as the zero set of a smooth function H

$$\dot{x} = \begin{cases} F_1(x), & H(x) > 0, \\ F_2(x), & H(x) < 0, \end{cases} \tag{2.27}$$

where $F_1(x) = F_2(x)$ if $H(x) = 0$. Suppose that the flow is uniformly discontinuous with degree m as in Definition 2.22. Then local to $H = 0$ we must be able to write

$$F_2(x) = F_1(x) + J(x)H(x)^{m-1}, \tag{2.28}$$

for some smooth function $J(x) \in \mathbb{R}^n$. To see this, note that H may locally be chosen as one of the co-ordinates close to the boundary and that a non-zero coefficient of $H(x)^k$ in the Taylor series expansion of $F_2 - F_1$, for $k < m-1$, means that the kth derivative of $F_2 - F_1$ does not vanish on Σ. Hence H^{m-1} must be a factor of $F_2 - F_1$. For example, for the bi-linear oscillator (2.24)–(2.26), which has $m = 2$, we have $H(x) = x_1$ and $J(x) = (0, -k_2 + k_1, 0)^T$.

2.2.3 Filippov systems

The case of systems a with uniform degree of smoothness one must be treated with great care since we have to allow the possibility of sliding motion. In order to define sliding, it is useful to think of a system (2.27) local to a discontinuity boundary between two regions defined by the zero set of a smooth function $H(x) = 0$; see Fig. 2.16.

Definition 2.23. *The* **sliding region** *of the discontinuity set of a system of the form (2.27) with uniform degree of smoothness one is given by that portion of the boundary of $H(x)$ for which*

$$(H_x F_1) \cdot (H_x F_2) < 0.$$

That is, $H_x F_1$ (the component of F_1 normal to H) has the opposite sign to $H_x F_2$. Thus, the boundary is simultaneously attracting (or repelling) from both sides.

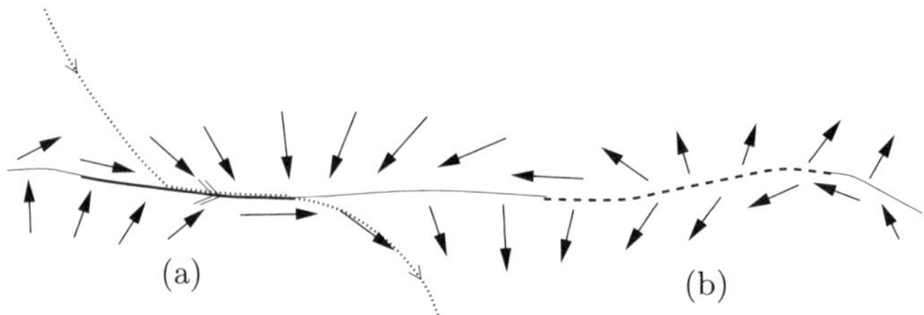

Fig. 2.16. A typical discontinuity boundary of a two-dimensional Filippov system showing the behavior of the vector fields on both sides. Bold and dashed regions represent (a) attracting and (b) repelling sliding motion, respectively. Dotted lines indicate three individual trajectory segments.

Note that the case of most interest is when the sliding region is attracting since, as is clear from Fig. 2.16, repelling sliding motion cannot be reached by following the system flow forward in time. However, attracting sliding motion can be reached in finite time. Henceforth, sliding will always be taken to mean 'attracting sliding' unless otherwise stated. Such motion leads to loss of information on initial conditions. Compare for example the two trajectories A and B of the two-dimensional flow represented in Fig: 2.16; they enter the sliding region at different points, but leave at the same point. Thus while they came from different initial conditions in the past, their future evolution is identical (the trajectory segment C). Thus, there is an infinite rate of attraction in forward time and the flow is not uniquely defined in reverse time. Another simple example of non-inevitability in mechanics is that of plastic impacts (e.g., imagine dropping a mature tomato on the floor!). Whatever the pre-impact velocity, the post-impact velocity is always zero.

As a consequence, any Poincaré mapassociated with trajectories that involve sliding motion will be noninvertible and have a multiplier that is zero (corresponding to the infinite rate of attraction). Now, the formalism of piecewise-smooth systems itself does not say how to define the evolution of the system as it undergoes sliding. One has to do something extra.

Two approaches exist in the literature for formulating the equations for flows that slide when written in the general form (2.27). These are *Utkin's equivalent control method* [257] and *Filippov's convex method* [100]. In Utkin's method one supposes that the system flows according to the sliding vector field F_{12}, which is the average of the two vector fields F_1 (in region S_1) and F_2 (in region S_2) plus a control $\beta(x) \in [-1, 1]$ in the direction of the difference between the vector fields:

$$F_{12} = \frac{F_1 + F_2}{2} + \frac{F_2 - F_1}{2}\beta(x). \tag{2.29}$$

Specifically the equivalent control is

$$\beta(x) = -\frac{H_x F_1 + H_x F_2}{H_x F_2 - H_x F_1}.$$

Filippov's method, by contrast, takes a simple convex combination of the two vector fields

$$F_{12} = (1 - \alpha)F_1 + \alpha F_2 \tag{2.30}$$

with $0 \leq \alpha \leq 1$, where

$$\alpha(x) = \frac{H_x F_1}{H_x(F_1 - F_2)}. \tag{2.31}$$

Sometimes, where there is no ambiguity, we shall write

$$F_{ij} := F_s$$

to represent the sliding vector field

Now it is a simple exercise to show that the above two methods are algebraically equivalent with $\beta = 2\alpha - 1$. (Note though that as shown in [257] there are some special cases where the two methodologies lead to subtly different results.) In both cases it is straightforward to show that the vector field F_s lies orthogonal to the direction H_x and so lies tangent to Σ. Utkin's method has the interpretation that β is precisely the control power that is needed to pull the flow back to being in a direction that is tangent to Σ; see Fig. 2.17(a). Another interpretation, from Filippov's method, is that just the right convex combination of the vector fields needs to be taken for the resulting field F_s to lie in Σ; see Fig. 2.17(b). A final interpretation is obtainable by separating the boundary to regions S_1 and S_2 slightly, within a hysteresis loop; see Fig. 2.17(c). That is, an initial condition in S_1 is allowed to evolve under flow F_1 until penetrating a small distance ε into S_2, then evolves under F_2 until passing back through Σ to a distance ε on the other side. (Thinking of the central heating example introduced in the Introduction, this would be where the temperature threshold for switching on the boiler is slighter greater than that for switching it off.) Then we can consider α to be proportion of time that a trajectory spends in the region S_1, in the limit $\varepsilon \to 0$.

Returning to the perfect sliding case, if the control $\beta(x) = -1$ (equivalently $\alpha = 0$), then the flow must be governed by F_1 alone, which must by definition be tangent to Σ at such a point. Similarly, $\beta = 1$ ($\alpha = 1$) represents a tangency of the flow F_2 with Σ. Hence we can define the sliding region as

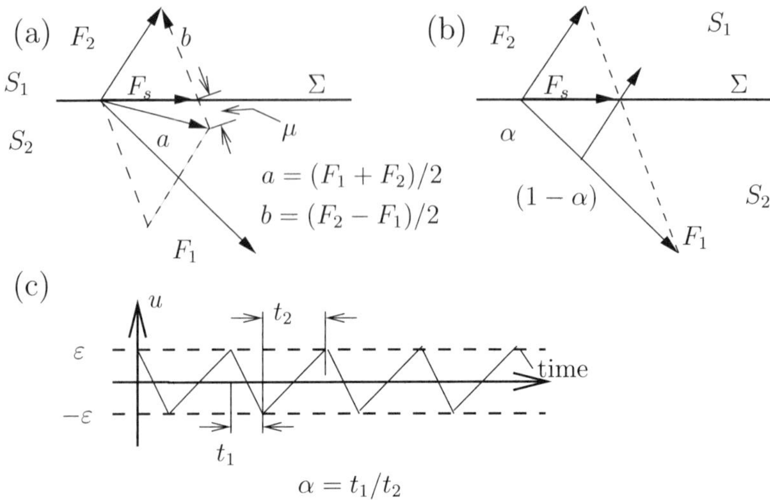

Fig. 2.17. The equivalent definitions of the sliding flow F_s, as defined in the text, illustrated in the two-dimensional case. In (c) the variable u is in the direction H_x orthogonal to Σ.

$$\widehat{\Sigma} := \{x \in \Sigma : -1 \leq \beta \leq 1\},$$

and the boundaries of the sliding region as

$$\partial\widehat{\Sigma}^{\pm} := \{x \in \Sigma : \beta = \pm 1\},$$

with tangency of one vector field or other occurring at the two different types of boundary.

2.2.4 Hybrid dynamical systems

Hybrid dynamical systems are combinations of maps and flows, giving rise to discontinuous, piecewise-smooth flows. They can arise both as models of impacting systems or in the context of the interaction between digital and analog systems. The notion of a *hybrid dynamical system* is a broad concept that encompasses a number of different formalisms in the literature. For example, *hybrid automata* [71, 183] are defined as dynamical systems with a discrete and a continuous part. The discrete dynamics can be represented as a graph whose vertices are the discrete states (or modes) and whose edges are transitions. The continuous states take values in $\mathbb{R}^n$ and evolve along trajectories, typically governed by ODEs or differential algebraic equations. The interaction between the discrete and the continuous dynamics takes place through *invariants* and *transition relations*. Each mode has an invariant associated with it, the violation of which as the system evolves says that a transition must take place. The transition relations describe conditions on the continuous state that enable the transition to occur and also the effect (or reset) that the transition will have on the continuous state. This formalism is really quite broad and covers a wide variety of possible systems both physical and virtual, and in fact all the other formulations we describe in this chapter can be seen as just a special case. The drawback with such a general description is that it does not necessarily allow much general information to be gleaned, which applies to all systems of this class. For more details, see the book by Van der Schaft and Schumacher [71].

In this book we shall reserve the name 'hybrid' for a specific kind of piecewise-smooth system that comprises a collection of different smooth flows and maps; see Fig. 2.18

Definition 2.24. *A* **piecewise-smooth hybrid system** *comprises a set of ODEs*

$$\dot{x} = F_i(x, \mu), \quad \text{if} \quad x \in S_i, \tag{2.32}$$

plus a set of reset maps

$$x \mapsto R_{ij}(x, \mu), \quad \text{if} \quad x \in \Sigma_{ij} := \bar{S}_i \cap \bar{S}_j. \tag{2.33}$$

Here $\cup_i S_i = \mathcal{D} \subset \mathbb{R}^n$ *and each* S_i *has a non-empty interior. Each* Σ_{ij} *is either an* $\mathbb{R}^{(n-1)}$*-dimensional manifold included in the boundary* ∂S_j *and* ∂S_i*, or is the empty set. Each* F_i *and* R_{ij} *are assumed to be smooth and well defined in open neighborhoods around* S_i *and* Σ_{ij}*, respectively.*

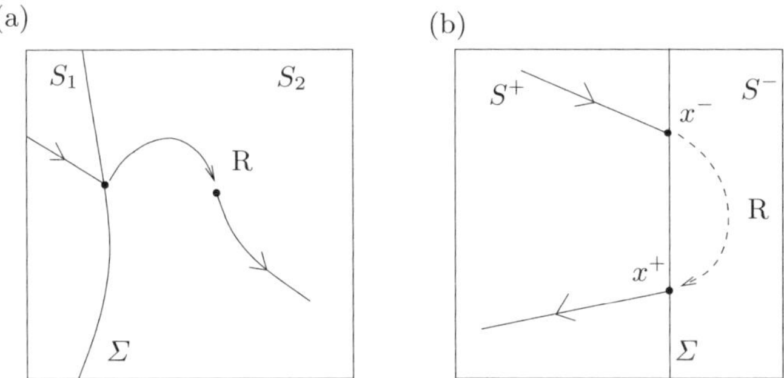

Fig. 2.18. (a) A hybrid system and (b) the impacting class of hybrid system.

In this book we will mostly study a special type of hybrid systems motivated by the impact oscillator example described in case study I. For such systems we generally consider surfaces Σ_{ij} that act as hard constraints, so that the reset R_{ij} maps the set Σ_{ij} back to itself.

Definition 2.25. *An* **impacting hybrid system** *is a piecewise-smooth hybrid system for which* $R_{ij} : \Sigma_{ij} \to \Sigma_{ij}$*, and the flow is constrained locally to lie on one side of the boundary; this is, in* $\overline{S}_i = S_i \cup \Sigma_{ij}$*.*

We shall often refer to the reset map R_{ij} in this context as being the *impact law* or *impact rule*. The discontinuity boundaries Σ_{ij} will be referred to as *impact surfaces* and the event of a trajectory intersecting Σ_{ij} as an *impacting event* or just an *impact*.

Throughout this book, we shall often consider a restrictive class of impacting hybrid systems that contain just one impact surface Σ. Suppose that such a surface Σ can be defined by the zero set of a smooth function $H(x)$,

$$\Sigma = \{x : H(x) = 0\}, \quad \text{and let} \quad S^+ = \{x : H(x) > 0\}, \tag{2.34}$$

with the dynamics constrained to the region S^+; see Fig. 2.19. Such systems can be thought of as describing the dynamics local to any impact surface in a general, multiple region system. Locally the dynamics may be written in the form

$$\dot{x} = F(x) \quad \text{if } H(x) > 0, \tag{2.35}$$

$$x \mapsto R(x) \quad \text{if } H(x) = 0, \tag{2.36}$$

for a smooth vector field F (which is well defined in a full neighborhood of Σ including for $H < 0$) and reset map R. Suppose an impact occurs at time t_0. Let x^- and x^+ represent the intersection of the flow with Σ both immediately before and immediately after the impact, so that $x^- = \lim_{t \to t_0^-}$, $x^+ = \lim_{t \to t_0^+}$. Hence we can write the impact surface as

$$x^+ = R(x^-). \tag{2.37}$$

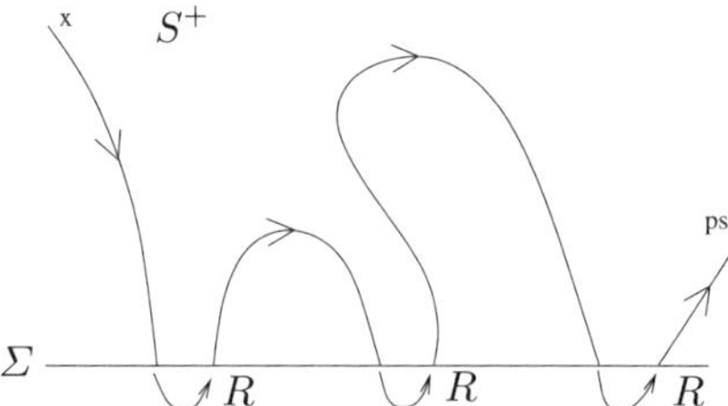

Fig. 2.19. The surface Σ and a multiple impacting trajectory for an impacting hybrid system with a single discontinuity boundary.

In order to be definite, we shall also assume a restrictive class of impact law that depends on the *normal velocity* $v(x)$ at which the trajectory approaches the impact manifold, given by

$$v(x) \equiv dH/dt = H_x F. \tag{2.38}$$

Specifically, we suppose that

$$R(x) = x + W(x)H_x F = x + W(x)v(x) \tag{2.39}$$

for a some smooth function $W(x) \in \mathbb{R}^n$. To motivate why (2.39) is a reasonable form to take, note that we would like an impact law that takes a grazing trajectory (one for which $v(x) = 0$) to itself and that is a smooth function of $v(x)$ otherwise. More complex forms of reset maps than (2.39) are required to deal with impacting mechanical systems with friction. For example, the so-called Painlevé paradox deals with mechanical systems that can both slide and impact; see, for example, [243, 174].

Given an impact rule of the form (2.39), the surface Σ can therefore be divided into three separate regions, Σ^-, Σ^+ and Σ^0 according to whether the normal velocity is, respectively, negative, positive or zero:

$$\Sigma^- = \{x \in \Sigma : v(x) < 0\}, \qquad \Sigma^+ = \{x \in \Sigma : v(x) > 0\},$$
$$\Sigma^0 = \{x \in \Sigma : v(x) = 0\}.$$

In general, if we write the impact law in the form (2.37), then we have $x^- \in \Sigma^-$ and $x^+ \in \Sigma^+$. In this case a flow in S^+ intersects Σ^-, is mapped to Σ^+ and then continues in S^+. The set Σ^0 is called the **grazing set**, and impacts close to it lead to subtle dynamics that we will analyze in detail in Chapter 6.

Example 2.7 (impact oscillator). Let us show that the impact oscillator with the simple coefficient of restitution law for impact, studied in case study I, fits into this framework. We can write the equations of motion (1.1) in the form

$$\begin{pmatrix} \dot{x}_1 \\ \dot{x}_2 \end{pmatrix} = \begin{pmatrix} 0 & 1 \\ -1 & -2\zeta \end{pmatrix} \begin{pmatrix} x_1 \\ x_2 \end{pmatrix} + \begin{pmatrix} 0 \\ \cos(\omega t) \end{pmatrix}, \tag{2.40}$$

together with the impact rule

$$\begin{pmatrix} x_1(t_j^+) \\ x_2(t_j^+) \end{pmatrix} = \begin{pmatrix} 1 & 0 \\ 0 & -1-r \end{pmatrix} \begin{pmatrix} x_1(t_j^-) \\ x_2(t_j^-) \end{pmatrix}, \tag{2.41}$$

which applies at times t_j for which $x_1 = \sigma$. Letting $x_3 = t$, we see that this fits into the above framework with $x = (x_1, x_2, x_3)^T$, $H = x_1 - \sigma$, $H_x F(x) = x_2$, and $W(x) = -(0, 1+r, 0)^T$.

Many more examples of hybrid systems of this form will be given in Chapter 6, in which the detailed dynamics of hybrid systems and their bifurcations will be analyzed.

Let us now consider the basic flow of the simple impacting system (2.35)–(2.39). Starting from an initial condition $x(0) = x_0$ in S^+, the ODE (2.35) generates a smooth flow $\Phi(x_0, t)$ up until the flow strikes Σ, at time t_0, say. Suppose that this impact is *transversal*, so that the normal velocity $v(x(t_0)) < 0$. Hence $x^- = x(t_0) \in \Sigma^-$. This point is then mapped instantaneously under the action of the reset map to the point $x^+ = R(x^-)$. If $v(x^+) > 0$, so that $x^+ \in \Sigma^+$, then the flow moves away from Σ back into the set S^+ and is described by the flow $\Phi(x^+, t)$. In principle, this scenario can repeat arbitrarily often, as illustrated in Figure 2.19.

However, this is not the only possible dynamics of the system. Consider a grazing point for which $v(x^-) = 0$, where the impact map becomes the identity. In order to understand what happens, it is useful to define the *normal acceleration* of the flow with respect to the boundary:

$$a(x) = d^2H/dt^2 = (H_x F)_x F = H_{xx} F F + H_x F_x F. \tag{2.42}$$

Now, in the case where $a(x^-) > 0$ at a grazing point, the curvature of the flow will cause the trajectory to immediately leave Σ. However, if $a(x) < 0$, then the flow will become **stuck** to the boundary, rather akin to the *sliding flow* of a Filippov system. Thus the **sticking subset** of the grazing set Σ^0 is determined by the conditions

$$\Sigma^0_- \equiv \{x : H(x) = 0, \quad v(x) = 0, \quad a(x) < 0\}.$$

The **sticking motion** evolves under the action of the vector field F, constrained to lie on the surface Σ. If we define the impact law according to (2.39), then it is possible to express the sticking vector field as

$$\dot{x} = F_s(x) = F(x) - \lambda(x)W(x), \tag{2.43}$$

where

$$\lambda(x) = \frac{a(x)}{(H_x F)_x W}. \tag{2.44}$$

To see that this corresponds to a sticking flow, note that in order to stick we require $H(x(t)) = v(x(t)) \equiv 0$. Differentiating the conditions $H(x) = 0$ and $v(x) = 0$ with respect to time, we have $H_x \dot{x} = 0$ and $v_x \dot{x} = 0$. The first of these conditions is satisfied identically when $H_x W = 0$, and the second condition if

$$0 = (H_x F)_x F - \lambda (H_x F)_x W = a(x) - \lambda (H_x F)_x W, \tag{2.45}$$

which defines λ according to (2.44). Note that (2.43)–(2.44) define a smooth flow $\Phi_s(x, t)$, which is also defined within a neighborhood of Σ, but for which the set $\Sigma\{x : H(x) = 0\}$ is invariant. For the hybrid system, the sticking flow ceases to apply when the trajectory leaves Σ^0_-. At such a point $a(x) = 0$, but $\frac{da(x)}{dt} := a_x(x)\dot{x} > 0$ and hence the system moves into S^+ where the original flow Φ applies. The condition that the vector field remains in the sticking region is $\lambda(x) > 0$. The formalism of complementarity systems described in the next section helps us understand the role played by this extra variable λ.

Typically, unlike the sliding motion in Filippov systems, impacting systems do not enter a sticking region directly, but via a **chattering sequence** (also known in control theory as a *Zeno phenomenon* [145]). Such a sequence begins if an impact occurs within Σ^-, close to the set Σ^0 with $v(x^+) \ll 1$ and

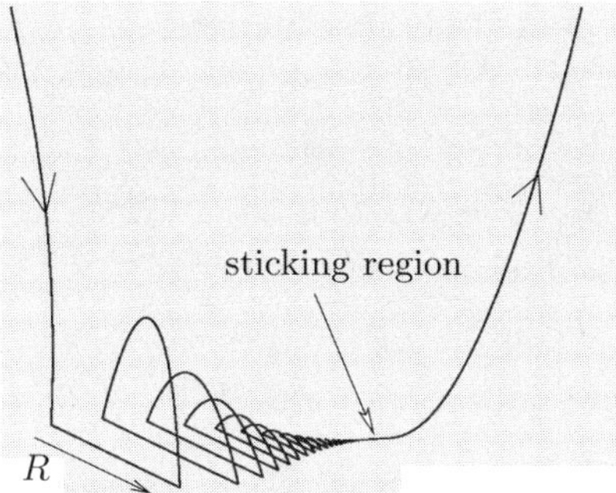

Fig. 2.20. A chattering sequence followed by sticking and release.

$a(x^+) < 0$; see Fig. 2.20. There follows an infinite sequence of impacts, of successively reduced velocity, which converges in *finite time*, onto a point in the sticking set [42, 203]. After the accumulation of such a sequence, the motion will evolve in the sticking set in the manner described above. We shall return to an analysis of chattering in Chapter 6. Chattering sequences are a commonly observed feature of hybrid systems and require special care when computing the flow numerically.

Hybrid systems, then, generally have state jumps. This should be contrasted with Filippov systems that have jumps in the vector field (time derivative of the flow) and piecewise-smooth continuous systems that have jumps in the second or higher derivative of the flow. Thus we can extend the notion of degree of smoothness in Definition 2.21 to say:

Definition 2.26. *A hybrid dynamical system that undergoes a jump in the system state* $\Phi(x,t) \mapsto R_{ij}(\Phi(x,t))$ *on a discontinuity boundary* Σ_{ij} *is said to have* **degree of smoothness zero**.

2.3 Other formalisms for non-smooth systems

The choice of formalism we choose in this book is essentially to deal with piecewise-smooth maps or with piecewise-smooth systems that have integer degree of smoothness across each of its boundaries Σ_{ij}. However, there is no guarantee that such a formulation leads to existence or uniqueness of solutions in all circumstances. Let us therefore briefly present several other formalisms for which more mature analytic theory is available.

2.3.1 Complementarity systems

Complementarity dynamical systems formalize the notion of a mechanical system with unilateral constraints. Such systems can be written most simply in the form of a differential algebraic equation plus inequality constraints (see for example the reviews by Brogliato and co-workers [38, 127] and references therein):

$$\dot{x} = f(x, \lambda), \tag{2.46}$$

$$g(x, w, \lambda) = 0, \tag{2.47}$$

$$0 \leq w \perp \lambda \geq 0, \tag{2.48}$$

$$\text{re-initialization rule } R \text{ for state } x. \tag{2.49}$$

Here $x \in \mathbb{R}^n$ is the system state, $g \in \mathbb{R}^m$ are a set of side relations, $\lambda \in \mathbb{R}^l$ is a so-called *slack variable* and $w \in \mathbb{R}^l$ is the corresponding signal or system output. The expression $w \perp \lambda$ means that the vector w is orthogonal to λ, whereas $\lambda, w \geq 0$ means that all components of λ and w are non-negative. Hence, if a component w_i is positive then the corresponding λ_i must be zero, and *vice versa*.

Let us consider the dynamics of a system written in the form (2.46)–(2.47). The set of m relations (2.47) implicitly defines the signal w in terms of the states and slack variables (often the relations can be written explicitly as $w = \tilde{g}(x, \lambda)$). The most important part of the system is the orthogonality (or '*complementarity*') relation (2.48). This should be understood component-wise. That is, for each i, *either* λ_i is zero and w_i is non-negative *or* λ_i is positive and w_i is zero. At transition points, that is at times t_j for which $\lambda_i(t_j)$ and $w_i(t_j)$ are both zero for some $i \leq m$, then one in general has to apply a rule (2.49) to reset the state $x(t_j^+) = R(x(t_j^-), w, \lambda)$. Complementarity systems may be seen as a special case of hybrid automata, where the discrete states are the λ_i, the particular set of w_i that are non-zero describe the invariants, and the state re-initialization rule (and choice of new set of non-zero w_i) gives the transition relations.

Example 2.8 (impact oscillator). We illustrate the complementarity framework with the impact oscillator, case study I, which can be written in the complementarity form

$$\begin{aligned}
\begin{pmatrix} \dot{x}_1 \\ \dot{x}_2 \end{pmatrix} &= \begin{pmatrix} 0 & 1 \\ -1 & -2\zeta \end{pmatrix} \begin{pmatrix} x_1 \\ x_2 \end{pmatrix} + \begin{pmatrix} 0 \\ \cos(\omega t) + \lambda \end{pmatrix}, \\
w &= x_1 - \sigma, \\
0 &\leq w \perp \lambda \geq 0, \\
\begin{pmatrix} x_1(t_j^+) \\ x_2(t_j^+) \end{pmatrix} &= \begin{pmatrix} 1 & 0 \\ 0 & -1 - r \end{pmatrix} \begin{pmatrix} x_1(t_j^-) \\ x_2(t_j^-) \end{pmatrix}.
\end{aligned} \tag{2.50}$$

For this example, there are two kinds of motion (active modes): *free motion* of the oscillator where $\lambda = 0$ and $w > 0$, which implies that $x_1 > \sigma$; and *sticking motion* where $\lambda > 0$ and $w = 0$; hence, $x_1 = \sigma$. Here, the slack variable λ should be interpreted as a *Lagrange multiplier*, namely the force being exerted by the obstacle on the particle to stop it from penetrating. Clearly if this force became negative, then the particle would be pulled into the obstacle; hence the requirement that $\lambda \geq 0$. In fact, it is possible to calculate explicitly the value that λ must take. Suppose that the particle is instantaneously in contact with the obstacle at some time τ. Then $x_1(\tau) = \sigma$. If it is to remain in contact, then we require that $x_1(t) \equiv \sigma$ for some time interval $t \in (\tau, \tau + \varepsilon)$. Hence $\dot{x}_1(\tau) = 0$ and $\ddot{x}_1(\tau) = 0$ also. The first of these conditions gives $x_2(\tau) = 0$, and the second gives

$$\lambda = \sigma - \cos(\omega t). \tag{2.51}$$

Hence, sticking motion can only occur for t-values such that $\cos(\omega t) < \sigma$.

In order to describe the motion completely we need to consider the transition times t_j when both λ and w are zero. Suppose first that the system is in free motion at t_j^- and reaches the constraint $x_1 = \sigma$. Then $w = 0$. Here, we apply the reset rule (2.50), which is just Newton's restitution law. Now, exceptions arise when the velocity $x_2 = 0$ at the impact point, so that grazing occurs. Then we have to look at the sign of $\dot{x}_2 = \cos(\omega t_j^-) - \sigma$. If this is positive, then we have a grazing trajectory, which immediately passes back into free motion again, since the reset rule gives $\dot{x}_1(t_j^+) = x_2(t_j^+) = 0$ but $\ddot{x}_1(t_j^+) = \dot{x}_2(t_j^+) > 0$ and so $x_1 > \sigma$ for $t = t_j + \varepsilon$ for some $\varepsilon < 0$. However, if $\dot{x}_2 < 0$, then a sticking motion ensues with a non-zero value of λ. As we have described above, the sticking region can only be entered after an infinite sequence of impacts (a chattering sequence). In contrast, the exit boundary from the sticking region is given by a zero of λ defined by (2.51). Hence, at this point we have that the first three time derivatives of x_1 are zero, but $\frac{d^3}{dt^3} x_1(t) = -\omega \sin(\omega t)$, which if negative implies that we are once again in the regime of free motion for small subsequent times. (In practice, this quantity will always be negative since the particle enters the sticking region at some time t such that $\cos(\omega t) < \sigma$ and leaves it at the first *later* time at which $\cos(\omega t) = \sigma$. Hence the angle ωt must be in the third or fourth quadrant, depending on whether $\sigma > 0$ or $\sigma < 0$. Hence $\sin(\omega t)$ must be positive.)

We can generalize this example by putting any piecewise-smooth ODE system into complementarity form, at least local to a single discontinuity boundary. For piecewise-smooth continuous systems in a neighborhood of a single uniformly discontinuous boundary, where $F_1(x, \mu) - F_2(x, \mu)$ is of the form (2.28), a corresponding complementarity formulation of (2.27) is

$$\begin{cases} \dot{x} = F_1(x, \mu) + \lambda^{m-1} J(x, \mu), \quad w = -H(x, \mu) + \lambda, \\ 0 \leq w \perp \lambda \geq 0, \end{cases} \tag{2.52}$$

for which there is no need for a reset rule. Table 2.1 shows the possible active modes of motion of the system.

In the Filippov case, i.e., for systems with degree of smoothness one, a different form of complementarity formulation is required. For example, given a two-zone system (2.27) with a single discontinuity boundary, we have

Table 2.1. The different possible active modes for the dynamics of the piecewise-smooth continuous ODE (2.52).

w and λ	dynamical system
$w = 0, \lambda > 0$	$\dot{x} = F_2 = F_1(x, \mu) + H(x, \mu)^{m-1} J(x, \mu)$
$w > 0, \lambda = 0$	$\dot{x} = F_1(x, \mu)$
$w = 0, \lambda = 0$	$\dot{x} = F_1(x, \mu)$ and $H(x, \mu) = 0$

$$\begin{cases} \dot{x} = F_1(x, \mu) + \lambda_1(F_2(x, \mu) - F_1(x, \mu)), \\ w_1 = -H(x, \mu) + \lambda_2, \\ w_2 = 1 - \lambda_1, \\ 0 \leq w_1 \perp \lambda_1 \geq 0, \\ 0 \leq w_2 \perp \lambda_2 \geq 0. \end{cases} \tag{2.53}$$

Table 2.2 gives the different possible dynamical regimes of such systems.

Table 2.2. The different possible active modes for the dynamics of the Filippov system (2.53).

C1	C2	dynamical system
$w_1 = 0,\ \lambda_1 = 1$	$w_2 = 0,\ \lambda_2 \geq 0$	$\dot{x} = F_2(x, \mu)$
$w_1 \geq 0,\ \lambda_1 = 0$	$w_2 = 1,\ \lambda_2 = 0$	$\dot{x} = F_1(x, \mu)$
$w_1 = 0,\ 0 \leq \lambda_1 \leq 1$	$0 \leq w_2 \leq 1,\ \lambda_2 = 0$	$\dot{x} = F_1(x, \mu) + \lambda_1(F_2(x, \mu) - F_1(x, \mu))$ $H(x, \mu) = 0$

Notice that the concept of the sliding vector field is embedded in the complementarity description of the system of interest. In fact, in the third case in Table 2.2, the dynamical system is a convex combination of the two original vector fields. The parameter λ_1 can be calculated directly from the requirement that $H(x, \mu) \equiv 0$ along such solutions. Hence

$$\frac{dH}{dt}(x, \mu) := H_x(x, \mu) \left[F_1(x, \mu) + \lambda_1(F_2(x, \mu) - F_1(x, \mu))\right] = 0. \tag{2.54}$$

Thus

$$\lambda_1 = \frac{H_x F_1}{H_x F_1 - H_x F_2},$$

which is the parameter α in Filippov's convex method introduced in (2.30). There thus seems an advantage of the complementarity framework over the piecewise-smooth one in this case. Checking the slack variables will automatically detect when sliding is occurring and when the sliding region is exited. That we had to differentiate the constraint once to obtain λ_1 means that the constraint and the differential equation have *relative degree one.* Equivalently the sliding mode of the complementarity system is an *index 1* differential algebraic equation (DAE). Note, in contrast, that the complementarity formulation for piecewise-smooth continuous systems (2.52) does not require differentiation of the constraint, since λ is given explicitly. Thus the constraint has relative degree zero, and the mode when $w = 0$ is a DAE with index 0, which is equivalent to just an ODE.

Finally, consider a hybrid system with a single impact boundary for which the reset map is written in the form (2.39). This can be written as the complementarity system

$$\begin{cases} \dot{x} = F(x,\mu) - \lambda W(x,\mu), \\ w = H(x,\mu), \\ 0 \leq w \perp \lambda \geq 0, \\ x(t^+) = x(t^-) + W(x(t^-),\mu) H_x F(x(t^-),\mu), \end{cases} \tag{2.55}$$

which is a generalization of the complementarity framework for the impact oscillator (2.50). Note that this has relative degree two in the sticking mode, since the value for λ is obtained by differentiating the constraint $H(x) = 0$ twice with respect to t.

The complementarity framework is not just restricted to problems with single discontinuity boundaries. In principle each of the above kinds of constraints and corresponding slack variables can be concatenated to take account of multiple boundaries.

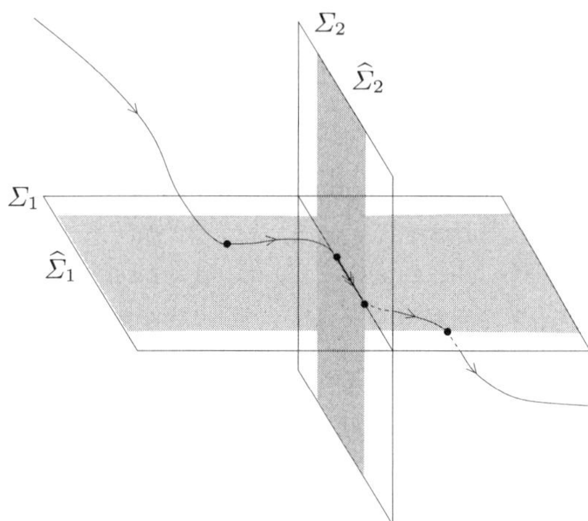

Fig. 2.21. Higher-order sliding occurring when two sliding regions $\widehat{\Sigma}_1$ and $\widehat{\Sigma}_2$ intersect.

For example, suppose a piecewise-smooth system is written in the form

$$\dot{x} = \begin{cases} F_1(x), & \text{if } H_1(x) > 0, \;\; H_2(x) > 0, \\ F_2(x), & \text{if } H_1(x) > 0, \;\; H_2(x) < 0, \\ F_3(x), & \text{if } H_1(x) < 0, \;\; H_2(x) > 0, \\ F_4(x), & \text{if } H_1(x) < 0, \;\; H_2(x) < 0; \end{cases}$$

see Fig. 2.21. This can be rewritten as the following complementarity system:

$$\dot{x} = \lambda_1\lambda_2 F_1(x) + \lambda_1(1-\lambda_2)F_2(x) + (1-\lambda_1)\lambda_2 F_3(x) + (1-\lambda_1)(1-\lambda_2)F_4(x), \tag{2.56}$$

$$\begin{cases} w_1 = -H_1(x,\mu) + \lambda_3, & w_2 = -H_2(x,\mu) + \lambda_4, \\ w_3 = 1 - \lambda_1, & w_4 = 1 - \lambda_2, \\ 0 \le w_1 \perp \lambda_1 \ge 0, & 0 \le w_2 \perp \lambda_2 \ge 0, \\ 0 \le w_3 \perp \lambda_3 \ge 0, & 0 \le w_4 \perp \lambda_4 \ge 0. \end{cases} \tag{2.57}$$

The most interesting case is when both w_1 and w_2 are zero, whereas w_3 and w_4 are both positive. Then, according to (2.57), we have $0 < \lambda_1 < 1$ and $0 < \lambda_2 < 1$ and the motion is constrained to a codimension-two set $\{H_1(x) = H_2(x) = 0\}$. The flow on this set we refer to as *higher-order sliding*, which we shall return to in Chapter 8, and is given by (2.56), where λ_1 and λ_2 are obtained from the pair of simultaneous equations that arise from differentiation of the constraints

$$\begin{aligned} H_{1x}\left[\lambda_1\lambda_2 F_1 + \lambda_1(1-\lambda_2)F_2 + (1-\lambda_1)\lambda_2 F_3 + (1-\lambda_1)(1-\lambda_2)F_4\right] &= 0, \\ H_{2x}\left[\lambda_1\lambda_2 F_1 + \lambda_1(1-\lambda_2)F_2 + (1-\lambda_1)\lambda_2 F_3 + (1-\lambda_1)(1-\lambda_2)F_4\right] &= 0. \end{aligned}$$

Once the problem has been formulated in the complementarity framework, it is possible to study its well-posedness using the analytic tools developed for unilaterally constrained optimization; see, for example, [40] for an extensive review. For example, any complementarity systems for which we can write

$$w = G(x) + D\lambda$$

for a smooth function G and invertible matrix D, is equivalent to a set of ODEs with degree of smoothness at least 1. If the matrix D is a so-called P-matrix (i.e., a matrix with positive principal minors), then it can be shown that the corresponding complementarity problem has a unique solution. This means that systems such as (2.52) where $D = 1$ and $G = H(x,\mu)$ have a unique solution for all parameter values μ.

Complementarity systems are also useful because they provide a general framework for describing systems with more than one (perhaps many thousands) of constraints. They come armed with a set of numerical solution techniques, that do not require the precise detection of the events t_j, where λ_i and w_i are both zero; see [41] for a review.

2.3.2 Differential inclusions

Another way of putting piecewise-smooth systems on a rigorous footing is to use a variational formulation. We shall not go into details, but the key notion is that of a *differential inclusion*. Here we allow the right-hand side of an ordinary differential equation $\dot{x} = f(x)$ to be not strictly a function (that is, returning a single value $f(x)$ for each x), but to be set-valued. For example, such set-valued functions arise in Coulomb dry friction laws encountered in mechanics. Specifically, Coulomb friction models objects in contact that slip over each other with velocity v only if their tangential contact force f_t exceeds some critical value. The function

$$f_t = C(v) = \alpha_0 \text{sgn}(v) - \alpha_1(v) + \alpha_2(v)^3 \tag{2.58}$$

occurring in case study IV is an example of such a law, see Fig. 2.22(a). The problem with (2.58) is that it does not specify what value f_t should take at $v = 0$. Using the notion of a differential inclusion, we rewrite f_t as a set-valued function

$$f_t(v) = \begin{cases} \{[-\alpha_0, \alpha_0]\}, & \text{if} \quad v = 0, \\ \alpha_0 \text{sgn}(v) - \alpha_1 v + \alpha_2 v^3, & \text{otherwise.} \end{cases}$$

So now, instead of $\ddot{y} + y = f_t(1 - \dot{y}) + a\cos(\nu t)$, we write

$$\ddot{y} + y - a\cos(\nu t) \in f_t(1 - \dot{y}),$$

because at $\dot{y} = 1$, f_t can take on a range of values. In [69], Deimling explains that to obtain a well-posed problem, one has to 'fill in' the gap between $[-\alpha_0, \alpha_0]$ at $v = 0$ (i.e., perform a so-called *convexification* of the problem).

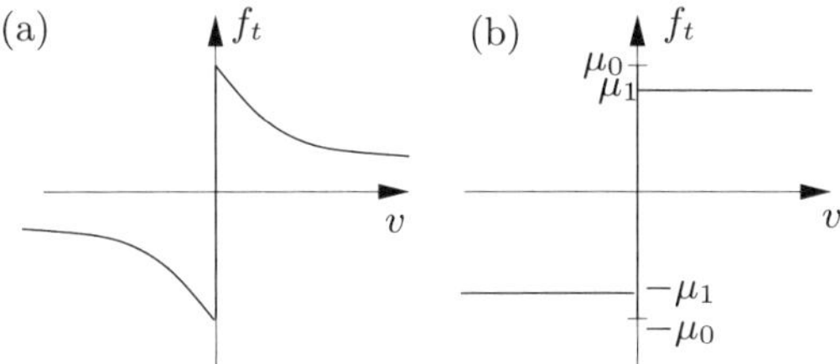

Fig. 2.22. Two idealized Coulomb friction characteristics showing the tangential force f_t as a function of velocity v.

In general, any Filippov system can be written as a differential inclusion. For example, a two-zone system can be written as

$$\dot{x} \in f(x), \quad \text{where} \quad f = \begin{cases} \{F_1(x)\}, & \text{if} \quad H(x) > 0, \\ \{F_2(x)\}, & \text{if} \quad H(x) < 0, \\ \{F_1(x) + \alpha(F_2 - F_1) | 0 \leq \alpha \leq 1\}, & \text{if} \quad H(x) = 0. \end{cases}$$

The concept of the inclusion is especially useful when we take more general Coulomb friction laws like the one in Fig. 2.22(b), where the static coefficient of friction is different from the dynamic one:

$$f_t(v) = \begin{cases} [-\mu_0, \mu_0], & \text{if} \quad v = 0, \\ \{\mu_1 \text{sgn}(v)\}, & \text{otherwise.} \end{cases}$$

In cases where the right-hand side $f(x)$ of an inclusion $\dot{x} \in f(x)$ satisfies some quite general properties (f is upper semi-continuous, non-empty, convex and compact for all x, and is bounded by an affine function of x), then there is a general theory that gives the existence of absolutely continuous solutions [69]. Unfortunately, many non-smooth systems when put into an inclusion form do not satisfy these hypotheses.

Example 2.9 (Lagrangian systems). Another way of writing a general impacting mechanical system in differential inclusion form is to use a Lagrangian approach. This leads to a second-order ODE system for generalized co-ordinates $q \in \mathbb{R}^n$. Consider such a system with mass matrix $M(q)$ and generalized non-contact forces $F(q, \dot{q}, t)$, in the absence of damping, that is constrained to the region of configuration space $S = \{h(q) \geq 0\}$. Its dynamics may be written as

$$M(q)\ddot{q} + F(\dot{q}, q, t) \in \partial\psi_S(q), \text{ if } \quad t \neq t_k, \tag{2.59}$$

$$\dot{q}(t_k^+) = R(\dot{q}(t_k^-)), \text{ if } \quad t = t_k, \tag{2.60}$$

where the $\{t_k, k = 1, 2, 3 \ldots\}$ are the *a priori* unknown set of impact times where $q \in \Sigma = \{h = 0\}$, and R is a reset rule. Here $\partial\phi(q)$ represents the *sub-differential* (the set of all possible one-sided limits $\lim_{t\to 0} \frac{\phi(q+tv)-\phi(q)}{t}$ for any vector v) and ψ_{S_1} is the *indicator function* of the set S_1, which is 0 for all points inside S_1 and infinite outside. Thus $\partial\psi_{S_1}$ is the empty set for all points outside S_1, is equal to the *normal cone* $N_K(x) = \{z | z \cdot (x - z) = 0$, for all $z \in S_1\}$ inside the boundary $\Sigma = \partial S$, and is 0 for points in the interior of S. This set is not compact, and so the general existence theory does not apply. A particular form of reset map R (corresponding to coefficient of restitution 0) is the so-called Moreau collision mapping that the velocity $\dot{q}(t_k^+)$ is in the so-called *polar cone* $V(q)$ for $q \in \Sigma$. Here $V(x) = \{z | z \cdot x \leq 0$, for all $x \in N_K(x)\}$, with the additional constraint that the jump in kinetic energy is minimized.

A way of dealing more generally with systems that have state jumps is via the formalism of *measure differential inclusions* introduced by Schatzman [230] and Moreau [192]. This is motivated by the idea that one would like a framework that allows the velocity jumps at impacts to be included explicitly in the differential equation.

Example 2.10 (impact oscillator without damping).

$$\ddot{u} = -u + f(t), \quad u > \sigma, \quad \text{plus the impact law.}$$

It is tempting to integrate, and write formally

$$\int \mathrm{d}\dot{u} = \int (-u + f(t))\mathrm{d}t + \int \mathrm{d}R(u),$$

where $\mathrm{d}R$ is a *measure* that is zero at all times other than t_k and gives the value of the jump in $\dot{u}$ at impact. This leads to a general formulation where one defines a measure $\mathrm{d}\mu = \mathrm{d}t + \Sigma_{k\geq 0}\delta(t_k)$, where $\delta(x)$ is the Dirac δ function, and we write

$$\frac{\mathrm{d}\dot{u}}{\mathrm{d}\mu} + (u - f(t))\frac{\mathrm{d}t}{\mathrm{d}\mu} \in F(u,t),$$

where

$$F(u,t) = \begin{cases} \{0\}, & \text{if} \quad t \neq t_k, \\ [0,\infty], & \text{if} \quad t = t_k. \end{cases}$$

Here we have introduced an example of a measure differential inclusion, which more generally can be written in the form

$$\frac{\mathrm{d}x}{\mathrm{d}\mu} + g(x(t^+),t^+)\frac{\mathrm{d}t}{\mathrm{d}\mu} \in F(x(t^+),t^+),$$

where μ is a positive measure on the time axis, and the set-valued function $F(x,t)$ satisfies the properties of being a cone for all x and t. The more general multi-degree-of-freedom mechanical system (2.59), (2.60) with impacts can also be put in this framework, upon writing

$$-M(q(t))\frac{\mathrm{d}\dot{q}}{\mathrm{d}\mu} - F(q(t),v(t))\frac{\mathrm{d}\dot{t}}{\mathrm{d}\mu} \in \partial\psi_{V(q(t))(\dot{q}(t^+)} \subseteq \partial\Psi_S\left(g(t)\right),$$

which is an example of a so-called *Moreau sweeping process*; see [166].

Many things can be proved about the dynamics of each of many subtly different classes of differential inclusions (either in measure form or not). They also have use in that they suggest natural ways to define numerical algorithms that preserve the properties of the inclusion that can be proved theoretically. However, we shall ignore such mathematical technicalities in this book and stick to a more pragmatic approach.

2.3.3 Control systems

Many concepts in non-smooth dynamical systems have a counterpart (often with different notation) in control theory. There, the goal is often to prove stability of some target state (such as an equilibrium point), or to design control laws in order to achieve such stability. See for example [240, 179, 255]. This book takes a rather different emphasis, which is to gain a qualitative understanding of complex dynamics via the (discontinuity-induced) transitions that can occur upon varying a parameter. It is nevertheless useful to spell out links with some of the ideas that arise in the control theory literature. For simplicity, we stick to the case of single-input single-output (SISO) systems. Here, the concept of *relative degree* is important; with the term having a rather different meaning to its use in complementarity systems, but nevertheless having a close link to our concept of degree of smoothness.

Consider a SISO linear system [240] given by

$$\begin{aligned} \dot{x} &= Ax + Bu, \\ y &= C^T x + Du. \end{aligned} \tag{2.61}$$

Here $x(t) \in \mathbb{R}^n$ is the *state vector*, $u(t) \in \mathbb{R}$ the *control input* and $y(t) \in \mathbb{R}$ the *output* of the system.

Definition 2.27. *The* **relative degree** *of the SISO linear control system (2.61) can be defined in terms of* **Markov parameters**

$$(M_0, M_1, M_2, M_3, \ldots) := (D, C^T B, C^T AB, C^T A^2 B, \ldots)$$

as the first index i for which M_i is nonzero.

Now it is easy to see how this concept is closely related to the degree of smoothness introduced in Definition 2.21. Take a relay control system, case study III, as a representative example. The system can be written as

$$\begin{aligned} \dot{x} &= Ax + Bu, \\ y &= C^T x + Du, \\ u &= -\mathrm{sgn}(C^T x). \end{aligned} \tag{2.62}$$

Then, according to Definition 2.27, this system has *relative degree 0* provided $D \neq 0$. In this case the discontinuity of the input u is translated into a discontinuity of the output y. Thus the relative degree is equal to one less than the degree of smoothness defined by Definition 2.21.

If instead $D = 0$ but $C^T B \neq 0$ in (2.62), then the *relative degree* is 1 and the output y is continuous but

$$\dot{y} = C^T \dot{x} = C^T Ax + C^T Bu$$

is discontinuous. Again the relative degree is one less than the degree of smoothness, which is 2 since the first derivative is the lowest differential of the state y having discontinuity. Similarly, if $D = 0$, $C^T B = 0$ but $C^T AB \neq 0$, then (2.62) has relative degree two because y and $\dot{y}$ are continuous, but

$$\ddot{y} = C^T \ddot{x} = C^T A^2 x + C^T ABu$$

is discontinuous. That is, the second derivative of the output is now discontinuous and the degree of smoothness is thus three.

These concepts extend to nonlinear control systems too. A general single-input single-output nonlinear system can be written as

$$\begin{aligned} \dot{x} &= f(x) + g(x)u, \\ y &= h(x). \end{aligned} \tag{2.63}$$

The system (2.63) is said to have *relative degree* r at a point x^* if

1. $\mathcal{L}_g \mathcal{L}_f^k h(x) = 0$ for all x in a neighborhood of x^* and all $k < r - 1$;
2. $\mathcal{L}_g \mathcal{L}_f^{r-1} h(x^*) \neq 0$.

Here we have introduced the following useful notation

Definition 2.28. *The* **Lie derivative** $\mathcal{L}_f$ *is the total time derivative along the direction of the flow governed by vector field* f*. Specifically, if* $f(x)$ *and* $g(x)$ *are smooth vector fields and* $h(x)$ *is a smooth scalar function, then we have*

$$\begin{aligned}
\mathcal{L}_f h(x) &:= \frac{\partial h}{\partial x} f(x), \\
\mathcal{L}_g \mathcal{L}_f h(x) &:= \frac{\partial(\mathcal{L}_f h)}{\partial x} g(x), \\
\mathcal{L}_g \mathcal{L}_f^k h(x) &:= \frac{\partial(\mathcal{L}_f^{k-1} h)}{\partial x} g(x), \\
\mathcal{L}_f^0 h(x) &:= h(x).
\end{aligned}$$

Consider, for example, a case where the relative degree is two at a point x^*. Here $r = 2$ and $k < 1$, which gives

$$\begin{aligned}
\mathcal{L}_g \mathcal{L}_f^0 h(x) &= \tfrac{\partial(\mathcal{L}_f^0 h)}{\partial x} g(x) = h_x g(x) = 0, \\
\mathcal{L}_g \mathcal{L}_f^1 h(x^*) &= \tfrac{\partial(\mathcal{L}_f^1 h)}{\partial x} g(x^*) = \tfrac{\partial(h_x f}{\partial x} g(x^*) = (h_{xx} f + h_x f_x) g(x^*) \neq 0;
\end{aligned}$$

or, showing the link to the linear SISO system (2.61)

$$\begin{aligned}
&\mathcal{L}_g \mathcal{L}_f^0 h(x) = h_x g(x) = AB = 0, \\
&\mathcal{L}_g \mathcal{L}_f^1 h(x^*) = (h_{xx} f + h_x f_x) g(x^*) = (0Ax^* + C^T A)B = C^T AB \neq 0,
\end{aligned}$$

where $Ax = f(x)$, $B = g(x)$ and $C^T x = h(x)$.

The Lie derivative will prove useful in Chapters 6, 7 and 8 for analyzing the flow near to grazing intersections. At various points in the book, we will borrow other concepts from control theory, where it is useful, such as observer canonical form, controllability and relay control.

2.4 Stability and bifurcation of non-smooth systems

The extension of well-established concepts for smooth systems to the case of non-smooth systems is still an open research area. We shall hence try to establish a pragmatic approach for studying the asymptotic and structural stability of our chosen classes of piecewise-smooth maps, flows and hybrid systems (Definitions 2.18, 2.20 and 2.24). Our aim is to come up with a utilitarian definition of a discontinuity-induced bifurcation (DIB) that allows us to explain the dynamical transitions that were observed in the case study examples introduced in Chapter 1. First we need to assess the notion of stability.

2.4.1 Asymptotic stability

It is a particularly cumbersome task to provide necessary and sufficient conditions that guarantee the asymptotic stability of an invariant set of a piecewise-smooth systems if that set straddles the boundary between two regions S_i and S_j; see, for example, [179] for a review. Even the problem of assessing the asymptotic stability of an equilibrium that rests on a discontinuity boundary is an open problem in general [36]. Let us focus on the problem for the special case of piecewise-linear systems, which will be of relevance to the material in Chapter 5.

Consider the piecewise-linear system

$$\dot{x} = \begin{cases} A^- x & \text{if } C^T x \le 0 \\ A^+ x & \text{if } C^T x \ge 0 \end{cases}, \tag{2.64}$$

where $A^\pm \in \mathbb{R}^{n \times n}$ and $c \in \mathbb{R}^n$. We assume that the overall vector field is continuous across the hyperplane $\{x : C^T x = 0\}$, but the degree of smoothness is uniformly one. This means that

$$A^- - A^+ = EC^T,$$

for some $E \in \mathbb{R}^n$. For the planar case, i.e., $n = 2$, a complete theory is possible and it can be shown that the equilibrium point $x = 0$ of (2.64) is asymptotically stable under certain strict conditions, provided the system obeys the property of *observability* often used in control theory.

Definition 2.29. *Two matrices $A \in \mathbb{R}^{n\times n}$ and $C^T \in \mathbb{R}^{p \times n}$ are said to be* **observable** *if the observability matrix, $\mathcal{O}$, defined as*

$$\mathcal{O} = \begin{pmatrix} C^T \\ C^T A \\ \vdots \\ C^T A^{n-1} \end{pmatrix}$$

has full rank. Equivalently, for single-output systems, where $V \in \mathbb{R}^{1\times n}$, observability implies det($\mathcal{O}$) $\neq 0$.

Theorem 2.6 ([49]). *Consider the system (2.64) with $n = 2$. Assume that the pair (C^T, A^-) is observable. Then*

1. *The origin is asymptotically stable if and only if*
   a) *neither A^- nor A^+ has a real non-negative eigenvalue, and*
   b) *if both A^- and A^+ have non-real eigenvalues, then $\sigma^-/\omega^- + \sigma^+/\omega^+ < 0$, where $\sigma^\pm \pm i\omega^\pm$ $(\omega^\pm > 0)$ are the eigenvalues of $A^\pm$.*
2. *The system (2.64) has a non-constant periodic solution if and only if both A^- and A^+ have non-real eigenvalues and $\sigma^-/\omega^- + \sigma^+/\omega^+ = 0$, where $\sigma^\pm \pm i\omega^\pm$ $(\omega^\pm > 0)$ are the eigenvalues of $A^\pm$. Moreover, if there is one periodic solution, then all other solutions are also periodic. Moreover any such periodic solution has period equal to $\pi/\omega^- + \pi/\omega^+$.*

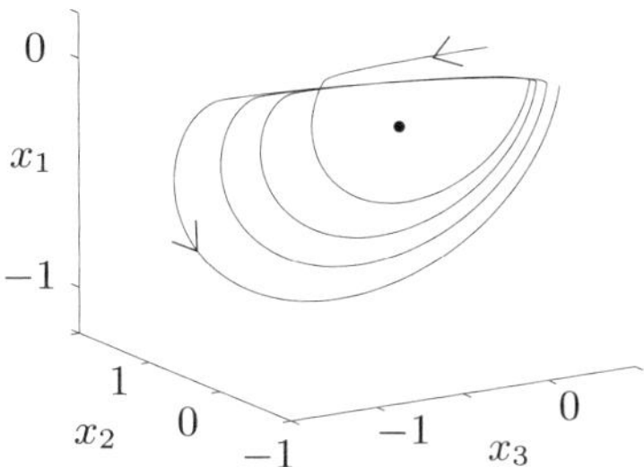

Fig. 2.23. A trajectory of the piecewise-linear system (2.64)–(2.66).

In higher dimensions, the problem becomes considerably more difficult. A seemingly paradoxical situation can occur whereby the origin of the individual systems $\dot{x} = A^- x$ and $\dot{x} = A^+ x$ is asymptotically stable, but is unstable for the combined system (2.64):

Example 2.11 (Unstable piecewise-linear system [51]). Consider the system (2.64) with

$$A^- = \begin{pmatrix} -1 & -1 & 0 \\ 1.28 & 0 & -1 \\ -0.624 & 0 & 0 \end{pmatrix}, \quad A^+ = \begin{pmatrix} -3.2 & -1 & 0 \\ 25.61 & 0 & -1 \\ -75.03 & 0 & 0 \end{pmatrix} \tag{2.65}$$

and

$$c = \begin{pmatrix} 1 \\ 0 \\ 0 \end{pmatrix}. \tag{2.66}$$

Now, the eigenvalues of A^- are $-0.2 \pm i$ and -0.6, whereas the eigenvalues of A^+ are $-0.1 \pm 0.5i$ and -3. Both sets are strictly in the left half-plane which would imply stability of the origin of each linear systems individually. Yet the combined piecewise-linear system has trajectories that tend to ∞; see Fig. 2.23.

In essence, the paradox is caused by the geometric relationship between the *eigenvectors* of the matrices A^- and A^+. Clearly if the eigenvectors of the two matrices were perfectly aligned, then stability of the matrices A^- and A^+ would be sufficient to establish stability of the piecewise-linear system. In fact, in certain other special cases, it is possible to establish conditions for stability for systems of the form (2.64) in three dimensions. For example, using the theory of invariant cones, Carmona *et al.* [51] have established the following result.

Theorem 2.7 ([51]). *Consider the system (2.64) with $n = 3$. Assume that the pair (C^T, A^-) is observable. Let $A^\pm$ and c be given by*

$$A^{\pm} = \begin{pmatrix} t^{\pm} & -1 & 0 \\ m^{\pm} & 0 & -1 \\ d^{\pm} & 0 & 0 \end{pmatrix}, \quad C = \begin{pmatrix} 1 \\ 0 \\ 0 \end{pmatrix}$$

(so-called observer canonical form). Suppose that the eigenvalues of the matrices $A^{\pm}$ are $\lambda^{\pm} \in \mathbb{R}$ and $\sigma^{\pm} \pm i\omega^{\pm}$, where $\omega^{\pm} > 0$. Also, suppose that

$$(\sigma^{-} - \lambda^{-})(\sigma^{+} - \lambda^{+}) < 0 \quad \textit{and} \quad (t^{+} - t^{-})(\sigma^{+} - \lambda^{+}) \le 0,$$

then the origin is an asymptotically stable equilibrium point if, and only if, $\lambda^{\pm}$ are both negative.

In the control theory literature, a more general tool has been proposed for the stability analysis of piecewise-smooth dynamical systems. Take, for example, the problem of establishing whether an equilibrium point in a discontinuity boundary of a piecewise-smooth dynamical system is asymptotically stable. One technique for proving such stability is to find a *common Lyapunov function*, that is, a function $V(x)$ that is *Lyapunov* (positive definite and decreasing along trajectories) for each of the vector fields defining the system dynamics in each of the phase space regions [179]. However, finding such functions in practice is at best difficult.

General progress toward understanding and classifying the dynamics of piecewise-smooth systems using such methods would appear hopeless. Drawing lessons from smooth dynamical systems theory, we advocate in this book a rather different approach. Instead of focusing on asymptotic stability of individual states or invariant sets, we focus instead on structural stability and bifurcation. Since *proving* stability from first principles can be hard, one should instead attempt to *classify* all the mechanisms that can lead to instability as a parameter is varied. Along with the classification should come techniques, both analytical and numerical, for identifying which case occurs in a particular example system and for understanding the nearby dynamics.

2.4.2 Structural stability and bifurcation

Consider a general invariant set of a piecewise-smooth dynamical system as defined in Definitions 2.18, 2.20 or 2.25. Bifurcations that involve invariant sets contained within a single region S_i for all parameter values of interest can be studied using smooth bifurcation theory. Also, it may be that the invariant set of a flow crosses several discontinuity boundaries, but nevertheless the Poincaré map associated with that invariant set is smooth. For example, in Sec. 2.5 below, we shall show that the Poincaré map associated with a periodic orbit that crosses all discontinuity sets Σ_{ij} transversally is smooth. Thus, *all the bifurcations discussed in Sec. 2.1.6 can also occur in piecewise-smooth systems.* However, other bifurcations are unique to piecewise-smooth systems. These typically involve non-generic interactions of an invariant set with a discontinuity boundary.

For piecewise-smooth systems such as (2.23) and (2.22) or (2.32) and (2.33), which define a dynamical system (this excludes the possibility of regions of phase space where there is repelling sliding or repelling sticking motion since this leads to no uniqueness in forward time), one can adopt the same notion of bifurcation as in Definition 2.16, applied to the entire system. However, we may wish to highlight other events that might not be a bifurcation of the entire system in this classical sense. In control systems for example, it may be important to identify whether a certain switch is activated. Or, in a mechanical system, we may need to know whether an attractor contains trajectories that impact or go beyond a certain threshold at which a bi-linear spring moves into its stiffer region. The transition that causes such an event will typically represent an invariant set forming a new crossing of a discontinuity boundary, as a parameter is varied. For example, at a parameter value $\mu = \mu_0$, a limit cycle of a piecewise-linear flow may become tangent to a discontinuity boundary Σ_{ij} at a *grazing point.* Alternatively, an equilibrium of a flow, or fixed point of a map, may approach a discontinuity boundary as $\mu \to \mu_0$. Now, if the degree of smoothness is sufficiently high, this will not affect the stability of these invariant sets and there will be no bifurcation in the sense of Definition 2.16. In the Russian literature, (e.g., [95, 98]), the term *C-bifurcation* has been adopted for such transitions that involve an invariant set doing something structurally unstable with respect to a discontinuity boundary. (The Russian character C, pronounced "S", stands for sewing, as one *sews* together two different trajectory segments on either side of the discontinuity boundary.) When the invariant set is the fixed point of a map, these have also been termed *border-collision bifurcations* [205].

Here we shall introduce the broader concept of a *discontinuity-induced bifurcation* [64, 79]. By this term we will identify qualitative changes to the topology of invariant sets with respect to the discontinuity boundaries. Specifically, we wish to single out parameter values at which the invariant set changes its *event sequence*; that is, the order and sense of interaction with the discontinuity boundaries. Such changes are typically brought about (or induced) through non-transversal interaction with a discontinuity boundary. However, in keeping with the qualitative theory of dynamical systems, we should like a definition of a discontinuity-induced bifurcation that is purely topological and does not refer to individual trajectories or invariant set. In order to come up with such a notion, we will need new definitions of structural stability and topological equivalence that call two dynamical systems non-equivalent if key invariant sets in the dynamics change their event sequence. We shall state this new definition of equivalence in the case of a hybrid dynamical system; corresponding definitions for piecewise-smooth maps and flows follow in an obvious manner.

Definition 2.30. *Let* $\{T, \mathbb{R}^n, \phi^t\}$ *and* $\{T, \mathbb{R}^n, \tilde{\phi}^t\}$ *be two hybrid piecewise-smooth dynamical systems (2.32), (2.33) defined by countably many different smooth flows* $\phi_i(x,t)$ *and* $\tilde{\phi}_i(x,t)$ *in finitely many phase space regions* S_i

and $\tilde{S}_i$*, respectively,* $i = 1 \dots N$*, with smooth resets* R_{ij} *and* $\tilde{R}_{ij}$ *applying, respectively, at each non-empty discontinuity boundaries* Σ_{ij} *and* $\tilde{\Sigma}_{ij}$*.*

Two such piecewise-smooth systems are called **piecewise-topological equivalent** *if:*

1. *They are topological equivalent; that is, there is a homeomorphism h that maps the orbits of the first system onto orbits of the second one, preserving the direction of time so that* $\phi^t(x) = h^{-1}(\tilde{\phi}^s(h(x)))$ *where the map* $t \to s(t)$ *is continuous and invertible.*
2. *The homeomorphism h can be chosen so as to preserves each of the discontinuity boundaries. That is, for each i and j,* $h(\Sigma_{ij}) = \tilde{\Sigma}_{ij}$*.*

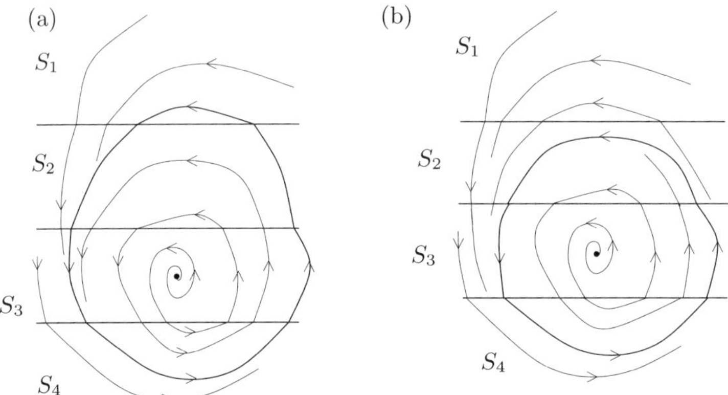

Fig. 2.24. Two phase potraits that are topological equivalent but not piecewise topological equivalent according to Definition 2.30. Note that the portrait in each separate region S_i, $i = 1, \dots, 4$ is topological equivalent between (a) and (b); yet in (a) there is a limit cycle that does not enter all four phase space regions, whereas in (b) the corresponding limit cycle does not visit region S_1. If a parameter is varied between these two cases, a DIB must occur, in this case, a grazing bifurcation of the limit cycle with the boundary Σ_{12}.

To motivate the second part of this definition, we want to call the two phase potraits illustrated in Fig. 2.24 non-equivalent, because in panel (a) there is a limit cycle that visits all four phase space regions, whereas in (b) the limit cycle visits only three of them. To see that this example fails the definition of equivalence, note that to transform from one phase portrait to another the limit cycle must be 'pulled through' the boundary Σ_{12}. Such a transformation cannot be achieved in a continuous way. In other words, the limit cycle and Σ_{12} are not in the same *general position* with respect to each other. This then leads us to our topological definition of a *discontinuity-induced bifurcation* (DIB) for parameterized piecewise-smooth dynamical systems. For example, we shall want to say that a discontinuity-induced bifurcation must occur if we continuously vary parameters between those used to obtain the two phase potraits in Fig. 2.24.

The definition of DIB proceeds as for the definition of smooth bifurcations. We start by saying what we mean by structural stability:

Definition 2.31. *A piecewise-smooth system is* **piecewise-structurally stable** *if there is an $\varepsilon > 0$ such that all C^1 perturbations of maximum size ε of the vector field (map) f, that leave the number and degree of smoothness properties of each of the boundaries Σ_{ij} unchanged, lead to piecewise-topological equivalent phase potraits.*

Definition 2.32. *A* **discontinuity-induced bifurcation** *(DIB) occurs at a parameter value at which a piecewise-smooth system is not piecewise-structurally stable. That is, there exists an arbitrarily small perturbation that leads to a system that is not piecewise-topological equivalent.*

Remarks

1. Note that we have been somewhat imprecise about what kind of perturbations are allowed in Definitions 2.31 and 2.32. One wants only to consider perturbed systems for which the partitioning of phase space into regions S_i remains topologically the same and that the degree of smoothness across each boundary does not change. We also want that the resets R_{ij} map boundaries Σ_{ij} to the equivalent parts of phase space. In fact, it remains an open problem to show that Definitions 2.30 and 2.32 are well defined mathematically. Strictly speaking, we need to define a topological space for each class of piecewise-smooth system in order to define topological equivalence correctly. The rigorous theory of DIBs is still in its infancy, and we shall not pursue this further here. Rather we shall treat Definition 2.32 as a working definition.
2. The concepts of *codimension* and *unfolding* can also be constructed, as in Definition 2.16 for bifurcations in smooth systems, but here one has to be even more careful to state what kinds of perturbation are allowed. Again we adopt a working definition of codimension that it is the 'degree of unlikeliness' of the discontinuity-induced bifurcation. That is, how many parameters would one expect to have to vary in order to correctly unfold the bifurcation?
3. Under Definition 2.32 classical bifurcations are also DIBs. However, our main focus in this book is the particular class of discontinuity-induced bifurcations that are caused by something structurally unstable happening with respect to a discontinuity set Σ_{ij}. Bifurcations that have nothing to do with discontinuity sets we shall refer to as *smooth bifurcations.* Most of the rest of this book will be about cataloging the various non-smooth transitions (particularly those of codimension-one) that can occur in piecewise-smooth systems. We shall also provide unfoldings of the ensuing dynamics and ways of calculating these unfoldings in examples. Moreover we shall seek to show how these DIBs explain the observed dynamics.

4. In the special case of Filippov systems in $\mathbb{R}^3$ there is now a rigorous structural stability theory; see the work by Filippov [100], Teixera [248], Simic & Johansson [239] and references therein.

2.4.3 Types of discontinuity-induced bifurcations

The main aim of rest of this book is devoted to a classification and analysis of the most commonly occurring types of DIBs. As we shall see, these lie at the heart of explaining what was observed in the case study examples introduced in Chapter 1. Let us list some of the most commonly occurring types of codimension-one DIBs (see Fig. 2.25):

Border collisions of maps. These are conceptually the simplest kind of DIB and occur when, at a critical parameter value, a fixed point of a piecewise-smooth map lies precisely on a discontinuity boundary Σ. For maps with singularity of order one (i.e., locally piecewise-linear continuous), there is now a mature theory for describing the bifurcation that may result upon varying a parameter through such an event. Remarkably, the unfolding may be quite complex. Even in one dimension, we saw in case study VIII that a period-1 attractor can jump to a period-n attractor for any arbitrary n, or to robust chaos without any periodic windows. In one and two dimensions, more or less everything is known. But in general n-dimensional maps, bifurcation information on only the simplest kinds of periodic points is known. This material is presented in Chapter 3. Chapter 4 then goes on to study border collision bifurcations in maps with other degrees of singularity, including the discontinuous and square-root cases from case studies VI and VII.

Boundary equilibrium bifurcations. The simplest kind of DIB for flows occurs when an equilibrium point lies precisely on a discontinuity boundary Σ. In Filippov systems and hybrid systems with sticking regions, there is also the possibility of *pseudo-equilibria*, which are equilibria of the sliding or sticking flow but are not equilibria of any of the vector fields of the original system. There are thus possibilities where the equilibrium lies precisely on the boundary between a sliding or sticking region and a pseudo-equilibrium turns into a regular equilibrium (either under direct parameter variation or in a fold-like transition where both exist for the same sign of the perturbing parameter). There is also the possibility that a limit cycle may be spawned under parameter perturbation of the boundary equilibrium, in a Hopf-like transition. This material is treated in Chapter 5.

Grazing bifurcations of limit cycles. One of the most commonly found DIBs in applications is caused by a limit cycle of a flow becoming tangent to (i.e., grazing) with a discontinuity boundary. One might naively think that this can be completely understood (upon taking an appropriate Poincaré section that contains the grazing point) as a border collision. However, as

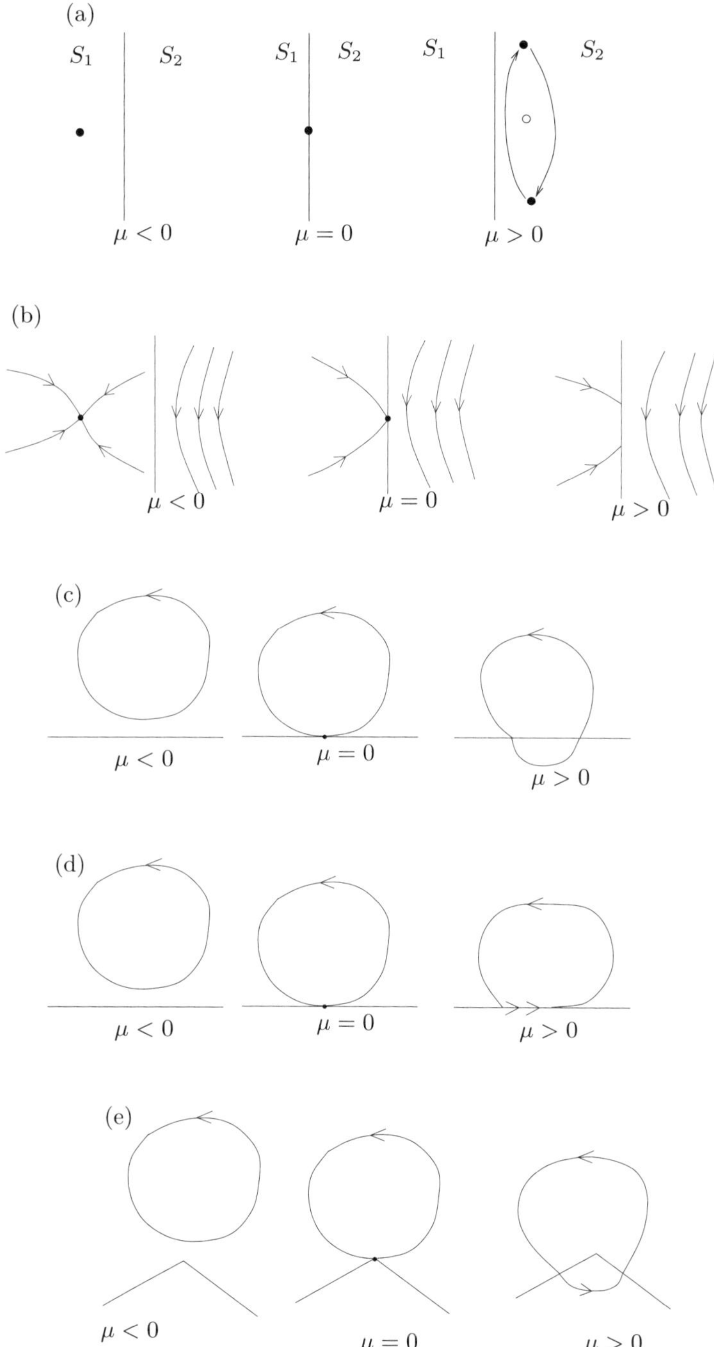

Fig. 2.25. Examples of DIBs: (a) a border collision in a map; (b) a boundary equilibrium bifurcation; (c) a grazing bifurcation of a limit cycle; (d) a sliding bifurcation in a Filippov system; (e) a boundary intersection crossing.

we shall see in Chapter 6 for hybrid systems and Chapter 7 for piecewise-smooth ODEs this is not necessarily the case. Instead one has to analyze carefully what happens to the flow in the neighborhood of the grazing point. In fact, one can derive an associated map (the, so-called, discontinuity map). But, the link between the singularity of the map and the degree of smoothness of the flow is a subtle one that also depends on whether the flow is uniformly discontinuous at the grazing point. This analysis explains what is observed at the grazing bifurcations in the impact and bi-linear oscillators, case studies I and II.

Sliding and sticking bifurcations. There are several ways that an invariant set such as a limit cycle can do something structurally unstable with respect to the boundary of a sliding region in a Filippov system. Chapter 8 is devoted to a careful unfolding of each of these. The Poincaré maps so derived have the property of typically being noninvertible in at least one region of phase space, owing to the loss of information backward in time inherent in sliding motion. This analysis helps to explain the dynamics observed in the relay control and dry friction examples described in case studies III and IV. In addition, in impacting systems, sticking regions can be approached by infinite *chattering* sequences of impacts, which we have seen already in case study I. Further details of such events will be given in Chapter 6 in the context of the single degree-of-freedom impact oscillator.

Boundary intersection crossing/corner collision. Another possibility for a codimension-one event in a flow is where an invariant set (e.g., a limit cycle) passes through the $(n-2)$-dimensional set formed by the intersection of two different discontinuity manifolds Σ_1 and Σ_2. In Chapter 7 we shall consider such intersection crossing in Filippov systems in the case where there is no sliding. We also consider there the special case where the jumps in vector field across Σ_1 and Σ_2 are such that their intersection can be considered as a 'corner' in a single discontinuity surface. This can explain the dynamics observed in the DC–DC converter, case study V.

Some possible global bifurcations. One example, which we shall mention in Chapter 5, involves a connection between the stable and the unstable manifolds of *pseudo-equilibria*, which are equilibria of a sliding flow but not of the individual flows either side of a discontinuity boundary.

Chapter 9 briefly treats extensions to the theory of DIBs, which are in each case motivated by a further case study example of practical significance, for which a detailed treatment is beyond the scope of the book. Topics include parameter and noise sensitivity; bifurcations that involve invariant tori grazing with a discontinuity surface; the similarity between grazing in piecewise-smooth flows and hybrid systems in the limit of large discontinuities; and codimension-two bifurcations.

2.5 Discontinuity mappings

The analysis of discontinuity-induced bifurcation in maps is relatively straightforward; one merely has to consider the fate of iterates that land either side of the discontinuity. DIBs in piecewise-smooth flows or hybrid systems are far harder to analyze, because one must establish the fate of topologically distinct trajectories close to the structurally unstable event that determines the bifurcation. In this section we introduce a key analytical tool that enables the study of DIBs involving limit cycles and other invariant sets that are more complex than mere equilibria. The concept is that of a *discontinuity map* (DM), a term first introduced by Nordmark [197]. This is a synthesized Poincaré mapthat is defined *locally* near the point at which a trajectory interacts with a discontinuity boundary. When composed with a global Poincaré map(for example around the limit cycle) ignoring the presence of the discontinuity boundary, one can then derive a (typically non-smooth) map whose orbits completely describe the dynamics in question.

To illustrate why discontinuity maps are both necessary and useful, consider the piecewise-smooth flow illustrated in Fig. 2.26(a),(b), for which there is a Poincaré surface Π lying in one of the regions S_i, which is intersected transversally at the point x_p by a periodic orbit $p(t)$ of period T.

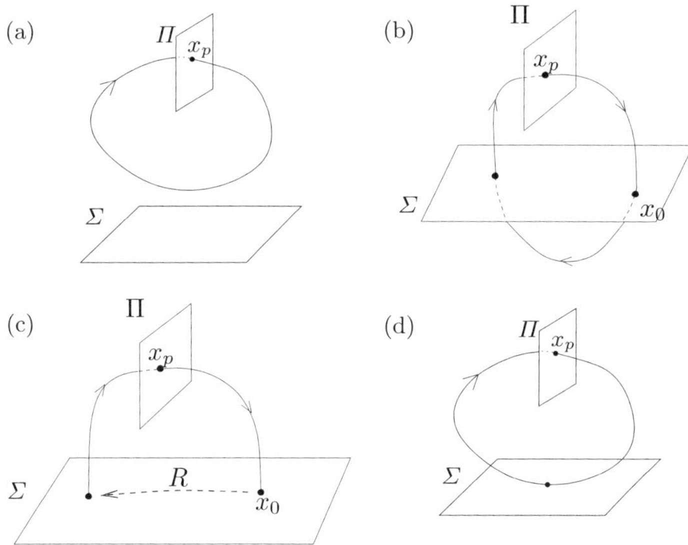

Fig. 2.26. (a) Simple periodic orbit $p(t)$ in piecewise smooth ODE that does not intersect any discontinuity surfaces. (b) Simple periodic orbit that intersects a single surface twice. (c) Equivalent to (b) but for an impacting hybrid system. (d) A grazing periodic orbit.

For points $\widehat{x} \in \Pi$ close to x_p, we may define a Poincaré map $P : \Pi \to \Pi$. It is natural to ask what form P takes when $\|x - x_p\|$ is small. The answer to this question takes three forms, and depends crucially upon the nature of the orbit $p(t)$.

If $p(t)$ lies wholly inside S_i, as in Fig. 2.26(a) then nearby orbits will also lie inside S_i. In this case the time-T map starting from x will be the smooth flow map $P(x) = \Phi_i(x, t)$, which has a well-defined Taylor series,

$$P(x) = N(x - x_p) + O\left(|x - x_p\|^2\right), \tag{2.67}$$

where $N = \Phi_{i,x}(x_p, T)$ is the Jacobian derivative with respect to x of the flow Φ_i around the periodic orbit, evaluated at $x = x_p$. More interesting things happen if the periodic orbit $p(t)$ intersects discontinuity surfaces Σ_{ij}.

Consider next the case illustrated in Fig. 2.26(b), where $p(t)$ has two *transverse* intersections with a discontinuity set Σ. In this case it is tempting to write that the linearization of the Poincaré maptakes the form $P(x) = N_1 N_2 N_3 (x - x_p)$, where N_1, N_2 and N_3 are linearizations of the flows Φ_1, Φ_2 and Φ_1, respectively, for the appropriate times for the trajectory starting at x_p to, respectively, reach Σ for the first time, to pass between the first and second intersections of Σ, and to pass from Σ back to Π. However, this is not the case because, as we shall see in Sec. 2.5.2, each time Σ is crossed transversally, one must apply a correction to the Poincaré map. This correction is necessary because the time taken for trajectories at points x close to x_p to reach the discontinuity boundary Σ will in general vary, and so a small error will be made in assuming that the linearization required is that of Φ_1 for a constant time. The correction to this error is the discontinuity map in this case. The effect of the DM on the matrix N_1 is to multiply it by a so-called *saltation matrix* [2, 194, 173] whose derivation we give below. A similar correction must be applied to the matrix N_2. Not introducing these corrections will in general result in wrong conclusions being made about the Floquet multipliers of the periodic orbit $p(t)$. Note in this case, provided the form of the jump in the vector fields upon crossing Σ is described by a smooth function, then the discontinuity mapping and the associated global Poincaré maparound $p(t)$ will both be smooth. Similar considerations apply to impacting hybrid systems where a periodic orbit $p(t)$ has a single impact with a discontinuity surface as in Fig. 2.26(c).

Now consider for a moment the special case where the velocity normal to Σ is zero, so that the periodic orbit *grazes* the discontinuity surface, as in Fig. 2.26(d). Note that the trajectories from some initial conditions $x \in \Pi$ near x_p do not intersect Σ at all, whereas others intersect Σ with a low normal velocity. The discontinuity mapping in this case is the identity for orbits that do not cross Σ, but is defined as the local correction that must be applied to initial conditions that *do* cross Σ, so that a Poincaré mapcan be applied as if Σ were not there. As we shall motivate briefly in Sec. 2.5.3 below, the effect of applying the DM to the map in (2.67) in this case is to introduce additional terms proportional to fractional powers of $\|x - x_p\|$, such as $\|x - x_p\|^{1/2}$ or

$\|x - x_p\|^{3/2}$. An analysis of the behavior of maps with fractional powers will be given in Chapter 4. Detailed derivations of DMs close to different kinds of DIBs, along with analyzing their dynamical consequences, form the main subject Chapters 6, 7 and 8.

In the case that trajectories intersect discontinuity boundaries transversally, then typically one still has to compute a discontinuity mapping in order to derive a globally correct Poincaré map. This is because even though the trajectory itself may be continuous (or in the case of a hybrid system, the trajectory's evolution would be defined by a continuous reset map), there is a correction that must be to the first and higher derivatives of the flow. This correction arises because the discontinuity boundary acts like a new Poincaré section that is distinct from the fixed time-t section that is implicitly defined flow.

2.5.1 Transversal intersections; a motivating calculation

Before deriving the general form of the *transverse discontinuity mapping* for an arbitrary piecewise-smooth or hybrid system, let us start with the motivating case of a simple impacting hybrid system of the form (2.35) and (2.36). Here we assume a smooth reset map R applies whenever the smooth flow Φ, governed by vector field F, impacts the discontinuity surface $\Sigma := \{x : H(x) = 0\}$ transversally; see Fig. 2.27. We shall analyze the dynamics of a trajectory with initial condition $\widehat{x}$ that is close to a reference trajectory x_p that impacts $\Sigma := \{x : H(x) = 0\}$ at a point x_*. It is perhaps most useful to think of x_p as belonging to a periodic orbit $p(t)$, for which x_* is the unique point of impact; although the analysis that follows shall be entirely local to a neighborhood of x_*. Let us define t_1 to be the time for which $\Phi(x_p, t_1) = x_*$ and let x_0 be the point reached by flowing for the same time from initial condition $\widehat{x}$, so that $\Phi(\widehat{x}, t_1) = x_0$. Note that, in general, x_0 will not lie in Σ. Instead, we must continue the flow for a small additional time δ (which may be positive or negative) until the trajectory intersects Σ. From this intersection point, the map R is applied to reach the point x_3 in Fig. 2.27. Now, the discontinuity map Q is defined to be the mapping that takes x_0 to x_4, which is the point obtained by flowing from x_3 through a time $-\delta$. That is,

$$Q(x_0) = \Phi(x_3, -\delta) = \Phi(R(\Phi(x_0, \delta)), -\delta).$$

Thus, Q maps x_0 to the appropriate point on the trajectory of the true hybrid flow which, without resetting the time variable, can be evolved forward under Φ as if the impact had occurred at time t_1. In the case that $x_0 \in \Sigma$, then note that by definition $\delta = 0$ and Q reduces to the reset map R. However, for general points x_0 close to x_*, Q contains an additional correction. We shall now proceed to derive an expression for the leading-order correction.

For a general flow described by the differential equation $\dot{x} = F(x)$, the solution for small subsequent times δ starting from the point x_0 can be expressed as a Taylor series in δ:

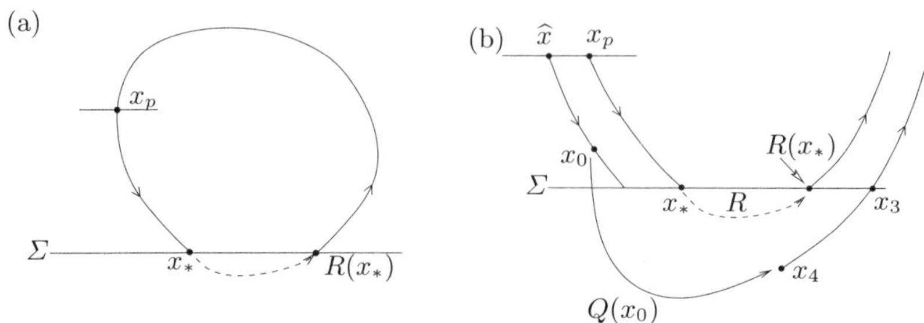

Fig. 2.27. (a) An impacting periodic orbit. (b) A blow-up near x_* defining the transverse discontinuity mapping $Q(x)$.

$$x(t) = \Phi(x_0, \delta) = x_0 + \delta F(x_0) + O(\delta^2).$$

Setting $x_0 = x_* + \Delta x$, we obtain

$$x(t) = x_* + \Delta x + \delta F(x_*) + O(\delta^2, \delta \Delta x, (\Delta x)^2).$$

We wish to find the time δ for which $H(x(\delta)) = 0$. Thus we require

$$H\left(x_* + \Delta x + \delta F(x_*) + O(2)\right) = 0. \tag{2.68}$$

where $O(2)$ refers to general quadratic terms in the small variables δ and Δx. Now, because $H(x_*) = 0$, we find for x close to x_*

$$H = H_x(x_*)(x - x_*) + O(\|x - x_*\|^2).$$

Hence, from (2.68) we seek a solution δ to the equation

$$0 = H_x(x*)\left[\Delta x + \delta F(x_*)\right] + O(2)$$

Thus, we find

$$\delta = -\frac{H_x(x_*)\Delta x}{H_x(x_*)F(x_*)} + O(2). \tag{2.69}$$

Note that this expression is only valid provided $H_x(x_*)F(x_*) \neq 0$, which is precisely the condition that the flow crosses Σ *transversally*, i.e., with non-zero velocity. Applying the reset map to the point $\Phi(x_0, \delta)$ gives for x_3:

$$x_3 = R[x_* + \Delta x + \delta F(x_*)] + O(2).$$

To obtain the discontinuity map we must find an expression for x_4 in the form

$$\begin{aligned} x_4 &= \Phi(x_3, -\delta) \\ &= x_3 - \delta F(x_3) + O(2) \\ &= R[x_* + \Delta x + \delta F(x_*)] - \delta F\left(R[x_* + \Delta x + \delta F(x_*)]\right) + O(2). \end{aligned}$$

Now, both R and F can be expanded as Taylor series about x_*. Hence, we obtain

$$x_4 = R(x_*) + R_x(x_*)[\Delta x + \delta F(x_*)] - \delta F(R_x(x_*)) + O(2).$$

Using the expression (2.69) for δ, we finally obtain

$$x_4 = R(x_*) + \frac{R_x(x^*) + [F(R(x_*)) - R_x(x_*)F(x_*)]H_x(x_*)}{H_x(x_*)F(x_*)}\Delta x + O(2).$$

Recalling that $\Delta x := x_0 - x^*$, note that $R(x^*) + R_x(x^*)\Delta x + O(2)$ is just the first the first two terms in the Taylor expansion of $R(x_0)$. Hence, the transverse discontinuity mapping is given, to leading order, by

$$x_0 \to Q(x_0) = R(x_0) + \frac{[F(R(x_*)) - R_x(x_*)F(x_*)]H_x(x_*)}{H_x(x_*)F(x_*)}(x_0 - x_*). \quad (2.70)$$

The second term in (2.70) is the leading-order correction to the reset map $R(x_0)$; note that this term is linear in $(x_0 - x_*)$. Hence, failure to apply this mapping when computing periodic orbits with impacts will in general lead to incorrect linearizations (Monodromy matrices), hence incorrect Floquet multipliers and (potentially) incorrect conclusions about stability of the periodic orbit.

2.5.2 Transversal intersections; the general case

Consider now a general hybrid system (2.32), (2.33) in $\mathbb{R}^n$, which we assume to have two phase space regions S_1 and S_2 as illustrated in Fig. 2.28, with corresponding flows Φ_1 and Φ_2, and a single reset map R applying on the boundary Σ between the two regions. Note that this covers both the case of a piecewise-smooth flow (2.23) and an impacting hybrid system (as in the previous calculation). In the case of a piecewise-smooth flow, the reset map R is the identity mapping. For the impacting hybrid system, R maps $\Sigma \to \Sigma$, and the flow Φ_1 applies after the impact, so that the flow Φ_2 becomes identically Φ_1 in what follows.

Suppose that a periodic orbit $p(t)$ crosses the discontinuity set Σ transversally at the two points $x = x_*$ and x_{**} as illustrated in Fig. 2.28. The key observation is that all nearby trajectories must then cross Σ transversally. Then, since R, Φ_1 and Φ_2 are smooth, the Poincaré map associated with this periodic orbit is smooth and has non-singular Jacobian. To compute this Jacobian, and indeed the entire Poincaré map, consider the flow map for the specific sequence of events that ensue from an initial condition x close to x_p in a Poincaré section Π.

Let us choose an origin of time such that the periodic orbit intersects the Poincaré section Π at $x_p \in S_1$ when $t = 0$ and intersects Σ at the two times $t_2 > t_1 > 0$. Trajectories close to the point x_* and the time t_1 are depicted in Fig. 2.28(b).

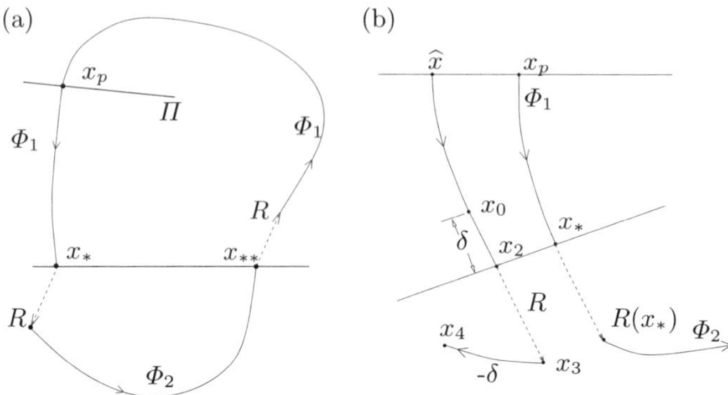

Fig. 2.28. (a) Defining the event sequence for a simple periodic orbit that generalizes the two cases in Fig. 2.26. (b) Construction of the transverse discontinuity mapping $Q : x_0 \mapsto x_4$.

Taking a nearby initial condition x to x_p and evolving the flow forward leads to a trajectory that intersects Σ at the point x_2 close to x_*, at time $t_1 + \delta$. If, in contrast, we evolve the flow from x for a fixed time t_1, we reach the point $x_0 = \Phi_1(x, t_1)$ in the figure. Applying the map R and then the flow Φ_2 to the point x_0 for $t > t_1$ gives an error as we have applied R at x_0 and $t = t_1$ rather than at x_2 and at $t = t_1 + \delta$. To correct this we can find the point x_4 such that the action of Φ_2 on the point x_4 for future times $t > t_1$ coincides with the action of Φ_2 on the point $x_3 = R(x_2)$ for $t > t_1 + \delta$. The correction $x_4 = Q(x_0)$ is indicated in the figure and is the *discontinuity mapping* in this case. This correction is applied to the point x_0 and can be defined theoretically by the expression

$$Q(x_0) = \Phi_2(R(\Phi_1(x_0, \delta)), -\delta) = \Phi_2(x_3, -\delta). \tag{2.71}$$

The points x_0, x_2, x_3 and x_4 are all indicated in Fig. 2.28(b). Note that the total elapsed time of the flow combination described by the discontinuity map is $\delta - \delta = 0$.

A similar map can be applied to the subsequent intersection with Σ at the point x_{**}. Then, the time-T map for the evolution of the overall piecewise-smooth flow around $p(t)$ becomes

$$P(x, T) = \Phi_1[Q(\Phi_2[Q(\Phi_1[x_p, t_1]), t_2 - t_1]), T - t_2],$$

which has Jacobian derivative

$$P_x(x, T) = \Phi_{1,x}[R(x_{**}), T - t^0]Q_x(x_{**})\Phi_{2,x}[R(x_*), t^0 - t_0]Q_x(x_*), \Phi_{1,x}(x_p, t_0) \tag{2.72}$$

where Q_x is the linearization of (2.71).

Definition 2.33. *The* **transverse discontinuity map** Q *for the transverse crossing of a discontinuity set* Σ_{ij} *in a piecewise-smooth flow (or hybrid system) is the extra mapping that the flow maps* Φ_i *and* Φ_j *must be composed with in order to get a description of the piecewise-smooth (hybrid) flow. Thus, if* Σ *is crossed in the sense of passing from region* S_i *to* S_j*, the correct flow map is* $\Phi_2 \circ Q \circ \Phi_1$*. The Jacobian derivative* Q_x *of* Q *is called the* **saltation matrix**.

We shall now extend the earlier calculation to derive an explicit expression for the discontinuity mapping Q, and its derivative, the saltation matrix Q_x. In order to do so, suppose that the discontinuity set can be written locally as

$$\Sigma = \{x \in \mathbb{R}^n : H(x) = 0\}$$

for some smooth function H. Consider again the local piece of the trajectory in Fig. 2.27 with initial condition x in a neighborhood of x_p. Evolving through a time t_1 we reach the point $x_0 = \Phi_1(x, t_1)$, which is in a small neighborhood of the point $x_* = \Phi_1(x_p, t_1)$. We suppose that $x_0 = x_* + \Delta x$, where $\|\Delta x\|$ is small and develop a Taylor series for $\Phi_1(x_0, \delta)$, for small times δ.

For a flow described by the differential equation $\dot{x} = F_1(x)$, the solution for subsequent times δ starting from the point x_0 is given by

$$\Phi(x_0, \delta) = x_0 + \delta F_1(x_0) + \frac{\delta^2}{2} F_{1,x}(x_0) F_1(x_0) + O(\delta^3).$$

If we now set $x_0 = x_* + \Delta x$ this expression takes the form

$$x(t) = x_* + \Delta x + \delta F_1(x_*) + \delta \Delta x F_{1,x}(x_*) + \frac{\delta^2}{2} + F_{1,x}(x_0) F_1(x_0) + O(3). \quad (2.73)$$

Here $O(3)$ refers to cubic terms in δ and Δx. The transversality of the intersection of $p(t)$ with Σ allows us to assume that these are of the same order.

The first step to computing $Q(x_0)$ is to find the time δ and the point x_2 at which $H(x_2) = H(\Phi_1(x_0, \delta)) = 0$. Thus

$$H\left[x_* + \Delta x + \delta F_1(x_*) + \delta \Delta x F_{1,x}(x_*) + F_{1,x}(x_*) F_1(x_*) \delta^2/2 + O(3)\right] = 0. \quad (2.74)$$

Now, as $H(x_*) = 0$, the function H can also be expanded in x about x_* as

$$H(x) = H_x(x_*)(x - x_*) + \frac{1}{2}(x - x_*)^T H_{xx}(x_*)(x - x_*) + O(3).$$

So (2.74) can also be expressed as a Taylor series and solved term by term for δ, under the assumption that the leading-order term

$$H_x(x_*) F_1(x_*) \neq 0. \quad (2.75)$$

As before, this is precisely the condition that $p(t)$ crosses Σ *transversally*. Specifically we obtain

$$\delta = -\frac{H_x(x_*)\Delta x}{H_x(x_*)F_1(x_*)} + O(2), \qquad x_2 = x_* + \Delta x + \delta F_1(x_*) + O(2),$$

The quadratic and all higher terms in these expressions can also be evaluated if higher-order expressions for the discontinuity mapping are required. In fact, the assumption (2.75) guarantees that the discontinuity map Q is an analytic function provided F_1, F_2 and R are also analytic.

According to (2.71), we now compute Q by applying the flow Φ_2 to the point

$$x_3 := R(x_2) = R\left[x_* + \Delta x + \delta F_1(x_*)\right] + O(2).$$

for a time $-\delta$. Now, Φ_2 can be expanded about the point $R(x_*)$ in the same way as Φ_1 was expanded about x_*; see (2.73). This gives

$$\begin{aligned} Q(x_0) &= x_4 = \Phi_2(x_3, -\delta) \\ &= R(x_2) - F_2(R(x_2))\delta + O(2) \\ &= R(x_0 + \delta F_1(x_*)) - F_2(R(x_0 + \delta F_1(x_*)))\delta + O(2). \end{aligned}$$

Furthermore, we will assume that the map R can be expanded about the point x_*, so that

$$R(x_0) = R(x_*) + R_x(x_*)\Delta x + \frac{1}{2}\Delta x^T R_{xx}(x_*)\Delta x + O(3).$$

Using this expression we have

$$\begin{aligned} Q(x_0) &= R(x_*) + R_x(x_*)\Delta x + R_x(x_*)F_1(x_*)\delta\Delta x - F_2(R(x_*))\delta\Delta x + O(2) \\ &= R(x_*) + \left[R_x + \frac{H_x(x_*)}{H_x(x_*)F_1(x_*)}\left[F_2(R(x_*)) - R_x(x_*)F_1(x_*)\right]\right](x_0 - x_*) \\ &\quad + O(\|x_0 - x_*\|^2). \end{aligned} \tag{2.76}$$

Thus, the saltation matrix Q_x in this general case is given by

$$Q_x(x_*) = R_x(x_*) + \frac{\left[F_2(R(x_*)) - R_x(x_*)F_1(x_*)\right]H_x(x_*)}{H_x(x_*)F_1(x_*)}.$$

We now consider examples where we can calculate the saltation matrix explicitly.

Example 2.12 (A two-zone Filippov system without sliding). For Filippov systems in which $R(x) = x$ and $F_1 \neq F_2$ the saltation matrix is given by the expression

$$Q_x = I + \frac{(F_2 - F_1)H_x}{H_x F_1}, \tag{2.77}$$

where I is the identity matrix. This expression was first derived in [2].

Example 2.13 (Impacting systems). For impacting systems with a single impact boundary written in the form (2.35), (2.36). the vector field F_2 should be identified with F_1, since the R maps Σ^- to Σ^+. Upon letting $F_1 = F_2 := F$, we find that

$$Q_x(x_*) = R_x(x_*) + \frac{[F(R(x_*)) - R_x(x_*)F(x_*)]H_x(x_*)}{H_x(x_*)F(x_*)}, \tag{2.78}$$

which is precisely the linearization of (2.70) derived earlier.

As a specific application, consider the one-dimensional impact type hybrid system considered in case study I, in which $x = (x_1, x_2, x_3)$, $H(x) = x_1 - \sigma$, $R(x) = (x_1, -rx_2, x_3)$ and $F(x) = (x_2, a, 1)$, with $a = \cos(\omega t)) - x_1 - 2\zeta x_2$ being the acceleration. (Note that the subscripts here refer to vector indices rather than to the points in the Fig. 2.27.) We therefore have

$$R_x = \begin{pmatrix} 1 & 0 & 0 \\ 0 & -r & 0 \\ 0 & 0 & 1 \end{pmatrix}, \quad H_x = (1, 0, 0).$$

If we set $v = H_x(x_*)F(x_*)$ to be the normal velocity immediately before impact and a^- and a^+ to be the normal accelerations immediately before and immediately after the impact, we obtain

$$\begin{aligned} Q_x &= \begin{pmatrix} 1 & 0 & 0 \\ 0 & -r & 0 \\ 0 & 0 & 1 \end{pmatrix} + \frac{1}{v}\left[\begin{pmatrix} -rv \\ a^+ \\ 1 \end{pmatrix} - \begin{pmatrix} 1 & 0 & 0 \\ 0 & -r & 0 \\ 0 & 0 & 1 \end{pmatrix}\begin{pmatrix} v \\ a^- \\ 1 \end{pmatrix}\right](1, 0, 0) \\ &= \begin{pmatrix} -r & 0 & 0 \\ \frac{a^+ + ra^-}{v} & -r & 0 \\ 0 & 0 & 1 \end{pmatrix} = \begin{pmatrix} -r & 0 & 0 \\ \frac{1+r}{x_2}(\cos\omega t - \sigma) & -r & 0 \\ 0 & 0 & 1 \end{pmatrix}, \end{aligned} \tag{2.79}$$

which is a result first derived by Fredriksson and Nordmark [107].

Example 2.14 (Onset of sliding in Filippov systems). Saltation matrices also apply for trajectories of Filippov systems that undergo a transition into sliding, as depicted in Fig. 2.29. Proceeding as above it is straightforward to derive that the saltation matrix for this case is

$$Q_x = I + \frac{(F_{12} - F_1)H_x}{H_x F_1},$$

where F_{12} is the sliding flow defined by (2.80).

2.5.3 Non-transversal (grazing) intersections

The above discontinuity mapping (2.76) was derived under the transversality condition (2.75). So-called *grazing* occurs when a trajectory becomes tangent

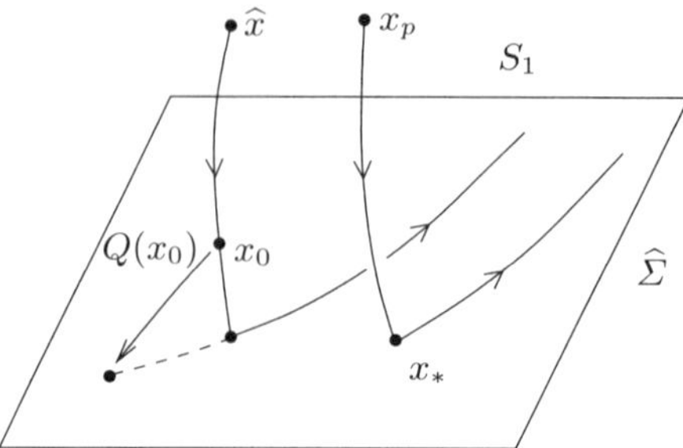

Fig. 2.29. Defining the saltation matrix for the onset of sliding.

to a discontinuity surface Σ. This occurs precisely when (2.75) is violated; that is, when $v := H_x(x_*)F_1(x_*) = 0$. Notice, from the expression (2.76) that the saltation matrix contains terms which are proportional to $1/v$ as $v \to 0$. Specifically, the coefficient of $1/v$ is $F_2 - F_1$, which would be zero in the case of flow with degree of smoothness two or higher. However, evaluation of subsequent terms in the Taylor expansion of the discontinuity map show that the factor $1/v$ enters at all orders, such that discontinuity with smoothness degree $n-2$ implies a singularity (proportional to $1/v$) of the nth derivative of the discontinuity map. Thus, the map is no longer analytic in the case of a grazing impact. In fact, detailed calculations which will be given in Chapters 6, 7 and 8 show that we should expect terms like $\sqrt{\Delta x}$ to occur in the expressions for the resulting discontinuity maps in this case. However, first we have to explain what we mean by a discontinuity map in the case of a grazing impact.

We illustrate the situation close to a grazing for an impacting hybrid system in Fig. 2.30. In this figure, which is analogous to Fig. 2.27, we show a distinguished trajectory locally lying entirely S_1. This trajectory we assume to graze with the discontinuity boundary Σ at the point x_* at time t_0.

To construct the discontinuity mapping, we need to know the fate of two different types of trajectory with initial conditions close to x_*. Some trajectories do not cross Σ locally; for these, the discontinuity mapping is the identity. In contrast, the discontinuity mapping will be non-trivial for the trajectory illustrated in Fig. 2.30 that passes through the point x_0 close to Σ at time t_1, hits Σ at the point x_2 at time $t_0 + \delta$, is mapped to the point x_3 by the map $\Phi_2(R(x_2), t_2 - t_1)$ and continues in S_1 from this point. Note that we allow here for both the impacting hybrid system case, in which Φ_2 is the identity, or the piecewise-smooth flow case, where R is in the identity. In the latter case, $t_2 - t_1$ is the time of flight of the trajectory until its second impact with Σ.

We shall describe two different ways of defining the non-trivial part of the discontinuity map. These are constructed either, like the DM for transversal trajectories defined above, such that the total elapsed time is zero—a so-called *zero-time discontinuity mapping (ZDM)*—or are defined with respect to a local *Poincaré* section—a *Poincaré-section discontinuity mapping (PDM)*.

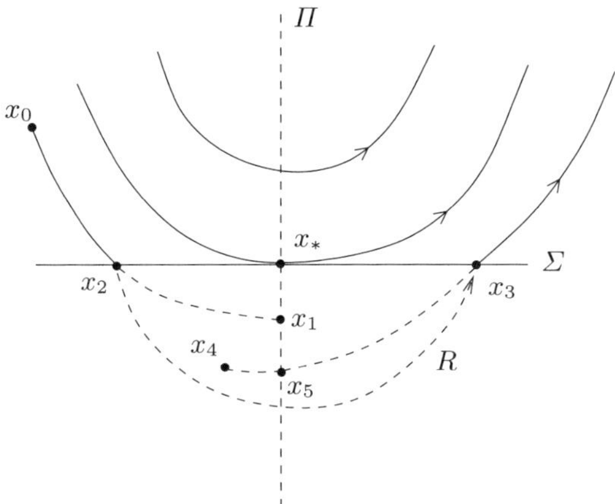

Fig. 2.30. A local illustration of the ZDM and PDM close to a grazing in an impacting hybrid system. In this figure the solid line represents the actual flow of the hybrid system, and the dashed line the extended flow. The ZDM is the map $x_0 \mapsto x_4$ and the PDM is the map $x_1 \to x_5$.

Our treatment is inspired by the analysis of n-dimensional impacting systems by [106], which extends earlier results in [236, 264, 197, 142].

To explain the difference between the ZDM and the PDM, consider in more detail the trajectory in Fig. 6.7 that passes through x_0. It intersects the Σ at x_2 and is mapped to x_3, where it subsequently evolves to the point x_6. By extending the smooth flow field $F_1(x)$ defined in the region $H(x) > 0$ (so that it lies above Σ in S_1) to the region $H(x) < 0$ (so that it now lies below Σ), we may continue the trajectory forward from x_2 under the action of the flow map Φ_1, or equally backward from x_3. As the point x_0 is close to x_g, then the smooth trajectory carried forward from x_2 under the action of Φ_1 will intersect the Poincaré surface

$$\Pi_N = \{x : v = H_x(x) F_1(x) = 0\}$$

at a point x_1 close to $x_g = 0$. Similarly, the backward continuation of the flow from x_3 will intersect the set Π_N at the point x_5. The mapping which takes x_1 to x_5 is the PDM.

Definition 2.34. *The* **Poincaré-section discontinuity mapping (PDM)** *near a grazing orbit is the discontinuity mapping defined on a suitable surface Π_N transverse to the flow, which contains the grazing set and intersects Σ transversally, that takes initial conditions on Π_N back to themselves. There is no requirement that this map take zero time.*

The same trajectory starting from x_3 can also be continued backward under the action of Φ_1 for a time $-\delta$ so that it passes through the point x_4 at the time t_0. We then define the ZDM as the map from x_0 to x_4.

Definition 2.35. *The* **zero-time discontinuity mapping (ZDM)** *near a grazing orbit is the discontinuity mapping in a neighborhood of the grazing point x_0 that takes zero time. That is, when this map is composed with the flow map of the non-impacting system in order to define a trajectory of the full system, the time taken is the same as for the flow map alone.*

In order to analyze *grazing bifurcations* of periodic orbits, we suppose that the trajectory passing through x_* is part of a limit cycle $p(t)$. Then, in order to unfold the dynamics, we need to combine a grazing discontinuity mapping (PDM or ZDM) with a Poincaré map defined around the limit cycle ignoring the grazing point. For example, zero time condition allows the ZDM to be incorporated in a natural way into the calculation of a fixed time-T Poincaré map P_S, sometimes called a *stroboscopic map*. For instance, for a grazing periodic orbit that is contained entirely within region S_1, the stroboscopic map can be written as

$$P_S = P_2 \circ ZDM \circ P_1,$$

where P_1 describes the evolution with flow Φ_1 through time t_1 and P_2 describes the Φ_1 through time $T - t_1$. The PDM may be preferable to use as an analytical tool for studying bifurcations of grazing limit cycles. It is also natural to apply the PDM for autonomous systems and the ZDM for time-periodically forced ones. The leading order terms of the ZDM and PDM generically have the same power, but the PDM correction takes non-zero time.

We do not give here the general forms of these maps. Unlike the case for transverse discontinuity mappings, there is no simple general expression valid for all cases of hybrid and piecewise-smooth systems. Indeed the detail evaluation of these mappings is rather lengthy in some cases; and forms the main thrust of Chapters 6, 7 and 8 in the cases of impacting hybrid systems, and piecewise-smooth systems with and without sliding, respectively. We shall also show in more detail how to combine the ZDM or PDM with the full flow of the system to produce an overall Poincaré mapand hence unfold the dynamics near a grazing limit cycle and other related DIBs.

2.6 Numerical methods

Many examples presented in this book rely on computations of orbits of piecewise-smooth and/or hybrid flows. For smooth flows, there are broadly speaking two classes of numerical methods for investigating the possible dynamics for a range of parameter values, namely: *direct numerical simulation*, and *numerical continuation* (also known as path-following). This classification also applies to piecewise-smooth systems. The rigorous numerical analysis of non-smooth dynamical systems remains a theory that is far from complete. Therefore, we shall take a practical approach in this book, since our goal is to use numerics to illustrate theory, rather than to analyze or derive optimal numerical algorithms.

2.6.1 Direct numerical simulation

When computing solutions to piecewise-smooth systems it is usually not possible to use general-purpose software packages directly, as most black-box numerical integration routines assume a high degree of smoothness of the solution. Accurate numerical computations must make special allowance for the non-smooth events that occur when a discontinuity boundary Σ_{ij} is crossed. Simulation methods for non-smooth systems fall broadly into two categories; *time-stepping* or *event-driven*. The former is most often used in many-particle rigid body dynamics written in complementarity form for which there can be perhaps millions of constraints and corresponding slack variables (Lagrange multipliers). For such problems, to accurately solve for *events* when one of the multipliers or constraint functions becomes zero within each time-step and to subsequently re-initiate the dynamics would be prohibitively computationally expensive. In contrast, the basic idea of time-stepping is to only check constraints at fixed times at intervals Δt. There are adaptations to standard methods for integrating ODEs and DAEs that are specifically designed for complementarity systems, some of which are based on linear complementarity problem solvers that have been developed in optimization theory. Clearly, errors are introduced by not accurately detecting the transition times, and therefore time-stepping schemes are often of low-order accuracy (i.e., with error estimates that $\sim O(\Delta t)^q$ for a low q) and can completely miss grazing events associated with low-velocity collisions. Several commercially available implementations of time-stepping algorithms are available, especially for the specific case of rigid body mechanics. These often have a variational formulation and are able to deal with the difficult problem of the collision of two rough bodies that may not have unique solutions. See the review by Brogliato and co-workers [41] and the Chapter by Abadie in [39, Ch. 2] for more details.

In this book we are largely concerned with low-dimensional systems with a small number of discontinuity boundaries (no more than say 10 of each). In that context, explicit *event-driven* schemes are feasible, fast and accurate. In these methods, trajectories within regions S_i are solved using standard numerical integration algorithms for smooth dynamical systems (e.g., Runge-Kutta, implicit solvers, etc.). Using these methods, the times at which a discontinuity boundary is hit are accurately solved for, and the problem is re-initialized there. Here we include the possibility of sliding or sticking flow by allowing portions $\tilde{\Sigma}_{ij}$ of discontinuity sets that are attracting to have the same status as open regions S_i, and to let the sliding vector field F_{ij} apply there. We then treat the boundary $\partial\tilde{\Sigma}$ as another discontinuity set. Similarly, in sticking regions for impact-hybrid systems, we can compute the explicit sticking flow that satisfies (2.43). Alternatively, one can use a DAE formulation, so that the Lagrange multipliers α or λ remain as part of the problem and a constraint $H_{ij} = 0$ is added which defines the discontinuity surface Σ_{ij}. There are now many reliable solvers for systems of differential algebraic equations, for example, DDASSL [218].

A key requirement for an event-driven method is the ability to define each discontinuity boundary as the zero set of a smooth function $H_{ij} = 0$. Also we have to carefully define a set of transition rules at each boundary that applies, if necessary, a reset rule R_{ij} and switches to the integration of a new dynamical system on the far side of the boundary. Thus, the time-integration of a trajectory of the dynamical system is reduced to the finding of a set of *event times* t_k and events $H_{ij}^{(k)}$ such that

$$H_{ij}^{(k)}(x(t_k)) = 0.$$

To achieve this we set up a series of *monitor functions*, the values of which are computed during each step of the time-integration. If one of these functions changes sign during a time step, then one needs to use a root finding method to accurately find where $H_{ij} = 0$. These ideas have been implemented in Matlab by Piiroinen & Kuznetsov [222].

Special care has to be taken to allow for the possibility of a sequence of event times converging onto a limit (for example, in a chattering sequence) followed by sticking or sliding. Clearly it is not appropriate to calculate all of the event times. To overcome this, it is typical to keep a record of the last few events. If it appears that the event times are converging to a limit, then this limit can be determined asymptotically, and then the procedure for a sliding or slicking solution applied; see [203] for details.

Let us now see how the event-driven method works specifically in the case of a two-zone Filippov system with sliding:

$$\dot{x} = \begin{cases} F_1(x), & \text{if} \quad H(x) > 0, \\ F_2(x), & \text{if} \quad H(x) < 0. \end{cases}$$

Note that the sliding vector field

$$F_s = (1 - \alpha)F_1(x) + \alpha F_2(x), \quad \text{where} \quad \alpha = \frac{H_x F_1}{H_x F_1 - H_x F_2} \tag{2.80}$$

is defined in a full neighborhood of $\Sigma = \{H = 0\}$. The flow is such that $H_x \equiv 0$, so that it is confined to level sets $H = \text{const}$. So, a small error in initial condition $H(x(t_j)) = \varepsilon$ will not in theory lead to flow precisely on Σ but on another manifold a small distance away from it. In fact, as is well known from constrained time-integration [10], a numerical approximation to such flows will cause $H(x(t))$ to slowly drift away from this manifold. One resolution to this [222] is to replace the sliding vector field with a regularized version

$$\widehat{F}_s(x) = F_s - CH_x(x)H(x),$$

where C is a positive constant. Note that $\widehat{F}_{12} = F_{12}$ on Σ, but away from the switching manifold, we have exponential attraction in the direction H_x onto it. See Fig. 2.31.

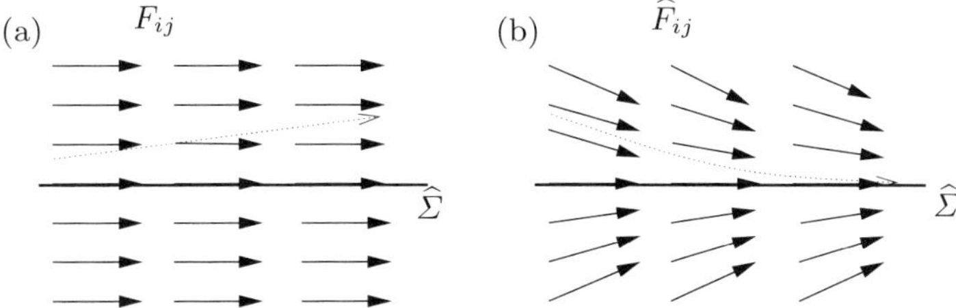

Fig. 2.31. (a) The sliding vector field F_{ij} and (b) the regularized version $\widehat{F}_{ij}$ near a discontinuity boundary Σ. Dashed lines indicate qualitatively what might happen to a numerical approximation to the flow.

One of the main uses of direct numerical simulation is to compute the bifurcation diagrams of the set of attracting solutions directly. In this process, for a fixed parameter value, a set of initial points is chosen and the flow from each point is determined. The flow is computed for a sufficiently long time for transients to decay and for the ensuing dynamics to be deemed to have converged onto an attractor. This dynamics is then recorded, perhaps in a suitable Poincaré section. The parameter is then changed slightly and the same process is repeated. Of course, one has to build up experience about the system in order to determine how long is a 'sufficiently long time'. However, an even more crucial question is to determine what set of initial conditions to take in order to converge to the various possible attractors. One approach here, which may minimize transient times, is to choose an initial condition for the new parameter value to be a point on the attractor at the previous parameter value. However, such an approach will necessarily miss the possibility of competing attractors present in the system. For example, consider the bifurcation diagram Fig. 1.26 for the DC–DC converter example, case study V, one sees several short intervals of the input voltage E for which in addition to the main bifurcation branch there are competing attractors (for example, a period-3 attractor around $E = 24$).

Thus, in general one should start from a range of different points within a suitably defined subset $\mathcal{D}$ of the phase space from which one has *a priori* knowledge that the attractors of the system must lie. But how to choose such points within this set? The number of points should of course be chosen to be as large as possible for the computational time available. One could start with a regular grid of points, but there are advantages in choosing the initial points at random. That is, at each fixed parameter value, use a *random* number generator to choose initial conditions in $\mathcal{D}$ uniformly. This way, the situation where attractors with small basins of attraction are missed consistently at each parameter value are likely to be avoided. We will refer to this method for computing bifurcation diagrams as a **Monte Carlo method**. Indeed, most of the bifurcation diagrams presented in this book were computed this way. The direct simulation method has many advantages in giving a quick and realistic picture of the bifurcation diagram of a system without assuming any *a priori* structure about the number or form of the attractors.

2.6.2 Path-following

While having the merits described above, direct simulation suffers from the two disadvantages that it does not accurately pinpoint bifurcation points, and it only computes stable invariant sets (attractors). In order to accurately locate bifurcations it is sometimes necessary to compute *unstable* invariant sets. For example, the collision of a limit cycle with an unstable equilibrium can cause the sudden disappearance of that limit cycle; or, one might want to detect the presence of an unstable limit cycle that at some subsequent parameter value may re-stabilize at a fold. Hence there is a complimentary need for *direct methods* for computing specific invariant sets of dynamical systems. These typically comprise methods for numerical path-following of these solutions as a parameter varies, for detecting codimension-one bifurcations, and possibly continuation of these bifurcation points in two or more parameters. These bifurcations might be regular bifurcations that can also occur in smooth systems, or they might be DIBs associated with the changing of the event sequence of the orbit. For smooth systems, there is a large literature on such methods; see, for example, [168, Ch. 10] or [232] for general explanations, and for example the software AUTO [88] and MatCont [74].

Let us illustrate the key ideas applied to the continuation, as a single parameter μ varies, of fixed points of maps $x \to f(x,\mu)$. This will be equivalent, via the numerical construction of Poincaré maps, to the problem of finding periodic orbits of non-smooth systems that have a given event sequence. It is also entirely equivalent to computing equilibrium solutions of a piecewise-smooth system $\dot{x} = f(x,\mu)$ within some fixed region S_i. Once we can do this, then we can use event detection to find parameter values where the event sequence of the periodic orbit changes, or where a fixed point or equilibrium hits a switching set Σ_{ij}. Hence we naturally have a method for the detection of DIBs.

The general setting is to find paths in the parameter space of smooth parameterized systems of n equations in n unknowns that take the form

$$G(x,\mu) = 0, \qquad G : \mathbb{R}^n \times \mathbb{R}^1 \to \mathbb{R}^n, \tag{2.81}$$

given some initial solution $x = x_0$ at $\mu = \mu_0$. For example, when computing fixed points of maps, we take

$$G(x,\mu) = x - f(x,\mu).$$

The key idea behind numerical continuation is based on an appeal to the Implicit Function Theorem to compute sequences of points at small intervals along the solution curve $x(\mu) \approx \{(x_i,\mu_i), i = 0 \dots N\}$.

The most commonly used method for solving systems of nonlinear equations is Newton's method, but it is well known that this requires a sufficiently good initial guess in order to converge [153]. There are many modifications to the above method and implementation details. Its strength is the local *quadratic convergence* guaranteed by Newton's method. Its drawback is the requirement to know the Jacobian matrix G_x. This is particular problematical in the case of periodic orbits.

When computing a periodic orbit of a non-smooth system that involves the crossing of a discontinuity boundary, one essentially needs a method to compute the Poincaré map$P(x, \mu)$ from some section $\Pi := \{x : \pi(x) = 0\}$ to itself, see Fig. 2.26. This can be done via a *shooting method.* (For alternative ways of computing periodic orbits in Filippov systems via concatenating different boundary-value problems for each trajectory segment, see [72]). The shooting approach takes an initial condition in Π and solves the flow, through the various regions S_i back to Π again, taking a total time $\tau(x)$. This defines the point $P(x)$, and the function to which we apply the continuation algorithm is thus

$$G = x - P(x, \mu),$$

a zero of which represents the existence of a periodic orbit. Therefore the Jacobian we need is $G_x = I - P_x$, so we therefore need an expression for the linearized Poincaré mapP_x. Here it is useful to apply the discontinuity mapping Q described in the previous section. Indeed, for the case of transversal intersections, the saltation matrix correction has been used successfully for path-following [66] and for the detection of limit cycles [1].

So, now that we have the notation, definitions and methods (both analytical and numerical) at our disposal; we are ready to embark on a tour of different discontinuity-induced bifurcations. We start, in the next two chapters, with the case of maps.

3

Border-collision in piecewise-linear continuous maps

This chapter concerns the analysis and classification of non-smooth bifurcations of fixed and periodic points of n-dimensional maps that are locally piecewise-linear and continuous. The majority of the chapter deals with maps composed of precisely two linear pieces. For such maps, a simple discontinuity-induced bifurcation occurs when a fixed point of one piece of the map reaches the discontinuity boundary, a so-called *border-collision* bifurcation. Techniques are described to classify the simplest resulting scenarios—namely persistence, non-smooth fold and non-smooth period-doubling—based on properties of the two component linear maps. In addition, bifurcation diagrams of remarkable complexity are found, including sudden transition from a stable fixed point to a fully developed *robust* (i.e., without periodic windows) chaotic attractor.

A complete classification of the behavior of the simplest orbits is given for one- and two-dimensional maps. Special attention is placed on the case of piecewise-linear maps that are nonintertible in one region, due to the presence of a zero eigenvalue. Such maps will be of importance to the grazing-sliding bifurcation studied in Chapter 8. Here we are able to prove the existence of robust chaos in a certain region of parameter space. Finally, we discuss briefly possible effects of nonlinear terms.

3.1 Locally piecewise-linear continuous maps

Throughout we focus attention on a local region $\mathcal{D} \subset \mathbb{R}^n$ of phase space that contains just one discontinuity boundary. That is, by an appropriate choice of local co-ordinates, the map under investigation can be described as

$$x \mapsto f(x,\mu) = \begin{cases} F_1(x,\mu), & \text{if } H(x,\mu) < 0, \\ F_2(x,\mu), & \text{if } H(x,\mu) > 0, \end{cases} \tag{3.1}$$

where $F_i : \mathbb{R}^n \times \mathbb{R} \mapsto \mathbb{R}^n$, $i = 1, 2$ and $H : \mathbb{R}^n \mapsto \mathbb{R}$ are sufficiently smooth and differentiable scalar functions of $x \in \mathbb{R}^n$. The surface described implicitly by the condition $H(x,\mu) = 0$ defines a smooth boundary

$$\Sigma := \{x \in \mathcal{D} : H(x,\mu) = 0\}, \tag{3.2}$$

which separates $\mathcal{D}$ into two regions

$$S_1 := \{x \in \mathcal{D} : H(x,\mu) < 0\}, \tag{3.3}$$
$$S_2 := \{x \in \mathcal{D} : H(x,\mu) > 0\}. \tag{3.4}$$

We assume that the map is continuous across the boundary; i.e., $F_1(x,\mu) = F_2(x,\mu)$ for all $x \in \Sigma$, in the sense that we must be able to write

$$F_2(x,\mu) = F_1(x,\mu) + E(x,\mu)H(x,\mu) \tag{3.5}$$

for some smooth function $E(x,\mu)$. [Indeed, for a point $\widehat{x} \in \Sigma$, we have $H(\widehat{x},\mu) = 0$ and so, from (3.5), $F_1(\widehat{x},\mu) = F_2(\widehat{x},\mu)$.]

In this chapter, we assume that functions F_1 and F_2 have well-defined linearizations along Σ. Then, intuitively, by choosing a small enough neighborhood of $\mathcal{D}$, the overall map (3.1) can be well approximated by a locally piecewise-linear map. More precisely, we will assume that the Jacobians $F_{1,x}$ and $F_{2,x}$ are well defined and non-singular throughout $\mathcal{D}$, but that $F_{1,x} \neq F_{2,x}$ for $x \in \Sigma$. The next chapter shall deal with maps that are not locally piecewise-linear; specifically maps with a jump (i.e., such that $F_1 \neq F_2$ on Σ); maps with infinite derivatives for which $F_{i,x}$ is undefined along Σ for $i = 1$ or 2; or maps with higher order discontinuity for which $F_{1,x} = F_{2,x}$ when $H(x,\mu) = 0$. Later in this chapter (Sec. 3.7) we shall also look at the case of locally piecewise-linear continuous maps that are noninvertible on one side, i.e., for which $F_{1,x}$ or $F_{2,x}$ is singular along Σ.

The aim is to derive a set of simple conditions on F_1 and F_2 to allow a classification to be made of the various types of discontinuity-induced bifurcations that a fixed point (or periodic point) $\widehat{x}$ of (3.1) can undergo as it approaches Σ under parameter variation. In what follows we shall assume without loss of generality that a simple fixed point within S_1 exists for parameter values $\mu < 0$ and that this approaches Σ as $\mu \to 0^-$. The goal is then to classify the dynamics that ensues for $\mu > 0$. The classification strategy will consist of three basic steps:

1. Linearize F_1 and F_2 about the bifurcation point, and where possible define a global co-ordinate transformation that places F_1 and F_2 in their simplest canonical forms.
2. Check if the linearized map satisfies a set of conditions identifying the existence and stability of certain periodic orbits.
3. Combine the conditions identified under step 2, and describe a bifurcation scenario of the simplest possible periodic orbits. Where no simple orbit is found to be stable, check conditions for the existence of more complex invariant sets such as robust chaotic attractors or invariant tori.

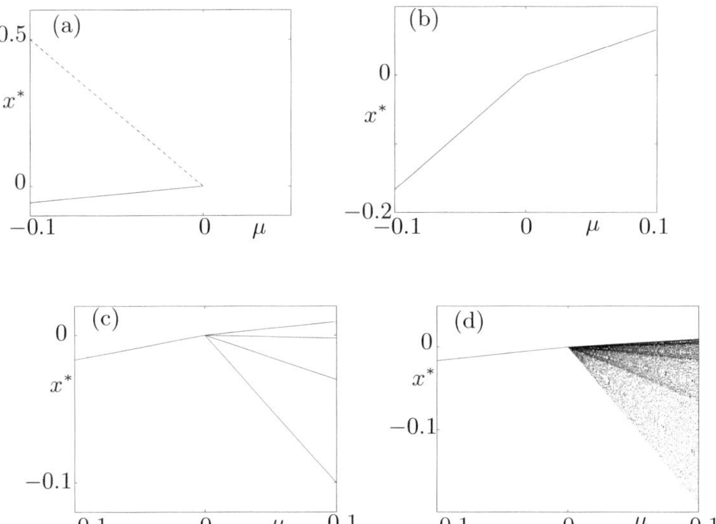

Fig. 3.1. Examples of Monte Carlo bifurcation diagrams of border collision bifurcations of a simple piecewise linear continuous maps. In each case, the attractor for $\mu < 0$ is a stable fixed point within region S_1, whereas for $\mu > 0$, a number of different attractors may be observed. For the specific details of the map, see Eq. 3.44. The parameter values taken are (a) $\nu_1 = -0.8$, $\nu_2 = 1.2$, (b) $\nu_1 = 0.4$, $\nu_2 = -0.5$, (c) $\nu_1 = 0.4$, $\nu_2 = -12$, and (d) $\nu_1 = 0.4$, $\nu_2 = -20$.

The classification of simple fixed points, steps 1 and 2, can be carried out in n-dimensional generality. This leads to the main theoretical result of this Chapter, Theorem 3.1 below. However, there would seem to be no general classification strategy for showing the existence of invariant circles, chaos, or more complex invariant sets in n-dimensional piecewise-smooth maps. So, for step 3, we restrict attention here to one- and two-dimensional maps (in Secs. 3.4 and 3.5, respectively), and even then are only able to give partial results. The advantage though of piecewise-linear maps is that they are remarkably easy to simulate, and, using a Monte Carlo approach, it is often possible to gain the correct qualitative information on the dynamics just by looking at pictures. For example, Figure 3.1 shows four such numerically obtained bifurcation diagrams of piecewise-linear maps, whose dynamics can be explained by the theory presented in this chapter, but for which the correct qualitative conclusion can be drawn by looking at these graphical outputs from an almost trivial computer code. Nevertheless theory, where available, provides much more information such as the identification of parameter regimes where certain behavior can or cannot occur. To begin, we start with some definitions.

3.1.1 Definitions

Definition 3.1. *We say that a point $x = x^*$ is an* **admissible** *fixed point of (3.1) if, for $i = 1$ or $i = 2$, $x^* = F_i(x^*, \mu)$ and $x^* \in S_j$ with $j = i$. We say instead that $x = \tilde{x}$ is a* **virtual** *fixed point of (3.1) if $\tilde{x} = F_i(\tilde{x}, \mu)$ and $\tilde{x} \in S_j$ with $i \neq j$.*

Definition 3.2. *A fixed point $x = x^*$ is a* **boundary fixed point** *$x \in \Sigma$; i.e., $F_1(x^*, \mu) = F_2(x^*, \mu)$ and $H(x^*, \mu) = 0$.*

Now, suppose $x^*(\mu)$ is a branch of fixed points of the map $x \mapsto F_i(x, \mu)$ that depends continuously on the parameter $\mu \in (-\varepsilon, \varepsilon)$. Suppose that x^* becomes a boundary fixed point for the piecewise-smooth map (3.1) when $\mu = 0$. That is, $H(x^*(0)) = 0$. Then, if the approach to Σ is transversal, the Implicit Function Theorem implies that $x^*(\mu)$ is an admissible fixed point of (3.1) for one sign of μ only. Without loss of generality, let us suppose that this sign is negative. Following [205], we then make the following definition:

Definition 3.3. *Such a fixed point x^* is a called a* **border-crossing fixed point** *if the branch $x^*(\mu)$ crosses Σ transversally as μ passes through zero. In so doing, the fixed point will be admissible for one and only one sign of μ, without loss of generality $\mu < 0$. Specifically, we require that (for either $i = 1$ or $i = 2$):*

1. $x^* = F_i(x^*, \mu) \in S_i$ *for* $-\varepsilon < \mu < 0$;
2. $x^* \in \Sigma$ *for* $\mu = 0$;
3. $x^* = F_i(x^*, \mu) :=\in S_j$, *where* $j \neq i$ *for* $0 < \mu < \varepsilon$;
4. $H_x F_{i,\mu} \neq 0$ *when* $\mu = 0$.

Note that condition 3 implies x^* changes continuously into a virtual fixed point of (3.1) for $\mu > 0$, whereas condition 4 ensures that the crossing of Σ is transversal.

Definition 3.4. *We say that a fixed point x^* undergoes a* **border-collision** *bifurcation for $\mu = 0$ if $x^*(\mu)$ is a border-crossing fixed point and linearizing the map about $x^*(0)$ in S_1 and S_2 yields*

$$\left.\frac{\partial F_1}{\partial x}(x^*, \mu)\right|_{\mu=0} \neq \left.\frac{\partial F_2}{\partial x}(x^*, \mu)\right|_{\mu=0}, \tag{3.6}$$

or equivalently, using condition (3.5)

$$\left.E(x^*, \mu)\frac{\partial H}{\partial x}(x^*)\right|_{\mu=0} \neq 0.$$

Note that condition (3.6) ensures that the Jacobian of the map (3.1) about the fixed point is discontinuous at the border-collision point.

Note that the above definitions can be easily extended to define admissible, virtual and boundary period-n points of a PWS map of interest [205] and their border-collision. This is because, the nth-iterate map of (3.1) when linearized within S_1 and S_2 separately about a period-n point lying within Σ will under appropriate non-degeneracy conditions, take exactly the same form as (3.1) for different functions F_1 and F_2. For example, we can give the following definition.

Definition 3.5. *A period-two point $(x_1^*(\mu), x_2^*(\mu))$ of (3.1) characterized by $F_1(x_1^*) = x_2$ and $F_2(x_2^*)$ is termed admissible if and only if $H(x_1^*, \mu) < 0$ and $H(x_2^*, \mu) > 0$. The point would be called a* **boundary period-two point** *if at least one of these conditions is replaced with equality and, otherwise, would be termed a virtual period-two point.*

3.1.2 Possible dynamical scenarios

At a border-collision, many different dynamical scenarios can be observed under parameter variation. For example, depending on the precise forms of F_1 and F_2 (admissible or virtual) fixed, period-two and more general period-n points may branch off the boundary fixed point for either sign of μ.

The three basic possible scenarios involving fixed and period-two points for general n-dimensional maps of the form (3.1) are as follows.

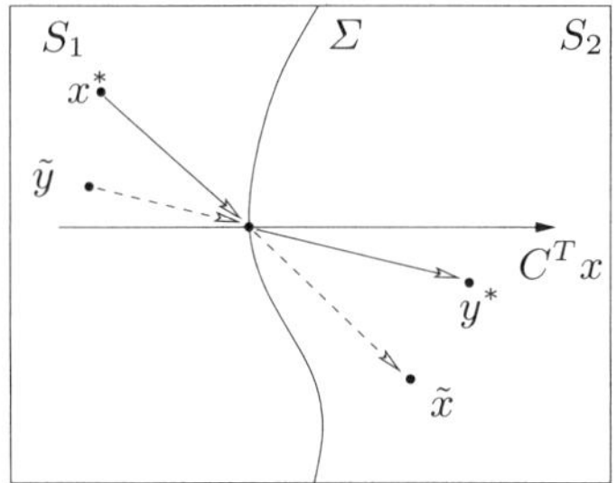

Fig. 3.2. Schematic planar phase space diagram of a border-collision leading to the *persistence* of the border-crossing fixed point. As the parameter μ is varied past the bifurcation point, the fixed point $x^* \in S_1$ crosses the boundary transversally changing continuously into the (admissible) fixed point $y^* \in S_2$.

Persistence: An admissible fixed point $x^* \in S_1$ and a virtual fixed point $\tilde{y} \in S_1$ hit the boundary at the border-collision point, turning into a virtual fixed point $\tilde{x} \in S_2$ and an admissible fixed point $y^* \in S_1$ respectively past the bifurcation point (see Fig. 3.2).

Non-smooth fold: Two coexisting admissible border-crossing fixed points, $x^* \in S_1$ and $y^* \in S_2$, hit the boundary and change continuously into two

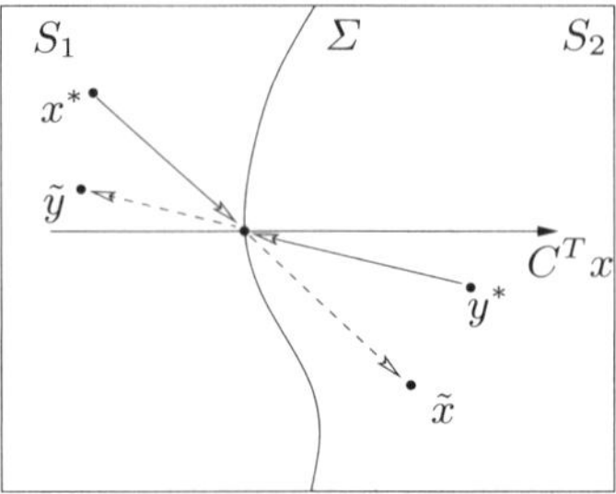

Fig. 3.3. Schematic planar phase space diagram of a border-collision leading to a *non-smooth fold*. As the parameter μ is varied past the bifurcation point, the two coexisting (admissible) fixed points $x^* \in S_1$ and $y^* \in S_2$ hit the boundary transversally changing continuously into two (virtual) fixed points. Thus, past the bifurcation point, the map has no admissible fixed point.

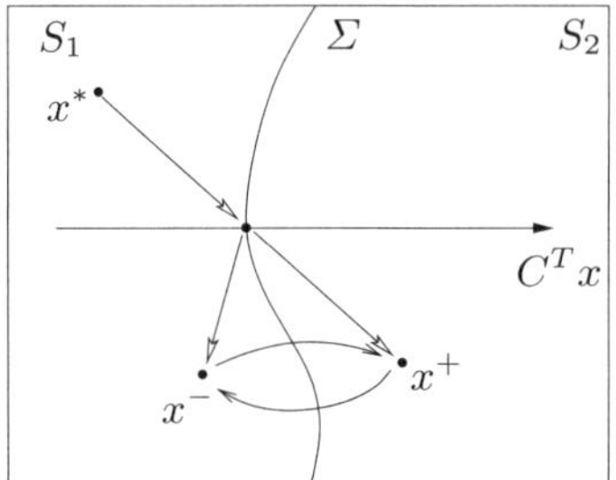

Fig. 3.4. Schematic planar phase space diagram of a border-collision leading to a *non-smooth period-doubling*. As the parameter μ is varied past the bifurcation point, the fixed point $x^* \in S_1$ crosses the boundary transversally. A period-two solution characterised by one iterate x^- in S_1 and the other x^+ in S_2 is involved in the bifurcation scenario.

virtual fixed points $\tilde{x} \in S_2$ and $\tilde{y} \in S_1$. Hence, no admissible fixed point exists past the bifurcation point (see Fig. 3.3).

Non-smooth period-doubling: A period-two orbit (x_1^*, x_2^*) characterized by having one iteration on each side of the boundary (e.g., $x_1 \in S_1$, $x_2 \in S_2$) branches off the boundary fixed point at the border-collision (see Fig. 3.4).

In fact, non-smooth period-doubling is just the simplest example of

Non-smooth period-multiplying: A period-m point for $m > 1$ branches off from the boundary fixed point. Typically, in order to be admissible, such orbits have one iterate on one side of the boundary, and $n-1$ on the other (e.g. $x_1 = F_2(x_n) \in S_1$, $x_2 = F_1(x_2) \in S_2$ and $x_j = F_2(x_{j-1}) \in S_2$, for $j = 3 \ldots n$).

In each of these simple scenarios, the fixed and period-m points could be stable or unstable, or exist for $\mu > 0$ or $\mu < 0$ depending on the precise form of F_1

and F_2. In what follows we shall find precise conditions for deciding which of these scenarios exist. For simplicity, we shall exclude detailed analysis of the period-multiplying bifurcation except for the period-doubling case $m = 2$.

3.1.3 Border-collision normal form map

As shown originally by Feigin [95, 96, 97, 98, 99] and later in a slightly different context by Nusse & Yorke [205, 206, 207] (see also [21]), a classification as intimated in the previous section is most simply carried out by studying an appropriate *normal form* map valid in a neighborhood of the border-collision bifurcation point. Under certain other non-degeneracy conditions [50], this map takes the form of a piecewise-linear map.

Suppose that a border-collision occurs for $\mu = \mu^*$ at $x = x^* \in \Sigma$. To derive an appropriate local normal form mapping, consider a sufficiently small neighborhood of the border-collision point in both state and parameter space.

First, we introduce a change of co-ordinates $\tilde{x} = x^* - x$, $\tilde{\mu} = \mu^* - \mu$ (and drop the tildes in what follows), so that the border-collision occurs at the point $x = 0$, when $\mu = 0$. Next, we expand the map (3.1), expressed in the new variables, about the bifurcation point $(x, \mu) = (0, 0)$, to obtain

$$x \mapsto \begin{cases} N_1 x + M_1 \mu + O(\|x\|^2, \|x\|\mu, \mu^2), & \text{if } C^T x + D\mu < 0, \\ N_2 x + M_2 \mu + O(\|x\|^2, \|x\|\mu, \mu^2), & C^T x + D\mu > 0, \end{cases} \tag{3.7}$$

where

$$N_1 = \frac{\partial F_1}{\partial x}, \qquad N_2 = \frac{\partial F_2}{\partial x},$$
$$M_1 = \frac{\partial F_1}{\partial \mu} \qquad M_2 = \frac{\partial F_2}{\partial \mu},$$
$$C^T = \frac{\partial H}{\partial x}, \qquad D = \frac{\partial H}{\partial \mu},$$

all evaluated at $x = 0, \mu = 0$.

Now, since the error terms are nonlinear, it is reasonable to assume that the dynamics local to the bifurcation point can be described by the piecewise-linear map obtained by neglecting higher-order terms in (3.7). Such a statement can be made rigorous using the terminology of topological equivalence introduced in Chapter 2, but we do not go into the details here. Clearly, from this point of view, the Implicit Function Theorem can be used in order to describe the behavior of simple period-m points correctly, provided each of the matrices $(I - N_i)$ are non-singular and that M_i is non-zero for $i = 1, 2$.

In what follows, for the sake of clarity, we will also assume that the boundary is independent of μ; that is, we assume

$$D = 0$$

in (3.7). Again, an appeal to the Implicit Function Theorem shows that such an assumption can be justified by a further co-ordinate transformation, provided certain extra non-degeneracy conditions are satisfied. Moreover, the assumption that H is smooth means that, close to the bifurcation point, we can always choose a suitable change of co-ordinates, which moves the boundary to the surface $\{x_1 = 0\}$ (see [205, 50, 21] for further details). Moreover, from the assumption of continuity across Σ for $\mu = 0$, it follows that the matrices N_1 and N_2 must then satisfy the condition

$$N_2 - N_1 = EC^T \tag{3.8}$$

for some constant vector E. That is, N_1 and N_2 differ by at most a matrix of rank 1. Specifically, given our choice of co-ordinate so that $C^T = (\,1 \;\; 0 \;\; \ldots \;\; 0\,)$, this means that N_1 and N_2 differ only in their first column. Moreover, continuity for non-zero μ implies that the vectors M_1 and M_2 in (3.7) must be the same. Thus, we set $M := M_1 = M_2$. Finally then we arrive at the piecewise-linear map

$$x \mapsto \begin{cases} N_1 x + M\mu, & \text{if } C^T x < 0, \\ N_2 x + M\mu, & \text{if } C^T x > 0, \end{cases} \tag{3.9}$$

where $N_{1,2}$ satisfy (3.8).

3.2 Bifurcation of the simplest orbits

For smooth bifurcations, particular unfoldings of the dynamics are associated with a set of non-degeneracy hypothesis and simple sign conditions on normal form coefficients. For example, there are two kinds codimension-one possibilities for Hopf bifurcation; either super- or sub-critical depending on the sign of the key cubic term in the normal form (the so-called *Lyapunov coefficient*), see Chapter 2 and [168, Chapter 2] for more details. For border-collisions, however, such a general classification of all possible unfoldings does not seem possible. Indeed, as we shall shortly see, (an infinite number of) different possible unfoldings can arise from the same normal form map (3.9), depending on the coefficients of the matrices N_1 and N_2.

In what follows, we present a strategy for the classification of only one aspect of the dynamics of this normal form map, namely the bifurcation behavior of period-one and two points. That is, we provide a method for distinguishing among the three simplest scenarios introduced in the previous section, as represented Figs. 3.2–3.4. This methodology is based on the original work of Feigin [98] and its later extensions [80].

3.2.1 A general classification theorem

Section 3.3 is devoted to a proof of the following fundamental result:

Theorem 3.1 ([95, 96]). *Let* $p_1(\lambda)$ *be the characteristic polynomial of matrix* N_1 *and* $p_2(\lambda)$ *the characteristic polynomial of* N_2 *in (3.9). Moreover, define*

$\sigma_1^+ :=$ *number of real eigenvalues of* N_1 (α_i) *greater than* 1;
$\sigma_2^+ :=$ *number of real eigenvalues of* N_2 (β_i) *greater than* 1;
$\sigma_1^- :=$ *number of real eigenvalues of* N_1 *less than* -1;
$\sigma_2^- :=$ *number of real eigenvalues of* N_2 *less than* -1.

Assume that the following non-degeneracy conditions are satisfied:

$$\begin{aligned} \det(I - N_1) &\neq 0, \\ \det(I + N_1) &\neq 0, \\ C^T(I - N_2)^{-1}M &\neq 0, \\ 1 - C^T(I - N_1)^{-1}E &\neq 0, \\ 1 - C^T(I + N_1)^{-1}E &\neq 0. \end{aligned}$$

Then, at a border-collision, we have the following scenarios:

Persistence *if either*

$$1 - C^T(I - N_1)^{-1}E > 0; \tag{3.10}$$

or, equivalently

$$p_1(1)p_2(1) > 0; \tag{3.11}$$

or

$$\sigma_1^+ + \sigma_2^+ \text{ is even.} \tag{3.12}$$

non-smooth fold *if*

$$1 - C^T(I - N_1)^{-1}E < 0; \tag{3.13}$$

or equivalently

$$p_1(1)p_2(1) < 0; \tag{3.14}$$

or

$$\sigma_1^+ + \sigma_2^+ \text{ is odd.} \tag{3.15}$$

non-smooth period-doubling *if*

$$1 + C^T(I + N_1)^{-1}E < 0; \tag{3.16}$$

or equivalently

$$p_1(-1)p_2(-1) < 0; \tag{3.17}$$

or

$$\sigma_1^- + \sigma_2^- \text{ is odd.} \tag{3.18}$$

Remarks

1. The conditions in this theorem can in principle be used to classify various different bifurcation scenarios as in Fig. 3.5. Note that there are three equivalent sets of conditions for each scenario, the proof of the equivalence of which forms the subject of the next section. Take, for example, the persistence case; then we have condition (3.10) based on matrices, or (3.11) based on characteristic polynomials, or (3.12) based on eigenvalues. Note that, whereas the latter two sets of conditions, were original derived by Feigin in the early 1970s [95], the matrix-based conditions are presented here for the first time (see also [79]). Despite the equivalence, in practice using eigenvalues for classification is probably most intuitive, since computations of the eigenvalues are required in order to assess the stability of the branching solutions. If, instead, it is important to assess the effects of an additional parameter (not μ) on the possible types of bifurcation scenarios observed at a border-collision, it might be easier to use the characteristic polynomials or the matrices directly.
2. As we are about to see, typically a combination of these conditions need to be used to derive specific bifurcation scenarios associated with a border-collision. For instance, condition (3.16) could be satisfied at the same time as either (3.10) or (3.13). Also, (3.16) does not say for which sign of μ the admissible period-two point exists. For instance, if for a non-smooth fold condition (3.16) additionally holds, then in principle the period-two point could exist for same sign of μ as the two period-one points, or for the other sign μ. Which of these occurs requires further analysis, which leads to cumbersome expressions in the general n-dimensional case. Later in this chapter we deal with detailed enumeration of all the various subcases only in the cases $n = 1$ or 2.
3. Similar conditions to those given by Theorem 3.1 can be given to characterize the fate of fixed points of higher-iterates of the map that interact with Σ at the border-collision. For example, suppose there is a period-two point that visits S_1 at least once (otherwise swap the labeling of S_1 and S_2). Then by looking at the eigenvalues of the matrices N_1N_2 and N_1N_1, we can use the above classification method to determine whether such period-two points persist, undergo a non-smooth fold or indeed generate a period-four orbit. For example, let $\sigma_{11}^+\sigma_{12}^+$ be the number of real eigenvalues of the matrices N_1N_1 and N_1N_2, respectively, which exceed 1. Then, the period-two point persists or folds depending on whether the quantity $\sigma_{11}^+ + \sigma_{12}^+$ is even or odd, respectively.
4. We have made generic assumptions on the form on the matrices involved, in particular that $(I - N_1)$ and $(I - N_2)$ are both non-singular. If either of these matrices were singular this would mean that the boundary fixed point would be non-hyperbolic 'from one side', that is the maps F_1 or F_2 would have a multiplier at unity. This would generically correspond to a fold bifurcation occurring at precisely the boundary equilibrium point, and its unfolding should be treated as a codimension-two border-collision/fold bifurcation. Some discussion of codimension-two discontinuity-induced bi-

furcations occurs in Chapter 9, but is beyond the scope of the current chapter. In Sec. 3.6 below, we shall also look at another special case that will be of relevance to the unfolding of sliding bifurcations in Chapter 8, in which either N_1 or N_2 is itself singular. Note that such a case does not violate any of the conditions of the above.

3.2.2 Notation for bifurcation classification

The proof of Theorem 3.1 forms the subject of Sec. 3.3. First let us explore some of its further consequences. Specifically, by combining the three elementary conditions in the Theorem, one can delineate all possible behaviors of period-one and two orbits; see Fig. 3.5. There, we indicate a stable admissible fixed point in region S_1 by the letter A, an unstable admissible fixed point in S_1 by a, a stable admissible fixed point in S_2 by B and an unstable admissible fixed point in S_2 by b. A stable period-two point, such as the one depicted in Fig. 3.4, is denoted by AB if it is stable and ab otherwise. Also, we use $\leftrightarrow$ to indicate the occurrence of a border-collision as μ passes through zero. Using this notation, a non-smooth fold, for instance, will be described by $A, b \leftrightarrow \varnothing$. Note that the $\leftrightarrow$ symbol does not necessarily imply that the indicated transition happens as μ increases, since the direction of bifurcation depends on the sign of $C^T M$. Finally note that we use the notation /, for example, A/a, to indicate fixed points whose stability is at present unknown and would require checking whether the map multipliers (eigenvalues of N_1 or N_2) are inside or outside the unit circle.

Let us look in more detail at some of the branches of the tree in Fig. 3.5. Suppose for example that

$$\sigma_1^- + \sigma_2^- \quad \text{is even } (\textit{no period doubling})$$

then we can conclude that no period-doubling occurs at the border-collision and the following cases are possible according to the specific sets of eigenvalues of N_1 and N_2:

1. If $\sigma_1^+ + \sigma_2^+$ is even then we have persistence of the fixed point at the border-collision; i.e.,
$$A/a \leftrightarrow B/b.$$
2. If $\sigma_1^+ + \sigma_2^+$ is odd then we have a non-smooth fold; i.e.,
$$A/a, b/B \leftrightarrow \varnothing.$$

The stability of the fixed points involved in the bifurcation scenario can be determined by looking at whether the eigenvalues of N_1 and N_2 are inside or outside the unit circle. For example, if $\sigma_1^+ = \sigma_2^+ = 1$ in the first case, we must have $a \rightarrow b$ since a real eigenvalue greater than unity implies instability of the fixed point. In contrast, in the second scenario above, we have that the number of eigenvalues inside the unit circle must change as we cross the

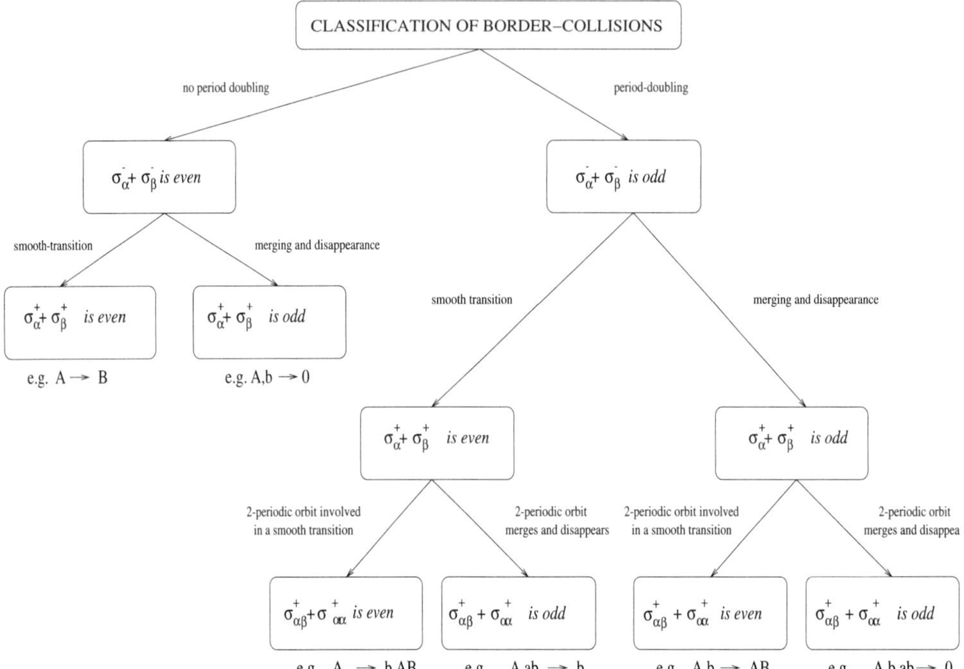

Fig. 3.5. Border-collision bifurcation tree generated by the application of the three elementary conditions (3.12), (3.15) and (3.17)

border-collision point. For example, if $\sigma_1^+ = 0$ and $\sigma_2^+ = 1$, we have a direct analog of the fold; that is, $A, b \leftrightarrow \varnothing$. However, note that there is no eigenvalue of N_1 that approaches unity as it would for a smooth fold bifurcation.

These represent the simplest possibilities, which might occur in locally piecewise-linear maps. In higher dimensions, other possibilities can occur. From the first case we could have $\sigma_1^+ - \sigma_2^+ = 2$ so that two eigenvalues of the map jump across the unit circle at the bifurcation point, or in the second scenario that $\sigma_1^+ - \sigma_2^+ = 3$ so that three eigenvalues jump. In addition, any number of complex pairs can jump without this affecting the signs of $\sigma_{1,2}^+$. This illustrates that classifying the possible scenarios in terms of the "jumping" of eigenvalues at a border-collision point is not the most helpful way of classifying this discontinuity-induced bifurcation in general n-dimensional maps.

Consider now another branch of the bifurcation tree in Fig. 3.5:

$$\sigma_1^- + \sigma_2^- \quad \text{is odd } (\textit{period-doubling occurs})$$

then there is period-doubling and a new stable (AB) or unstable (ab) period-two orbit will be involved in the border-collision in a way that can be determined by looking at the quantity $\sigma_{11}^+ + \sigma_{12}^+$. Specifically we have the following possibilities:

1. If $\sigma_1^+ + \sigma_2^+$ is even (*persistence of the border-crossing fixed point*) then
   (a) If $\sigma_{11}^+ + \sigma_{12}^+$ is even (*continuous transition to a fixed point and a coexisting period-two orbit*) which gives one of the following scenarios:

$$\begin{aligned} A &\leftrightarrow b, AB, \\ A &\leftrightarrow b, ab, \\ a &\leftrightarrow B, ab, \\ a &\leftrightarrow b, AB, \\ a &\leftrightarrow b, ab. \end{aligned}$$

   To determine which scenario happens for a particular map, the specific eigenvalues of N_1, N_2 and N_1N_2 must be computed to check for the stability of the fixed and period-two points involved.

   (b) If $\sigma_{11}^+ + \sigma_{12}^+$ is odd (*the period-two orbit disappears*)

$$\begin{aligned} A, ab &\leftrightarrow b, \\ A, ab &\leftrightarrow B, \\ a, AB &\leftrightarrow b, \\ a, AB &\leftrightarrow B. \end{aligned}$$

2. If $\sigma_1^+ + \sigma_2^+$ is odd (*non-smooth fold of the border-crossing fixed point*), then
   (a) if $\sigma_{11}^+ + \sigma_{12}^+$ is even (*continuous transition to a period-two orbit*) we would get one of the following possible scenarios:

$$\begin{aligned} A, b &\leftrightarrow AB, \\ A, b &\leftrightarrow ab, \\ a, b &\leftrightarrow AB, \\ a, b &\leftrightarrow ab, \end{aligned}$$

   Note that these scenarios are not possible for one- or two-dimensional maps, and it is conjectured in [95, 96] that this is true also in arbitrary dimensions, although we are aware of no immediate proof for n-dimensional maps.

   (b) if $\sigma_{11}^+ + \sigma_{12}^+$ is odd (*period-two orbit disappears*) we get one of the following cases:

$$\begin{aligned} A, b, ab &\leftrightarrow \varnothing, \\ a, b, ab &\leftrightarrow \varnothing, \\ a, b, AB &\leftrightarrow \varnothing. \end{aligned}$$

Example 3.1 (border-collision classification in a two-dimensional map). To better illustrate the classification strategy described so far, consider a general two-dimensional map of the form (3.24) where

$$N_1 = \begin{pmatrix} d_{11} & 1 \\ d_{21} & 0 \end{pmatrix}, \quad N_2 = \begin{pmatrix} d_{12} & 1 \\ d_{22} & 0 \end{pmatrix}, \quad M = \begin{pmatrix} 1 \\ 0 \end{pmatrix}. \tag{3.19}$$

We shall take two different representative cases characterized by different values of the coefficients of the matrices N_1 and N_2.

Case 1. Assume $d_{11} = 1.45$, $d_{21} = -0.5$, $d_{12} = -1.7$ and $d_{22} = -0.5$ in (3.19). With this choice of parameters, the eigenvalues of matrix N_1 are $\lambda_{11} = 0.89, \lambda_{12} = 0.56$ whereas those of matrix N_2 are $\lambda_{21} = -1.32, \lambda_{22} = -0.38$. Thus, using the notation introduced above, we have

$$\sigma_1^- + \sigma_2^- = 1 \quad \text{(odd)}, \quad \sigma_1^+ + \sigma_2^+ = 0 \quad \text{(even)}.$$

Hence, according to the classification tree presented in Fig. 3.5, a nonsmooth period-doubling will occur at the border-collision. In order to identify the right scenario, we need then to look at the eigenvalues of $N_1 N_1$ and $N_1 N_2$. In particular, we have $\sigma_{11}^+ + \sigma_{12}^+$ is even. Therefore the period-two orbit coexists with the A/a orbit and we have one of the scenarios:

$$A, ab \leftrightarrow b, \quad A, ab \leftrightarrow B, \quad a, AB \leftrightarrow b, \quad a, AB \leftrightarrow B.$$

As the eigenvalues of N_1 are inside the unit circle and those of N_2 and $N_1 N_2$ are not, we have that the fixed point A is stable, whereas b and the period-two point ab are not:

$$A, ab \leftrightarrow b.$$

Note that this is one of those cases where, because of its general n-dimensional nature, the classification strategy is able to predict only that after the border-collision there will not exist any stable fixed or period-two point. We need therefore to use other tools to establish what attractor, if any, is indeed observed after the bifurcation event. In two dimensions, further classification is possible; see Sec. 3.5 for details. A numerical simulation of the bifurcation diagram in this case is shown in Fig. 3.6(a) in which we see the transition from a stable fixed point A to what appears to be a chaotic attractor.

Case 2. Keeping all the parameters as in the previous case, we now vary just the element d_{12} of N_2 in (3.19) from -1.7 to the value -1.45. In this case, the eigenvalues of matrix N_1 are $\lambda_{11} = 0.89, \lambda_{12} = 0.56$, whereas those of matrix N_2 are $\lambda_{21} = -0.89, \lambda_{22} = -0.56$. Thus, using the notation introduced above, we have:

$$\sigma_1^- + \sigma_2^- = 0 \quad \text{(even)}, \quad \sigma_1^+ + \sigma_2^+ = 0 \quad \textit{(even)}.$$

Hence, according to the classification method, no period-doubling will occur at the border-collision. As $\sigma_1^+ + \sigma_2^+$ is even, we can have one of the following scenarios:

$$A \leftrightarrow B, \quad A \leftrightarrow b, \quad a \leftrightarrow b.$$

Now, since N_1 and N_2 have all eigenvalues inside the unit circle, the scenario we predict using Feigin's strategy is:

$$A \leftrightarrow B.$$

That is, at the border-collision, the admissible border-crossing fixed point $A \in S_1$ will move to the other side, changing continuously into the admissible stable fixed point $B \in S_2$. The numerical bifurcation diagram is shown in Fig. 3.6(b), which confirms this but illustrates the dramatic change in slope of the locus of fixed points versus parameter.

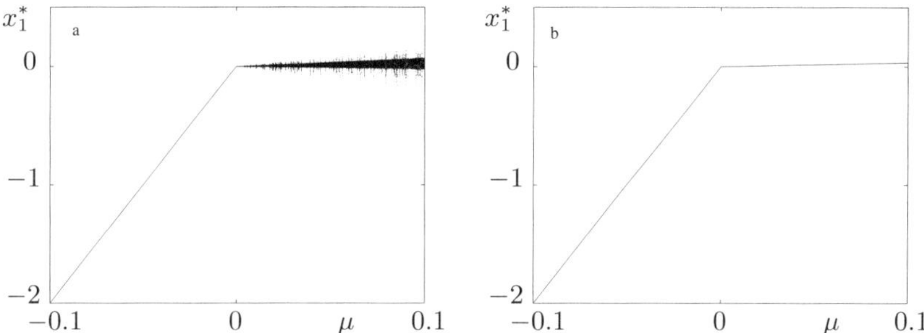

Fig. 3.6. Monte Carlo bifurcation diagrams for (a) case 1 and (b) case 2 of Example 3.1.

Example 3.2 (A three-dimensional example). To further illustrate the simplicity of using Theorem 3.1 for classification purposes, we move now to the case of a higher-dimensional map. Assume that the local mapping associated with a border-collision in a system of interest is derived to be

$$x \mapsto \begin{cases} N_1 x + B\mu, & \text{if } C^T x < 0, \\ N_2 x + B\mu, & \text{if } C^T x > 0, \end{cases} \tag{3.20}$$

where

$$N_1 = \begin{pmatrix} 0.6 & 0 & 1 \\ 0 & -0.2 & 0.8 \\ 0.5 & -0.8 & -0.2 \end{pmatrix}, \quad N_2 = \begin{pmatrix} -1 & 0 & 1 \\ 1.25\rho & -0.2 & 0.8 \\ 0.5 & -0.8 & -0.2 \end{pmatrix},$$

with ρ being a tunable parameter and

$$B = \begin{pmatrix} 1 \\ 0 \\ 0 \end{pmatrix}, \qquad C^T = \begin{pmatrix} 1 & 0 & 0 \end{pmatrix}.$$

Theorem 3.1 can now be used to determine the fixed and period-two points branching off the boundary fixed point at the border-collision. For example, using the characteristic polynomials for classification purposes, we have that in this case:

$$p_1(\lambda) = \lambda^3 - 0.2\lambda^2 - 0.06\lambda - 0.508,$$
$$p_2(\lambda) = \lambda^3 + 1.4\lambda^2 + 0.58\lambda + 0.58 + \rho.$$

Then, we have

$$p_1(1)p_2(1) = 0.826 + 0.232\rho,$$
$$p_1(-1)p_2(-1) = -0.6592 - 1.648\rho,$$

and, by varying ρ, we can expect to see different bifurcation scenarios at the border-collision. Specifically, we look at the following cases.

Case 1: $\rho = 0$. Here, condition (3.11) is satisfied, and therefore, the bifurcating fixed point will persist at the border-collision. As condition (3.17) is also satisfied, a period-two point will also branch off. The eigenvalues of the matrices N_1, N_2, determine the stability of the fixed points on either sides of the boundary, whereas the eigenvalues of $N_1 N_2$ will determine the stability of the period-two point. In this case, we find that N_1 and $N_1 N_2$ have all eigenvalues inside the unit circle, whereas N_2 is unstable. From such calculations we find the transition from a stable fixed point in region S_1 to an unstable one lying in region S_2 coexisting with a stable period-two point. This confirms the numerical observations in Fig. 3.1(a).

Case 2: $\rho = 4$. In this case, conditions (3.11) and (3.17) are still satisfied but now one of the eigenvalues of $N_1 N_2$ is outside the unit circle, thus we expect the transition from a stable fixed point to an unstable period-two point coexisting with an unstable fixed point. Thus, other stable attractors might exist past the bifurcation point. Indeed, as shown in Fig. 3.1(b), we observe the sudden transition from a fixed point to a stable chaotic attractor. (Conditions for the existence of chaos branching off the border-collision point will be given in Secs. 3.4 and 3.5 below.)

Case 3: $\rho = -0.8$. Now condition (3.17) is no longer satisfied; hence, we have the persistence scenario. As the eigenvalues of both N_1 and N_2 are now within the unit circle, the transition is from a stable fixed point in region S_1 to a stable fixed point in region S_2 as depicted in Fig. 3.1(c).

Case 4: $\rho = -3.57$. In this case condition (3.14) holds, whereas (3.17) is no longer satisfied. Hence we have a non-smooth fold at the border-collision. The eigenvalues of N_1 are within the unit circle, whereas the eigenvalues of N_2 are outside. Hence, we observe the transition from a stable admissible fixed point in region S_1 coexisting with an unstable fixed point in region S_2 to no admissible fixed or period-two point past the bifurcation point, as depicted in Fig. 3.1(d).

3.3 Equivalence of border-collision classification methods

This section provides a proof of Theorem 3.1, in particular focusing on the equivalence of the three alternative sets of conditions in the theorem statement. We start by showing that the normal form map (3.9) can be recast in a canonical form through an appropriate similarity transformation, which makes the analysis simpler.

3.3.1 Observer canonical form

Using a classical result from linear algebra, commonly exploited in control theory [240], it is possible to perform a further similarity transformation that puts the matrices N_1 and N_2 in a form that is more amenable for the classification of border-collision bifurcations. Note that the characteristic polynomials, and hence the eigenvalues, of all the matrices involved are invariant under such transformations.

Specifically, we suppose that the matrices $(I - N_i)$, $i = 1, 2$, are non-singular and the associated *observability matrices* defined as

$$O_i := \begin{pmatrix} C^T & C^T N_i & \dots & C^T N_i^{n-1} \end{pmatrix}^T \tag{3.21}$$

have full rank. Under these non-degeneracy conditions, we can perform a further co-ordinate transformation

$$\tilde{x} = Wx, \qquad \text{with} \qquad W = T_i O_i, \quad i = 1 \text{ or } 2, \tag{3.22}$$

where O_i is the observability matrix given by (3.21) and T_i is the matrix

$$T_i = \begin{pmatrix} 1 & 0 & \dots & 0 \\ d_{i1} & 1 & \dots & 0 \\ \vdots & \vdots & \dots & \vdots \\ d_{i(n-1)} & d_{i(n-2)} & \dots 1 & \end{pmatrix},$$

with d_{ij} being the coefficients of the characteristic polynomials of N_i given by:

$$p_i(\lambda) = \lambda^n + d_{i1}\lambda^{n-1} + \ldots + d_{i(n-1)}\lambda + d_{in}. \tag{3.23}$$

Under such a transformation, (3.7) can be put in the so-called *observer canonical form*

$$x \mapsto \begin{cases} \tilde{N}_1 x + \tilde{M}\mu, & \text{if } C^T x < 0, \\ \tilde{N}_2 x + \tilde{M}\mu, & \text{if } C^T x > 0, \end{cases} \tag{3.24}$$

where $\tilde{N}_i = W N_i W^{-1}, \tilde{M} = WM$ have the form:

$$\tilde{N}_i = \begin{pmatrix} -d_{i1} & 1 & 0 & \cdots & 0 \\ -d_{i2} & 0 & 1 & \cdots & 0 \\ \vdots & \vdots & \vdots & \ldots & \vdots \\ -d_{i(n-1)} & 0 & 0 & \cdots & 1 \\ -d_{in} & 0 & 0 & \cdots & 0 \end{pmatrix}, \qquad i = 1, 2, \tag{3.25}$$

$$\tilde{M} = (\, m_1 \quad m_2 \quad \ldots \quad m_n \,)^T, \qquad C^T = (\, 1 \quad 0 \quad \cdots \quad 0 \,).$$

The peculiarity of the co-ordinate transformation W is that, because matrices N_1 and N_2 satisfy (3.8) and $C^T = (\, 1 \quad 0 \quad \ldots \quad 0 \,)$, the transformation takes the same form for $C^T x > 0$ and $C^T x < 0$. That is we use the *same* transformation in regions S_1 and S_2. Henceforth in this section we shall drop the tildes on N_i, M when it is clear that the system is already in observer canonical form.

Remarks

1. The co-ordinate change described above is a generalization to n-dimensions of the one proposed by Nusse & Yorke [205] for the two-dimensional case. In two-dimensional it is possible to choose an appropriate co-ordinate transformation so that the choice of axis is independent of the parameter. In so doing, (3.7) becomes the normal form

$$\tilde{N}_1 = \begin{pmatrix} \tau_1 & 1 \\ \delta_1 & 0 \end{pmatrix}, \qquad \tilde{N}_2 = \begin{pmatrix} \tau_2 & 1 \\ \delta_2 & 0 \end{pmatrix}$$

and

$$\tilde{B} = \begin{pmatrix} 0 \\ 1 \end{pmatrix}, \qquad C^T = (\, 1 \quad 0 \,). \tag{3.26}$$

Here the parameters τ_i and δ_i, $i = 1, 2$ have the convenient interpretation as the traces and determinants, respectively, of the matrices N_i.
2. The normal form map given by (3.24) is particularly amenable for carrying out the classification according to Theorem 3.1 because the characteristic polynomials evaluated at $+1$ and -1, respectively, can easily read off from the first columns of N_1 and N_2 in (3.24). Specifically, the characteristic polynomials of these matrices are given by

$$p_i(\lambda) = \lambda^n - d_{i1}\lambda^{n-1} + \ldots - d_{i(n-1)}\lambda - d_{in},$$

where the $d_{i1}, d_{i2}, \ldots, d_{in}$ are just the elements of the first columns of the transformed matrices N_i. Thus, the quantities involved in the classification of border-collision in higher-dimensional maps required by Theorem 3.1 can be read off directly:

$$p_i(+1) = 1 - \sum_{k=1}^{n} d_{ik}, \tag{3.27}$$

and

$$p_i(-1) = (-1)^n - \sum_{k=1}^{n} (-1)^{n-k} d_{ik}. \tag{3.28}$$

For example, map (3.20) can be recast in the canonical form (3.24) by considering the change of variables (3.22) with

$$W = \begin{pmatrix} 1 & 0 & 0 \\ 0.4 & 0 & 1 \\ 0.68 & -0.8 & 0.2 \end{pmatrix}.$$

¡ This gives the transformed matrices of the normal form map:

$$\tilde{N}_1 = WN_1W^{-1} = \begin{pmatrix} 0.2 & 1 & 0 \\ 0.06 & 0 & 1 \\ 0.508 & 0 & 0 \end{pmatrix},$$

$$\tilde{N}_2 = WN_2W^{-1} = \begin{pmatrix} -1.4 & 1 & 0 \\ -0.58 & 0 & 1 \\ -0.58 - \rho & 0 & 0 \end{pmatrix}.$$

The characteristic polynomials of the two matrices can then be read off directly from their first columns.

Recall from linear algebra the definition of the adjoint of an $n \times n$ matrix A. Written adj(A), the adjoint is defined to be the transpose of the matrix of co-factors, where the co-factor of an element a_{ij} is $(-1)^{i+1}(-1)^{j+1}$ times the determinant of the $(n-1) \times (n-1)$ matrix obtained from A by removing the ith row and the jth column. The structure of the dynamical matrices N_1 and N_2 given by (3.25) together with the continuity condition (3.8) yield the following useful expressions involving the adjoint of $(I \pm N_i)$, where $i = 1, 2$.

Lemma 3.1. *Suppose* $E = \begin{pmatrix} e_1 & e_2 & \dots & e_n \end{pmatrix}^T$ *is a generic vector such that* $N_2 = N_1 + EC^T$ *then; we have*

$$C^T \text{adj}(I - N_1)E = C^T \text{adj}(I - N_2)E = \sum_{i=1}^{n} e_i := \eta \tag{3.29}$$

and

$$C^T \text{adj}(I + N_1)E = C^T \text{adj}(I + N_2)E = \sum_{i=1}^{n} (-1)^{i-1} e_i := \theta. \tag{3.30}$$

Moreover, the characteristic polynomials, $p_1(\lambda), p_2(\lambda)$ *of* N_1 *and* N_2 *verify the properties:*

$$p_2(\lambda) = p_1(\lambda) - \lambda^{n-1} e_1 - \lambda^{n-2} e_2 + \ldots - e_n, \tag{3.31}$$

$$p_2(1) = p_1(1) - \eta, \tag{3.32}$$

$$p_2(-1) = p_1(-1) - \theta. \tag{3.33}$$

Proof. Expressions (3.29) and (3.30) can be obtained immediately by observing that

$$I \pm N_i = \begin{pmatrix} 1 \mp d_{i1} & \pm 1 & 0 & \cdots & 0 \\ \mp d_{i2} & 1 & \pm 1 & \cdots & 0 \\ \vdots & \vdots & \vdots & \cdots & \vdots \\ \mp d_{i(n-1)} & 0 & 0 & \cdots & \pm 1 \\ d_{in} & 0 & 0 & \cdots & 1 \end{pmatrix}$$

Thus, the first row of the adjoint of the matrix $(I \pm N_i)$, given by $C^T \mathrm{adj}(I \pm N_i)$, is simply the vector $(1 \quad \mp 1 \quad 1 \quad \mp 1 \ldots)$ obtained by considering the cofactors of $(I \pm N_i)$, computed neglecting the first column. As this is the only column that differs between matrix N_1 and N_2, it also follows that $C^T \mathrm{adj}(I \pm N_1) = C^T \mathrm{adj}(I \pm N_2)$ as stated.

Moreover, we have

$$N_2 = \begin{pmatrix} -d_{11} + e_1 & 1 & 0 & \cdots & 0 \\ -d_{12} + e_2 & 0 & 1 & \cdots & 0 \\ \vdots & \vdots & \vdots & \cdots & \vdots \\ -d_{1(n-1)} + e_n & 0 & 0 & \cdots & 1 \\ d_{1n} & 0 & 0 & \cdots & 0 \end{pmatrix}.$$

Using (3.23) we find the characteristic polynomial to be

$$p_2(\lambda) = \lambda^n + (d_{11} - e_1)\lambda^{n-1} + \ldots + (d_{1n} - e_n).$$

Hence, (3.32) and (3.33) immediately follow.

3.3.2 Proof of Theorem 3.1

Without loss of generality, we adopt the observer canonical form (3.24) since the similarity transformation (3.22) from a general form (3.9) preserves characteristic polynomials.

We begin by considering conditions (3.10) and (3.13), the persistence and non-smooth fold scenarios, respectively. Let x_1^* and x_2^* be fixed points of the sub-mappings $\Pi_1 := N_1 x + M\mu$ and $\Pi_2 := N_2 x + M\mu$, respectively. That is,

$$x_1^* = N_1 x_1^* + M\mu, \quad C^T x_1^* < 0, \tag{3.34}$$
$$x_2^* = N_2 x_2^* + M\mu, \quad C^T x_2^* > 0. \tag{3.35}$$

Assuming $N_1 - I$ and $N_2 - I$ to be invertible, from (3.34) we find

$$x_1^* = (I - N_1)^{-1} M\mu. \tag{3.36}$$

Moreover, using (3.8), from (3.35), we get

$$x_2^* = N_1 x_2^* + M\mu + EC^T x_2^*,$$

or equivalently

$$x_2^* = (I - N_1)^{-1} M\mu + (I - N_1)^{-1} EC^T x_2^*. \tag{3.37}$$

Substituting (3.36) into (3.37), we then have

$$x_1^* = \left[I - (I - N_1)^{-1} EC^T\right] x_2^*. \tag{3.38}$$

Now, let $\delta_1 = C^T x_1^*$ and $\delta_2 = C^T x_2^*$. From (3.38) we obtain

$$\delta_1 = \left[1 - C^T (I - N_1)^{-1} E\right] \delta_2. \tag{3.39}$$

For each fixed point to be admissible, we then require $\delta_1 < 0$ and $\delta_2 > 0$. Thus, the two solutions will be admissible for opposite signs of μ (persistence) if

$$\left[1 - C^T (I - N_1)^{-1} E\right] > 0.$$

In contrast, the fixed points will be both of the same type—admissible or virtual—for one sign of μ (a non-smooth fold) if

$$\left[1 - C^T (I - N_1)^{-1} E\right] < 0,$$

as claimed by (3.10) and (3.13), respectively.

Using Lemma 3.1, it is simple to show the equivalence of these conditions to (3.11) and (3.14) given in terms of the characteristic polynomials of the matrices N_1 and N_2. In particular, we observe that

$$(I - N_1)^{-1} = \frac{\operatorname{adj}(I - N_1)}{\det(I - N_1)} = \frac{\operatorname{adj}(I - N_1)}{p_i(1)}$$

and

$$\det(I - N_1) = p_1(1).$$

Hence, we have

$$\begin{aligned} \left[1 - C^T (I - N_1)^{-1} E\right] &= 1 - \frac{C^T \operatorname{adj}(I - N_1) E}{p_1(1)}, \\ &= \frac{p_1(1) - \eta}{p_1(1)}. \end{aligned}$$

Now, using Lemma 3.1, we obtain

$$\left[1 - C^T (I - N_1)^{-1} E\right] = \frac{p_2(1)}{p_1(1)}.$$

Thus, checking the sign of $\left[1 - C^T (I - N_1)^{-1} E\right]$ yields the same result as checking the sign of $p_2(1)/p_1(1)$ or equivalently $p_1(1)p_2(1)$.

We now move onto condition (3.16). Suppose that a period-two point branches off the boundary equilibrium at the border-collision. The period-two solution involved in the bifurcation is characterized by two fixed points

x_1^*, x_2^* of the second iterate map $f^{(2)}$. Specifically $f(x_1^*) = x_2^*$ and $f(x_2^*) = x_1^*$. For such a period-two point to be admissible, we require that x_1^* and x_2^* be in opposite half-planes S_1 and S_2, respectively. Hence, we require

$$x_2^* = N_1 x_1^* + M\mu, \tag{3.40}$$

$$x_1^* = N_2 x_2^* + M\mu. \tag{3.41}$$

Using (3.8), from (3.41), we have

$$x_1^* = N_1 x_2^* + M\mu + EC^T x_2^*.$$

Moreover, from (3.40) we find

$$x_2^* - N_1 x_1^* = M\mu.$$

From simple algebraic manipulations, we thus obtain

$$x_1^* = \left[I + (I + N_1)^{-1} EC^T\right] x_2^*. \tag{3.42}$$

Now, letting $\delta_1 = C^T x_1^*$ and $\delta_2 = C^T x_2^*$, from (3.42) we have

$$\delta_1 = \left[1 + C^T (I + N_1)^{-1} E\right] \delta_2. \tag{3.43}$$

Thus, the period-two point will be admissible if δ_1 and δ_2 have opposite signs for the same value of μ; i.e.,

$$\left[1 + C^T (I + N_1)^{-1} E\right] < 0.$$

As before, we can recast this condition in terms of the characteristic polynomials of the matrices N_1 and N_2 as follows. Namely, simple linear algebra shows that

$$(I + N_1)^{-1} = -\frac{\mathrm{adj}(I + N_1)}{p_1(-1)}.$$

Thus, we get

$$1 + C^T (I + N_1)^{-1} E = 1 - \frac{C^T \mathrm{adj}(I + N_1) E}{p_1(-1)}.$$

Hence, using Lemma 3.1, we finally have

$$1 + C^T (I + N_1)^{-1} E = \frac{p_2(-1)}{p_1(-1)}.$$

Therefore we find that checking the sign of $1 + C^T (I + N_1)^{-1} E$ is equivalent to checking the sign of $p_1(-1)p_2(-1)$. Hence we have established the condition (3.17).

Finally, note that the three elementary conditions (3.11), (3.14) and (3.17) are written in terms of products of the characteristic polynomials of the matrices N_i evaluated at ± 1. We now show that it is possible to recast the same conditions in terms of the signs of eigenvalues of the matrices and hence prove conditions (3.12), (3.15) and (3.18). Specifically, let $\{\alpha_i\}_{i=1,2,...,n}$ and $\{\beta_i\}_{i=1,2,...,n}$ be the eigenspectra of N_1 and N_2, respectively; then, by definition,

$$p_1(\lambda) = (\lambda - \alpha_1)(\lambda - \alpha_2)...(\lambda - \alpha_n),$$
$$p_2(\lambda) = (\lambda - \beta_1)(\lambda - \beta_2)...(\lambda - \beta_n).$$

Consider now the sign of $p_1(1)$. The first thing to note is that pairs of complex conjugate eigenvalues, $\eta \pm j\gamma$ will always generate quadratic factors in $p_1(1)$ of the form $(1-\eta)^2 + \gamma^2$ that have a positive sign. Therefore, the overall signs of $p_1(1)$ [and, of course, $p_2(1)$] will depend solely on the number of real eigenvalues α_i (respectively, β_i) greater than 1.

In fact, a similar argument to that above shows that any complex roots contribute a positive factor of $p_i(\lambda)$ for any real λ. Therefore, the signs of $p_1(-1)$ and $p_2(-1)$ similarly depend on the number of real eigenvalues less than -1.

The three conditions (3.11), (3.14), and (3.17) then imply

1. $p_1(1)p_2(1) > 0 \Leftrightarrow \sigma_1^+ + \sigma_2^+$ is even (persistence);
2. $p_1(1)p_2(1) < 0 \Leftrightarrow \sigma_1^+ + \sigma_2^+$ is odd (non-smooth fold);
3. $p_1(-1)p_2(-1) < 0 \Leftrightarrow \sigma_1^- + \sigma_2^-$ is odd (non-smooth period-doubling).

3.4 One-dimensional piecewise-linear maps

The border-collision bifurcation scenarios we have seen so far involve fixed points and period-two points of n-dimensional PWS maps. In general, higher periodic attractors or chaos might also branch off a boundary fixed point at a border-collision. For example, some of the scenarios listed in Sec. 3.2.2 (e.g., $A \leftrightarrow b$ and $A, b \leftrightarrow \varnothing$) predict the absence of any attracting fixed point after the bifurcation. In this case, we need to investigate if other types of attractors (e.g., higher-period periodic points or chaos) can be born at the border-collision. To address this issue, the next stage in Feigin's method is to investigate possible period-two orbits AB or ab. In principle this could be continued up to period-N points for arbitrary N. Unfortunately, given the n-dimensional nature of the strategy, it is algebraic cumbersome to go beyond period-two points. A more complete classification can instead be given in the case of one- and two-dimensional maps. We start with the case of a simple one-dimensional piecewise-linear map. The results presented here summarize and extend those presented in [205, 206, 207, 21, 80, 16, 115, 204, 90].

We study a simple normal form for one-dimensional piecewise-linear maps, depending on three real parameters. We show how the simple classification strategy presented above can be extended to reveal a complete understanding of the bifurcation structure in the parameter plane, including the behavior of all possible periodic orbits. New effects include the instantaneous generation

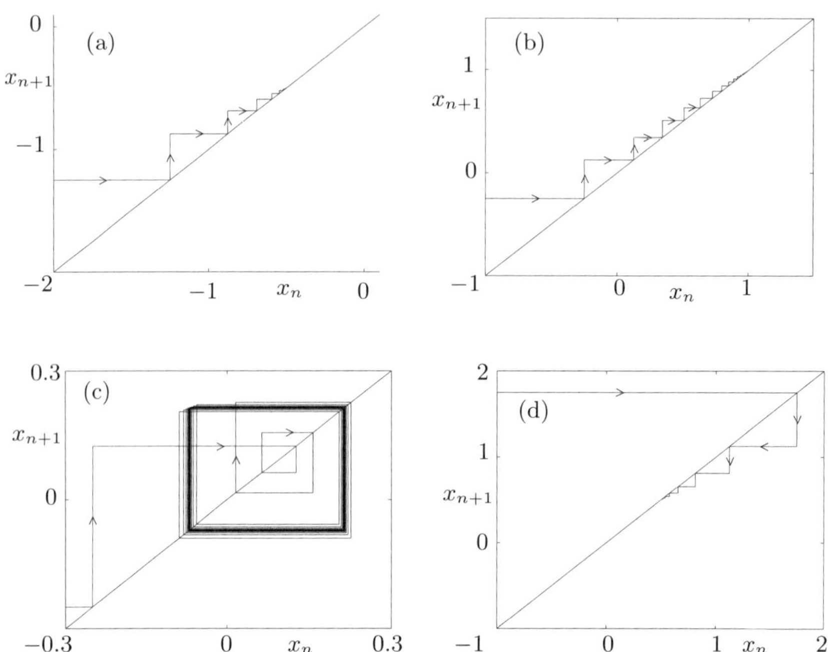

Fig. 3.7. Cobweb diagrams depicting iterates of the map (3.44) for (a) $\mu = -0.25$, $\nu_1 = 0.5$, $nu_2 = 1.5$; (b) $\mu = 0.25$, $\nu_1 = 0.5$, $\nu_2 = 0.75$; (c) $\mu = 0.25$, $\nu_1 = 0.5$, $\nu_2 = -1.5$; and (d) $\mu = 0.25$ $\nu_1 = -1.5$, $\nu_2 = 0.5$.

of an arbitrary number of orbits of increasing period from a single stable solution, and a mechanism for the sudden jump to a *robust* chaotic attractor that has often been associated with piecewise-smooth systems, such as the case study examples introduced in Chapter 1.

Consider a mapping from $\mathbb{R}$ to $\mathbb{R}$, depending on three real parameters μ, ν_1, ν_2:

$$x \mapsto \begin{cases} \nu_1 x + \mu, & \text{if } x \leq 0, \\ \nu_2 x + \mu, & \text{if } x > 0, \end{cases} \tag{3.44}$$

which is such that, as μ passes through zero, the fixed point $x = 0$ undergoes a border-collision. Note that this map is written in the most general form (3.9) in one dimension. A simple rescaling $x \to |\mu| x$, shows that without loss of generality we may assume $\mu = \pm 1$ if μ is non-zero. However, since we are interested in the passage of μ through zero, we shall not apply such a transformation explicitly, but remember that any dynamics for a particular sign of μ will be topologically the same no matter what is the magnitude of μ. Figure 3.7 depicts the dynamics of the map for negative μ for different values of ν_1 and ν_2. Note from panels (a) and (b) that, if ν_1 and ν_2 have the same sign, then the dynamics must remain somewhat trivial. That is, the only form of allowed attractor is a stable fixed point. In particular no period-m points can exist for $m > 1$. Hence in what follows we shall focus exclusively on the case that ν_1 and ν_2 have opposite signs. Now, there is a further transformation under which the map (3.44) is invariant:

$$x \to -x, \quad \mu \to -\mu, \quad S_1 \leftrightarrow S_2.$$

In fact, Fig. 3.7(c),(d) depicts the dynamics of the map for two cases that are mapped into each other under this transformation. Thus, we can assume without loss of generality that

$$\nu_1 > 0, \qquad \nu_2 < 0.$$

3.4.1 Periodic orbits of the map

We start our analysis by seeking the domains of existence and stability of fixed and periodic points. For convenience, we define the two submappings

$$\Pi_1 : x \mapsto \nu_1 x + \mu, \quad \text{if } x \leq 0, \tag{3.45}$$

$$\Pi_2 : x \mapsto \nu_2 x + \mu, \quad \text{if } x > 0. \tag{3.46}$$

Consider first simple fixed points of the map. Clearly the map Π_1 has fixed point

$$x_1^* = \frac{\mu}{1-\nu_1},$$

which to be admissible must satisfy $x_1^* < 0$. So an unstable a mode exists for $\mu > 0$ if $\nu_1 > 1$, and a stable A for $\mu < 0$ if $\nu_1 < 1$. Similarly, the map Π_2 has the simple fixed point given by

$$x_2^* = \frac{\mu}{1-\nu_2},$$

which must be positive to be admissible. So the B/b mode exists only for $\mu > 0$. In addition, the B/b mode has eigenvalue ν_2 and hence is stable (B) for $\nu_2 > -1$ and unstable (b) for $\nu_2 < -1$. From these calculations, it is easy to see that (in the notation of the previous section) that if $\nu_1 > 1$, then $\sigma_1^+ + \sigma_2^+ = 1$, so A/a and B/b undergo a non-smooth fold. Alternatively, if $\nu_1 < 1$, then $\sigma_1^+ + \sigma_1^+ = 0$, and so there is persistence; i.e., a continuous transition from A/a to B/b. In addition, for $\nu_1 < 1$ and $\nu_2 < -1$, we find $\sigma_1^- + \sigma_2^- = 1$, which implies that a period-doubled mode AB/ab branches off at $\mu = 0$. In contrast, for $\nu_2 > -1$ there is no period doubling, since $\sigma_1^- + \sigma_2^- = 0$. So, for $\nu_2 > -1$, the simplest border-collision bifurcation scenarios are

$$\nu_1 < 1, \nu_2 > -1 \Rightarrow A \leftrightarrow B,$$

$$\nu_1 > 1, \nu_2 > -1 \Rightarrow \varnothing \leftrightarrow a, B.$$

Now consider a general periodic orbit. Such an orbits can be of numerous different types, which are characterized by different numbers of iterations lying in regions S_1 and S_2, respectively. Extending the notation from the previous section, we will label such a periodic point by a concatenation of arbitrary strings of symbols of A and B (or a and b if unstable), e.g., $AABBA$ or $aabba$, which means a particular period-five orbit with $x \in S_1$, $f(x) \in S_1$, $f^{(2)}(x) \in S_2$, $f^{(3)}(x) \in S_2$, $f^{(4)}(x) \in S_1$ and $f^{(5)}(x) = x$. We shall in what follows keep track of all possible periodic orbits of the form A^iB (or a^ib for an unstable orbit) with i iterates in region S_1 followed by just one in S_2 before repeating periodically. We shall show shortly that only orbits of this type may ever be stable, given our assumption on the signs of ν_1 and ν_2. We will show that there exist $\nu_{1,2}$-values that give existence of border-collision bifurcation scenarios such as

$$
\begin{aligned}
A &\leftrightarrow b, ab, \ldots, a^{k-2}b, A^{k-1}B, \\
A &\leftrightarrow b, ab, \ldots, a^{k-2}b, a^{k-1}b, \\
\varnothing &\leftrightarrow a, b, ab, \ldots, a^{k-1}b,
\end{aligned}
$$

for any $k \geqslant 2$. Specifically we have

Theorem 3.2 ([80]). *The $A^{k-1}B/a^{k-1}b$ mode exists for $\mu > 0$ if and only if*

$$
\nu_2 < \frac{1 - \nu_1^{k-1}}{\nu_1^{k-1} - \nu_1^{k-2}} := \psi_k(\nu_1). \tag{3.47}
$$

Moreover, $A^{k-1}B/a^{k-1}b$ is stable if and only if

$$
\nu_2 > -\frac{1}{\nu_1^{k-1}} =: \phi_k(\nu_1). \tag{3.48}
$$

Proof. Clearly the $A^{k-1}B/a^{k-1}b$ mode exists if and only if there is a periodic sequence

$$
x_1^\star \geqslant 0, \quad x_2^\star \leqslant 0, \quad \ldots, \quad x_k^\star \leqslant 0
$$

of the map (3.45),(3.46), where

$$
x_2^\star = \nu_2 x_1^\star + \mu, \tag{3.49}
$$

$$
x_j^\star = \nu_1 x_{j-1}^\star + \mu, \qquad \text{for all } 3 \leqslant j \leqslant k, \tag{3.50}
$$

$$
x_1^\star = \nu_1 x_k^\star + \mu. \tag{3.51}
$$

Suppose that $x_k^\star \leqslant 0$; then we have

$$
x_{k-1}^\star = \frac{x_k^\star - \mu}{\nu_1} < 0,
$$

since $x_k^\star$ is negative and μ and ν_1 are both positive. Thus, by induction $x_j^\star < 0$ for all $2 \leqslant j \leqslant k$. In addition

$$
x_1^\star = \frac{x_2^\star - \mu}{\nu_2} > 0
$$

as $x_2^\star < 0$, $\mu > 0$ and $\nu_2 < 0$.

We may find $x_k^\star$ by backwards and forward substitution of (3.49), (3.50) and (3.51). First we obtain an expression for $x_1^\star$:

$$x_1^\star = \frac{1 + \nu_1 + \ldots + \nu_1^{k-1}}{1 - \nu_1^{k-1}\nu_2}\mu,$$

and then an expression for $x_k^\star$:

$$x_k^\star = \frac{1 + \nu_1 + \ldots + \nu_1^{k-2} + \nu_1^{k-2}\nu_2}{1 - \nu_1^{k-1}\nu_2}\mu. \tag{3.52}$$

So, since $\mu > 0$ and $1 - \nu_1^{k-1}\nu_2 > 0$, we have $x_k^\star < 0$ if and only if the numerator of equation (3.52) is negative; that is,

$$\nu_2 < -\left(1 + \frac{1}{\nu_1} + \frac{1}{\nu_1^2} + \ldots \frac{1}{\nu_1^{k-2}}\right).$$

Note that, on summing the geometric progression, we have

$$\begin{aligned} -\left(1 + \frac{1}{\nu_1} + \frac{1}{\nu_1^2} + \ldots \frac{1}{\nu_1^{k-2}}\right) &= -\frac{1 + \nu_1^{1-k}}{1 - \nu_1^{-1}} \\ &= \frac{-\nu_1^{k-1} + 1}{\nu_1^{k-1} - \nu_1^{k-2}} \\ &= \psi_k(\nu_1). \end{aligned}$$

Hence we obtain (3.47).

The eigenvalue of the $A^{k-1}B/a^{k-1}b$ orbit is

$$\nu_1^{k-1}\nu_2$$

and so $A^{k-1}B/a^{k-1}b$ is stable if and only if

$$\nu_2 > -\frac{1}{\nu_1^{k-1}} =: \phi_k(\nu_1)$$

3.4.2 Bifurcations between higher modes

Let us consider the two conditions (3.47) and (3.48) for the existence and stability of the $A^{k-1}B/a^{k-1}b$ mode. We shall be interested only in $\nu_2 < -1$, since there is no period doubling for $\nu_2 > -1$. Recall that $A^{k-1}B/a^{k-1}b$:

exists for $\mu > 0$ if and only if $\nu_2 < \psi_k(\nu_1)$;
is stable for $\mu > 0$ if and only if $\nu_2 > \phi_k(\nu_1)$.

The existence of the $A^{k-1}B/a^{k-1}b$ mode implies that all of $a^j b$ for $1 \leqslant j \leqslant k-2$ exist and are unstable, since, for all $1 \leqslant j \leqslant k-1$

$$\nu_2 < \psi_k(\nu_1) < \psi_j(\nu_1) \Rightarrow A^j B/a^j b \text{ exists for } \mu > 0,$$
$$\nu_2 < \psi_k(\nu_1) < \phi_j(\nu_1) \Rightarrow a^{j-1}b \text{ is unstable.}$$

Therefore, we can classify the following border-collision scenarios in the (ν_1, ν_2)-plane as depicted in Fig. 3.8:

Case 1 $\nu_1 > 1$ **and** $\nu_2 < -1$. In this case all modes are unstable. Modes a and b both exist for $\mu > 0$. If $\psi_k(\nu_1) > \nu_2 > \psi_{k+1}(\nu_1)$, $a^{k-1}b$ exists for $\mu > 0$, and is the highest such mode to exist, and so the simplest possible bifurcation structure is

$$\varnothing \leftrightarrow a, b, ab, \ldots, a^{k-1}b.$$

Thus, as ν_2 decreases for fixed ν_1, we see a sequence of increasingly complex border-collision scenarios such as:

$$\varnothing \leftrightarrow a, b, ab,$$
$$\varnothing \leftrightarrow a, b, ab, a^2b,$$
$$\varnothing \leftrightarrow a, b, ab, a^2b, a^3b,$$
$$\vdots$$

Case 2 $\nu_1 < 1$ **and** $\nu_2 < -1$ Here A/a exists for $\mu < 0$ and is stable, and B/b is unstable and exists for $\mu > 0$ together with higher periodic modes. The highest periodic mode in any region may be either stable or unstable, depending on whether condition (3.48) is verified. To better understand this condition, consider the function $\phi_k(\nu_1) - \psi_k(\nu_1)$, which is monotone increasing and equal to $-\infty$ at $\nu_1 = 0$ and $k-2$ at $\nu_1 = 1$. Thus $\phi_k(\nu_1) - \psi_k(\nu_1)$ has a unique zero $\nu_{1k} \in (0, 1)$. In addition, $\psi_{k+1}(\nu_1) < \phi_k(\nu_1)$. Hence, as ν_2 is decreased from -1, for $0 < \nu_1 < \nu_{1k}$, the (stable) $A^{k-1}B$ mode is produced ($\nu_2 = \psi_k(\nu_1) > \phi_k(\nu_1)$), then becomes unstable ($\nu_2 = \phi_k(\nu_1) > \psi_{k+1}(\nu_1)$), followed by the generation of the A^kB/a^kb mode ($\nu_2 = \psi_{k+1}(\nu_1)$). Alternatively, if $\nu_{1k} < \nu_1 < 1$, the unstable $a^{k-1}b$ mode is produced without the prior existence of the stable mode. Thus, as ν_2 is decreased, we see bifurcation structures such as

$$A \leftrightarrow b, AB,$$
$$A \leftrightarrow b, ab,$$
$$A \leftrightarrow b, ab, A^2B,$$
$$A \leftrightarrow b, ab, a^2b,$$
$$A \leftrightarrow b, ab, A^3B,$$
$$A \leftrightarrow b, ab, a^3b,$$
$$\vdots$$

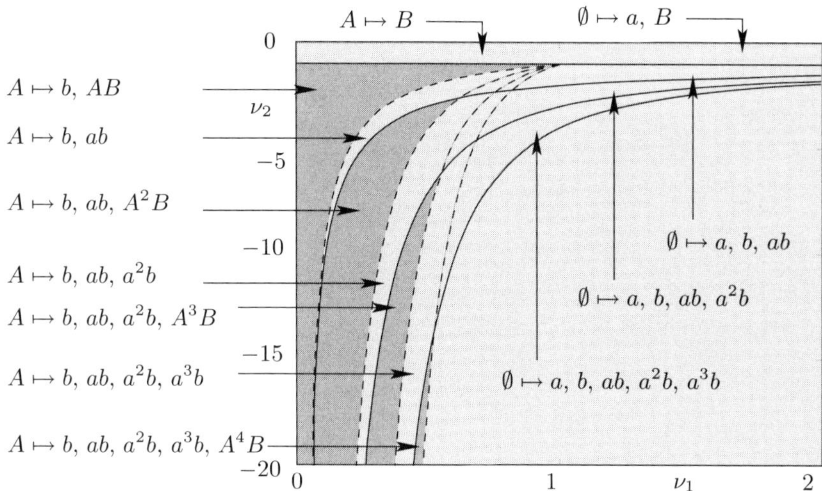

Fig. 3.8. Simplest possible bifurcation structure of the map (3.44) in each region of parameter space. Solid lines denote ψ_k (existence boundaries, which are called f_k in the figure) and dashed lines denote ϕ_k (stability boundaries). Regions in which the highest periodic mode is stable are shaded.

The question remains as to what happens between the regions where stable periodic orbits exist. Figure 3.9 shows bifurcation diagrams of the map at different parameter values, showing different bifurcation scenarios associated with the border-collision occurring at $\mu = 0$. We observe that when the classification strategy predicts the absence of any stable attractor past the bifurcation point, the sudden transition to a stable chaotic attractor is observed as shown in Fig. 3.9(d). The existence of such an attractor can be proved, and its features can be further characterized in the case of one-dimensional maps as is shown in the next section.

The subtle changes in the attractor for $\mu > 0$ as ν_1 and ν_2 vary are better illustrated by the bifurcation diagram in Fig. 3.10 obtained by fixing $\nu_1 = 0.4$ and decreasing ν_2 for $\mu > 0$. As expected, we can observe orbits of increasing periodicity interleaved with regions of chaotic motion.

3.4.3 Robust chaos

Let us focus on the case $0 < \nu_1 < 1$, $\nu_2 < -1$. Under these conditions, we expect the scenario $A \leftrightarrow b, AB/ab$ at the border-collision. Figure 3.8 shows that upon decreasing ν_2, we observe bifurcation scenarios of increasing complexity associated with the generation of stable higher-periodic points. In particular, the shaded regions in Fig. 3.8 denote areas of parameter space where stable A^jB solutions with $j = 1, 2, 3, \ldots, n$ exist. Between these shaded regions we numerically observe chaotic dynamics for $\mu > 0$.

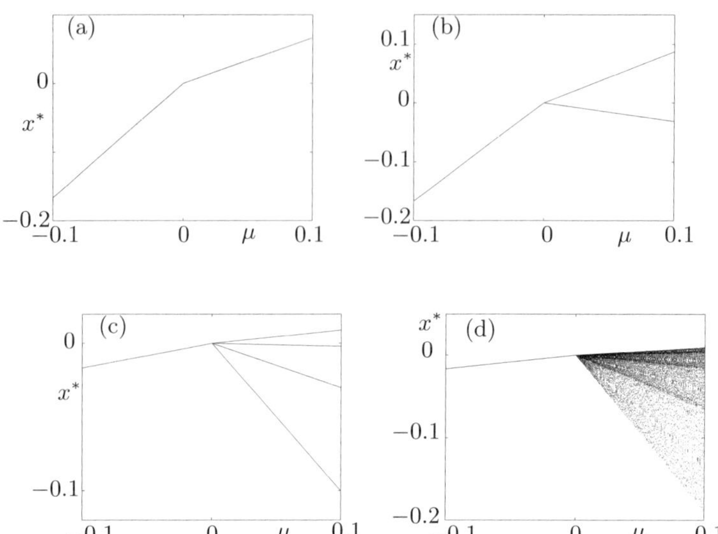

Fig. 3.9. Monte Carlo bifurcation diagrams of the piecewise-linear map for $\nu_1 = 0.4$ and (a) $\nu_2 = -0.5$; (b) $\nu_2 = -1.5$; (c) $\nu_2 = -12$; (d) $\nu_2 = -20$ showing the occurrence of border-collisions leading to the formation of orbits of type $A^{k-1}B$ and chaos. In particular, we observe the following scenarios: (a) $A \leftrightarrow B$; (b) $A \leftrightarrow b, AB$; (c) $A \leftrightarrow b, ab, a^2b, A^3B$; (d) $A \leftrightarrow b, ab, a^2b, a^3b, \ldots$.

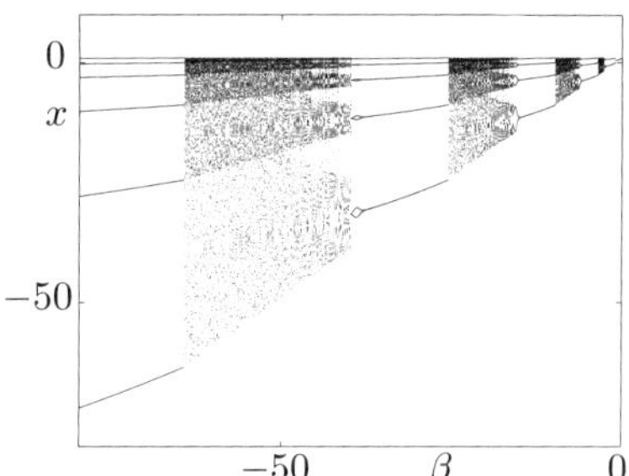

Fig. 3.10. Monte Carlo bifurcation diagram of (1.33) for $\mu = 1$, $\nu_1 = 0.4$ and $\nu_2 \in (-80, 0)$.

The following theorem for the simplifying case of one-dimensional maps has an elementary proof.

Theorem 3.3 (robust chaos in one-dimensional piecewise-smooth maps). *Consider the map (3.44) for* $0 < \nu_1 < 1$, $\nu_2 < -1$. *Suppose that*

$$\psi_{k+1}(\nu_1) < \nu_2 < \psi_k(\nu_1) \quad \textit{and} \quad \nu_2 > \phi_k(\nu_1), \tag{3.53}$$

for some $k > 2$, *where* ψ_k *and* ϕ_k *were defined in (3.47) and (3.48). Then there exists a chaotic attractor for all* $\mu > 0$. *Moreover the attractor is* robust *in the sense that there exist no stable periodic orbits for any* $\mu > 0$ *or any* ν_1 *and* ν_2 *values satisfying the bounds (3.53).*

Proof. We shall explain the existence of chaos for $\mu > 0$ by proceeding in two stages. First, we show that there must exist an attractor. Then, we argue that such an attractor cannot be a stable periodic orbit so that a chaotic attractor is the only possibility. This argument applies uniformly *throughout* the parameter region (3.53); hence, we have the stated robustness. Note that robustness with respect to μ is trivial because the dynamics is scale invariant; that is, invariant under a rescaling $x \to x/\mu$ replacing μ by 1 in the map (3.44). Hence all dynamics for $\mu > 0$ can be trivially related to that for $\mu = 1$. Without loss of generality, we then suppose $\mu = 1$ in what follows.

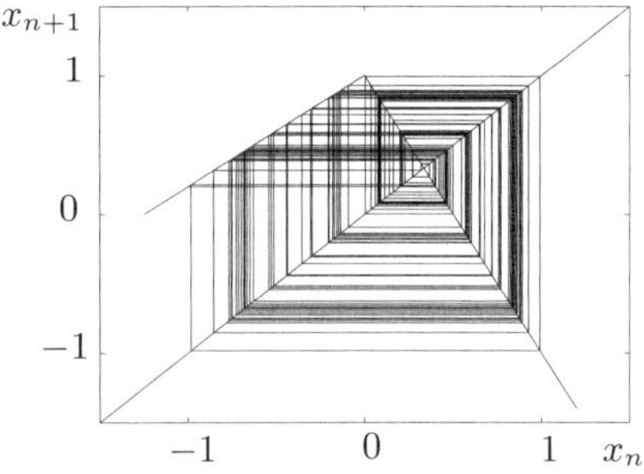

Fig. 3.11. Cobweb diagram showing the iterates of the map (3.44) for $\mu = 1$, $\nu_1 = 0.8$ and $\nu_2 = -2$, which parameters satisfy the hypothesis of Theorem 3.3, and the existence of the trapping region.

To show **existence of an attractor**, consider the dynamics of (3.44) for $\mu > 0$; see Fig. 3.11. Suppose x is large and positive ($x > 1$). Clearly, as $\nu_2 < -1$, such an initial condition gets mapped in one iteration to a point $x_1 < (1 + \nu_2)$ in region S_1. The condition that $\nu_2 < \psi_k(\nu_1)$ implies that such a point gets mapped under Π_1 to a further point in S_1, and since $0 < \nu_1 < 1$, further iterations of x then decrease under the action of Π_1. Therefore, finitely many further iterates of map Π_1 keep x inside S_1 until an iterate gets mapped into a point within the range $1/\nu_1 < x \leq 0$. In the next iterate, such a point gets mapped into S_2 at a point x_2 that is bounded below by 1.

Now, either (with probability zero) x_2 corresponds with the fixed point b, i.e., $x_2 = 1/(1 - \nu_2)$—in which case, all further iterates remain there for all time—or we reach a point with $-1/\nu_2 < x < 1$ (possibly after finitely many further iterates inside S_2). This point now gets mapped into S_1 again with $x > (1 + \nu_2)$. Further application of the above arguments show that we now have that $(1 + \nu_2) \leq x \leq 1$ for all further iterates of the map.

Similar arguments apply to any initial condition with $x < (1+\nu_2)$. Therefore we have shown that the interval $I := \{x : (1 + \nu_2) \leq x \leq 1\}$ is a trapping region, which all initial conditions must eventually enter and remain for all time. Namely, successive iterations of the map generate a bounded sequence, $\{f^n(x)\}_{n=1}^{\infty}$, and therefore there must be an attractor. That is, there must be points in I that are accumulation points of this sequence.

To show **existence of chaos**, we can appeal to the fact that the only attractors that can exist in one-dimensional maps are stable periodic orbits or chaotic attractors. Other possibilities, such as quasi-periodic motion (invariant circles) require at least two dimensions (or the state space to be a circle). So, if we can show that there cannot be a stable periodic orbit, we have proved that there must be a chaotic attractor. Let us consider possible stable periodic sequences. Now since the map is linear in S_1 we have that the multiplier associated with an iterate in this region is always ν_1. Similarly the multiplier associated with an iterate in S_2 is ν_2. So if we have a periodic sequence composed of a finite number of symbols 1 and 2 (e.g., a period-five orbit 11122), we can immediately calculate its multiplier as $\nu_1^{k_1}\nu_2^{k_2}$, where k_1 is the number of iterates within S_1 and k_2 the number within S_2 (e.g. $\nu_1^3\nu_2^2$ for the above period-five orbit). Now, the inequality, $\nu_2 < \phi_k(\nu_1) = -\nu_1^{-k+1}$, implies we must have

$$k_1 > (k-1)k_2 \tag{3.54}$$

for the multiplier to be inside the unit circle and hence for the periodic orbit to be stable. We shall now show that any such periodic orbit cannot exist given the inequality $\psi_{k+1}(\nu_1) < \nu_2$.

To show this, consider a possible periodic sequence that starts at a point x_2 within $I \cap S_2/b$. That is, $-1 < x_2 < 0$. This is without loss of generality since any periodic orbit must visit region S_2. Supposing that $x_2 \neq 1/(1-\nu_2)$, the fixed point b, we have that either immediately or after finitely many further iterates within S_2 such a point is mapped to S_1. Clearly, the maximum number of iterates that can now be spent within S_2 is determined by the image of the point $x_2 = -1$. After $k-1$ iterates, we have $f^{k-1}(-1) = (1+\nu_1(1+\nu_1(\ldots\nu_1(1+\nu_2))))$ where ν_1 appears $k-2$ times. The inequality $\nu_2 < \psi_k(\nu_1)$ implies that $f^{k-1}(-1) < 0$. However, the inequality $\nu_2 > \psi_{k+1}(\nu_1)$ similarly implies that $f^k(-1) > 0$. Hence $(k-1)$ is the maximum number of iterates that can be spent inside S_2. Thus we have shown that for any periodic sequence for *each* iterate in S_2 we can have at most $(k-1)$ iterates in S_1. That is, $k_1 \leq (k-1)k_2$. Thus we have shown that (3.54) cannot hold for any periodic orbit, and hence there cannot be any stable periodic orbit.

Remarks

1. A casual glance at Fig. 3.10 suggests a contradiction to this theorem. It seems that upon reducing ν_2, chaotic attractors are destroyed, although some kind of sudden *crisis* as the stable period-k orbit is created (upon crossing curves ψ_k). However, further reduction of ψ_k results in the loss of

stability of this orbit, upon crossing ϕ_k, in what appears to be a period-doubling type bifurcation. This would suggest a bifurcation on varying ν_2 of $A^{k-1}B \to a^{k-1}b, A^{k-1}BA^{k-1}B$, where the latter stable sequence is a period-$2k$ orbit. However, this would contradict the above theorem. In fact, such a period-k orbit cannot exist, let alone be stable. This is because the map, when confined to orbit sequences that are repetitions of $1^{k-1}2$ (using the notation of the proof), is completely linear. Hence there cannot exist any period-$2k$ solution with this sequence.
Even if such an orbit did exist, it must be unstable owing to the inequality $\nu_2 < \psi_k$. However, note that as ν_2 approaches ψ_k from above, the multiplier of the $A^{k-1}B$ orbit approaches -1. Hence we should expect long transients of a flip-flop nature (like a period-doubled motion around the orbit). These transients get longer and longer as ν_2 approaches ψ_k. For ν_2 just greater than ψ_k, the orbit becomes weakly unstable, with the transients causing slow growth towards the chaotic attractor. This is what we are seeing in the Monte Carlo bifurcation diagram in Fig. 3.10. The 'ghost' of the period-doubling is due to the finite length of the transient data that is thrown away before the points are plotted. In contrast, the periodic orbits born at the ends of the chaotic regions in the plot are born in border-collision type events, and so when they occur they have a multiplier that is bounded away from ± 1. Hence, in this case, we do not see the effects brought about by transients.

2. bifurcation diagrams showing border-collision bifurcations leading to the sudden onset of chaos in parameter regions described by the theorem are depicted in Fig. 3.12. We first consider the scenario depicted in Fig. 3.12(a),(c). Here $\nu_1 = 0.5$. Then, to obey the inequalities in the above theorem for $k = 3$, from (3.47) and (3.48), we get $\psi_3 = -3$ and $\phi_3 = -2$. Thus, to be in the chaotic region, we choose $-3 < \nu_2 < -2$. The bifurcation diagram shown in Fig. 3.12 (c) was obtained for $\nu_2 = -2.2$ and in (a) for $\nu_2 = -2.8$. Note that for the lower-ν_2 value the attractor appears to be arranged around $2(k-1)$ pieces. In this case 4. For higher ν_2 values within the band, the four pieces have merged into a connected chaotic attractor. For further increase of μ within this chaotic region, this becomes a six-piece chaotic attractor, as we get the pseudo period-doubling cascade as mentioned in remark 1, above. Similarly in Fig. 3.12(b) and (d) we depict one- and six-piece chaotic attractors obtained within the band of chaos for $k = 4$. More details of the transitions to chaos due to border-collision bifurcations in one-dimensional piecewise-smooth maps can be found in [246, 247].

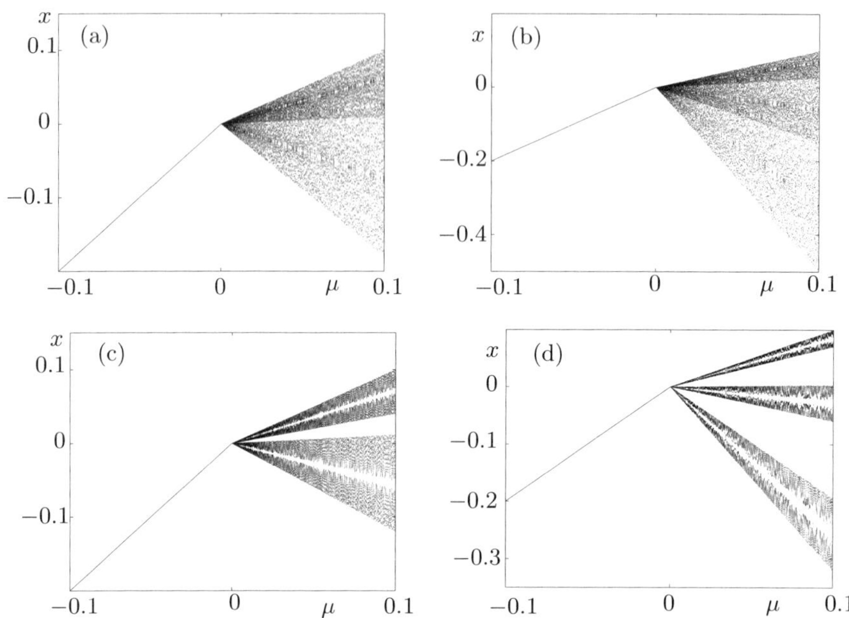

Fig. 3.12. Bifurcation diagrams for $\nu_1 = 0.5$ representing border-collisions to chaotic attractors that (a,b) have a single connected component, or (c,d) are organised in four or six separate pieces. Values of ν_2 are, respectively, -2.8, -6, -2.2 and -4.2

3.5 Two-dimensional piecewise-linear normal form maps

We consider now border-collision bifurcations of hyperbolic fixed points in planar piecewise-linear maps. Here there is the possibility of yet more intricate behavior emerging from a border-collision point; see [21, 205, 22, 80].

As mentioned in Sec. 3.1.3, under appropriate non-degeneracy assumptions, any locally piecewise-linear continuous map can be transformed using linear changes of co-ordinates into normal form (3.24). In the planar case, this normal form can be written as

$$\begin{pmatrix} x_1 \\ x_2 \end{pmatrix} \mapsto \begin{cases} \begin{pmatrix} \tau_1 & 1 \\ -\delta_1 & 0 \end{pmatrix} \begin{pmatrix} x_1 \\ x_2 \end{pmatrix} + \begin{pmatrix} m_1 \\ m_2 \end{pmatrix} \mu, & \text{if } x_1 \le 0, \\ \begin{pmatrix} \tau_2 & 1 \\ -\delta_2 & 0 \end{pmatrix} \begin{pmatrix} x_1 \\ x_2 \end{pmatrix} + \begin{pmatrix} m_1 \\ m_2 \end{pmatrix} \mu, & \text{if } x_1 > 0, \end{cases} \tag{3.55}$$

where we have changed notation to emphasize that τ_i is the *trace* (sum of the diagonal terms) of the matrix N_i, and δ_i is its determinant. Moreover, it is possible to make a further change of co-ordinates to normalize the coefficients m_1 and m_2. In particular, setting $\tilde{x}_1 = x_1$, $\tilde{x}_2 = x_2 - m_2\mu$ and $\tilde{\mu} = \mu(m_1 + m_2)$, dropping the tildes we obtain

$$\begin{pmatrix} x_1 \\ x_2 \end{pmatrix} \mapsto \begin{cases} \begin{pmatrix} \tau_1 & 1 \\ -\delta_1 & 0 \end{pmatrix} \begin{pmatrix} x_1 \\ x_2 \end{pmatrix} + \begin{pmatrix} 1 \\ 0 \end{pmatrix} \mu, & \text{if } x_1 \leq 0, \\ \begin{pmatrix} \tau_2 & 1 \\ -\delta_2 & 0 \end{pmatrix} \begin{pmatrix} x_1 \\ x_2 \end{pmatrix} + \begin{pmatrix} 1 \\ 0 \end{pmatrix} \mu, & \text{if } x_1 > 0. \end{cases} \tag{3.56}$$

Note that, because traces and determinants are invariant under the coordinate transformation, it is enough to calculate these quantities for the map of interest *without* putting it into observer-canonical form and then apply the classification of border-collisions that we present below.

In what follows, as is often the case in applications, we restrict attention to dissipative systems, for which δ_1 and δ_2 are both positive and less than unity. A similar analysis can be carried out to classify the possible border-collision scenarios if this is not the case (see, for example, [21] for the case of negative δ_1 and δ_2).

3.5.1 Border-collision scenarios

We start with the classification of the simplest bifurcation scenarios; those involving fixed and period-two points. The fixed points of the map (3.56) are

$$A/a = \left[\frac{\mu}{1 - \tau_1 + \delta_1}, \; -\frac{\delta_1 \mu}{1 - \tau_1 + \delta_1} \right]$$

and

$$B/b = \left[\frac{\mu}{1 - \tau_2 + \delta_2}, \; -\frac{\delta_2 \mu}{1 - \tau_2 + \delta_2} \right] \quad \text{for} \quad \frac{\mu}{1 - \tau_2 + \delta_2} > 0.$$

In the notation of Theorem 3.1, we have $p_1(\lambda) = \lambda^2 - \tau_1 \lambda + \delta_1$ and $p_2(\lambda) = \lambda^2 - \tau_2 \lambda + \delta_2$. Thus,

$$p_1(1)p_2(1) = (1 - \tau_1 + \delta_1)(1 - \tau_2 + \delta_2), \tag{3.57}$$

$$p_1(-1)p_2(-1) = (1 + \tau_1 + \delta_1)(1 + \tau_2 + \delta_2), \tag{3.58}$$

and therefore, according to conditions (3.11), (3.14) and (3.17), we have the following possible scenarios:

persistence if either

$$\tau_1 > 1 + \delta_1 \quad \text{and} \quad \tau_2 > 1 + \delta_2,$$

or

$$\tau_1 < 1 + \delta_1 \quad \text{and} \quad \tau_2 < 1 + \delta_2.$$

non-smooth fold if either

$$\tau_1 > 1 + \delta_1 \quad \text{and} \quad \tau_2 < 1 + \delta_2,$$

or

$$\tau_1 < 1 + \delta_1 \quad \text{and} \quad \tau_2 > 1 + \delta_2.$$

non-smooth period-doubling if either

$$\tau_1 > -(1+\delta_1) \qquad \text{and} \qquad \tau_2 < -(1+\delta_2),$$

or

$$\tau_1 < -(1+\delta_1) \qquad \text{and} \qquad \tau_2 > -(1+\delta_2).$$

To complete the classification, we need to assess the stability of the fixed and period-two points branching off at the border-collision. Note that, as the map is linear on both sides of the discontinuity boundary, any period-two point must be characterized by one iteration in region S_1 and one in S_2.

The eigenvalues of matrix N_1 can be written as

$$\lambda_{11,12} = \frac{1}{2}(\tau_1 \pm \sqrt{\tau_1^2 - 4\delta_1}),$$

and those of N_2 as

$$\lambda_{21,22} = \frac{1}{2}(\tau_2 \pm \sqrt{\tau_2^2 - 4\delta_2}),$$

from which we can easily work out parameter regions where A and B are stable. More specifically, we have:

Case 1: $\tau^2 - 4\delta < 0$. The eigenvalues of the fixed points w_i, $i = 1$ or 2 are complex conjugate; i.e.,

$$\lambda_{i1,i2} = \frac{1}{2}(\tau_i \pm j\sqrt{4\delta_i - \tau_i^2}),$$

and we have $|\lambda_{i1,i2}| = \sqrt{\delta_i}$, which is less than unity. Hence, both fixed points are stable.

Case 2: $\tau^2 - 4\delta > 0$. In this case, the eigenvalues are real. In particular,

$$|\lambda_{i1}| = \frac{1}{2}(\tau_i - \sqrt{\tau_i^2 - 4\delta_i}),$$
$$|\lambda_{i2}| = \frac{1}{2}(\tau_i + \sqrt{\tau_i^2 - 4\delta_i}),$$

and we find that each equilibrium w_i is stable if $|\tau_i| < 1+\delta_i$, and unstable if $|\tau_i| > 1+\delta_i$.

So, both equilibria A and B are stable when

$$-1 - \delta_i < \tau_i < 1 + \delta_i. \tag{3.59}$$

Period-two points are characterized by one iteration in S_1 and one in S_2; hence their stability is determined by the eigenvalues of the matrix $N_1 N_2$. Thus, we find the eigenvalues of the period-two point to be

$$\frac{1}{2}\left[\tau_1\tau_2 - \delta_1 - \delta_2 \pm \sqrt{(\tau_1\tau_2 - \delta_1 - \delta_2)^2 - 4\delta_1\delta_2}\right].$$

Hence, if $(\tau_1\tau_2 - \delta_1 - \delta_2)^2 - 4\delta_1\delta_2 < 0$, such eigenvalues are complex conjugate and it is easy to show that they must lie inside the unit circle if $\delta_1\delta_2 < 1$, which is always the case in the parameter region being considered here. If, instead, $(\tau_1\tau_2 - \delta_1 - \delta_2)^2 - 4\delta_1\delta_2 > 0$, then the eigenvalues are real and are within the unit circle if:

$$1 + \delta_1 + \delta_2 + \delta_1\delta_2 - \tau_1\tau_2 < 0,$$
$$1 - \delta_1 - \delta_2 + \delta_1\delta_2 + \tau_1\tau_2 > 0.$$

3.5.2 Complex bifurcation sequences

In general, higher-periodic orbits and chaotic attractors can also branch off the boundary fixed point at a border-collision. Figure 3.13 shows a selection of different possible cases, which are computed by Monte Carlo simulation of the map. This selection shows that transitions from fixed or higher-periodic points to chaos are possible, as well as transitions between different types of periodic orbits or transitions to no attractor. A key feature of the two-dimensional piecewise-affine maps, which was not seen in the one-dimensional piecewise-affine maps, is the coexistence of different attractors. It should be mentioned here that the mechanism leading to border-collisions involving multiple attractors has not been studied as yet in the literature and that its explanation remains an open question.

As the map is planar, it is possible to use geometric arguments to give conditions for the existence of chaotic attractors and/or higher-periodic points involved in the border-collision scenario of interest. For example, conditions for the existence and the stability of higher periodic points can be given by following the same strategy as the one followed for period-two points in the previous section.

One thing to note about the two-dimensional map (3.56) is that in the limit $\delta_1, \delta_2 \to 0$ we recover the one-dimensional normal form (3.44) with $\nu_1 = \tau_1$ and $\nu_2 = \tau_2$. Hence, for small δ_1 and δ_2, the region $\tau_2 < -(1 + \delta_2)$, $0 < \tau_1 < (1 + \delta_1)$ is likely to contain all the complex dynamics highlighted in the previous section. That is, we expect to see regions where there exist stable high-period periodic orbits interspersed with parameter regimes of robust chaos. However, the proof of the robustness of the chaos does not necessarily apply here. This is because the argument no longer applies that no periodic orbit can ever be stable if $A^{k-1}B$ is not stable (where $(k-1)$ is the highest number of iterates allowed within S_1). The stability of possible longer periodic chains (e.g., $A^{k-2}BA^{k-1}B$) requires more careful treatment since, it does not necessarily follow that if C and D are matrices with all eigenvalues less than 1 in modulus, then this must also be true of their product CD.

For the two-dimensional map, another parameter region that has attracted interest is a region where a transition from no attractor for $\mu < 0$ to a robust chaotic attractor for $\mu > 0$ occurs. It is clear from the plausible argument in [21, 24] that additional (possibly generic) conditions must be true in order to

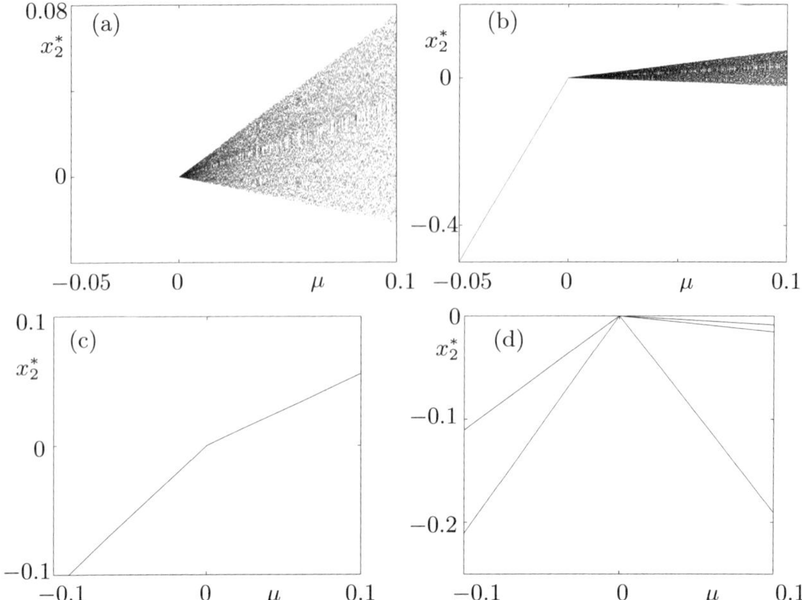

Fig. 3.13. Monte Carlo bifurcation diagrams illustrating different cases of border-collision bifurcations in the planar map (3.56) (after [21]). (a) From no attractor to chaos, parameter values are set to $\tau_1 = 1.5$, $\delta_1 = 0.3$, $\tau_2 = -1.6$, and $\delta_2 = 0.4$; (b) from a fixed point attractor to chaos; parameter values are set to $\tau_1 = -1.6$, $\delta_1 = 0.4$, $\tau_2 = 1.2$, and $\delta_2 = 0.3$ (c) from a fixed point attractor to a fixed point attractor, parameter values are set to $\tau_1 = 0.4$, $\delta_1 = 0.3$, $\tau_2 = -0.5$, and $\delta_2 = 0.3$; and (d) from period-two attractor to period-three attractor, parameter values are set to $\tau_1 = -1.2$, $\delta_1 = -0.3$, $\tau_2 = -1.2$, and $\delta_2 = 1.6$

guarantee that such a scenario definitely occurs, and the precise enumeration of the parameter region in which a chaotic attractor occurs remains an open problem. In the next section, we shall deal with a special case when one of the δ_i is zero, where it *is* possible to provide precise information on the existence of robust chaos in (3.56).

Under other conditions, it is possible that the two-dimensional piecewise-linear continuous map (3.56) generates quasi-periodic motion corresponding to the existence of stable invariant circles [280, 282, 27]. Again, we are unaware of a general classification of where in the four-dimensional $(\delta_i, \tau_i, \delta_i, \tau_i)$-parameter space such behavior should be observed.

In higher dimensions, more or less any dynamical behavior may be observed in piecewise-linear continuous maps (see, e.g., [241, 21, 148, 189, 90, 99, 152]), and a general classification, such as we attempted for the one-dimensional case would appear impossible.

3.6 Maps that are noninvertible on one side

In this section, we consider piecewise-linear continuous maps of the form studied in this chapter, that are invertible only in one of their two linear pieces. More precisely, we assume that in one region, say S_2, the matrix representing the dynamics has co-rank one; that is, it has a single zero eigenvalue. The importance of the study presented here will become clear in Chapter 8, where it will be shown that locally noninvertible normal form maps arise from the local analysis of grazing-sliding bifurcations in Filippov flows. noninvertible piecewise-linear maps have been studied in a number of papers in the literature (see for example [187, 188, 190, 191, 158]).

Specifically, consider a map of the form (3.24) and assume now that, although N_1 is non-singular, N_2 is such that $N_1 B = 0$ for some non-zero matrix $B \in \mathbb{R}^1 \times \mathbb{R}^n$ with $C^T B = 0$. Hence N_2 has corank 1; that is, $\det(N_2) = 0$ and N_2 has a single zero eigenvalue with right eigenvector B. The dynamics of such a family of mappings is trivial if $n = 1$, so we focus here on the two-dimensional case (as would arise from Poincaré maps of three-dimensional flows) [158]. In two-dimensions, after co-ordinate transformations as introduced earlier in the chapter, we arrive at a normal form map of the form:

$$\begin{pmatrix} x_1 \\ x_2 \end{pmatrix} \mapsto \begin{cases} \begin{pmatrix} \tau_1 & 1 \\ -\delta_1 & 0 \end{pmatrix} \begin{pmatrix} x_1 \\ x_2 \end{pmatrix} + \begin{pmatrix} 1 \\ 0 \end{pmatrix} \mu, & \text{if} \quad x_1 \le 0, \\ \\ \begin{pmatrix} \tau_2 & 1 \\ 0 & 0 \end{pmatrix} \begin{pmatrix} x_1 \\ x_2 \end{pmatrix} + \begin{pmatrix} 1 \\ 0 \end{pmatrix} \mu, & \text{if} \quad x_1 \ge 0. \end{cases} \tag{3.60}$$

The bifurcation scenario that results from a border-collision is explained using the same classification approach for two-dimensional maps in Sec. 3.5. Specifically, note that the main requirement of Theorem 3.1 is that $(I - N_1)$ and $(I - N_2)$ be non-singular. Hence the fact that N_2 itself is singular does not invalidate that theorem. Existence of higher-periodic orbits and chaos can proven in the planar case we are treating by means of simple geometric arguments (see [158] for an exhaustive list of possible cases). Here, though, we restrict attention to the existence of robust chaotic attractors.

3.6.1 Robust chaos

We focus on the chaotic attractor created for the map (3.60) in the parameter region where

$$\tau_1 > -1 - \delta_1, \quad \tau_2 < -1, \quad |\tau_1| < 2.$$

Under these conditions, period-two points branch off the border-collision bifurcation point. Note that in our current case a fixed point A and (or its virtual counterpart) $\tilde{A}$ can lie either above or below the x_1-axis, whereas the fixed point b (or $\tilde{b}$) must lie on the axis; see Fig. 3.14

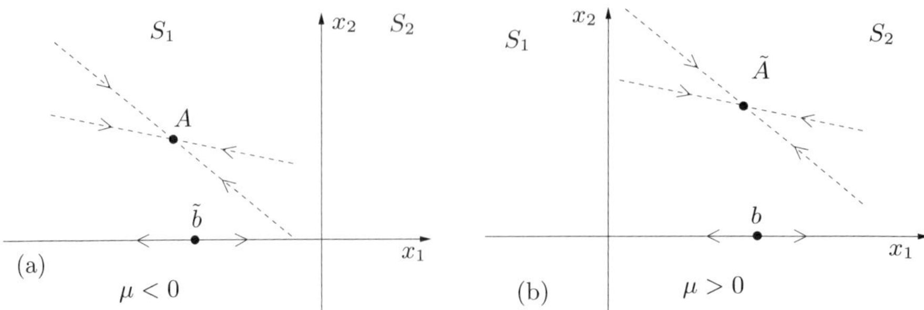

Fig. 3.14. Schematic representation of the fixed points of (3.60) before (a) and after (b) the border-collision bifurcation from an admissible stable fixed point to an admissible saddle. Lines with arrows schematically depict eigendirections of both fixed points.

Consider first the situation when $\mu < 0$, Fig. 3.14(a). Note that A is an attracting fixed point for both regions S_1 and S_2. Consider an initial condition in S_2. After one iterate, it is mapped onto the x-axis and, after at most one further iteration, into S_1. For $\mu = 0$, the fixed point A moves to the origin where it remains an attractor.

Consider next what happens for $\mu > 0$; see Fig. 3.14(b). The stable admissible fixed point A becomes a virtual fixed point $\tilde{A}$ and the unstable virtual saddle $\tilde{b}$ becomes the unstable admissible fixed point b. By continuity, for $\mu > 0$, $\tilde{A}$ will attract points from S_1, so all initial conditions in S_1 will eventually be mapped into S_2. The eigenstructure of the fixed point b of S_2 will govern subsequent behavior. Simple calculations reveal that b is a flip saddle (that is, it has a negative multiplier). Therefore all points in S_2 will be mapped immediately onto a segment of the x_1-axis and eventually, after a finite number of iterations, onto the part of the axis with $x_1 < 0$. Hence, a trapping region is formed which consists of part of the x_1-axis and the forward iterates of this part. The immediate question arises as to what type of attractors can exist within this trapping region. We break the classification into the two cases as follows.

Theorem 3.4 (Two-piece chaotic attractor). *If the border-collision bifurcation from an admissible fixed point attractor to an admissible flip-saddle is exhibited by map (3.60) under the variation of μ and in addition conditions*

$$\tau_1 < -\frac{1}{1+\tau_2}, \tag{3.61}$$

$$\tau_1(\tau_2+1) - \delta_1(1+\frac{1}{\tau_2}) < 0, \tag{3.62}$$

$$\tau_1(\tau_2+1) - \delta_1(1+\frac{1}{\tau_2}) + 1 > 0, \tag{3.63}$$

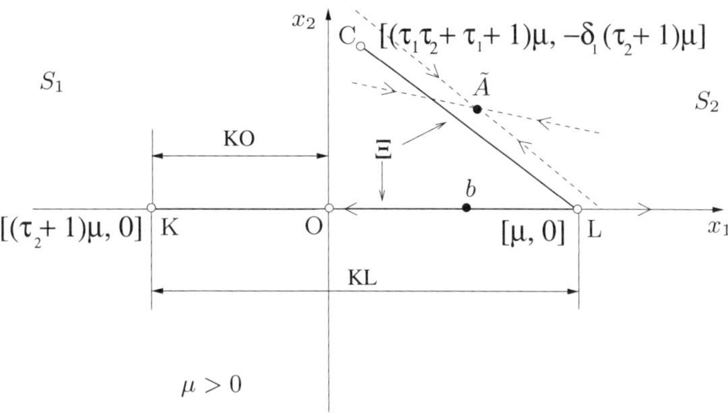

Fig. 3.15. An example of a limit set of (3.60) after the border-collision bifurcations from a stable fixed point to a flip saddle.

hold, then there exists an attractor for $\mu > 0$ *that lies within the piecewise-linear continuous invariant segment* KLC*, where* $K = [(\tau_2 + 1)\mu,\ 0]$, $L = [\mu,\ 0]$ *and* $C = [(\tau_1\tau_2 + \tau_1 + 1)\mu,\ -\delta_1(\tau_2 + 1)\mu]$ *(see Fig. 3.15).*

If, in addition,

$$|\tau_1\tau_2 - \delta| > 1, \tag{3.64}$$

then this attractor is chaotic and robust.

Proof. Let us focus on Fig. 3.15. As mentioned, for $\mu > 0$, points from both regions S_1, S_2 are eventually mapped onto a portion of the negative x_1-axis. Let us denote this segment KO (see Fig. 3.15) and consider what happens to this segment under forward iteration. The origin O is mapped to the point $L : (\mu,\ 0)$, and by continuity, the image of KO joins the x_1-axis at this point, but as it cannot lie on the axis itself, the image of the entire x_1-axis must exhibit a non-differentiable corner here. The image of KO is denoted LC in Fig. 3.15.

If (3.61) holds LC does not cross the x_2-axis. Suppose this situation to be the case, and consider the image of LC under the map. L is then mapped to the point $((\tau_2 + 1)\mu,\ 0)$, which is how we define the point K. Let us suppose that C is mapped into the interior of the segment KL. This holds under the inequality constraints (3.62), (3.63). Hence K is a pre-image of C and so the piecewise-linear continuous invariant segment KLC is an invariant set that must contain all the long-term dynamics. Let us denote this invariant set by Ξ; see Fig. 3.15.

Because of the geometry of the set Ξ, one of the period-two points must lie on KO and the other on LC, thus, we observe switchings between S_1 and S_2. The period-two points can be either stable or unstable depending on whether the quantity $\tau_1\tau_2 - \delta_1$ lies within the unit circle. Assuming it does, the period-two point is the global attractor of the system.

A more intriguing scenario can be observed if the period-two point is unstable, i.e., when $\tau_1\tau_2 - \delta_1$ lies outside of the unit circle. In this case, we shall first show that there are no other stable periodic points within Ξ. We can analyze the dynamics on Ξ by considering a map that maps KL back onto itself. Such a map will be discontinuous and consist of two linear pieces. If the absolute values of both slopes of the map are greater than unity, then no stable periodic point can live on Ξ. To obtain the functional form of such a one-dimensional map, consider the first two forward iterates of L, O and K. After straightforward calculation [158], we obtain the map

$$\pi(x;\mu) = \begin{cases} -1(\tau_2\tau_1 - \delta_1)x + ((\tau_2\tau_1 - \delta_1) + 1)(\tau_2 + 1)\mu, & \text{for} \quad x \le 0, \\ \tau_2 x + \mu, \quad \text{for} \quad x \ge 0, \end{cases} \tag{3.65}$$

from which we can see that the slope is strictly greater than unity in absolute value everywhere. Thus there can be no stable periodic orbits.

To show that the dynamics on Ξ (or on a subset of) must be chaotic, we appeal to the Smale–Birkhoff homoclinic theory and show that there exists a transverse homoclinic intersection between the stable and the unstable manifolds of b. These sets form locally straight lines in $\mathbb{R}^2$, joined by continuous, but non-differentiable corners. The unstable manifold of b is formed by the set Ξ itself. The stable manifold of b, up to the first intersection with the x-axis, is formed by the set of points that in one iteration are mapped onto b. In S_2, its functional form can be given by

$$S_b : x_2 = -\tau_2 x_1 + \frac{\tau_2}{1-\tau_2}\mu.$$

We now seek to find whether S_b intersects LC. The intersection point, say X_{Cr}, is given by

$$X_{Cr} = \left(\frac{\tau_2\tau_1 + \tau_2\delta_1 - \delta_1}{1-\tau_2}, -\frac{\tau_2^2\delta_1}{(1-\tau_2)(\tau_2\tau_1 - \delta_1}) \right)^T \mu.$$

It can be then checked that indeed, in the parameter range of interest, X_{Cr} belongs to LC.

Finally, note that the chaotic attractor must be robust, since within the open parameter region stated in the Theorem, the invariant set Ξ persists, there are no stable periodic orbits on Ξ and the homoclinic intersection remains transverse.

Remark. We should note here that we have not proved that the chaotic attractor covers the whole of the set Ξ. In simulations we find the chaotic dynamics is typically organized around the unstable period-two points and depending on precise parameter values, can form a four, two, or one-piece chaotic attractor.

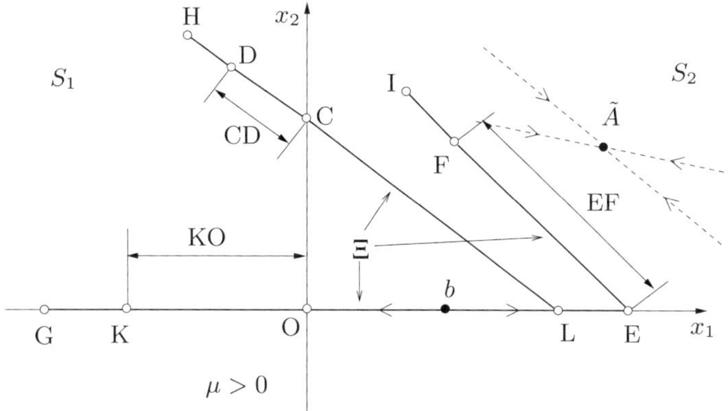

Fig. 3.16. An example of a limit set of (3.60) after the border-collision bifurcations from a stable fixed point to a flip saddle.

Theorem 3.5 (Three-piece chaotic attractor). *If the border-collision bifurcation from an admissible fixed point attractor to an admissible flip-saddle is exhibited by map (3.60) under the variation of μ and in addition conditions*

$$\tau_1 > -\frac{1}{1+\tau_2}, \tag{3.66}$$

$$\tau_1^3(\tau_2+1)+\tau_1+1-\delta_1(1+\frac{2}{\tau_2}) > 0, \tag{3.67}$$

$$\tau_1^3(\tau_2+1)+\tau_1^2(1+\tau_2\delta_1)-\delta_1\tau_1(1+\frac{1}{\tau_2})-\frac{\delta_1}{\tau_2}+1 > 0, \tag{3.68}$$

hold then there exists an attractor born in the border-collision bifurcation that must necessarily lie within a piecewise-linear continuous invariant set built from segments GEI and LH.

Proof. Under the hypotheses of this theorem, the image of KO now crosses the x_2-axis; see Fig. 3.16. The image of KO, denoted by LD, contains a segment CD, which requires one additional iteration to get mapped into S_2. Segment LC is mapped in one iteration onto the x_1-axis. By continuity, the image of CD, denoted EF, must join the image of LC. Note that the point E must lie outside of the segment KL but in region S_2 and that the maximum value of x_1 at E is greater than the maximum value of x_1 of the pre-image of KO, from the hypothesis of the theorem.

From (3.66) it also follows that $\tau_1 > 0$. Then, if (3.67) holds in forward iteration, the piecewise-linear segment OEF is mapped within some segment, say GE. [Condition (3.67) ensures that the value of x_1 of the image of F is less than that of x_1 at E; see Fig. 3.16.] Now we need to consider forward iterations of GO only (OE is mapped onto GO). The image of GO will cross the x_2-axis and is denoted in the figure by LH. Finally, the image of LH will form a piecewise-linear continuous segment KEI lying partly on the x_1-axis (the part KE) where at point E it is joined by segment EI (see Fig. 3.16). Our main concern is to determine the image of EI. If condition (3.68) is satisfied, then the image of EI lies within GE and we have obtained a bounded set, say Ξ, which must contain the ω-limit set. The set Ξ will be formed by *three* linear segments, GE, EI and LH.

Remarks

1. The possible ω-limit sets on Ξ are higher-periodic points or a chaotic attractor. Similarly to the previous case, these different scenarios can be distinguished by determining the stability properties of higher-periodic points. By obtaining a one-dimensional map that maps GE back onto itself, it can be shown that, provided the period-three point is unstable, there are no other possible stable periodic points on Ξ. A homoclinic intersection between the stable and the unstable manifolds of b can be shown. Hence a chaotic attractor is born on Ξ. Similarly as in the two-piece case, robustness can be also shown. Because of the structure of Ξ, we might observe the onset of a six-, three- or one-piece chaotic behavior.
2. In principle, we could also consider the possibility of the existence of a Ξ set formed by a higher number of piecewise-linear segments, four, five and so on. For example if EI (see Fig. 3.16) is not mapped within GE (violation of (3.68)) or if the maximum value of x of the image of OEF is greater than at E (violation of (3.67)), then, if there exists a bounded set, it must necessarily contain a higher number of piecewise-linear segments than three. The dynamics on such sets can be analyzed in a likewise manner.

3.6.2 Numerical examples

Example 3.3 (Two-piece chaotic attractor).

Consider the map derived in [86] at which a grazing sliding bifurcation was found to lead to the onset of chaos. The map in question falls into the class (3.60) with

$$N_1 = \begin{pmatrix} 0.8540 & 1 \\ -0.009 & 0 \end{pmatrix}, \qquad N_2 = \begin{pmatrix} \tau_2 & 1 \\ 0 & 0 \end{pmatrix}. \tag{3.69}$$

We will vary the parameter τ_2 and show that for different values of this parameter we can observe a border-collision as μ increases that creates a stable period two point, then a four-piece chaotic attractor, that merges into a two-piece attractor and finally a one-piece chaotic attractor.

For $\tau_2 < -1$ the map is characterized by a stable fixed point, say $A/\tilde{A}$ and a saddle point $b/\tilde{b}$ given by

$$A/\tilde{A} = \left(\frac{\mu}{0.1550}, \ 0.0581\mu\right)^T, \quad \text{and} \quad b/\tilde{b} = \left(\frac{\mu}{1-\tau_2}, \ 0]\right).$$

The eigenvalues of $A/\tilde{A}$ are 0.8433 and 0.0107, and those of $b/\tilde{b}$ are τ_2 and 0. For these values, a period-two point is born in the border-collision. To see a transition to a stable period two point we need the eigenvalues of N_1N_2 to lie within the unit circle. In our case the eigenvalues of N_1N_2 are $\lambda_1 = 0$, $\lambda_2 = 0.854\tau_2 - 0.009$. So, for $\tau_2 = -1.1$, for example, we find a stable period-two point after the bifurcation; see Fig. 3.17(a).

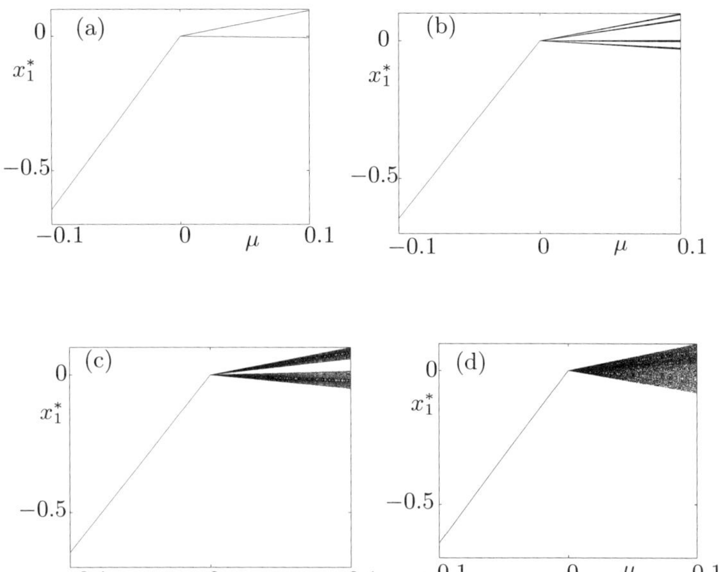

Fig. 3.17. Bifurcation diagrams showing birth of attractors lying on a two-piece invariant set Ξ after a border-collision bifurcation for (3.69) with (a) $\tau_2 = -1.1$, (b) $\tau_2 = -1.3$, (c) $\tau_2 = -1.5$ and (d) $\tau_2 = -1.85$.

Under further reduction of τ_2, the period-two points become unstable. For example, for $\tau_2 = -1.3$ [Fig. 3.17(b)] it can be easily shown that all the inequality conditions in the statement of Theorem 3.4 are satisfied and that we get a chaotic attractor lying on a two-piece invariant set Ξ. In this case, since the period-two points are only weakly unstable, the chaotic attractor is organized near the these periodic points (flipping on either side of each one) and so has four pieces. With yet further decrease of τ_2, the limit set covers a bigger part of Ξ, with first pairs of pieces of the attractor merging [Fig. 3.17(c) and eventually, for $\tau = -1.85$, forming a single piece chaotic attractor (Fig. 3.17(d)]. Figure 3.18 shows the corresponding invariant sets, confirming that in each case the attractor lies on a two-piece invariant set Ξ.

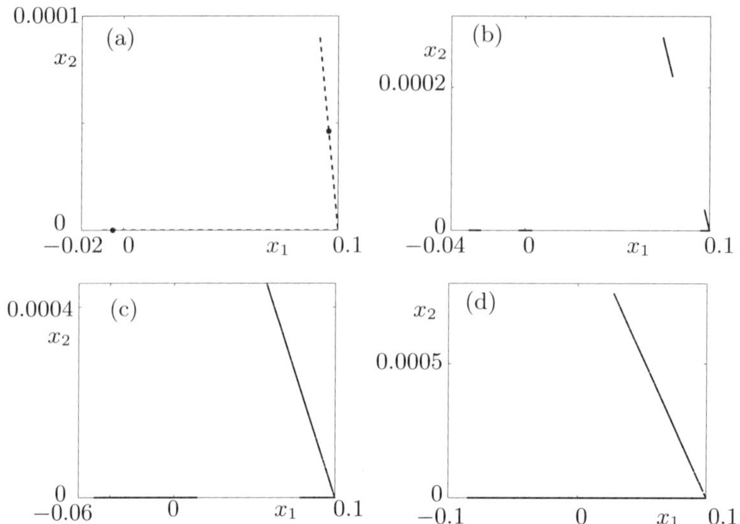

Fig. 3.18. Invariant sets corresponding to each panel in Fig. 3.17 for $\mu = 0.1$

Example 3.4 (Three-piece attractor).

Similarly to the previous example, consider a piecewise-linear map of the form (3.60), but with

$$N_1 = \begin{pmatrix} 0.5 & 1 \\ -0.06 & 0 \end{pmatrix}, \quad N_2 = \begin{pmatrix} \tau_2 & 1 \\ 0 & 0 \end{pmatrix}. \tag{3.70}$$

We will vary τ_2 starting from $\tau_2 = -3.5$ and show different cases of border-collision bifurcation scenarios under the variation of the bifurcation parameter μ, all linked with the existence of the three-piece piecewise-linear continuous set Ξ (see Fig. 3.16). It is straightforward to check that for $\tau_2 = -3.5$, (3.70) satisfies all the conditions of Theorem 3.5. Moreover, we find that for this τ_2 value, the eigenvalues of $N_1N_1N_2$ lie within the unit circle. Thus, under the variation of the bifurcation parameter μ, we should expect border-collision bifurcations from a fixed point to a period-three attractor; see Fig. 3.19(a).

If we decrease τ_2, the period-three point will become unstable (at $\tau_2 = -5.105$), but the set Ξ will retain its properties. Thus we see the onset of a robust chaotic attractor with, at first six pieces [Fig. 3.19(b)] then three [Fig. 3.19(c)], and finally a single piece for $\tau_2 = -6.5$ [Fig. 3.19(d)]. The corresponding ω-limit sets are depicted in Fig. 3.20, from which we see that they all lie on a three-piece invariant set Ξ.

We now provide an example of a border-collision bifurcation from two virtual saddles to two admissible saddles. This leads to the spontaneous creation of a chaotic attractor from no local attractor before the bifurcation. A similar proof of robust chaos may be constructed in this case as in Theorems 3.4

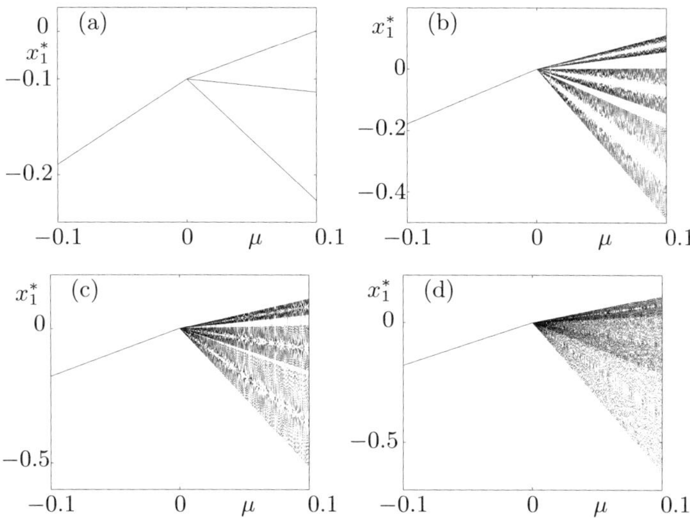

Fig. 3.19. Bifurcation diagram depicting the border collision bifurcation for (3.70) with $\tau_2 =$ (a) -3.5, (b) -5.25, (c) -5.5, and (d) -6.5

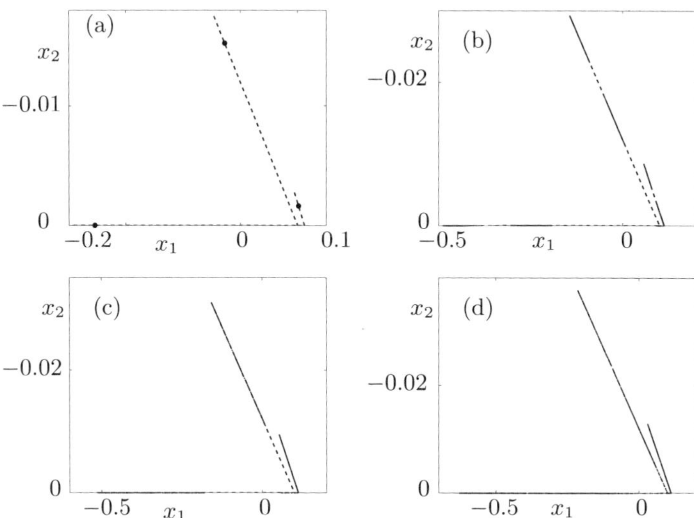

Fig. 3.20. Invariant sets corresponding to each panel in Fig. 3.19 for $\mu = 0.1$

and 3.5 for the flip-saddle case. For brevity, though, we omit the details and restrict attention to an example system only.

Example 3.5 (Sudden onset of chaos in saddle case). We now consider the case where

$$N_1 = \begin{pmatrix} 1.2448 & 1 \\ 0 & 0 \end{pmatrix}, \quad N_2 = \begin{pmatrix} -4.5502 & 1 \\ -0.0698 & 0 \end{pmatrix}, \tag{3.71}$$

for which it is straightforward to calculate that for $\mu < 0$ we have two virtual fixed points, which become admissible for $\mu > 0$ in a non-smooth fold type border-collision. For $\mu \leq 0$ there is no attractor, as all initial conditions other than the fixed points eventually get mapped towards $-\infty$ along the negative x_1-axis. For $\mu > 0$ we observe the birth of admissible fixed points a and b which are unstable. Other periodic points can be born also. These, however, must also be unstable. More information on the dynamics can be inferred if we consider the geometry of the stable and unstable manifolds of a and b (see Fig. 3.21). First, we find the slopes of the unstable eigenvector of a (labeled U_a in Fig. 3.21) are such that the two manifolds form a homoclinic tangle via the point of intersection $x_d = (\mu, 0)^T$ in the figure. We next note that the stable

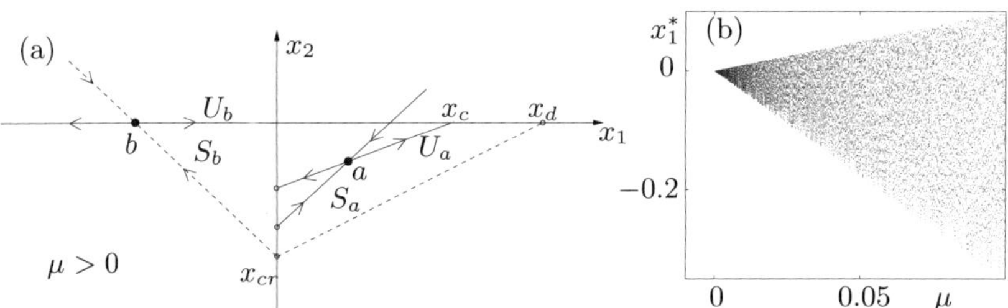

Fig. 3.21. (a) Fixed points of affine map (3.71) for $\mu > 0$. (b) Bifurcation diagram of the map under variation of μ.

manifold of b (labeled S_b in Fig. 3.21)), forms a seperatrix for points with $x_1 < 0$. That is, all points with $x_1 < 0$ to the left of the stable manifold S_b are mapped towards $-\infty$, whereas points to the right are eventually mapped into the half-plane $x_1 > 0$. We also find that the right-hand branch of U_b does not intersect S_b; therefore, we can consider the triangle formed by the points b, the point x_d, and x_c, where S_b intersects the x_2-axis. It is straightforward then to show that this triangle is a trapping region which must contain an attractor, which must be chaotic since no periodic point is stable. This is confirmed by numerical iteration of the map (3.71) in Fig. 3.21(b).

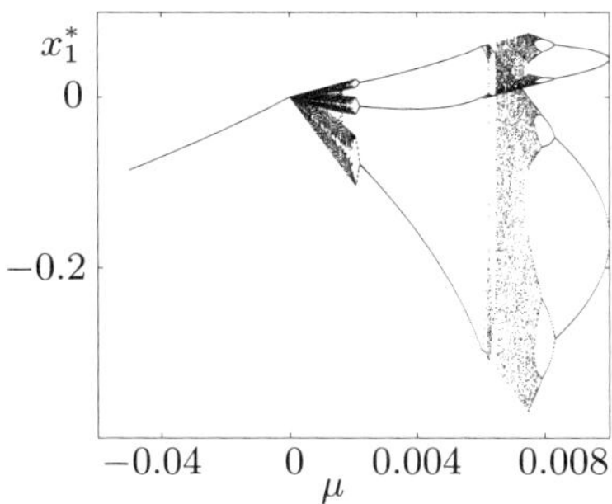

Fig. 3.22. Monte Carlo bifurcation diagram representing sudden jump to chaos followed by a further sequence of bifurcations for the nonlinear map (3.72).

3.7 Effects of nonlinear perturbations

As discussed in the previous two sections, one of the most striking features of border-collision bifurcations in piecewise-linear maps is the possible transition to a robust chaotic attractor. As we have said, robust chaos is characterized by the fact that no periodic windows can be found within the parameter regime where a piecewise-linear map exhibits chaotic dynamics. In this sense, the attractor is robust to parameter variations. However, all our analysis so far has been for exactly piecewise-linear maps. In most applications, for example, when derived as Poincaré maps of flows, such maps arise only as leading-order approximations to a full map that contains nonlinear terms also. Therefore it is relevant to assess whether the chaotic attractors we have constructed are also robust to nonlinear perturbations.

Currently, there is no general theory to account for this case, so to illustrate this issue we have merely computed numerically the bifurcation diagram of the following map:

$$x \mapsto \begin{cases} 0.5x + x^2 + \mu, & \text{if } x \leq 0, \\ -6x + x^2 + \mu, & \text{if } x > 0, \end{cases} \tag{3.72}$$

which is a perturbed version of map (3.44) for $\nu_1 = 0.5$, $\nu_2 = -6$. For these parameters, close to the origin, we therefore have to leading order a piecewise-linear continuous map that we know undergoes a border-collision bifurcation from a stable fixed point attractor to a robust chaotic attractor. We expect that this bifurcation scenario will be preserved in the neighborhood of $\mu \approx 0$ also for (3.72). However, for sufficiently large $\mu > 0$, the quadratic term might break the robustness of the chaotic attractor.

The bifurcation diagram of (3.72) is depicted in Fig. 3.22, which confirms these predictions. We indeed observe a sudden transition to chaos from a fixed point under increase of μ through 0. However, unlike the case where the map is purely piecewise-linear (see Fig. 3.12), we observe periodic windows embedded within the attractor, for example the period-three periodic window for

$\mu \approx 0.04$ in Fig. 3.22. Although a detailed analysis of the effects of nonlinear perturbations is beyond the scope of this book, it would seem reasonable that chaotic attractor is robust for small $\mu > 0$, since the construction of chaos relies on transverse intersections between certain manifolds that are likely to be preserved under small nonlinear perturbations.

4

Bifurcations in general piecewise-smooth maps

The previous chapter considered the nature of discontinuity-induced bifurcations that arise at the border collisions of fixed and periodic points of maps that are both piecewise-linear and continuous. However, the range of systems described by such maps does not cover the wide variety of cases that we shall derive in subsequent chapters from Poincaré maps of non-smooth flows, or other mappings that arise directly from applications such as the heart attack map in case study VI. Accordingly, we now extend the range of maps considered to include those that are piecewise-smooth, but are not necessarily continuous across a discontinuity boundary, and those that are not locally linearizable on either side. Again we will see a rich bifurcation behavior, but with subtle distinctions from the locally piecewise-linear continuous case.

4.1 Types of piecewise-smooth maps

Recall from Chapter 2, the Definition 2.18 of a piecewise-smooth map. Clearly, the set of all possible such maps is vast, so to give our discussion some structure, we will restrict our attention to maps with a single discontinuity surface. That is, we assume that there is a set $\mathcal{D} \subset \mathbb{R}^n$ and a codimension one subset $\Sigma \subset \mathbb{R}^{n-1}$ of $\mathcal{D}$. In general, Σ will be assumed to be a smooth manifold defined by a smooth function $H(x)$ so that

$$\Sigma = \{x \in \mathcal{D} : H(x) = 0\}. \tag{4.1}$$

We identify two open subsets S_1 and S_2 of $\mathcal{D}$ so that $H(x)$ takes opposite signs in each of these two regions. We then consider a map $f : \mathcal{D} \to \mathcal{D}$, which is defined for all of $\mathcal{D}$ but takes different functional forms in S_1 and S_2:

$$f(x) = \begin{cases} F_1(x), & \text{if } x \in S_1, \\ F_2(x), & \text{if } x \in S_2. \end{cases} \tag{4.2}$$

Although, formally, f may not be uniquely defined for points in Σ. We will assume that each function $F_{1,2}$ is *smooth* over its domain of definition, but that the overall function $f(x)$ is not. The lack of smoothness may arise in many ways, for example through a lack of continuity of f as x varies through Σ or a lack of continuity of a first or higher derivative. In this chapter, we shall extend the analysis of border-collisions in piecewise-linear continuous maps presented in the previous chapter to more general piecewise-smooth maps. To be concrete, we will only consider three cases, of successively increasing *order of singularity* (see Def. 2.19) of the map at Σ.

Discontinuous maps. An important class of maps that arise in many applications, for example the heart attack model that we introduced as case study VI in Chapter 1, are discontinuous in the system state. That is, F_1 and F_2 are functions with a bounded derivative as $x \to \Sigma$ (so that they are well approximated by linear maps close to Σ), but the map F is *discontinuous* as x passes through Σ. That is, if $x \in \Sigma$, $x_n^+ \in S_2 \to x$ and $x_n^- \in S_1 \to x$, then

$$\lim F_2(x_n^+) \neq \lim F_1(x_n^-). \tag{4.3}$$

In Sec. 4.2 we will show that even an analysis of one-dimensional dicontinuous maps, under the simplification that F_1 and F_2 are completely linear leads to a rich complexity of dynamics.

Square-root maps. In this case we let f be continuous but allow F_1 or F_2 to have a square-root form as x approaches Σ, so that f is proportional to $\sqrt{|H(x)|}$ for small $H(x)$ of one sign. These maps arise naturally as local approximations to the Poincaré maps associated with grazing bifurcations in impacting systems, as we have already seen in case studies I and VII in Chapter 1 and will be further elaborated in Chapter 6. The key aspect of these maps is that the square-root form leads to infinite local stretching on at least one side of Σ. This can cause an infinite sequence of transitions in the dynamics arbitrarily close to the border collision point, and as we shall show in Sec. 4.3, the creation of a large number of high period periodic orbits.

Higher-order maps. In Sec. 4.4 we then consider maps for which the nth derivative is continuous at Σ but the $(n+1)$st is not. As an example, we can suppose that f and its first derivative be continuous but allow F_2 and or F_1 to behave as $|H(x)|^{3/2}$ as x approaches Σ. The second derivative of the map therefore becomes unbounded in this limit. Maps of this type will be shown to arise in various kinds of grazing and sliding bifurcations of piecewise-smooth flows in Chapters 7 and 8. One thing to note about such maps is that, as they are continuous across the boundary, then a border-collision will not change the existence *or stability* of a fixed point crossing the boundary. That is, the Implicit Function Theorem applies locally and there is no immediate change in the dynamics at the border-collision point. Thus, any consequent change in the dynamics must be delayed until *after* the discontinuity-induced bifurcation.

The rest of this chapter provides a detailed description of the dynamics that occurs when a simple fixed point undergoes a border collision in each of these three classes of map. Many technicalities are specific to each case; nevertheless, we will see many of the characteristic features of the bifurcations associated with piecewise-linear continuous maps that were observed in the last chapter, including persistence and non-smooth folds, period-adding, robust chaos, and so on. To see why this is so, at least for one-dimensional continuous maps, recall the 2-parameter plot in Fig. 3.8 from Chapter 3 which represents different outcomes for the one-dimensional piecewise-linear continuous map when the main bifurcation parameter μ is varied. Recall that the co-ordinate axes in this plot, ν_1 and ν_2 represent the slopes of the two linear pieces of the map.

Now, consider instead a continuous map that is not locally piecewise-linear, specifically one composed of two pieces, one of which is linear, the other of which is $O(x^\gamma)$ for $\gamma \neq 1$; e.g.,

$$x \mapsto \begin{cases} \nu_1 |x|^\gamma + \mu_1, \text{ if} & x \leq 0, \\ \text{if } \nu_2 x + \mu_2 & x > 0. \end{cases} \tag{4.4}$$

If $\mu_1 = \mu_2 = \mu$, then this map is continuous. Let $x^*(\mu)$ be the fixed point of this map in the region where the map is nonlinear. Then, under variations of the parameter μ, $x^*(\mu)$ travels up the nonlinear piece of the map and the local slope of the map will be given by $\tilde{\nu}_1 = -\nu_1 \gamma |x|^{\gamma-1}$. So, effectively, μ-variation also leads to variation of the slope parameter $\tilde{\nu}_1$ in Fig. 3.8. Thus, we might expect to see a cascade of bifurcations under variations of μ that only occurs for the piecewise-linear map if we allow the slopes ν_1 and ν_2 to vary. Hence, *period-adding cascades* (either interspersed with chaos or not) arise naturally out of the maps we study, because μ-variation causes the map solutions to move across the parameter space boundaries identified in Fig. 3.8 where the attractor transforms from being of period n to $n+1$. Clearly, the detailed scaling of the period-adding cascades (the size of periodic 'windows' in parameter space, extent of the attractor in phase space, etc.) will be specific to the value of the exponent γ in simple maps such as (4.4).

Assuming γ to be positive, important distinctions occur between the cases $\gamma < 1$ and $\gamma > 1$. We deal only with the indicative cases $\gamma = 1/2$ ('square-root maps') in Sec. 4.3 below and $\gamma = 3/2$ or 2 ('higher-order maps') in Sec. 4.4. These choices are motivated by the forms of discontinuity maps that arise at discontinuity-induced bifurcations in flows, as will be derived in Chapters 6, 7 and 8. Even considering just maps that have a single discontinuity and are linearizable on one side of it, the presentation that follows will be far from exhaustive. In the case of square-root maps, there is an almost complete theory known in n dimensions and because of its relevance to grazing bifurcations in impacting systems, we shall dwell on this case in some detail. Most of the rest of the material will concern one-dimensional maps only. Moreover, as in the previous chapter, we restrict attention to a local region where the map can be approximated by its leading term only on each side of the discontinuity.

First, though, we present results for piecewise-smooth discontinuous maps (*maps with a gap*), which we can think of as the case $\gamma \to 0$ in (4.4). There, we will see an additional phenomena, that of a *Devil's staircase* of periodic points of the map that have a *Farey tree* structure where between the parameter values at which there is a period-m orbit and that at which there is a period-n there is a period-$(n+m)$ orbit. This structure is well known in quasi-periodic systems described by circle maps (see e.g.,[9]), as will be observed in what follows.

4.2 Piecewise-smooth discontinuous maps

In this section we will look at maps that are smooth and have a well-defined derivative up to the discontinuity boundary, but they are discontinuous across the boundary itself. piecewise-smooth discontinuous maps of the form (4.1)–(4.2) satisfying (4.3) arise naturally in their own right, for example, in the heart attack problem described in the introduction (case study VI in Chapter 1) and other models for the general behavior of excitable media [118]. Also, discontinuous maps have been used in descriptions of various switching phenomena in electrical circuits [143] and in simple models of the firing of neurons [37]. In addition, such maps will arise as Poincaré maps associated with systems of impact oscillators with multiple impacts (see Chapter 6). In general, little work has been reported on the analysis of discontinuous maps, which are becoming the subject of increasing scientific interest. Recent work include [180], [157], [164], [46] where such systems were shown to possess period-adding bifurcations, [226] where the presence of multiple devil's staircases was demonstrated and [227] where so-called type V intermittency was found. A quadratic map with a gap defined on the interval was studied in [15, 16]. More recently, extensions of Feigin's classification strategy to piecewise-linear discontinuous maps were presented in [137].

The simplest type of piecewise-smooth discontinuous maps are those that are locally piecewise-linear. We will focus our discussion on this type of maps. We first look at fixed and period-two points of piecewise-linear maps in general dimensions, and we will then look at the periodic and chaotic orbits of such maps in one-dimension, identifying period-adding behavior.

4.2.1 The general case

We start with the discontinuous version of the piecewise-linear map studied in Chapter 3 and look at conditions for the existence and bifurcation of fixed points and period-two points. Here, we extend the classification strategy presented in Chapter 3 to the discontinuous case in the manner presented in [137]. All through this chapter, we use the same notation for bifurcation classification presented in Sec. 3.2.2.

Consider a discontinuous piecewise-linear map of the form

$$x \mapsto \begin{cases} N_1 x + M\mu, & \text{if } C^T x < 0, \\ N_2 x + M(\mu + l), if & C^T x > 0, \end{cases} \tag{4.5}$$

where $\mu, l \in \mathbb{R}$, N_1 and N_2 are real $n \times n$ matrices, $M \in \mathbb{R}^{n\times 1}$ and $C \in \mathbb{R}^{n\times 1}$. We assume that the map is smooth for $C^T x \neq 0$, and so we suppose that the matrices N_1 and N_2 still satisfy condition (3.8). To motivate this choice of system, note that not all discontinuous maps can be put into this kind of normal form, for example period-adding of a more general class of maps is studied in [46]. Nevertheless, there is a strong connection with the form (3.7) from the previous chapter, except that there is a finite *gap* lM between the value of the map on either side of the discontinuity surface $H(x) = C^T x = 0$.

This map has potentially two fixed points given by

$$x_1^* = (I - N_1)^{-1} M\mu, \quad \text{if} \quad C^T x_1^* < 0,$$

$$x_2^* = (I - N_2)^{-1} M(\mu + l), \quad \text{if} \quad C^T x_2^* > 0.$$

A border-collision occurs if either of these two fixed points passes through the plane $C^T x = 0$, which will occur at two values of μ, namely $\mu = 0$ and $\mu = -l$. In this spirit we can adapt the notation $A \leftrightarrow B$ and so on, for the continuous map to describe the fate of the fixed points under the pair of transitions. For example, we will use the notation $A \leftrightarrow A, B \leftrightarrow B$ to indicate that at the border-collision the transition is observed from one fixed point to two fixed points and then to the other fixed point. (Here we use $\leftrightarrow$ to give the indication that μ may actually increase or decrease in any realization of each scenario.) To this end, it is possible to rework the derivation presented in Chapter 3, in order to extend to this case the classification theory on the behavior of the simplest orbits in a border-collision. In particular, we have:

Theorem 4.1 ([137]). *Let $p_1(\lambda)$ be the characteristic polynomial of matrix N_1 and $p_2(\lambda)$ the characteristic polynomial of N_2 in (4.5). Moreover, define*

$$\sigma_1^+ := \textit{number of real eigenvalues of } N_1 \ (\alpha_i) \textit{ greater than } 1$$
$$\sigma_2^+ := \textit{number of real eigenvalues of } N_2 \ (\beta_i) \textit{ greater than } 1$$

Assume that the matrices $(I - N_1)$ and $(I - N_2)$ are invertible, then at a border-collision, according to the value of $l \neq 0$, we can have one of the following scenarios:

Case 1. *If*

$$p_1(1)p_2(1) < 0; \quad \textit{or, equivalently} \quad \sigma_1^+ + \sigma_2^+ \textit{ is odd,}$$

then as μ varies we have the possibility that either

$$\textit{no fixed point} \leftrightarrow \textit{one fixed point} \leftrightarrow \textit{two fixed points}$$

(e.g., $\varnothing \leftrightarrow B \leftrightarrow a, B$), or

$$\textit{two fixed points} \leftrightarrow \textit{one fixed point} \leftrightarrow \textit{no fixed points.}$$

Case 2. *If*

$$p_1(1)p_2(1) > 0, \quad \textit{or, equivalently} \quad \sigma_1^+ + \sigma_2^+ \textit{ is even,}$$

then we have either

one fixed point ↔ two fixed points ↔ other fixed point

(e.g., $A \leftrightarrow A, B \leftrightarrow B$), or

one fixed point ↔ no fixed point ↔ other fixed points.

Remarks

1. Note that when $l = 0$ these two cases reduce to non-smooth fold and persistence scenarios, respectively. For $l \neq 0$, the major difference is the intermediate stage where one, two or no fixed points can exist.
2. It is possible to refine the above conditions to say more about which specific bifurcation scenarios can occur, including stability of the various orbits, under conditions on the signs of l and μ and on the eigenvalues of various matrix products (see [137]). We omit the details here. Instead, we focus for the remainder of this section on the more illustrative one-dimensional case.

4.2.2 One-dimensional discontinuous maps

Consider the one-dimensional version of (4.5) given by

$$x \mapsto \begin{cases} \nu_1 x + \mu, & \text{if } x \le 0, \\ \nu_2 x + (\mu + l), & \text{if } x > 0. \end{cases} \tag{4.6}$$

Note that one-dimensional discontinuous maps of this form have formed a key role in the historical development of chaotic dynamics. For example, if we set $\nu_1 = \nu_2 = 1$ then under a simple rescaling, (4.6) becomes the well studied rotation (or twist) map [129], and it is closely related to the Bernoulli shift map [89]. As will be shown later, in general the case $\nu_1 = \nu_2 = \nu$ is simpler to analyze and yet exhibits very different behavior from that observed in piecewise-linear continuous maps when $l \neq 0$. Without loss of generality, we will assume that, by an appropriate rescaling, $l = -1$, $l = 1$ or $l = 0$.

For convenience, we define the two submappings

$$\Pi_1 : x \mapsto \nu_1 x + \mu \qquad x\text{if } \le 0 \tag{4.7}$$

$$\Pi_2 : x \mapsto \nu_2 x + (\mu + l) \qquad \text{if } x > 0, \tag{4.8}$$

and start our analysis by seeking the domains of existence and stability of fixed and periodic points.

The map Π_1 has one fixed point x_1^* given by

$$x_1^* = \frac{\mu}{1 - \nu_1}$$

which is admissible if $x_1^* \leq 0$ and stable for $|\nu_1| < 1$, whereas the map Π_2 has the fixed point

$$x_2^* = \frac{\mu + l}{1 - \nu_2},$$

which must be positive to be admissible and is stable for $|\nu_2| < 1$.

For $l = +1$, if $0 < \nu_1 < 1$, $\nu_2 < 1$ then x_1^* is the only admissible stable fixed point (A) when $\mu < 0$; x_1^* and x_2^* are both stable and admissible (A,B) when $0 < \mu < 1$ and x_2^* is the only stable admissible fixed (B) point when $\mu > 1$. Thus, at the border-collision we have the scenario $A \leftrightarrow A, B \leftrightarrow B$ predicted by Theorem 4.1. The resulting bifurcation diagram has the form illustrated in Fig. 4.1.

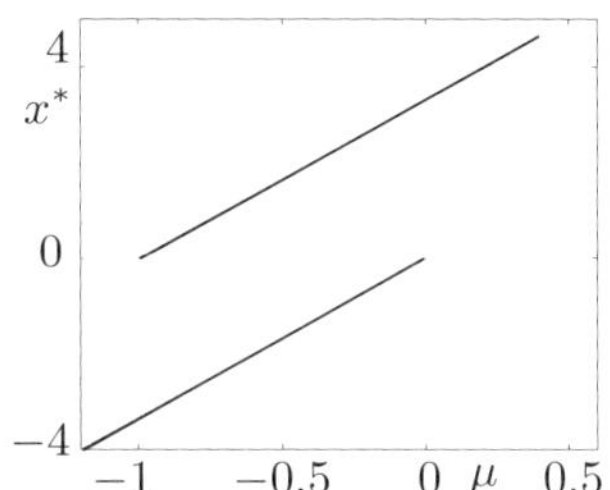

Fig. 4.1. The bifurcation diagram of the discontinuous map (4.6) for $l = 1$, $\nu_1 = \nu_2 = 0.7$ showing the scenario $A \rightarrow A, B \rightarrow B$ with the two coexisting fixed points in the gap region.

The case of $l = -1$ generally gives rise to much more complicated dynamics. The fixed point x_1^* is exactly as before, but the other fixed point exists only for $\mu > 1$. So for $-1 < \nu_2 < 0$, the simplest bifurcation scenarios are

$A \leftrightarrow \varnothing \leftrightarrow B$, if $0 < \nu_1 < 1$;
$\varnothing \leftrightarrow a \leftrightarrow a, B$, if $0 < \nu_1 < 1$;

and for $\nu_2 < -1$,

$A \leftrightarrow \varnothing \leftrightarrow b$, if $0 < \nu_1 < 1$;
$\varnothing \leftrightarrow a \leftrightarrow a, b$, if $0 < \nu_1 < 1$.

For example, the scenario $A \leftrightarrow \varnothing \leftrightarrow B$ is shown in Fig. 4.2.

Using the same approach presented in Chapter 3, it is possible to study next the existence and stability of period-two solutions by looking at the second-iterate of map (4.5). Also, by a simple extension of the arguments used in Chapter 3, it is straightforward to show [137] that, for $l = 0, \pm 1$, an $A^{k-1}B, a^{k-1}b$ solution exists for $\mu > 0$ if and only if the following two conditions are satisfied:

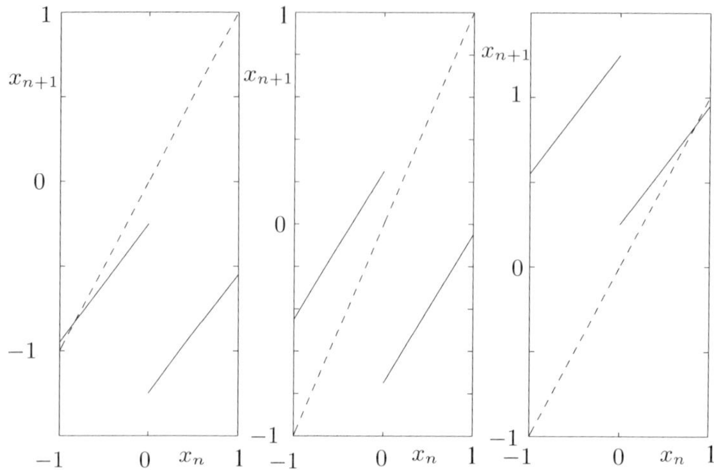

Fig. 4.2. The piecewise-linear discontinuous map (4.6) with $l = -1$ and (from left to right) $\mu < -1$, $-1 < \mu < 0$ and $\mu > 0$.

$$0 < \left(1 + \frac{l}{\mu} + \frac{1}{\nu_1} + \frac{1}{\nu_1^2} + \ldots + \frac{1}{\nu_1^{k-1}}\right) = \frac{1}{\mu} + \frac{1 - \nu^k}{\nu^k - \nu^{k-1}}, \tag{4.9}$$

$$\nu_2 < \left(1 + \frac{l}{\mu} + \frac{1}{\nu_1} + \frac{1}{\nu_1^2} + \ldots + \frac{1}{\nu_1^{k-2}}\right) = \frac{l}{\mu} + \frac{1 - \nu^{k-1}}{\nu^{k-1} - \nu^{k-2}}. \tag{4.10}$$

Moreover, it is clear that for all l the eigenvalue of the linearization of the map about the $A^{k-1}B, a^{k-1}b$ solution is $\nu_1^{k-1}\nu_2$, and so it is stable if and only if

$$\nu_2 > -\frac{1}{\nu_1^{k-1}}. \tag{4.11}$$

As in the case of piecewise-linear continuous maps, using conditions (4.9)–(4.11), it is possible to plot existence and stability curves of higher-periodic solutions in the (ν_1, ν_2) parameter space. Then, the possible border collision scenarios observed under variations of the parameter μ can be classified. Two representative diagrams for $l = +1$ and $l = -1$ are shown in Fig. 4.3.

A more extensive investigation for a map of this form can be found, for example, in the works of Jain & Banerjee [143], Hogan *et al.* [137] and Budd & Piiroinen [46]. The main results are summarized in Table 4.1. A complete study of all the cases is lengthy. The bare bones can be obtained by following the same steps used in Chapter 3 for the case of a piecewise-linear continuous map. In addition, Jain & Banerjee [143] compute a series of numerical bifurcation diagrams, indicating what dynamics is possible beyond simple periodic points. We mention also the extensive computations by Avrutin & Schantz [14, 17] for a closely related family of maps. The work of Keener [152] on non-smooth circle maps is also relevant for understanding the global dynamics.

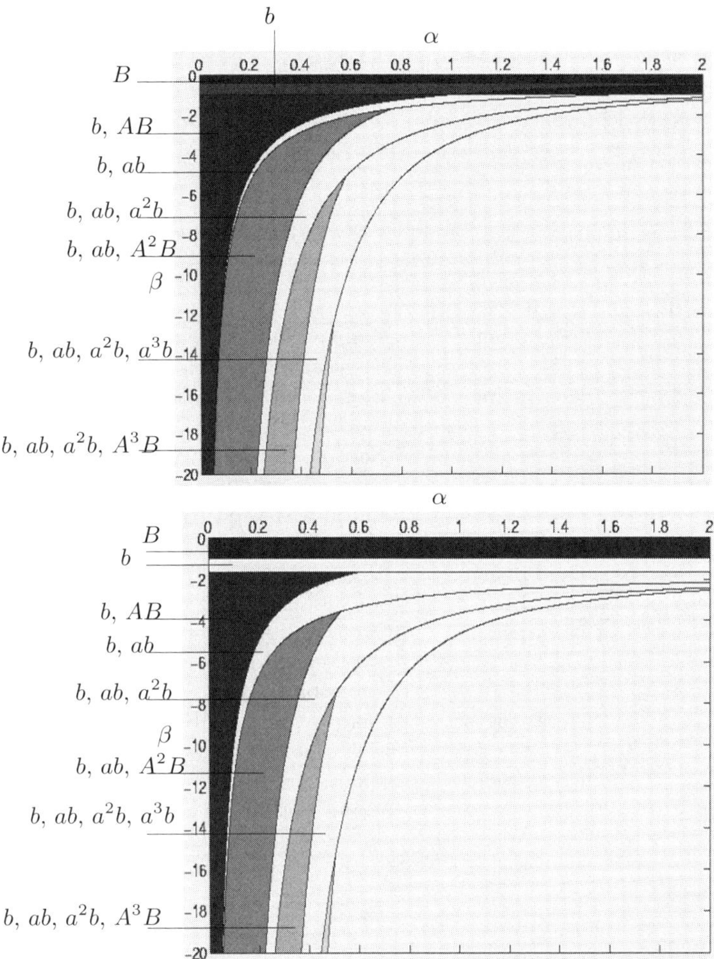

Fig. 4.3. The two-parameter bifurcation diagrams of the discontinuous map when $l = 1, \mu = \frac{3}{2}$ and $l = -1, \mu = \frac{5}{3}$ showing the existence and stability boundaries of higher-periodic orbits. (Reprinted from [137] with permission from the Royal Society).

In what follows, we shall analyze more details of the global dynamics arising in the more interesting case $l = -1$. We note that an important distinction occurs in the consequent dynamics depending on whether the origin lies inside the *discontinuity interval* $I = [\mu - 1, \mu]$. In the case $0 < \nu_1, \nu_2 < 1$ and $l = -1$, then the set $[\mu - 1, \mu]$ is invariant under the action of the map if $0 < \mu < 1$ and no fixed point exists. The map is one-to-one but not onto (an injection but not a subjection) on this set. As we shall see in Sec. 4.2.3 which follows, in this case, we find that only periodic orbits occur for all but a possible set of μ-values of zero measure.

In contrast, if we have $1 < \nu_1, \nu_2 < 2$, $l = -1$, then the set $[\mu - 1, \mu]$ is invariant only if $1 - 1/\nu_1 < \mu < 1/\nu_2$. (Indeed, if $\mu > 1/\nu_2$, then orbits can escape to infinity.) The map is onto in this case but not one-to-one and chaotic behavior is observed. Finally, if $0 < \nu_1 < 1 < \nu_2$, then we see both periodic and chaotic behavior. We treat these cases in Sec. 4.2.4.

Table 4.1. Summarizing the dynamics of (4.6) in the case $l = -1$.

Case	Ranges of $\nu_{1,2}$	$\mu < 0$	$0 < \mu < 1$	$\mu > -1$
1	$0 < \nu_1 < 1$, $0 < \nu_2 < 1$	fixed point	Many periodic orbits (Farey Tree)	fixed point
2	$0 < \nu_1 < 1$, $\nu_2 > 1$	fixed point	Many periodic orbits, 2-cycle and chaos	No attractor
3	$0 < \nu_1 < 1$, $-1 < \nu_2 < 0$	fixed point	Many periodic orbits (period-adding)	Coexisting fixed point and 2-cycle.
4	$0 < \nu_1 < 1$, $\nu_2 < -1$,	fixed point	High periodic orbits and chaos	2-cycle or chaos
5	$\nu_1 > 1$, $0 < \nu_2 < 1$,	No attractor	Chaos, 2-cycle and many periodic orbits	fixed point
6	$\nu_1 > 1$, $\nu_2 > 1$	No attractor	Chaos	No attractor
7	$\nu_1 > 1$, $-1 < \nu_2 < 0$	No attractor	Chaos or many periodic orbits	fixed point
8	$\nu_1 > 1$, $\nu_2 < -1$	No attractor	Chaos	Chaos
9	$-1 < \nu_1 < 0$ $0 < \nu_2 < 1$	Coexisting periods one and two	many periodic orbits	No attractor
10	$-1 < \nu_1 < 0$ $\nu_2 > 1$	Periods-1 and 2/period-1	High periodic orbit/chaos	No attractor

4.2.3 Periodic behavior: $l = -1$, $\nu_1 > 0$, $\nu_2 < 1$

This case is relatively easy to analyze. Note first that, as shown earlier, there is a stable fixed point x_1^*, which is admissible if $\mu < 0$. Similarly, if $\mu > 1$ then x_2^* is an admissible fixed point that may or may not be stable. A bifurcation diagram for this case is presented in Fig. 4.4(a), which shows that for $0 <$

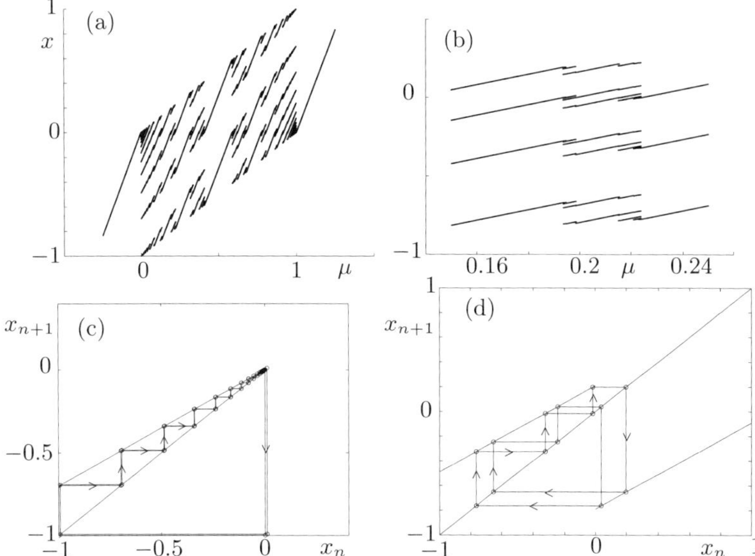

Fig. 4.4. (a) The bifurcation diagram of (4.6) for $l = -1$ and $\nu_1 = \nu_2 = 0.7$. (b) Zoom showing the transition from a period-3 to a period-4 orbit separated by an interval in which there are period-7 and other more complex orbits. (c) The limiting homoclinic orbit as $\mu \to 0^+$. (d) A periodic orbit with symbol sequence A^2BA^3B.

$\mu < 1$ purely periodic behavior is observed, but there is a rich variety of such behavior.

Indeed, for small $\mu > 0$ there is a sequence of high periodic orbits with symbol sequence $A^{k-1}B$ with $k \to \infty$ as μ decreases to zero, in which limit a where a homoclinic orbit is observed [see Fig. 4.4(c)]. Similar behavior is observed as μ is increased to 1, where the sequence is of the form AB^{k-1}.

A careful analysis of this intricate structure reveals an interesting relationship between nearby periodic orbits within the cascade.In particular, between each period-k and period-$(k+1)$ window, we see more complicated periodic motions with symbol sequences that involve a concatenation of those of the neighboring orbits, e.g., an orbit of type $A^{2k-1}B^2 \equiv A^{k-1}BA^kB$ between a solution of type $A^{k-1}B$ and one of type A^kB. These are illustrated in Fig. 4.4(b) in which we see a period-7 orbit between period-3 and period-4 orbits as well as more complex orbits (of period 10 and 11, respectively) between the period-7 orbit and the period-3, and between the period-7 orbit and the period-4. This concatenation of the symbol sequences leading to an orbit of period equal to the sum of the periods of the two neighboring orbits shows a striking resemblance with the well-known series in number theory proposed by Farey in 1816 [94]. Therefore, this phenomenon is referred to as *Farey arithmetic* and the complete family of orbits as a *Farey tree*. We now proceed to give a brief analysis of this behavior.

The Farey tree; devil's staircases and period-adding

Under our assumptions, if $\mu = 0^+$, then there is a fixed point at $x = 0$. This coexists with a (noninvertible) homoclinic orbit of the form $A^\infty B$. This orbit, illustrated in Fig. 4.4(c), is the limit of the orbits of the form $A^{k-1}B$ as $\mu \to 0$ and $k \to \infty$ and its existence can be explained as follows. Suppose $x_1 = -1 \in S_1$; then for $\mu = 0$, we have a sequence $x_n = -\nu_1^{n-1} \in S_1$, which converges to 0 in S_1 as we have assumed that $0 < \nu_1 < 1$. Considering now 0 as a point in S_2, this point is then mapped back in a single iterate to $x = l = -1$, leading to the orbit illustrated in Fig. 4.4(c).

More generally, for $0 < \mu < 1$, we consider a sequence of iterates $\{x_n\}$ of type $A \in S_1$ and $B \in S_2$. Clearly, if $x_{n-1} \in S_1$, the maximum value of $x_n = f(x_{n-1})$ is given by $\mu \in S_2$. Then, provided that $\mu < 1/(1+\nu_2)$, we have $f(x_n) < \mu - 1 + \nu_2\mu < 0$ so that $x_{n+1} \in S_1$. Thus, if $\mu < 1/(1+\nu_2)$, the sequence must comprise strings A^{k-1} separated by single iterates B. The simplest possible sequence of this type is the (maximal) periodic orbit $A^{k-1}B$.

To calculate such an orbit and obtain conditions for its existence, we set $\bar{x}_n = x_n - \mu/(1-\nu_1)$. If $x_n \in S_1$, so that $\bar{x}_n < -\mu/(1-\nu_1)$, we then have

$$\bar{x}_{n+1} = \nu_1 \bar{x}_n = \nu_1^{n-1}\bar{x}_1.$$

Similarly, if $x_n \in S_2$ so that $\bar{x}_n > -\mu/(1-\nu_1)$, we have (after rescaling)

$$\bar{x}_{n+1} = \nu_2\bar{x}_n - 1 + \mu\frac{(\nu_2-\nu_1)}{(1-\nu_1)}.$$

Now assume that k is such that

$$\bar{x}_{k-1} = \nu_1^{k-2}\bar{x}_1 < -\frac{\mu}{1-\nu_1} < \nu_1^{k-1}\bar{x}_1 = \bar{x}_k. \tag{4.12}$$

We then have a periodic orbit of period k provided that

$$\bar{x}_1 = f(\bar{x}_k) = \nu_2\bar{x}_k - 1 + \mu\frac{(\nu_2-\nu_1)}{(1-\nu_1)} = \nu_2\nu_1^{k-1}\bar{x}_1 - 1 + \mu\frac{(\nu_2-\nu_1)}{(1-\nu_1)}. \tag{4.13}$$

Similarly, rearranging (4.13), we must have

$$\bar{x}_1 = \frac{-1+\mu(\nu_2-\nu_1)/(1-\nu_1)}{(1-\nu_2\nu_1^{k-1})} \tag{4.14}$$

or equivalently,

$$x_1 = \frac{\mu}{(1-\nu_1)} + \frac{-1+\mu(\nu_2-\nu_1)/(1-\nu_1)}{(1-\nu_2\nu_1^{k-1})}. \tag{4.15}$$

Equation (4.15) determines the form of the orbit, and condition (4.12) determines whether it exits. The same sequence is repeated as $\mu \to 1$ if we make the transformation $\mu \to 1-\mu$ and swap A and B, S_1 and S_2 in the above expressions.

Note that the above expressions simplify considerably in the special case (and exemplifies the special role played by the map discontinuity (or gap) since, as shown in Chapter 3, the dynamics of piecewise-linear continuous maps with $\nu_1 = \nu_2$ is trivial)

$$\nu_1 = \nu_2 := \nu \tag{4.16}$$

as the term involving μ in (4.14) vanishes and

$$\bar{x}_1 = -\frac{1}{1-\nu^k}.$$

This orbit then satisfies the existence condition provided that, from (4.15), we have

$$\frac{\nu^{k-1}(1-\nu)}{1-\nu^k} < \mu < \frac{\nu^{k-2}(1-\nu)}{1-\nu^k}. \tag{4.17}$$

Continuing with the assumption (4.16), note that for $\nu < 1$ the orbit (when it exists) is necessarily stable. Hence, as $k \to \infty$, we have approximately that the orbit exists and is stable for

$$\nu^{k-1}(1-\nu) < \mu < \nu^{k-2}(1-\nu),$$

so that the window width scales geometrically. Note further that, from (4.17), the maximum value of μ in the period-k window is $\nu^{k-2}(1-\nu)/(1-\nu^k)$ and the minimum value in the period $k-1$ window is $\nu^{k-2}(1-\nu)/(1-\nu^{k-1})$ so the windows do not intersect. In the gap between the period k and $k-1$ orbits, that is for

$$\frac{\nu^{k-2}(1-\nu)}{(1-\nu^k)} < \mu < \frac{\nu^{k-2}(1-\nu)}{(1-\nu^{k-1})},$$

more complex behavior is observed. In terms of the symbol sequence the period k and $k-1$ orbits have the form $A^{k-1}B$ and $A^{k-2}B$, respectively. The simplest orbit in the gap is the one generated by concatenating their symbol sequences; i.e., the orbit of type $A^{2k-3}B^2 \equiv A^{k-2}BA^{k-1}B$, for which

$$\bar{x}_1 = \nu^k(\nu^{k-1}\bar{x}_1 - 1) - 1;$$

so that

$$x_1 = \frac{-(1+\nu^k)}{(1-\nu^{2k-1})},$$

with associated consistency conditions. This is the origin of the period-7 orbit between the period-3 and 4 orbits in Fig. 4.4(d). In fact, such a procedure of concatenating symbol sequences can be applied recursively so that we get orbits of arbitrarily long symbol sequences between any two finite length orbits, thus revealing some of the subtlety of the Farey tree.

Another way of describing these orbits, is via the *rotation number* ρ of a periodic orbit. A period-k orbit with r iterates in S_2 (and $k-r$ in S_1) is defined to have rotation number $\rho = r/k$. For example, an orbit of type A^nB has rotation number $\rho = 1/(n+1)$. As μ increases, spanning the Farey tree, the value of ρ increases monotonically from 0 to 1 in the manner of a *Cantor function*, with intervals of values of μ over which it is constant and takes rational values. These intervals correspond precisely to periodic motions. There is an uncountable Cantor set of measure zero over which ρ takes irrational values. On this set the iterates of the map lie on a non-periodic invariant set. A typical example of the form of ρ as a function of μ is given in Fig. 4.5(a), which is known as *devil's staircase* diagram.

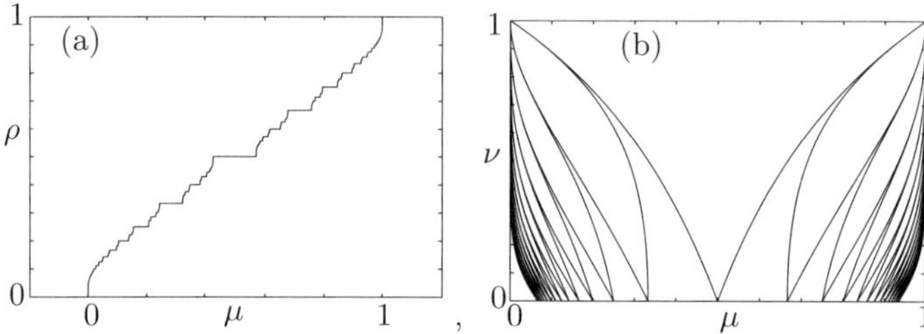

Fig. 4.5. Representation of the global dynamics of the map (4.6) in the case $l = -1$ and $\nu_1 = \nu_2 = \nu$. (a) The rotation number ρ for $\nu = 0.75$. (b) The (μ, ν) existence regions for the (simple) period-m orbits in the case.

A simple analysis shows that the interval of μ-values over which we see period-k orbits of the form $A^{k-1}B$ or AB^{k-1} is given by

$$\mu \in \left[\frac{1}{k} - \frac{k-1}{2k}\varepsilon + \mathcal{O}(\varepsilon^2), \frac{1}{k} - \frac{k-3}{2k}\varepsilon + \mathcal{O}(\varepsilon^2)\right],$$

where $\varepsilon = 1-\nu$, with a similar pattern for $1-\mu$. This gives the (μ, ν) existence regions presented in Fig. 4.5(b). Note that these regions shrink to isolated points at $x = \{1/k\}$ as $\nu \to 1$. [These existence regions closely resemble Arnold tongues of smooth circle maps [89] (which arise when $\nu = 1$), but here the widths of all regions exhibit linear growth as $\nu \to 1$.]

Finally, if we relax the assumption (4.16) and allow $\nu_1 \neq \nu_2$, then the analysis is more difficult as additional terms involving μ appear in the formulation. This can lead to a less clear separation of the windows. A nice example arises when $\nu_2 = 0$ so that we have the map $x \to \nu_1 x + \mu, x < 0, x \to \mu - 1, x > 0$. The first iterate in S_1 is then always $\mu - 1$, and hence, a period-k orbit is observed if

$$\frac{\mu}{1-\nu_1} + \nu_1^{k-1}\left(\mu - 1 - \frac{\mu}{1-\nu_1}\right) < 0 < \frac{\mu}{1-\nu_1} + \nu_1^{k}\left(\mu - 1 - \frac{\mu}{1-\nu_1}\right),$$

leading to the bifurcation diagram illustrated in Fig. 4.6(a). Note that, upon summing the geometric series, we see that this expression is precisely equivalent to (4.9)–(4.10) derived earlier for the case $l = -1$. Here, we see a different phenomenon, namely, period-adding. This differs from the Farey tree scenario highlighted earlier, because we no longer observe complex transitions between one solution and the other but simply the overlapping of their regions of existence. A detailed investigation of period-adding scenarios in one-dimensional maps with a gap can be found in [14, 17].

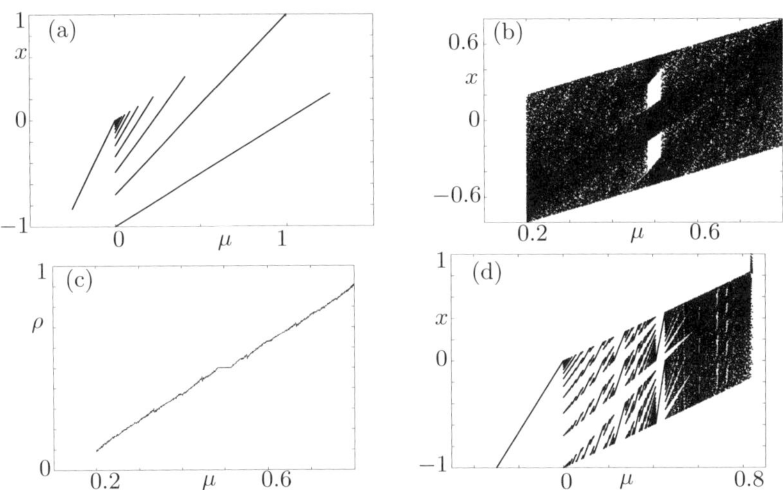

Fig. 4.6. Bifurcation diagrams of the map (4.6) for $l = -1$. We detect (a) period-adding when $\nu_1 = 0.7$, $\nu_2 = 0$, and (b) chaotic behavior when $\nu_1 = \nu_2 = 1.2$; (c) the rotation number ρ is plotted against μ; and (d) Farey tree and chaos when $\nu_1 = 0.7$, $\nu_2 = 1.2$.

4.2.4 Chaotic behavior: $l = -1$, $\nu_1 > 0$, $1 < \nu_2 < 2$

Consider first the case where $1 < \nu_1, \nu_2 < 2$. For this range of parameters, any fixed point or periodic cycle that exists must necessarily be unstable. The interval $[\mu - 1, \mu]$ is invariant under the action of the map provided that $1 - 1/\nu_1 < \mu < 1/\nu_2$. Hence the map is onto (i.e., surjective) when restricted to this interval.

If $\mu > 1/\nu_2$, then the iterations of the map become unbounded. An example of this case is given in Fig. 4.6(b). Figure 4.6(c) shows the plot of the corresponding rotation number ρ over a range of different initial conditions

x for each value of μ. Here we see that, in contrast to the previous case, in general ρ depends on x and so is not a uniquely defined function of μ.

Next, suppose that $0 < \nu_1 < 1$, but $1 < \nu_2 < 2$. Then, for $\mu < 0$ the fixed points x_1^* and x_2^* both exist, although only x_1^* is stable. As μ increases through zero, then initially we see a period-adding cascade similar to the case where both $\nu_{1,2} < 1$. Orbits of the form $A^n B/a^n b$ will be stable if and only if $\nu_1^n \nu_2 < 1$. As the value of μ is increased, we therefore reach a threshold value μ_1 such that all periodic sequences for $\mu > \mu_1$ become unstable, and we see instead chaotic behavior. The point of transition is precisely when the map changes from being one-to-one to being onto. A simple calculation shows that this arises when

$$\mu_1 = (1 - \nu_1)/(\nu_2 - \nu_1).$$

If $\mu > \mu_2 = 1/\nu_2\text{m}$ then the chaotic attractor becomes unstable and high-periodic orbits cease to exist. All trajectories then go monotonically to ∞. A representative bifurcation diagram in this case is given in Fig. 4.6(d) for $\nu_1 = 0.7$ $\nu_2 = 1.2$; hence $\mu_1 = 0.6$ and $\mu_2 = 5/6$.

Remark. Note that the dynamics of the one-dimensional discontinuous map presented above can be embedded in a more general setting without assuming that the map is locally piecewise-linear (but that it does have a well-defined derivative on either side of the discontinuity boundary). In particular, under certain assumptions, such maps can be considered as discontinuous maps of the unit circle to itself, which were analyzed by Keener in [152]. These mappings take the form

$$x_{n+1} = F(x_n) \mod 1, \quad x_n \in [0, 1]$$

with discontinuity at $x_n = \theta$ such that

$$\lim_{x_n \to \theta+} F(x_n) = 1, \qquad \lim_{x_n \to \theta-} F(x_n) = 0.$$

When viewed on the unit circle, F is actually continuous at θ, the true discontinuity being at $x = 0$ provided $F(0) \neq F(1)$.

In [152], it is shown that there are two fundamental cases to be considered depending on the sign of the quantity $\Delta := F(1) - F(0)$. If Δ is positive, the map is said to be non-overlapping. That is, the map of the unit circle is one-to-one but not onto. However, if $\Delta < 0$ the map is said to be overlapping since the image of the unit circle covers the entire circle, but is not one-to-one, because there is a region of the circle that has two pre-images. It is possible to show that overlapping dynamics produces chaotic dynamics, whereas non-overlapping dynamics produces only periodic motion (see [152] for further details). This was precisely the distinction observed in our simple piecewise-linear map with a gap.

To further illustrate the relationship between discontinuous piecewise-linear maps and circle maps we look now at a representative example.

Example 4.1 (A simple neuron model). Bressloff and Stark [37] proposed a discontinuous circle map to model the dynamics of a single neuron within a network. The model is inspired by the physiology of true brain cells and yet leads to possibilities for new kinds of artificial neural networks that take on real rather than discrete values and whose dynamics can be described by coupled discontinuous circle maps. This idea was later extended in [60] where the dynamics are studied of coupled circle maps, modeling a simple network of such neurons.

The neuron is modeled by a single scalar quantity, the value of its *action potential* (a measure of the internal voltage and hence the excitation level of the neuron). For simplicity it is supposed that the neuron can only fire at a set of discrete times $t_n = t_{n-1} + t_d$ for some fixed t_d that is related to the recovery time of the neuron. Let x_n be the value of action potential of the neuron at time t_n relative to its threshold value h for firing. Thus $x_n > 0$ implies that the neuron has fired and $x_n < 0$ corresponds to not firing this time around. Now, let us suppose that the neuron receives input I_n at each time instant; then we arrive at a map

$$x_{n+1} = [I_n + (1-\delta)h + w\Theta(x_n)] \exp\left(\frac{-w\Theta(x_n)}{S}\right), \tag{4.18}$$

where $\Theta(x)$ is the Heaviside step function which is zero for $x < 0$ and 1 for $x > 0$. Here, $0 < \delta < 1$ represents a voltage leakage factor during time interval t_d, and the constants w and w/S represent weights (degree of influence of the firing of one neuron on another) associated with self-interaction; w being a constant weight and w/S being the coefficient of a term proportional to the excess action potential above threshold at time t_{n-1}. Note that both S and w can take either sign depending on whether the neuron is self-excitatory or self-inhibitory.

If such a neuron is embedded in a network, in general I_n would be a sum of many terms proportional to weights w_{1j} times Heaviside functions that determine whether each neuron j in the network has fired at time t_n. Also there would be further exponential factors with weights w_{ij}/S_{ij}. If one wants to consider the dynamics of the single neuron, though, we can assume that at each time t_n the action potential receives a constant input $I_n = I$. Then we can express the map (4.18) more simply as

$$x_{n+1} = f(x_n) = \begin{cases} \nu_1 x_n + a, & \text{if} \quad x_n < 0, \\ \nu_2 x_n + b, & \text{if} \quad x_n \geq 0, \end{cases} \tag{4.19}$$

where $\nu_1 = \delta$, $\nu_2 = \delta e^{-w/S}$, $a = I - (1-\delta)h$, $b = (a+w)e^{-w/S}$. Note that a, b and ν_2 can take either sign, whereas $0 < \nu_1 < 1$.

Clearly, upon letting $\mu = a$ and $l = (b - a)$, then this map fits into the framework of the above theory. Indeed, the cases of Farey trees and of chaos were observed in the numerical simulations presented in [37].

4.3 Square-root maps

We consider now a class of continuous maps characterized by a square-root singularity on one side of the discontinuity boundary Σ with linear behavior (associated with the matrix N) on the other side. Such maps arise naturally in the study of grazing bifurcations of hybrid and piecewise linear flows studied theoretically in Chapters 6 and 7, and experimentally in Chapter 9. The border-collision of a fixed point of the map with Σ then corresponds to a *grazing bifurcation* of the flow. We will start by studying one-dimensional maps, with the main result being a classification theorem for the border-collision, describing various scenarios including period-adding and robust chaos. We then generalize this analysis to maps of higher dimension treating separately the case where N has positive real eigenvalues, which closely resembles the one-dimensional case. We then present quite general conditions for an orbit of a given symbol sequence to be born in the border collision. Finally, we restrict attention to the two-dimensional maps, where we show a rich variety of possible bifurcating stable behavior. Note that by being so general for the case of square-root maps (which is justified as they play such a key role in unfolding the dynamics near grazing bifurcations), the derivation of the results we present in this section are somewhat technical and we recommend that the details are skipped on a first reading.

4.3.1 The one-dimensional square-root map

In case study VII in the Introduction, we considered a simple continuous square-root map described by

$$x \mapsto f(x) = \begin{cases} F_1(x) = \sqrt{\mu - x} + \nu\mu, & \text{if} \quad H(x,\mu) \equiv x - \mu < 0, \\ F_2(x) = \nu x, & \text{if} \quad H(x,\mu) \equiv x - \mu > 0, \end{cases} \tag{4.20}$$

where $\Sigma = \{x : H(x,\mu) = 0\}$. This form of the map is motivated by the behavior of the simple impact oscillator near to grazing and, therefore, the region $H(x,\mu) < 0$ will be denoted the *impacting region*. Note that (4.20) has the form (4.4) with $\gamma = 1/2$ if we make the simple co-ordinate transformation $x \to x - \mu$. Following the notation introduced in Chapter 3, we will use a to denote an iteration of the map in the region for which $H(x,\mu) < 0$ (an *impacting iteration*) and b an iteration in the region for which $H(x,\mu) > 0$ (a *non impacting iteration*).

If $\mu < 0$, then the map has a single (non impacting) , stable fixed point at $x^* = 0$ for which $H(x^*,\mu) = -\mu > 0$ and $e := dH(x^*,\mu)/d\mu = -1$. Moreover, if $0\nu < 1$, this fixed point is stable. A border-collision bifurcation of this fixed point occurs when $\mu = x^* = 0$. At this value of μ we have $F_{1,x}(0^-) = -\infty$ and $F_{2,x}(0^+) = \nu$. We show that increasing μ through zero leads to the instant creation of an infinite number of new, unstable, periodic orbits. The specific stable scenario that will be observed for $\mu > 0$ however, depends on the value of $\nu \in \mathbb{R}$. The possible bifurcation scenarios were investigated by various authors and can be summarized by the following classification theorem, which describes how the form of the bifurcation depends upon the (damping) parameter ν.

Theorem 4.2 (Border-collision in the one-dimensional square-root map [199, 54, 43, 103]). *Consider the one-dimensional square-root map (4.20) with stable fixed point for $\mu < 0$ with positive eigenvalue $0 < \nu < 1$. Then the dynamics for $\mu > 0$ can be characterized as follows:*

1. **Weakly stable case** $2/3 < \nu < 1$. *As μ increases through zero, there is an immediate creation of robust chaotic motion, for which the chaotic attractor has size proportional to $\sqrt{\mu}$.*
2. **Intermediate case** $1/4 < \nu < 2/3$. *There is a* period-adding *cascade of stable period-m orbits of the form $B^{m-1}A$ (using the previous notation) with $m \to \infty$ as $\mu \to 0$. Moreover in between each period-m and period-$(m+1)$ window, there is an interval of values of μ for which we see a chaotic attractor. The width and the location of these windows decrease geometrically with asymptotic ratio ν^2 as $\mu \to 0$.*
3. **Strongly stable case** $0 < \nu < 1/4$. *There is again a period-adding cascade, accumulating on $\mu = 0$ as $m \to \infty$. However, this time there is no chaos and adjacent periodic windows overlap, giving multiplicity of attractors for some parameter values.*

To illustrate this result, we present in Fig. 4.7 plots of the bifurcation diagrams in each of the three cases.

Proof. Consider the simple square-root map given by (4.20) illustrated in Fig. 4.8 If x is just less than μ, then $f_x \equiv F_{1,x}$ is negative and approaches $-\infty$ as $x \to \mu$, leading to a significant degree of stretching. In contrast, if $x > \mu$, then $0 < f_x \equiv F_{2,x} = \nu < 1$. The function $f(x)$ becomes equal to its minimum value $f(x) = \nu\mu$ when $x = \mu$. Let V be the interval defined as $V \equiv \{x : \nu\mu < x < \mu\}$, in which $|f_x|$ is large. We proceed to show that, under the action of f, a discontinuous map F is then induced from V to itself, which can, in turn, be approximated by a single map G that is invariant under the rescaling $\mu \to \nu^2\mu$. By studying the fixed points of G, it is then possible to determine the form of the periodic orbits of f. We call the interval V the *trapping region* of the map.

To construct the map G from a generic initial condition $x_0 \in V$, we proceed as follows. Let $x_0 \in V$ and $x_1 = f(x_0)$. If μ is sufficiently small, then apart from a small interval close to μ, the stretching associated with f implies that $x_1 > \mu$. As $\nu < 1$, then there will be a (possibly large) number, say $m(x_0, \mu)$, of iterations x_n of $f(x)$ such that $x_{n+1} = f(x_n) < x_n$ and $x_n > \mu$ if $n < m$. It is easy to see that

$$x_n = \nu^{n-1}\sqrt{\mu - x_0} + \nu^n\mu.$$

Let us now define x_m to be the first of these iterates which again lies in V. The number $m(x_0, \nu)$ of iterations required to re-enter V takes its *maximal* value,

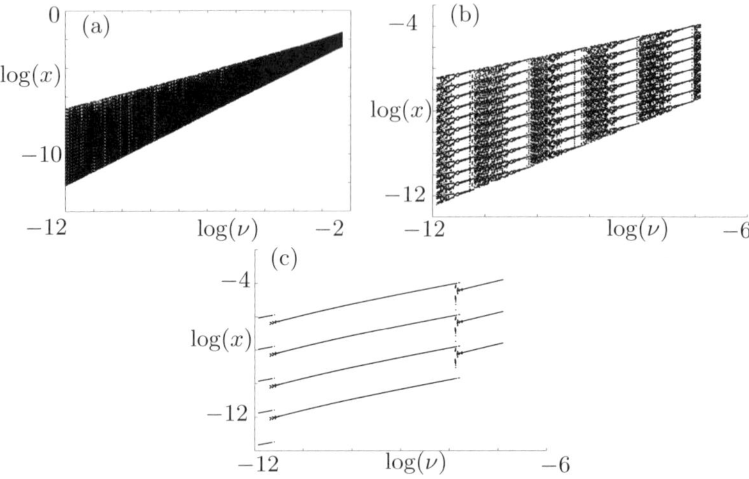

Fig. 4.7. Log-log plot of the dynamics of (4.20) in each of the three cases of Theorem 4.2. In each successive panel we have, respectively, (a) robust chaos if $\nu = 0.8$, (b) period-adding plus chaos if $\nu = 0.6$ and (c) period-adding if $\nu = 0.15$.

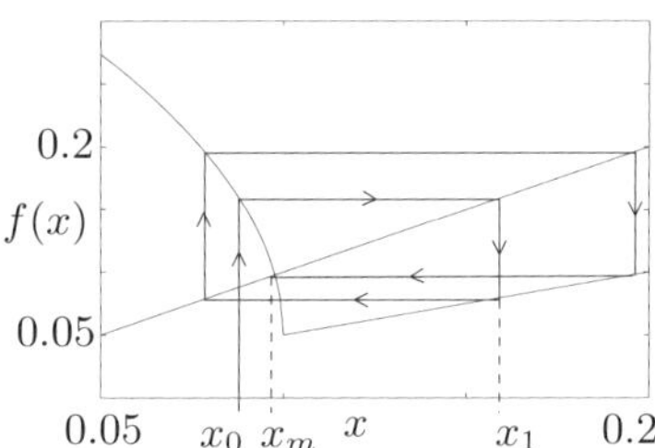

Fig. 4.8. The square-root map and the line $y = x$ for $\nu = 0.5$ and $\mu = 0.1$, illustrating the trapping region $V := \{x : \nu\mu < x < \mu\}$.

$M(\mu)$, when $x_0 = \nu\mu$ and decreases monotonically as x_0 increases towards μ. We can then define the induced map from $V \to V$ as

$$x_{m(x_0,\mu)} = \widehat{F}(z) \equiv \nu^{m-1}\sqrt{\mu}\sqrt{1-z} + \nu^m\mu \quad \text{where} \quad z = x_0/\mu. \tag{4.21}$$

The map $\widehat{F}(z)$ is continuous on intervals over which $m(x_0, \mu)$ is constant and is discontinuous across the boundaries of such intervals. These intervals are given by

$$V_m = \{z : \mu\nu < \nu^{m-1}\sqrt{\mu}\sqrt{1-z} + \nu^m\mu < \mu\}. \tag{4.22}$$

In particular $\widehat{F}$ has $M(\mu)$ continuous branches, where $M = m(\mu\nu, \mu)$ so that

$$\widehat{F}(\nu) = \nu^{M-1}\sqrt{\mu}\sqrt{1-z} + \nu^M\mu \in [\nu\mu, \mu]. \tag{4.23}$$

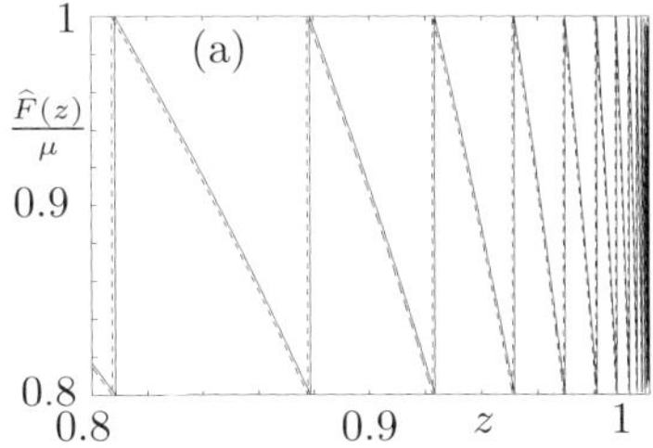

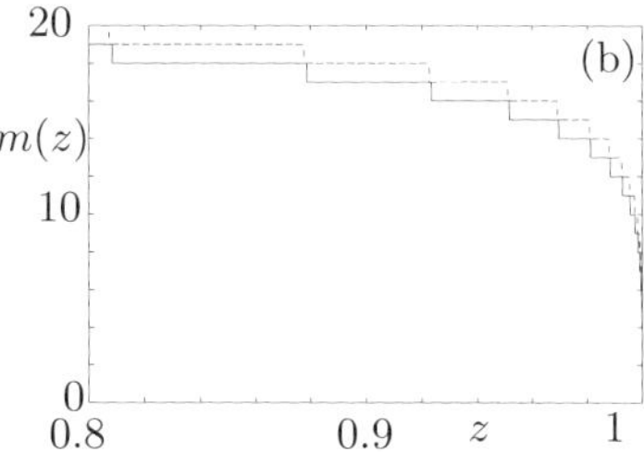

Fig. 4.9. (a) The induced map $\widehat{F}(z)/\mu$ illustrating its discontinuous form together with the intervals over which it is continuous and (b) the associated values of $m(z)$. In each figure we take $\nu = 0.8$ and $\mu = 10^{-4}$ (solid) and $\mu = \nu^2 \times 10^{-4}$ (dashed). These figures show the near equivalence of the function $\widehat{F}(z)/\mu$ in these two cases and the increase of $m(z)$ by one as μ is decreased by the factor ν^2.

Figure 4.9(a) illustrates the map $\widehat{F}(z)/\mu$ for two values of μ in the ratio of $1 : \nu^2$. In particular, we see the map when $M = 19$ and $M = 20$, respectively, and Fig. 4.9(b) shows the variation of $m(z)$ with z. It is apparent from these figures that the form of the map is similar in both cases and that m increases by one. Crucial to the development of the structure of the period-adding cascade is understanding this strong degree of self-similarity in the map $\widehat{F}$ under rescaling in μ. Indeed, if μ is small compared to $1 - \nu$, then $\widehat{F}(z)/\mu$ is closely approximated by $\nu^{m-1}\sqrt{1-\nu}/\sqrt{\mu}$. If we change μ to $\nu^2\mu$, then this approximation to $\widehat{F}(z)/\mu$ does not change on each subinterval provided that m is increased by one. When this occurs, M is also increased by one and an extra branch is added to the map $\widehat{F}$ close to $z = 1$. However, the local form of $\widehat{F}$ on the sub-branches does not change significantly. To make this calculation more precise, we set $k = M - m$ and introduce the parameter

$$\lambda = \left(\frac{\nu}{(\widehat{F}(\nu)/\mu)} \right)^2 \in [\nu^2, 1]. \tag{4.24}$$

It follows from (4.23) that on each interval, the map $\widehat{F}(z)/\mu$ takes the form

$$\frac{\widehat{F}(z)}{\mu} = \frac{\nu^{1-k}\sqrt{1-z}}{\sqrt{1-\nu}\sqrt{\lambda}} + \nu^{M-k}\left(1 - \frac{\sqrt{1-z}}{\sqrt{1-\nu}}\right).$$

Now, if M is large compared to k, then we have

$$\widehat{F}(z)/\mu \approx G(z, \lambda),$$

where the function $G(z, \lambda)$ is defined for all $k = 0, 1, 2, 3, 4, \ldots$ by

$$G(z, \lambda) = \frac{\nu^{1-k}\sqrt{1-z}}{\sqrt{1-\nu}\sqrt{\lambda}}, \tag{4.25}$$

with $G(\nu,\lambda) = \nu/\sqrt{\lambda}$ and k taking constant values on sub-intervals $z \in I_k \subset [\nu,1]$. Now, if λ is held constant, then the map $\widehat{F}(z)/\mu$ which maps the interval $J \equiv [\nu,1]$ to itself, is closely approximated by the function $G(z,\lambda)$. It follows from (4.23) that, if μ is small, then λ is constant precisely when μ is reduced by a factor of ν^2 and M is increased by one. This is the origin of the self-similar structures inherent in the period-adding cascade. The map $G(z,\lambda)$ has a series of branches of increasing slope as z increases. A simple calculation shows that

$$|G_z(z,\lambda)| > |G_z(\nu,\lambda)| = \frac{\nu}{2(1-\nu)\sqrt{\lambda}}. \tag{4.26}$$

Furthermore G has an infinite number of branches and a corresponding *infinite number of fixed points* $z_k \in I_k$. The stability of each such point can be determined from the formula

$$G_z(z_k,\lambda) = -\frac{z_k}{2(1-z_k)}. \tag{4.27}$$

Note that $|G_z(z_k,\lambda)|$ increases as z_k increases, indicating that it is the first fixed point (with $k=0$) that is most likely to be stable and observed. This point satisfies the equation

$$z_0 = \frac{\nu\sqrt{1-z_0}}{\sqrt{1-\nu}\sqrt{\lambda}}. \tag{4.28}$$

Now let us treat each subcase of Theorem 4.2 in turn.

1. If $2/3 < \nu < 1$, then for *all* $\lambda \in (\nu^2, 1]$, it follows from (4.26) that $|G_z(z,\lambda)| > 1$ for all z. Thus *no stable periodic points* exist for G. The only other possibility is chaotic motion confined to the set J. The maximum amplitude of the iterates of the original map $f(x)$ in this chaotic motion is given by the image of the point $\nu\mu$. Hence the strange attractor of $f(x)$ lies in the interval $(\nu\mu, \sqrt{\mu}\sqrt{1-\nu}+\nu\mu)$.
2. If $\nu < 2/3$, then for each fixed M it is possible for the first fixed point z_0 to be *stable* provided that $|G_z(z_0,\lambda)| < 1$. For the original map $f(x)$, such a fixed point corresponds to a periodic *maximal M-orbit* of the symbolic form AB^{M-1}, which has one iterate in the region where the map has a square-root form and $(M-1)$-iterates in the region where the map is linear. As M is kept fixed and λ varies, then there is typically a range of values of $\lambda \in [\nu^2, 1]$ where this orbit exists and is stable. When $\lambda = 1$ and $\nu < 2/3$ we have $z_0 = \nu$ and $|G'(z_0)| = \nu/2(1-\nu) < 1$. As λ decreases, then z_0 and $|G'(z_0)|$ increase, until $z_0 = 2/3$ when $G'(z_0) = -1$ and there is a period-doubling bifurcation, leading to an instability of the fixed point. This arises when
$$\lambda = \lambda_{PD} = \frac{3}{4}\frac{\nu^2}{(1-\nu)}.$$
If $\nu > 1/4$, then $\lambda_{PD} > \nu^2$ and the fixed point z_0 corresponding to the M-orbit exists and is stable if $\lambda \in [\lambda_{PD}, 1] \subset [\nu^2, 1]$. For $\lambda \in (\nu^2, \lambda_{PD})$ we expect to see more complex, and possibly chaotic, motion.

We can now see how *period-adding* arises. In terms of the original variable μ the M-orbit exists and is stable for a range $\mu \in [\mu_{PD}(M), \mu_e(M)]$. If μ is small, then we can estimate λ by

$$\lambda \approx \frac{\mu}{\nu^{2(M-2)}(1-\nu)},$$

so that

$$\mu_e(M) \approx (1-\nu)\nu^{2(M-2)}, \quad \mu_{PD}(M) \approx \frac{3}{4}\nu^{2(M-1)}$$

and there is a periodic M-maximal orbit in the interval

$$\mu \in P_M = (\mu_{PD}(M), \mu_e(M)) \approx \left(\frac{3}{4}\nu^{2(M-1)}, (1-\nu)\nu^{2(M-2)}\right).$$

Because $1/4 < \nu < 2/3$, these intervals do not overlap and have width in geometric progression so that $|\mathcal{C}_{M+1}| = \nu^2|\mathcal{C}_M|$. There are gaps between these intervals, and in these gaps more complex, indeed chaotic motion, will exist.

3. We now finally look at the case of $\nu < 1/4$. In this case, there are no period-doubling bifurcations at which the maximal periodic M-orbit loses stability. Instead, the M-orbit persists as ν is reduced and the regions for existence of the M and $M+1$ orbits overlap. This is because the second fixed point z_1 of the map $G(z,\lambda)$ becomes stable for values of λ for which z_0 is also stable.

4.3.2 Quasi one-dimensional behavior

The one-dimensional square-root map of the previous section is a special case of the more arbitrary n-dimensional square-root type maps that arise in the unfolding of grazing bifurcations in hybrid and piecewise-smooth flows (see Chapters 6 and 7). Generically, these maps combine a linearizable smooth map with a correction on the far side of a discontinuity boundary $\Sigma \equiv \{H(x,\mu) = 0\}$ that to leading order is proportional to $y = \sqrt{|H|}$.

For the sake of simplicity, in what follows we will consider a semi-linearization of such a general map. All of the stated results are easily generalizable under the inclusion of higher-order terms. Accordingly, we now consider square-root maps of the form $x \to f(x,\mu)$, $x \in \mathbb{R}^n$, $\mu \in \mathbb{R}$, where

$$f(x,\mu) = \begin{cases} F_1(x,\mu) = Nx + M\mu + Ey, & \text{if} \quad H(x,\mu) < 0, \\ F_2(x,\mu) = Nx + M\mu, & \text{if} \quad H(x,\mu) > 0. \end{cases} \tag{4.29}$$

Here we will take

$$H(x,\mu) = C^T F_2(x) + D\mu = C^T(Nx + M\mu) + D\mu, \quad y = \sqrt{-H(x,\mu)}, \tag{4.30}$$

and take N, M, E, C, and D to be general matrices of appropriate dimensions. We will refer to the region $H(x,\mu) < 0$ as the impacting region and denote such iterates by a. The slightly cumbersome form of the function $H(x,\mu)$ is motivated by the derivation of square-root maps for hybrid systems that we will describe in Chapter 6. The derivative of the map F_1 with respect to x is given by

$$F_{1,x} = N - \frac{EC^T N}{2y}.$$

Hence terms in this derivative become unbounded as $y \to 0$, which implies a high degree of stretching in appropriate directions. However, the map is still *dissipative* (so that it contracts area) if $|\det(F_{1,x})| < 1$. Maps derived from dissipative flows must have this property. A necessary condition for this to be true as $y \to 0$ is that the operator N is dissipative and that

$$C^T E = 0. \tag{4.31}$$

As we shall see, in Chapter 6, maps with this restriction arise naturally in the analysis of stable orbits undergoing a grazing impact in hybrid systems. In fact, we shall now assume that (4.31) holds in what follows. The map (4.29)–(4.31) was first introduced by Nordmark [197], and in [54] is referred to as the *Nordmark map*, although the map analyzed there is a linear transformation of (4.29).

The non-impacting fixed point x^* of the map F_2 (4.29)–(4.31) can be computed easily and we have

$$x^* = (I - N)^{-1} M\mu. \tag{4.32}$$

The condition for x^* to also be an admissible fixed point of f (i.e., to *not* impact) is that

$$0 < H(F_2(x^*),\mu) = H(x^*,\mu) = C^T(I-N)^{-1}M\mu + D\mu := e\mu. \tag{4.33}$$

However, if x^* is not a fixed point of f, it can still act as an attractor of the iterates of the map F_2. The form that this attractor takes then largely determines the overall dynamics of f. In the simplest case, the fixed point of F_2 is a *stable node* that attracts iterates along a one-dimensional stable manifold. Such behavior arises when the leading eigenvalue ν_1 of the matrix N is *real, positive and less than one.* In this case the iterates of the map f may approach x^* in the region $H(x,\mu) > 0$ until there is a first point when $H(x,\mu) < 0$ and the map F_1 applies. It is this transition that leads to interesting dynamics that can be understood in terms of the behavior of the one-dimensional map.

The situation is more subtle when the operator N has *complex* eigenvalues. In this case rather different behavior from that of the one-dimensional map is observed, with delicate transitions between various types of periodic orbit. We will consider this case in the next section, after we have shown the following direct analog of Theorem 4.2 for higher dimensional maps.

Theorem 4.3 (Quasi-one-dimensional border-collision in the n-dimensional square-root map [106, 171]). *Consider the map (4.29)–(4.30).*

Suppose that for all $n \geq 1$ *we have* $C^T N^n E > 0$ *and that* $0 < \nu_1 < 1$ *is the leading eigenvalue of* N *(that is, the unique real eigenvalue having the largest modulus), then as* μ *passes through zero, the stable fixed point* x^**, which occurs when* $e\mu > 0$ *evolves into one of the following scenarios when* $e\mu < 0$*:*

1. *If* $2/3 < \nu_1 < 1$*, there is a robust chaotic attractor close to the origin for* all *small negative values of* $e\mu$*. The chaotic attractor has size proportional to* $\sqrt{-e\mu}$*. Moreover, the attractor comprises a fixed number,* m*, of thin fingers (almost one-dimensional sets) in the positive* $N^n E$ *directions, where* $0 \leq n \leq m$*.*
2. *If* $1/4 < \nu_1 < 2/3$*, then for* all *small negative values of* $e\mu$ *there is an alternating series of chaotic and stable periodic motions. The periodic motions accumulate in a period-adding cascade as* $\mu \to 0$ *in a sequence of windows that are mapped into each other by a factor asymptotically proportional to* ν_1^2*. The periodic bands have width that increases from zero as* ν_1 *decreases from* $2/3$ *to* $1/4$*.*
3. *If* $0 < \nu_1 < 1/4$*, then there is a period-adding cascade in which the periodic bands overlap and increase in period as* $e\mu \to 0^-$*.*

Remark. In Case 1, it can be shown that the principal Lyapunov exponent on the chaotic attractor scales as $1/\log(-e\mu)$ as $\mu \to 0$ [171]. Moreover, for larger values of $-e\mu$, the attractor may terminate in a period-subtracting cascade from an initial maximal orbit.

Proof. If the effect of impacts are ignored, then the map F_2 has a fixed point $x^*(\mu)$. We assume that $H(x^*, \mu) = e\mu = -\sigma$ with $\sigma > 0$. To study the dynamics close to x^*, when σ is small, we set $x = x^* + z$. On rescaling, we have

$$z \to f(z) = \begin{cases} Nz, & C^T Nz > \sigma, \\ Nz + E\sqrt{\sigma - C^T Nz} & C^T Nz < \sigma. \end{cases}$$

We identify two regions: Region I, for which $C^T Nz < \sigma$, and Region II for which $C^T Nz > \sigma$. As N is contracting, and as $C^T N^n E > 0$ for all n, it follows that the origin in the new co-ordinate system is a global attractor. Hence, Region I is *a trapping region* for the map into which all iterates must eventually fall. Given a starting point z_0 in Region II, there is typically a series of (linear) iterates of $z_n = f^n(z_0) = N^n z_0$ in Region II that asymptotically approach the origin. Eventually an iterate z_n enters Region I. Let M be the first iterate to enter Region II, so that $z_n = N^n z_0$ is in Region I for $0 \leq n \leq M - 1$. Now, in Region I the high derivative of the square-root term implies that a small neighborhood of $z = 0$ is stretched along the direction of the vector E with the most significant local stretching occurring when $C^T N z_M$ is close to σ. For σ sufficiently small then, as $C^T NE > 0$, point $f(z_m)$ re-enters Region II in the next iteration of f, and so such an event sequence repeats. See Fig. 4.10.

In Region II we have linear dynamics: $z \to Nz$. Suppose that, corresponding to the eigenvalue ν_1, N has right eigenvector ψ_1 normalized so that

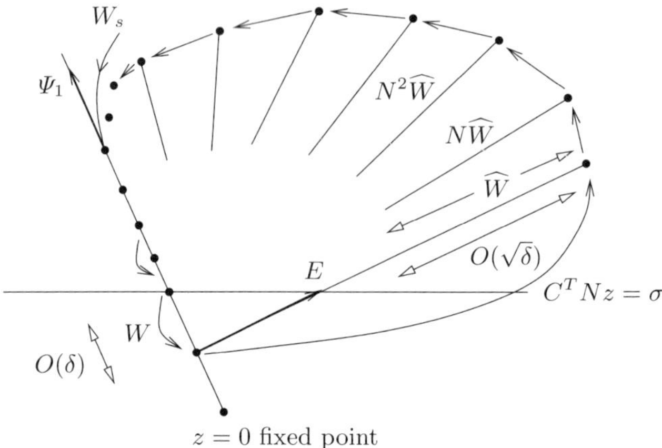

Fig. 4.10. Illustration of the proof of Theorem 4.3.

$C^T\psi_1 > 0$, and left eigenvector ϕ_1. That is, $N\psi_1 = \nu_1\psi_1$, and $N^T\phi_1 = \nu_1\phi_1$. If z_0 is a point in Region II, then for sufficiently large n, we have

$$z_n = N^n z_0 \approx \nu_1^n(\phi_1 z_0)\psi_1$$

so that the iterates approach the origin along the direction ψ_1. Along this line there will be a first value of n for which $C^T N z_n < \sigma$ but $C^T z_n > \sigma$.

Let $\gamma > 0$ and set $z = \sigma\gamma\psi_1$. We next consider the one-dimensional interval

$$W = \{\sigma\gamma\psi_1 : \nu_1\gamma_b < \gamma < \gamma_b\} \quad \text{with} \quad \gamma_b = 1/(\nu_1(C^T\psi_1)).$$

Here the range of values of γ is chosen so that W lies in Region I, but $N^{-1}W$ lies in Region II. For large n, the iterates of z_0 approaching the origin in the direction of ψ_1, enter the set W and then experience stretching in the direction of E. The image $\widehat{W}$ of the set W under the action of the square-root map is given by

$$\widehat{W} = F_1(W) = \{\gamma\nu_1\sigma\psi_1 + E\sqrt{\sigma}\sqrt{1-\gamma/\gamma_b}\}. \tag{4.34}$$

The set W is thus stretched into the set $\widehat{W}$, which is almost tangent to E for $\sigma \ll 1$ and is of 'length' $\sqrt{\sigma}\sqrt{1-\nu_1}|E|$. As $C^T N E > 0$, then, nearly all points of $\widehat{W}$ will be in Region II unless γ is very close to γ_b. Subsequent iterations of the map of the form $N^n\widehat{W}$ will comprise a series of 'fingers' which lie in Region II, and return to Region I (owing to the fact that the origin is an attracting fixed point) after M iterations; see Fig. 4.10.

The value of M will be largest when $\gamma = \nu_1\gamma_b$. Hence, we can estimate M by solving the equation

$$C^T N^M E\sqrt{\sigma}\sqrt{1-\nu_1} = \sigma.$$

Noting that for large M we have $C^T N^M E \approx \nu_1^M \alpha$ where $\alpha > 0$ is a constant, we get

$$\nu_1^M \approx \sqrt{\sigma}/(\alpha\sqrt{1-\nu_1}),$$

so that M varies as $\log(\sigma)/2\log(\nu_1)$. During the iterations, the orbits are bounded by a set of size proportional to $\sqrt{\sigma}$. More generally, the points in $\widehat{W}$ will return to W after a number $m(\gamma, \sigma)$ of iterations, giving an induced map F from W to itself parameterized by σ The function $m(\gamma, \sigma)$ is a locally constant discontinuous function of γ, which, to leading order, is constant over the interval

$$W_m = \{\gamma : \nu_1 \gamma_b \sigma < \alpha \nu_1^m \sqrt{\sigma}\sqrt{1-\gamma/\gamma_b} < \gamma_b \sigma.\}.$$

This behavior of the induced map F is entirely equivalent to that of the one-dimensional map described in (4.22), and the rest of the proof follows that of Theorem 4.2.

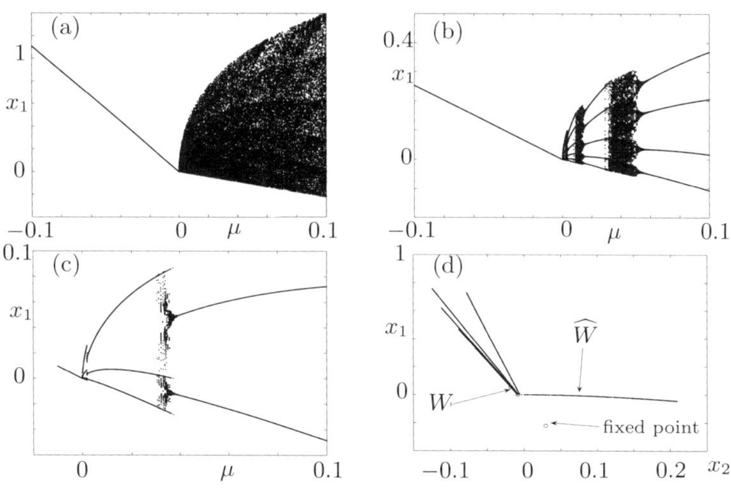

Fig. 4.11. Bifurcation diagrams for the two-dimensional maps in example 4.2: (a) Immediate jump to chaos when $N = N_1$, (b) period-adding cascade interspersed with chaos when $N = N_2$, (c) overlapping period adding cascade when $N = N_3$ and (d) chaotic attractor for case (a) when $\mu = 0.02$ showing the series of fingers given by the iterates of the set W. See text for explanation of labels

Example 4.2. We illustrate this behavior by taking three specific two-dimensional square-root maps with the matrix $N = N_i$ given by one of the following three cases:

$$N_1 = \begin{pmatrix} 1.4635 & 3.8396 \\ -0.2063 & -0.3935 \end{pmatrix},$$

$$N_2 = \begin{pmatrix} 0.7833 & 1.6660 \\ -0.0895 & -0.0175 \end{pmatrix}, \quad N_3 = \begin{pmatrix} 0.3252 & 0.7244 \\ -0.0389 & -0.0252 \end{pmatrix}.$$

These operators have eigenvalues $\{\nu_1, \nu_2\}$ given, respectively, by $\{0.8,\ 0.27\}$, $\{0.5,\ 0.27\}$, and $\{0.2,\ 0.1\}$. We take

$$M = (-0.5337,\ -0.2277), \quad E = (0,\ 1), \quad C^T = (1,\ 0), \quad D = 0,$$

and set $x = (x_1, x_2)$. The resulting bifurcation diagrams for iterations of the Nordmark map with $N = N_1, N_2, N_3$ are given in panels (a)–(c) of Fig. 4.11. In these plots we see, respectively, an immediate jump to chaos, a period-adding cascade with periodic windows separated by regions of chaos and a period-adding cascade with overlapping periodic orbits. In all cases the attractor grows like $\sqrt{\mu}$ for positive μ, in perfect agreement with Theorem 4.3. Note in panel (b) we see evidence of orbits of periods 4, 5 and 6 within windows whose width decreases by a factor of $0.25 = 0.5^2$ as $\mu \to 0$.

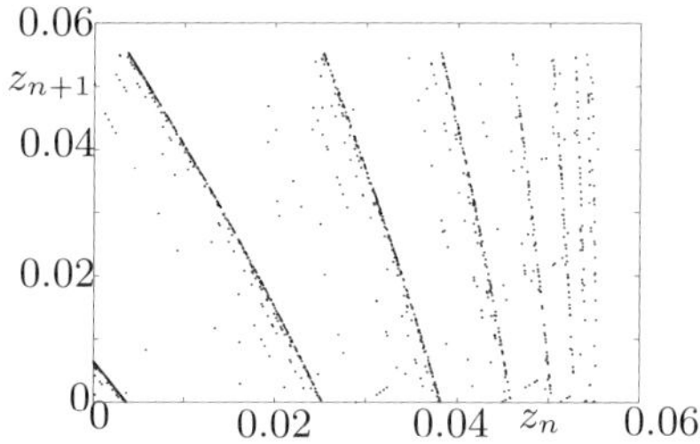

Fig. 4.12. The computed return map from the set W to itself, plotting successive iterates z_1. Note the strong similarity between this map and the map $G(x, \lambda)$ illustrated in Fig. 4.9.

The chaotic attractor plotted in the original co-ordinates for $\mu = 0.02$ in Fig. 4.11(a) with $N = N_1$ is given in Figure 4.11(d). In this case we have $\psi_1 = [0.9854 \ -0.1704]$ and the fixed point is given by

$$\bar{x} = (I - N_1)^{-1} M\mu = (-11.0653\ 1.474)\mu$$

so that $e = -11.06$ and $\sigma = -11.06\mu$. Note the star shaped appearance of the attractor with iterates lying along the sets $N_1^n E$. The sets W and $\widehat{W}$ are indicated together with the inadmissible fixed point of the linear portion of the map. A direct calculation shows that the set W is given by $x \in [0, 0.0553]$. Using this, we may calculate the first return map from the set W to itself directly from the iterates of the map. To do this we find those iterates for which $x_1 \in W$ and plot each iterate against the previous such iterate. The resulting map is plotted in Fig. 4.12. The almost piecewise (discontinuous) nature of the map is immediate from the figure, which we can compare directly with Fig. 4.9.

4.3.3 Periodic orbits bifurcating from the border-collision

For more general forms of the matrix N, particularly when it has complex eigenvalues, there is much less we can say about the overall dynamics of the square-root map. Instead, as in Sec. 3.2, we shall focus on classifying the behavior of the simplest kind of periodic orbits, first dealing with existence, and then stability. We will establish precise conditions for the existence of periodic orbits of arbitrary period and then consider the case of two-dimensional maps in detail. The presentation is based on the work of Nordmark in [199].

To do this we will slightly generalize our earlier discussions and consider a general square-root map of the following form, which agrees with the earlier map (4.30) to leading order:

$$f(x,\mu) = \begin{cases} F_1(x,\mu) = F_2(x,\mu) + E(F_2(x,\mu), y, \mu)y, & \text{if} \quad H(x,\mu) < 0, \\ F_2(x,\mu) = Nx + M\mu, & \text{if} \quad H(x,\mu) > 0, \end{cases} \tag{4.35}$$

where

$$H(x,\mu) = C^T F_2(x) + D\mu + O(|x,\mu|^2) \quad \text{and} \quad y = \sqrt{-H(x,\mu)} > 0, \tag{4.36}$$

and E is a generals smooth function whose leading-order term $E(0,0)$ is the matrix E defined earlier.

It follows immediately, from an application of the Implicit Function Theorem, that the fixed point $x^* = F_2(x^*, \mu)$ of the map F_2 exists provided that $(I-N)^{-1}$ exists. This is a small perturbation of the point $(I-N)^{-1}M\mu$ and is a fixed point of f provided that

$$H(x^*, \mu) \equiv e\mu > 0.$$

Our aim is now to establish similar necessary conditions for the existence of more complex periodic orbits of f, which have the symbolic form

$$b^{n_1-1}ab^{n_2-1}a \ldots b^{n_m-1}a.$$

Typically these will have a given *impacting sequence* in which starting with a point x_1 the system has $n_1 - 1$ non-impacting iterations of f followed by an impact for which $y = y_1$ leading to a new state x_2 with a further $n_2 - 1$ non-impacting iterations and so on. A *necessary* set of equations satisfied by such a periodic orbit is

$$\begin{aligned} x_2 - E(F_2^{n_1}(x_1,\mu), y_1, \mu)y_1 - F_2^{n_1}(x_1,\mu) &= 0, \\ H(F_2^{n_1-1}(x_1,\mu)) + y_1^2 &= 0, \\ &\vdots \\ x_1 - E(F_2^{n_m}(x_m,\mu), y_m, \mu)y_m - F_2^{n_m}(x_m,\mu) &= 0, \\ H(F_2^{n_m-1}(x_m,\mu)) + y_m^2 &= 0. \end{aligned} \tag{4.37}$$

For this periodic orbit to exist, it must also satisfy the compatibility conditions for the non-impacting iterates given by

$$H(F_2^k(x_i,\mu),\mu) > 0 \text{for} 0 \le k \le n_i - 1 \quad \text{and} \quad y_i > 0. \tag{4.38}$$

We initially consider the linearized system obtained from (4.37) by setting $F_2(x,\mu) = Nx + M\mu$ and $H(x,\mu) = C^T F_2(x) + D\mu = C^T(Nx + M\mu) + D\mu$ and $E(F_2(x), y, \mu) = E(0,0,0)$. This gives a set of linear equations, subject to linear constraints, acting on the vector $w \in \mathbb{R}^{m\times(n+1)}$ defined as

$$w = \begin{bmatrix} x_1 \; y_1 \; \cdots \; x_m \; y_m \end{bmatrix}^T.$$

In this linearized system, the solution $w(\mu)$ is directly proportional to μ.

More generally, if we consider the set of nonlinear equations $\mathcal{F}(w) = 0$ defined by (4.37), then the solution $w(\mu)$ is given uniquely, for small μ, by the Implicit Function Theorem, provided that $\frac{\partial \mathcal{F}}{\partial w}$ is nonsingular at $(0,0)$ and (4.38) hold. To simplify the calculation of this solution, we rescale the system of equations (4.37) by setting $x_i = x^*(\mu) + e\mu X_i$, and $y_i = e\mu Y_i$. A periodic solution, which solves the rescaled system, can then be described by the vector

$$W = \begin{bmatrix} X_1 \; Y_1 \; \cdots \; X_m \; Y_m \end{bmatrix}^T.$$

A period-k non-impacting sequence of iterations of F_2 then has the form

$$F_2^k(x_1,\mu) = F_2^k(x^*(\mu) + e\mu X_1, \mu),$$

which gives on expansion, up to an error of $O(\mu^2)$,

$$F_2^k(x_1,\mu) \approx F_2^k(x^*(\mu),\mu) + F_{1,x_1}^k e\mu X_1 = x^*(\mu) + e\mu N^k X_1 \tag{4.39}$$

Similarly, expressing H in terms of the new co-ordinates we have

$$H(F_2^k(x_1,\mu),\mu) \approx H(x^*(\mu)) + e\mu C^T N^k X_1 = e\mu(1 + C^T N^k X_1). \tag{4.40}$$

Taking the first line of (4.37) with (4.39) and (4.40) gives

$$e\mu X_2 - Ee\mu Y_1 - e\mu N^{n_1} X_1 + O(\mu^2) = 0,$$

and the second line of (4.37) yields

$$e\mu + e\mu C^T N^{n_1} X_1 + O(\mu^2) = 0.$$

Similar expressions for the rest of the impacting sequence can be derived and put into matrix form. For the impacting sequence $b^{n_1-1}ab^{n_2-1}a\ldots b^{n_m-1}a$, the Jacobian matrix of the system defined by (4.37), say $J = \mathcal{F}_W(0,0)$, can then be shown to have a block-banded structure of the form

$$J = \frac{\partial \mathcal{F}}{\partial W} = \begin{bmatrix} -N^{n_1} & -E & I & 0 & 0 & \dots & 0 & 0 \\ C^T N^{n_1} & 0 & 0 & 0 & 0 & \dots & 0 & 0 \\ 0 & 0 & -N^{n_2} & -E & I & \dots & 0 & 0 \\ 0 & 0 & C^T N^{n_2} & 0 & 0 & \dots & 0 & 0 \\ \\ \dots & \dots & \dots & \dots & \dots & \dots & \dots & \dots \\ I & 0 & 0 & 0 & 0\dots & -N^{n_m} & -E & \\ 0 & 0 & 0 & 0 & 0\dots & C^T N^{n_m} & 0 & \end{bmatrix}. \quad (4.41)$$

The entire linearized system can then be written as

$$JW(\mu) = \begin{bmatrix} 0 \\ -1 \\ \vdots \\ 0 \\ -1 \end{bmatrix} + \mathcal{O}(\mu). \quad (4.42)$$

This system of equations must be augmented with the compatibility conditions (4.38), which give

$$\operatorname{sign}(e\mu)(1 + C^T N^k X_i(\mu) + 1) + O(\mu) > 0 \text{ for } 1 \leq k \leq n_i - 1 \quad (4.43)$$

and

$$\operatorname{sign}(e\mu) Y_i(\mu) > 0 \quad (4.44)$$

derived as above. Here we divide by the modulus of $e\mu$ to preserve the dependence on the sign of $e\mu$. The linear set of equations defined by (4.42) can then be solved for W in all cases. A solution with impacting sequence $(n_1, n_2, ..., n_m)$ exists for small μ for one of $\mu > 0$ or $\mu < 0$ if the compatibility conditions (4.43) and (4.44) hold for the solution to (4.42). The conditions $\det(I - N) \neq 0$, $e \neq 0$ and $\det J \neq 0$ ensure that this calculation is correct to $\mathcal{O}(\mu^2)$ for the more general system. If any expression on the left-hand side of the compatibility condition (4.43) or (4.44) are negative then no solution exists for that sequence of impacts. Close to the border-collision at $\mu = 0$, the term e can be taken to be a constant, so that the left-hand sides of (4.43) and (4.44) change sign as μ passes through zero. Hence the necessary conditions cannot simultaneously hold for both signs of μ. So, if all conditions in (4.43) and (4.44) are of the same sign and positive for a given impacting sequence, then the orbit exists for the same sign of μ as the simple non-impacting fixed point. Conversely, if all the conditions are of the same sign but negative, then the orbit exists on the opposite side to the non-impacting orbit. Finally, if conditions (4.43) and (4.44) are of both signs for a particular orbit sequence $(n_1, n_2, ..., n_m)$, then no periodic orbit of that type can be either created or destroyed in the bifurcation.

Let us now discuss the *stability* of the bifurcating period orbits. Whereas the stability of a non-impacting orbit is determined by (the eigenvalues of) the matrix N, the stability of an impacting orbit is strongly affected by the stretching implicit in the map F_1. To estimate the nature of any such (in)stability, consider the map associated with a sequence of $n-1$ non-impacting iterations followed by one impact. In the notation of the previous section, this is given by

$$Q(X_1, y_1) \equiv N^n X_1 + E y_1, \quad \text{with} \quad C^T N^n X_1 + y_1^2 + e\mu = 0.$$

If we make the perturbations $X_1 \to X_1 + \Delta X_1, y_1 \to y_1 + \Delta y_1$ we have

$$\Delta Q = N^n \Delta X_1 + E \Delta y_1, \quad \text{with} \quad C^T N^n \Delta X_1 + 2y_1 \Delta y_1 = 0.$$

Eliminating Δy_1 between these two expressions gives

$$\Delta Q = (N^n - [C^T N^n E]/2y_1)\Delta X_1.$$

Hence the dominant growth factor in ΔQ is given by $-C^T N^n E/2y_1$. As y_1 is of order μ as $\mu \to 0$ at the bifurcation, this will lead to unbounded growth in this term; hence, close to the bifurcation the orbit corresponding to this impact sequence is unstable. If an orbit has more impacts, then this growth factor increases. As $|\mu|$ (and hence y_1) increases away from the bifurcation point at $\mu = 0$, the impacting orbit may stabilize if $|C^T N^n E/2y_1| < 1$. Clearly this is most likely if all the eigenvalues of N have modulus less than unity, n is large and only one impact occurs. periodic orbits with a single impact and period n for large n take the symbolic form $b^{n-1}a$. These orbits have the same form as the orbits of the one-dimensional square-root map and if n takes its largest possible value M for a given set parameters. We shall term these *maximal* orbits [54]. From the above discussion we see that of all possible periodic orbits these are the most likely to be stable in a neighborhood of the bifurcation point. Determining the maximum value of n in terms of the system parameters thus gives important insights into the resulting dynamics. We restrict attention to three special cases:

Example 4.3 (Bifurcation of single impact period-one orbit). The simplest periodic impacting orbit is a *ba* orbit. To leading order, the system in (4.42) can be written as

$$(I - N)X_1 - EY_1 = 0, \quad 1 + C^T N X_1 = 0.$$

Rearranging these equations and converting back into the original co-ordinates with $y_1 = e\mu Y_1$ yields

$$y_1 = -\frac{e\mu}{C^T N(I-N)^{-1}E}, \quad x_1 = x^* + (I-N)^{-1}E y_1.$$

For a general period-n orbit, we define the function $s(n)$ by

$$s(n) := C^T N^n (I - N^n)^{-1} E. \tag{4.45}$$

As $C^T E = 0$, this is identical to

$$s(n) = C^T(I - N^n)^{-1}E.$$

This linearized expression for the single impact period-one orbit then takes the form

$$y_1 = -\frac{e\mu}{s(1)}, \qquad x_1 = x^* - (I - N)^{-1}E\frac{e\mu}{s(1)}. \tag{4.46}$$

To satisfy the compatibility condition, we require $y_1 > 0$, so that $e\mu$ must have the opposite sign to $s(1)$. If we assume that the non-impacting orbit arises when μ is small and $\mu < 0$, then we must have $e < 0$, so that the sign of $e\mu$ is $+1$. It follows that if $s(1)$ is *negative*, then the impacting periodic orbit exists in the same range ($\mu < 0$) as the non-impacting periodic orbit and coalesces with it when $\mu = 0$ in a non-smooth fold. Conversely, if $s(1)$ is *positive*, then the impacting orbit exists in the opposite range ($\mu > 0$, for small μ) from the non-impacting orbit and hence we have a persistence scenario, where the period-one non-impacting orbit continuously evolves into the single-impact orbit as μ passes through 0. More formally, applying the Implicit Function Theorem, we deduce the existence of the period one impacting orbit of the full nonlinear system provided that $e, s \neq 0$ and $(I - N)^{-1}$ exists. Note further that y_1, and hence $x_1 - x^*$, is proportional to μ so that the typical bifurcation picture takes one of the two forms illustrated in Figs. 4.13 in which we observe either super- or sub-critical behavior.

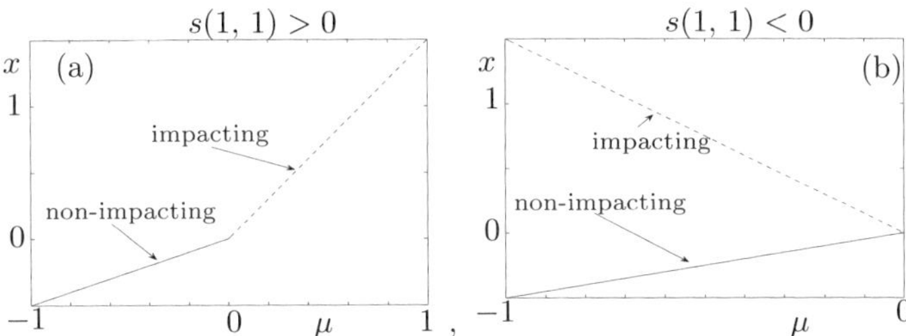

Fig. 4.13. The two kinds of discontinuity induced bifurcation of period-one orbits in the general square root map (4.29)–(4.31): (a) super-critical (persistence) if $s(1) > 0$ and (b) sub-critical (non-smooth fold) if $s(1) < 0$.

Example 4.4 (Bifurcation of single-impact period-two orbit). We now consider impacting orbits with symbol sequence b^2a. To leading order, these satisfy the linear equations

$$(I - N^2)X_1 - EY_1 = 0, \quad 1 + C^T N^2 X_1 = 0,$$

together with the compatibility condition

$$e\mu(1 + C^T N X_1) > 0.$$

Combining the two equations we have

$$e\mu C^T N(I-N)X_1 > 0 \quad \text{and} \quad X_1 = (I-N)^{-1}(I+N)^{-1}EY_1.$$

Hence this orbit exits provided that

$$e\mu C^T N(I+N)^{-1}EY_1 > 0, \quad \text{with} \quad Y_1 = -\frac{1}{s(2)}, \quad y_1 = -\frac{e\mu}{s(2)} > 0,$$

where $s(n)$ was defined by (4.45). As $e\mu Y_1 > 0$, the compatibility condition reduces to

$$C^T N(I+N)^{-1}E > 0.$$

The existence of such an orbit is thus guaranteed for some choice of sign of $e\mu$ provided that $s(2) \neq 0$, $C^T N(I+N)^{-1}E > 0$ and $(I+N)^{-1}$ and $(I-N)^{-1}$ both exist. Note that as $y_1 = -e\mu/s(2) > 0$, the b^2a orbit exists when $e\mu$ and $s(2)$ take opposite signs.

Example 4.5 (Bifurcations of maximal orbits). The two examples considered above are potentially both examples of maximal orbits of the form $b^{n-1}a$. As discussed, we are particularly interested in those maximal orbits for which n takes the largest possible for a given set of parameters. Such orbits satisfy the (linearized) equation

$$(I-N^n)X_1 - EY_1 = 0, \quad 1 + C^T N^n X_1 = 0,$$

with the compatibility conditions

$$e\mu(1 + C^T N^k X_1) > 0, \quad k = 1 \dots n-1, \quad e\mu Y_1 > 0.$$

For certain ranges of the eigenvalues of N there is indeed a maximum value of $n = M$ for which such an orbit exists. To see this, suppose that the leading eigenvalue of N is λ_1; then asymptotically, there is a constant a for which

$$1 + C^T N^k X_1 = 1 + a\lambda_1^k.$$

As $1+C^T N^n X_1 = 0$, then asymptotically $1+C^T N^k X_1 = 1-\lambda_1^{k-n}$. If k and n are both large, and if λ_1 is real and positive, then this expression has a constant sign. In this case we may see orbits of arbitrary period. In fact precisely this behavior was observed earlier in Theorem 4.3, closely associated with the existence of a chaotic attractor. In contrast, if λ_1 is negative or complex, then $1 + C^T N^k X$ cannot keep a constant sign for large values of k. In this case we see a maximal orbit with a largest value M of n for which the compatibility conditions can hold. If the eigenvalues of N change under parameter variation, then we would expect to see changes in the value of the maximal value M. In particular $M \to \infty$ as the imaginary part of $\lambda_1 \to 0$. We shall see this in the next sub-section when we look at maps in two-dimensions. Furthermore, if as parameters in the system vary, λ_1 changes from being complex to real, then we see a transition from a large-period maximal orbit, to chaotic behavior in the post-bifurcation regime. This transition was investigated by Chin *et al.* [54]. They observed that stable maximal orbits can arise in (smooth) fold bifurcations $\mu = \mu_{SN}$ close to the border-collision at $\mu = 0$.

In general it is not easy to analyze such fold bifurcations for the full nonlinear map as higher-order terms in the expressions for the orbits must be taken into account. In cases where the square-root maps arise as Poincaré maps of flows, these higher-order terms must also include detailed calculations of the flow that are often difficult to determine. However, an analysis of the fold structure is possible in the case of the Nordmark map described exactly by (4.29)–(4.31). If we include the quadratic terms and pose the system in the original co-ordinates, we find that y_1 satisfies the quadratic equation

$$e\mu + s(n)y_1 + y_1^2 = 0, \tag{4.47}$$

so that

$$y_1 = -\frac{s(n)}{2} \pm \frac{1}{2}\sqrt{s(n)^2 - 4e\mu}$$

with

$$\begin{aligned} F_2^{n-1}x_1 &= (I-N)^{-1}M\mu + N^{n-1}(I-N^n)^{-1}Ey, \\ x_1 &= NF_2^{n-1}x_1 + M\mu + Ey_1, \\ &\text{etc.} \end{aligned} \tag{4.48}$$

It is immediate that such orbits can only occur close to the bifurcation point for the values of μ for which $-e\mu/s(n) > 0$, however, for larger values of μ, they exist when $-e\mu > 0$. A saddle-node bifurcation then arises in the case of $s(n) < 0$ at the point

$$\mu_{SN} = \frac{s(n)^2}{4e}.$$

Of course, we must also check the compatibility conditions for any given example system. We give one such example in the following section.

4.3.4 Two-dimensional square-root maps

The conditions for periodic orbits derived in Sec. 4.3.3 can in principle be used to classify any border-collision in a square-root map. However, they are in general too unwieldy to use in practice. In fact, since only maximal orbits are likely to be stable, much more practical information can be obtained by considering only such orbits. We explain how to use maximal orbits in such a classification, restricting attention to the case of *two-dimensional maps*, for which we can develop a much more complete theory. Such maps are important in applications. Indeed, two-dimensional square-root maps arise naturally in the study of impacting systems of the form described in case study I in the Introduction, where the underlying system is described by a second-order, non-autonomous differential equation. This leads to a flow in a three-dimensional phase space, and the natural Poincaré map associated with this flow is two-dimensional. The link between this map and the map described in this chapter will be made precise in Chapter 6, but the key feature of this link is that nearly all of the parameter space of the possible two-dimensional maps can be explored by considering impacting systems with various combinations of coefficients.

A simple co-ordinate transformation, such as that described for the two-dimensional piecewise-linear continuous maps described in Chapter 3, transforms the matrices N E and C^T into the normal form given by

$$N = \begin{pmatrix} \tau & 1 \\ -\delta & 0 \end{pmatrix}, \quad E = \begin{pmatrix} 0 \\ 1 \end{pmatrix} b\gamma, \quad C^T = [1\ 0], \tag{4.49}$$

where $b > 0$. Note that the condition $C^T E = 0$ is satisfied automatically for such a normal form. Now suppose that the eigenvalues of N are given by λ_1, λ_2. Then $\tau = \lambda_1 + \lambda_2$ is the trace of N and $\delta = \lambda_1 \lambda_2$ is the determinant, and we can consider the different types of behavior that arise as τ and δ vary. A maximal orbit arises for those values of n, X and Y for which

$$(I - N^n)X - EY = 0, \quad C^T N^n X + 1 = 0, \quad Y = -1/C^T N^n (I - N^n)^{-1} E.$$

Because $C^T E = 0$, we have

$$C^T (I - N^n) X = 0, \quad \text{and hence} \quad C^T X + 1 = 0.$$

The compatibility conditions are then given by

$$e\mu(C^T N^k X + 1) > 0 \quad \text{and} \quad e\mu Y > 0,$$

these simplify to

$$e\mu b_{n,k} > 0 \quad \text{and} \quad e\mu\gamma c_n > 0,$$

where

$$b_{n,k} = C^T N^k X + 1, \quad \text{and} \quad c_n = C^T (I - N^n) X.$$

A key feature of the analysis of the two-dimensional square-root map is that N^n can be calculated in terms of its eigenvalues, and hence the fixed points and periodic orbits can be calculated explicitly. For example, the vector X is given by

$$X = \left(-1, \frac{(\lambda_1^{n+1} - \lambda_2^{n+1}) - (\lambda_1 - \lambda_2)}{\lambda_1^n - \lambda_2^n} \right).$$

Given this value for X, the curves along which $b_{n,k} = 0$ represent transitions (called *iterate boundaries* in [199]) across which the value of n must change along the maximal orbit. Curves $c_n = 0$ are lines in which the *direction* of the bifurcation changes. Observe that for a given X, the possible existence of the maximal n-orbit then depends upon the signs of $e\mu$ and γ. Using the expression for X we can calculate $b_{n,k}$ and c_n entirely in terms of the values of λ_i so that

$$b_{n,k} = \frac{-(\lambda_1\lambda_2)(\lambda_1^{n-k} - \lambda_2^{n-k}) - (\lambda_1^k - \lambda_2^k) + (\lambda_1^n - \lambda_2^n)}{\lambda_1^n - \lambda_2^n},$$

$$c_n = \frac{-(\lambda_1 - \lambda_2)(\lambda_1^n - 1)(\lambda_2^n - 1)}{\lambda_1^n - \lambda_2^n}.$$

As the parameters in the problem vary, the eigenvalues of N change and hence the possible periodic orbits also change. The above expressions for $b_{n,k}$ and c_n allow the existence regions for the various types of orbit to be determined entirely in terms of the eigenvalues of N, and hence in terms of τ and δ. Suppose we assume that the pre-impacting system is stable so that $0 < \delta \leq 1$. Following [197], [54] we can identify five regions of parameter space corresponding to qualitatively different forms of behavior. These arise from different configurations of the eigenvalues of N and the nature of the fixed point x^* of the non-impacting system.

1. $\tau > \delta + 1$; **saddle-point**: $0 < \lambda_2 < 1 < \lambda_1$.
2. $\tau - 1 < \delta < \tau^2/4$; **stable-node**: $0 < \lambda_2, \lambda_1 < 1$.
3. $\tau > \tau^2/4$; **stable-spiral**: $\lambda_{1,2} = \rho e^{\pm i\theta}$, $0 < \rho < 1$.
4. $-\tau - 1 < \delta < a_1^2/4$; **flip node**: $-1 < \lambda_2 < \lambda_1 < 0$.
5. $\delta < -\tau - 1$; **flip saddle**: $\lambda_2 < -1 < \lambda_1 < 0$.

In Regions 1, 2, 4 and 5 the system is 'over-damped' and the eigenvalues are real. In Region 3 the system is 'under-damped' and the eigenvalues are complex. We now briefly examine the dynamics of the full impacting system in Regions 1, 2 and 3. The dynamics in Regions 4 and 5 are similar to that in Regions 2 and 1, respectively, when we consider the second iterate of the map. Hence the fundamental bifurcation of the period-one orbit becomes more akin to a period-doubling than to a fold or persistence. Typically we see chaotic behavior and/or period-adding in Regions 1 and 2, similar to that of the quasi-one-dimensional map. In contrast, in Region 3 we typically see periodic behavior and complex transitions between orbits of high period.

Regions 1 and 2: real positive eigenvalues. In these regions the map has the behavior of the quasi-one-dimensional map described in Sec. 4.3.2, and we can now explore this in more detail. If the eigenvalues λ_i are both positive, then the sign of $b_{k,n}$ is independent of k and n and is the same as the sign of $(\lambda_1 - 1)(1 - \lambda_2)$ [199]. A similar result holds for c_n, with both c_n and $b_{n,k}$ changing sign as λ_1 passes through 1. Hence, provided that $0 < \lambda_i$, all orbits are possible.

In Region 2, we have $b_{n,k}$ and c_n both negative. Hence $e\mu$ must be negative and γ must be positive. In Region 2, all periodic orbits can exist, but most are unstable. The transitions between the various types of bifurcation diagram occur when $\lambda_1 = 2/3$ and $\lambda_1 = 1/4$. As $\tau = \lambda_1 + \lambda_2$ and $\delta = \lambda_1\lambda_2$ these boundaries form the two lines

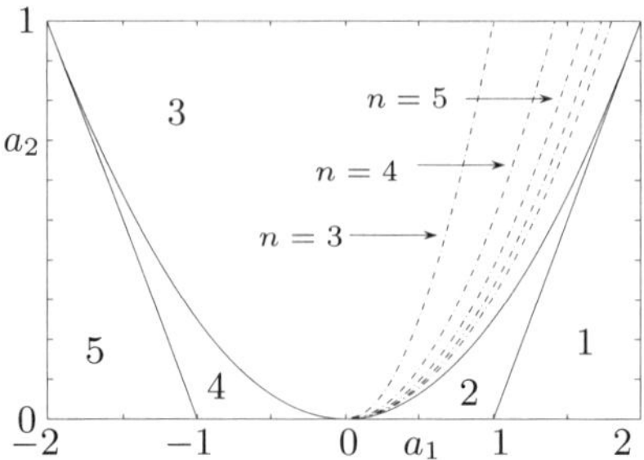

Fig. 4.14. The five regions (labeled by numbers from 1 to 5) of the two-dimensional square root map (4.29)–(4.31) as functions of τ and δ, corresponding to the qualitatively different eigenvalue configurations for the matrix N. Dashed lines indicate the boundaries inside Region 3 where the number n changes for the maximal orbit, see text for details.

$$\delta = \frac{2}{3}\left(\tau - \frac{2}{3}\right) \quad \text{and} \quad \delta = \frac{1}{4}\left(\tau - \frac{1}{4}\right).$$

Suppose now that λ_i has associated right eigenvector ϕ_i and left eigenvector ψ_i; then

$$C^T N^n E = \lambda_1^n (C^T \phi_1)(\psi_1 E) + \lambda_2^n (C^T \phi_2)(\psi_2 E) \equiv p\lambda_1^n + q\lambda_2^n.$$

As $0 < \lambda_2 < \lambda_1$ the condition $C^T N^n E$ is true for all n provided it is true for $n = 1$, which is true since we can choose

$$\phi_{1,2} = \left(1, -\frac{\tau}{2} \pm \frac{\sqrt{\tau^2 - 4\delta^2}}{2}\right)^T, \quad \psi_{1,2} = \left[-\frac{\tau}{2} \mp \frac{\sqrt{\tau^2 - 4\delta^2}}{2}, 1\right],$$

so that $(C^T \phi_i) = (\psi_i E) = 1$. Hence we can apply Theorem 4.3, and the bifurcation sequences will be much like those for the one-dimensional square-root map.

Region 3: Complex Eigenvalues. In Region 3, we see very different behavior from the quasi-one-dimensional case. For a fixed set of parameter values τ, σ periodic orbits of period up to a maximum value of $n \le M(\tau, \sigma)$ can exist and there are many internal boundaries corresponding to the change in type of the maximal orbit, i.e., where $b_{k,n}$ and c_n change sign for different n. The first few such curves are depicted as dashed lines in Fig. 4.14. They accumulate, as $M \to \infty$ on the curve $\delta = \tau^2/4$. This curve marks the transition between real and complex eigenvalues. If $n = 1$, then an orbit exists for $e\mu > 0, \gamma < 0$ and for $e\mu < 0, \gamma > 0$ when $\tau < \delta + 1$. If $n = 2$ we have $c_n = -|\lambda_1^2 - 1|^2/\tau$, which has a constant sign if τ has a constant sign, similarly for $b_{k,n}$. If $\tau > 0$ this orbit exists if $e\mu < 0$ and $\gamma > 0$. More generally, if $\lambda_{1,2} = \rho e^{\pm i\theta}$, then

$$\operatorname{sign}(b_{n,k}) = \operatorname{sign}\left(\frac{-r^n \sin((n-k)\psi_1) + r^{n-k}\sin(n\theta) - \sin(k\theta)}{\sin(n\psi_1)}\right)$$

and

$$c_n = -\frac{\sin(\theta)}{\sin(n\theta)}|1-\lambda_1^n|^2.$$

Hence

$$\operatorname{sign} c_n = -\operatorname{sign}(\sin(n\theta)).$$

An internal boundary therefore occurs when $n\theta = \pi$. As $\tau = 2r\cos(\theta)$ and $\delta = r^2$, we have an internal boundary when

$$\tau = 2\sqrt{\delta}\cos(\pi/n)$$

and existence of the maximal n-orbit provided that

$$2\sqrt{\delta}\cos(\pi/n) < \tau < \delta + 1.$$

This orbit exists for $e\mu < 0$ and $\gamma > 0$. It continues to exist for $\tau > \delta + 1$ if $e\mu, \gamma > 0$.

A complete analysis of all of the existence regions for the various orbits in Region 3 is missing, although various conjectures are made in [54, 199] for the existence regions of general period-n orbits, which rely on subtle number theoretic relations between n and $\theta/2\pi$.

Example 4.6 (Fold bifurcation of a period-three impacting orbit bifurcating from a non-impacting orbit.). To illustrate some of this behavior we consider a particular example of the map (4.29)–(4.31) in which

$$N = \begin{pmatrix} 0.4663 & 1.4337 \\ -0.227 & -0.1713 \end{pmatrix}, \quad M = \begin{pmatrix} -0.5337 \\ -0.2277 \end{pmatrix}, \quad E = \begin{pmatrix} 0 \\ 1 \end{pmatrix},$$

$$C^T = (1\ 0), \quad D = 0.$$

For this system, we have $\lambda_{1,2} = 0.1475 \pm 0.4731i$, $\tau = 0.295$ and $\delta = 0.2456$. In the notation introduced in the examples above, we have

$$e = -1, \quad s(1) = 1.5067, \quad s(2) = 0.2883, \quad s(3) = -0.1894.$$

We find, by numerical calculation, that an unstable period-3 maximal orbit is created at a (sub-critical) discontinuity-induced bifurcation when $\mu = 0$, and this coexists with a non-impacting orbit close to $\mu = 0$ for $\mu < 0$. The period-3 orbit stabilizes at the saddle-node bifurcation, which occurs when

$$\mu_{SN} = s(3)^2/4e = -0.0090$$

and appears to be globally attracting when $0 < \mu < 0.15$. For $\mu > 0.15$ it appears to coexist with either a chaotic motion or a period-two motion. In Fig. 4.15(a) we present a Monte Carlo bifurcation diagram of the iterates of the map over a broad range of values of μ. In panel (b) of the same figure, we show the result of a numerical continuation of the period-3 orbit, indicating both where it is stable *and* where it is unstable.

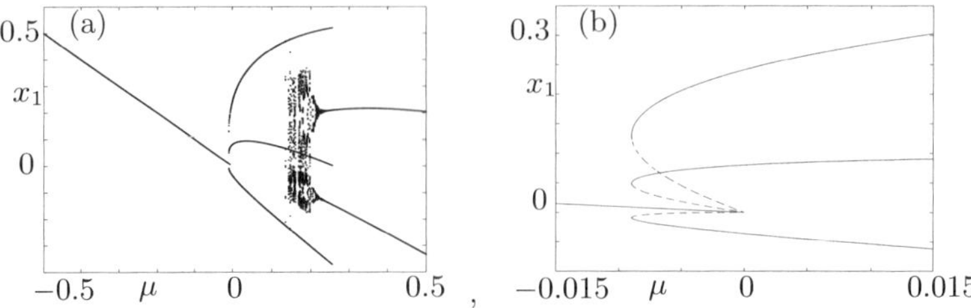

Fig. 4.15. (a) Monte Carlo bifurcation diagram of the two-dimensional Nordmark map example 4.6. (b) Numerical continuation of the period-three orbit; dashed lines indicating stability and solid lines stability.

4.4 Higher-order piecewise-smooth maps

We conclude this chapter by studying border-collisions in piecewise-smooth maps that are locally differentiable at the bifurcation point but have additional nonlinear terms acting on one side only so that the higher derivatives of these maps are not smooth. As will be discussed in Chapters 7 and 8, these types of map arise as normal form maps of grazing bifurcations in piecewise-smooth continuous flows, for example the Chua Circuit, or of sliding bifurcations in Filippov systems.

In general such maps will be shown in Chapters 7 and 8 to take the form:

$$x \mapsto \begin{cases} Nx + M\mu & : \quad C^T x \leq 0 \\ Nx + M\mu + E(C^T x)^\gamma & : \quad C^T x > 0, \end{cases}$$

for some $\gamma > 1$. Typically, we see local bifurcations of such maps that are the same as we see in smooth maps (because of the differentiability of the map), but close to the bifurcation point the behavior is very different. For example, we often see saddle-node bifurcations occurring close to the primary bifurcation which arise because of the non-smooth terms in the map. To illustrate the kind of analysis required to study the dynamics of such maps and the possible bifurcation scenarios they can exhibit, we shall focus exclusively on the one-dimensional case, which we can write without loss of generality in the form

$$x \mapsto \begin{cases} \nu x - \mu & x \leq 0 \\ \nu x + \eta x^\gamma - \mu & x > 0 \end{cases} \tag{4.50}$$

where $\nu \in \mathbb{R}$, $\eta = \pm 1$ $\gamma > 1$ are fixed parameters, and $\mu \in \mathbb{R}$ is the primary bifurcation parameter. Motivated by the normal forms that follow, we shall focus exclusively on the cases $\gamma = 3/2$ and $\gamma = 2$. As for piecewise-linear maps, we shall consider the border-collision at $\mu = 0$ of the simple fixed point that exists within $C^T x > 0$ and then classify the fate of the simplest orbits of period-one and two. Maps of this type were first studied by Halse *et al* [125], upon whose analysis most of the rest of this chapter is based.

4.4.1 Case I: $\gamma = 2$

We label as x_1^* the fixed point in the linear region S_1 and as $x_2^{*\pm}$ those in the nonlinear region S_2. It is easy to see that x_1^* will be admissible if

$$\frac{\mu}{\nu - 1} < 0. \tag{4.51}$$

The fixed points $x_2^{*\pm}$ are given by

$$x_2^{*\pm} = \frac{-(\nu - 1) \pm \sqrt{(\nu - 1)^2 + 4\eta\mu}}{2\eta} > 0, \tag{4.52}$$

which must take positive real values to be admissible. The inequalities (4.51) and $(4.52)^\pm$ thus define regions of parameter space where different combinations of x_1^*, x_2^{*+} and x_2^{*-} may exist.

Clearly, the stability condition for x_2^* is given by $-1 < \nu < 1$. The slope s of the map linearized about $x_2^{*\pm}$ is given by equation

$$s = \nu + 2\eta x_2^{*\pm} = 1 \pm \sqrt{(\nu - 1)^2 + 4\eta\mu}.$$

Therefore, if x_2^{*+} is admissible, then its eigenvalue must be greater than unity. Hence x_2^{*+} is always unstable. Similarly, if x_2^{*-} is admissible, its eigenvalue will be less that $+1$, so the condition for stability is that this eigenvalue is greater than -1, or equivalently:

$$(\nu - 1)^2 + 4\eta\mu < 4.$$

Period-two points (AB/ab) can be analyzed similarly. Adopting the notation used in Chapter 3, we start by considering orbits with a positive ($B > 0$) and a negative iterate ($A < 0$). Therefore

$$A = \nu B + \eta B^2 - \mu, \qquad B = \nu A - \mu,$$

which implies

$$A_{1,2}/a_{1,2} = \frac{-(\nu^2 - 1) \pm \sqrt{(\nu^2 - 1)^2 + 4\nu\eta\mu(\nu + 1)}}{2\nu\eta},$$

and

$$B_{1,2}/b_{1,2} = \frac{2\nu\eta\mu - (\nu^2 - 1) \pm \sqrt{(\nu^2 - 1)^2 + 4\nu\eta\mu(\nu + 1)}}{2\nu^2\eta}.$$

For the solutions A_1B_1/a_1b_1 and A_2B_2/a_2b_2 to be admissible we must have real solutions, that is,

$$(\nu^2 - 1)^2 + 4\nu\eta\mu(\nu + 1) > 0,$$

and also $A_{1,2}/a_{1,2} < 0$ and $B_{1,2}/b_{1,2} > 0$.

The eigenvalues of the period-two points are

$$1 \pm \sqrt{(\nu^2 - 1)^2 + 4\nu\eta\mu(\nu + 1)}.$$

Clearly, whenever such points exist, the square-root term is positive so $a_1 b_1$ must always be unstable whereas $A_2 B_2$ is stable if

$$(\nu^2 - 1)^2 + 4\nu\eta\mu(\nu + 1) < 4.$$

It is also possible to have period-two orbits existing with two positive iterates; i.e.,

$$B_1 = \nu B_2 + \eta B_2^2 - \mu, \quad B_2 = \nu B_1 + \eta B_1^2 - \mu,$$

with $B_1, B_2 > 0$. However, these orbits exist entirely in the nonlinear region of the map and are not affected by the discontinuity in the second derivative at $\mu = 0$. Thus, they will not be influenced by the occurrence of discontinuity-induced bifurcations in the system.

These existence and stability conditions derived so far make it possible to describe several different possible border-collision scenarios occurring as μ passes through 0; see Fig. 4.16. Note that, as the map is differentiable at $\mu = 0$, there is no actual bifurcation in the classical sense at $\mu = 0$. That is, the discontinuity-induced bifurcation does not imply an immediate change in the number or stability of periodic points. However, as we shall argue at the end of this chapter, classical bifurcations such as saddle-node bifurcations can be *induced* by the border-collision to occur a finite distance away in parameter space.

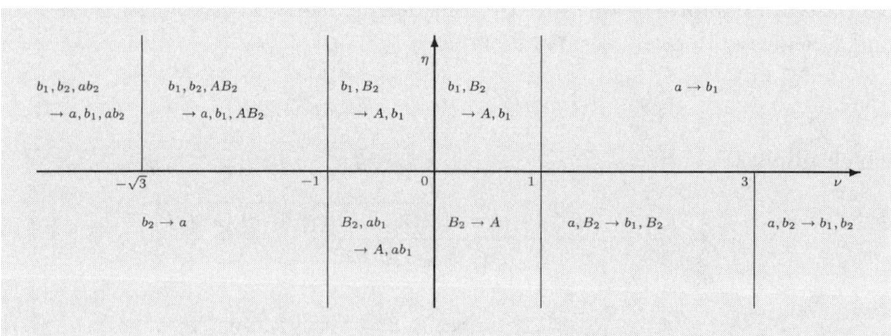

Fig. 4.16. Different discontinuity-induced transitions at $\mu = 0$ in different regions of the (ν, η) parameter space when $\gamma = 2$.

4.4.2 Case II: $\gamma = 3/2$

We can repeat the above analysis for the important case $\gamma = 3/2$ that we shall see in Chapter 7 arises in the unfolding of grazing bifurcations in piecewise-smooth continuous systems.

The condition for the existence (4.51) and stability ($|\nu < 1|$) of fixed points x_1^* remain unchanged. Fixed points, x_2^*, on the nonlinear side must satisfy

$$0 = m^3 + \frac{\nu - 1}{\eta} m^2 - \frac{\mu}{\eta}, \tag{4.53}$$

where $m = \sqrt{x_1^*}$. This equation has up to three solutions that can be found using standard formulae for cubic equations. Labeling the solutions m_1, m_2 and m_3, we have

$$\begin{aligned} m_1 &= \frac{h^{1/3} + 4(\nu-1)^2 h^{-1/3} + 2(\nu - 1)}{6\eta}, \\ m_{2,3} &= \frac{-h^{1/3} - 4(\nu-1)^2 h^{-1/3} + 4(\nu-1) \pm i\sqrt{3}(h^{1/3} - 4(\nu-1)^2 h^{-1/3})}{12\eta}, \end{aligned}$$

where

$$h = 108\mu\eta^3 - 8\nu^3 + 24\nu^2 - 24\nu + 8 + 12\eta^2 \sqrt{\frac{3\mu(-12\nu + 12\nu^2 - 4\nu^3 + 4 + 27\mu\eta^3}{\eta}}.$$

Moreover, since

$$0 = (m - m_1)(m - m_2)(m - m_3)$$

and there is no linear term in (4.53), we must have

$$0 = (m_1 m_2 + m_2 m_3 + m_1 m_3).$$

If there were three positive real solutions, this would lead to a contradiction. Therefore at most there can be two positive real solutions, and hence at most two admissible solutions $x_2^{*\pm}$. To fully determine the conditions for existence, it is necessary to consider separate regions of parameter space bounded by curves on which the number of real positive solutions to (4.53) changes. For the sake of brevity we omit such fine detail here. However, it can be easily shown that the stability of fixed points $x_2^{*\pm}$ is ensured if and only if

$$-1 < \nu + \frac{3\eta\sqrt{x_1^*}}{2} < 1. \tag{4.54}$$

For a period-two orbit AB, we must have

$$A = \nu B + \eta B^{3/2} - \mu, \quad B = \nu A - \mu,$$

which implies that $B = \nu^2 B + \nu\eta B^{3/2} - \mu$. This leads to three pairs of solutions (A_i, B_i) for $i = 1, 2, 3$; again at most two of which will have B_i positive as required. As before, for stability we require

$$-1 < \nu^2 + \frac{3\nu\eta}{2}\sqrt{B_i} < 1,$$

where B_i are the positive iterates of the period-two points in question.

The conditions for existence and stability can be examined in all areas of parameter space. The behavior is exactly the same as in the case $\gamma = 2$ (Fig. 4.16), because the map is still differentiable at the bifurcation point.

4.4.3 Period-adding scenarios

Let us now consider in detail maps of type (4.50) with parameter values

$$0 < \nu < 1, \quad \eta = -1, \quad \mu < 0, \quad \gamma > 1. \tag{4.55}$$

With these choices of parameters the map can be shown to exhibit an infinite period-adding sequence, which for small enough ν involves periodic orbits characterized by only one iterate on the nonlinear side. Specifically, we seek conditions for the existence and stability of period-k orbits of the form $A^{k-1}B$.

Since we suppose that the orbit in question has one iterate, say x_1, on the nonlinear side and $k-1$ iterations on the linear one, we thus obtain that x_1 must satisfy

$$x_1^\gamma + \frac{1-\nu^k}{\nu^{k-1}}x_1 + \frac{\mu(1+\nu+\nu^2+...+\nu^{k-1})}{\nu^{k-1}} = 0, \tag{4.56}$$

together with the compatibility conditions

$$x_1 > \nu^{\frac{1}{(\gamma-1)}} \tag{4.57}$$

and

$$x_1 < -\mu. \tag{4.58}$$

Moreover, for stability this solution must also satisfy the condition

$$x_1 < \left(\frac{\nu^k+1}{\nu^{k-1}\gamma}\right)^{\frac{1}{\gamma-1}}. \tag{4.59}$$

We can use these implicit expressions to compute the regions of existence for periodic windows for any value of γ. For example, the regions of existence of periodic windows for $\gamma = 2$ are shown in Fig. 4.17.

Note that in addition to stable orbits that have only one iteration in the nonlinear region, other types of orbit may also exist; for example, orbits of the form $A^{k-2}B^2$. A complete analysis of all stable period k orbits for arbitrary $\gamma > 1$ is lengthy and would require the derivation of conditions of existence for all different periodic sequences.

For the cases of primary interest, we have the following

Theorem 4.4. *Consider the map (4.50) with* $\eta = -1$.

1. *With* $\gamma = 2$ *the boundaries of existence of stable* k*-periodic orbits of the form* $A^{k-1}B$ *are given by*

$$\frac{-4 + (1 - \nu^k)^k}{4\nu^{k-1}(1 + \nu + \nu^2 + \ldots + \nu^{k-1})} < \mu < \frac{-(1 + \nu + \nu^2 + \ldots + \nu^{k-1})}{\nu^{k-2}}.$$

2. *With* $\gamma = 3/2$ *the boundaries of existence of stable* k*-periodic orbits of the form* $A^{k-1}B$ *are given by*

$$\frac{-8(\nu^k + 1)^3 - 12(1 - \nu^k)(1 + \nu^k)^2}{27\nu^{2(k-1)}(1 + \nu + \nu^2 + \ldots + \nu^{k-1})} < \mu < -\left(\frac{\nu + \nu^2 + \ldots + \nu^k}{\nu^{k-1}}\right)^2.$$

These boundaries are plotted for the cases $\gamma = 2$ in Figs. 4.17(a).

Remark. If $0 < \nu < 1$ is sufficiently small, then as $k \to \infty$ we see a form period-adding sequence in which a period-k orbit is observed for values of μ in a window, which scales geometrically as $1/\nu$ if $\gamma = 2$ and as $1/\nu^2$ if $\gamma = 3/2$. Typically these windows are bounded by regions in which we see chaotic behavior.

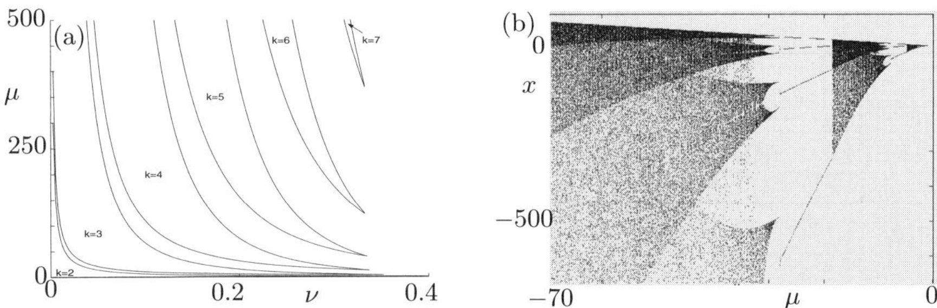

Fig. 4.17. (a) Regions of existence of stable period-k orbits of the form $A^{k-1}B$ orbits in map (4.50) with $\eta = -1$ and $\gamma = 2$. (b) Monte Carlo numerical bifurcation diagram for the case $\eta = -1$, $\nu = 0.27$, $\gamma = 2$ and $\mu \in [-72, 0]$. This shows both the periodic window and the bounding chaotic regions.

Proof. We consider first the case $\gamma = 2$ for boundaries of the existence and the stability of orbits of the forms $A^{k-1}B$. The existence conditions are given by (4.57) and (4.58). In this case with $\gamma = 2$ firstly we have

$$x_1 > \nu. \tag{4.60}$$

The second condition is again

$$x_1 < -\mu. \tag{4.61}$$

In this quadratic case we can solve (4.56) explicitly for x_1, giving

$$x_1 = \frac{(\nu^k - 1) + \sqrt{(1-\nu^k)^2 - 4\nu^{k-1}\mu(1+\nu+\nu^2+\ldots+\nu^{k-1})}}{2\nu^{k-1}}. \tag{4.62}$$

The alternative solution of the quadratic can be neglected as $\nu < 1$, which gives the solution a negative value.

We can substitute this expression for x_1 into the existence conditions (4.60) and (4.61), which, after some manipulation, gives

$$\mu < \frac{-\nu}{1+\nu+\nu^2+\ldots+\nu^{k-1}}, \tag{4.63}$$

and

$$\mu < \frac{-(1+\nu+\nu^2+\ldots+\nu^{k-1})}{\nu^{k-2}}. \tag{4.64}$$

We can see (4.64) implies (4.63) so we only need consider the latter.

The stability condition is given by (4.59). We can substitute our expression (4.62) for x_1 to get the requirement for stability that

$$\mu > \frac{-4+(1-\nu^k)^k}{4\nu^{k-1}(1+\nu+\nu^2+\ldots+\nu^{k-1})}.$$

Consider now the case $\gamma = 3/2$. For existence we have to satisfy conditions (4.57) and (4.58), which here are

$$x_1 > \nu^2 \tag{4.65}$$

and

$$x_1 < -\mu. \tag{4.66}$$

For stability the solution must satisfy (4.59), which here becomes

$$x_1 < \left(\frac{\nu^k+1}{\nu^{k-1}\gamma}\right)^2. \tag{4.67}$$

The first condition (4.65), when substituted into the implicit expression (4.56), leads to the inequality

$$\mu < \frac{-\nu^{k+2}+\nu^{k+4}-\nu^4}{1+\nu+\nu^2+\ldots+\nu^{k-1}}. \tag{4.68}$$

The second condition (4.66) leads to the inequality

$$\mu < -\left(\frac{\nu+\nu^2+\ldots+\nu^k}{\nu^{k-1}}\right)^2. \tag{4.69}$$

Substituting the stability condition (4.67) we have

$$\mu > \frac{-8(\nu^k+1)^3 - 12(1-\nu^k)(1+\nu^k)^2}{27\nu^{2(k-1)}(1+\nu+\nu^2+\ldots+\nu^{k-1})}. \tag{4.70}$$

It again holds that (4.69) implies (4.68).

4.4.4 Location of the saddle-node bifurcations

There is no immediate bifurcation at $\mu = 0$ for the maps of the form (4.50) for $\gamma > 1$. However, there is still a *discontinuity-induced bifurcation* because simple orbits can change their symbol sequence with respect to visits of Regions S_1 and S_2. This in turn may cause there to be a saddle-node or a period-doubling bifurcation at parameter values away from $\mu = 0$. However, if we derive the piecewise-smooth map (4.50) as a normal form close to a limit cycle bifurcation in a non-smooth flow, we should remember that the full Poincaré mapwill have additional (smooth) nonlinear terms. So, there is little point in analyzing maps of the form (4.50) if the nonlinear terms of the full Poincaré would cause there to be other nearby saddle-node or period-doubling bifurcations. What we shall now show is that, at least in the case $\gamma = 3/2$, we should expect that the saddle-node bifurcation induced by the border-collision occurs nearer to $\mu = 0$ than would have been expected for a smooth map.

Consider a map of the form (4.50) for which

$$0 < \nu < 1, \quad \gamma = 3/2.$$

Then a saddle-node bifurcation occurs at a double root $x = y$ of a fixed point of the map, which implies that

$$y(1-\nu) + \mu - \eta y^{3/2} = 0,$$
$$y(1-\nu) + \mu - \frac{3\eta}{2} y^{1/2} = 0.$$

From these equations we obtain the μ-value of the saddle-node bifurcation to be

$$\mu_{\rm SN} = -\frac{4(1-\nu)^3}{3\eta^2}. \tag{4.71}$$

This should be compared to the case of an analytic map with Taylor series coefficient of x^2 being η, which would exhibit a saddle-node bifurcation at the parameter value

$$\mu_{\rm smoothSN} = -\frac{(1-\nu)^2}{4\eta}. \tag{4.72}$$

On comparing (4.71) with (4.72) we notice that the power of $(1-\nu)$ in the numerator of the right-hand side is higher for the non-smooth map than for the analytic map. Thus, if a border-collision takes place for a map with multiplier ν close to unity, then the saddle-node would occur asymptotically closer than it would for to the corresponding smooth map. Also, as we shall from examples in Chapter 7, often the coefficient η can be large compared to ν. Then, the factor η^2 in the denominator of (4.71) can mean that that the saddle-node of the non-smooth map occurs extremely close to the parameter value at which the border-collision occurs. In fact, these two parameters can be almost indistinguishable, so that simulations would falsely suggest that the fold occurs precisely at the border-collision point.

5

Boundary equilibrium bifurcations in flows

We have seen so far that a rich variety of dynamical scenarios can occur when a fixed point of a non-smooth map undergoes a border-collision. This chapter is concerned with a closely related class of discontinuity-induced bifurcations, involving equilibria of n-dimensional piecewise-smooth flows. Specifically, in this Chapter we will study transitions studied that occur when a *boundary equilibrium*, that is one lying within a discontinuity manifold, is perturbed. We will show that such equilibria can either persist or disappear in non-smooth fold transitions when the system parameters are varied. We will also present partial results on the creation of other attractors (e.g. limit cycles) at boundary-equilibrium bifurcations (BEB). A complete analysis of what attractors may then arise in n-dimensions is unknown, but we will take care to show where specific results are known in special cases, such as planar systems.

In the three sections that follow, we treat in turn piecewise-smooth continuous systems (with degree of smoothness equal to 2), Filippov systems (degree one) and hybrid systems (degree zero), each of which leads to subtly different dynamics.

5.1 Piecewise-smooth continuous flows

Let us first focus on locally piecewise-smooth continuous systems, i.e. systems with a single discontinuity boundary Σ with degree of smoothness across Σ equal to 2. We restrict attention to a region of phase space, say $\mathcal{D}$, where the system can be described in terms of a local set of co-ordinate as introduced in Chapter 2:

$$\dot{x} = \begin{cases} F_1(x,\mu), & \text{if } H(x,\mu) < 0, \\ F_2(x,\mu), & \text{if } H(x,\mu) > 0, \end{cases} \tag{5.1}$$

where $x \in \mathbb{R}^n$, $\mu \in \mathbb{R}$, $F_1, F_2 : \mathbb{R}^{n+1} \mapsto \mathbb{R}^n$ and $H : \mathbb{R}^{n+1} \mapsto \mathbb{R}$ are sufficiently smooth functions of both their arguments throughout $\overline{\mathcal{D}}$, and Σ is defined by $H = 0$. Owing to the continuity assumption, we can define

$$F_2(x,\mu) = F_1(x,\mu) + J(x,\mu)H(x,\mu), \tag{5.2}$$

for some smooth function $J : \mathbb{R}^{n+1} \mapsto \mathbb{R}^n$, so that when $H(x,\mu) = 0$ then $F_1 = F_2$ as required. Locally, Σ divides $\mathcal{D}$ in the two regions S_1 and S_2 where the system is smooth and defined by the vector fields F_1 and F_2 respectively:

$$S_1 = \{x \in \mathcal{D} : H(x,\mu) < 0\},$$
$$S_2 = \{x \in \mathcal{D} : H(x,\mu) > 0\}.$$

We can identify different types of equilibria of system (5.1):

Definition 5.1. *We term a point $x \in \mathcal{D}$ an* **admissible equilibrium** *of (5.1) if x is such that either*

$$F_1(x,\mu) = 0 \text{ and } H(x,\mu) < 0$$

or

$$F_2(x,\mu) = 0 \text{ and } H(x,\mu) > 0\,.$$

Alternatively, we say that a point $y \in \mathcal{D}$ is a **virtual equilibrium** *of (5.1) if either*

$$F_1(y,\mu) = 0 \text{ but } H(y,\mu) > 0\,,$$

or

$$F_2(y,\mu) = 0 \text{ but } H(y,\mu) < 0\,.$$

For some value of the system parameters, it is possible for an equilibrium to lie on the discontinuity boundary.

Definition 5.2. *We term a point $z \in \mathcal{D}$ a* **boundary equilibrium** *of (5.1) if*

$$F_1(z,\mu) = F_2(z,\mu) = 0 \textit{ and } H(z,\mu) = 0\,.$$

Finally, we define a boundary equilibrium bifurcation as follows.

Definition 5.3. *The PWS system (5.1) is said to undergo a* boundary equilibrium bifurcation *(BEB) at $\mu = \mu^*$ if there exists a point x^* such that, for both $i = 1$ and 2:*

1. *$F_i(x^*,\mu^*) = 0$.*
2. *$H(x^*,\mu^*) = 0$.*
3. *$F_{i,x}(x^*,\mu^*)$ is invertible (equivalently $det(F_{i,x}) \neq 0$).*
4. *$H_\mu(x^*,\mu^*) - H_x(x^*,\mu^*)\left[F_{i,x}^{-1}F_{i,\mu}\right](x^*,\mu^*) \neq 0$.*

The first two requirements in this definition are the defining conditions for the bifurcation, namely that x^* is a boundary equilibrium at $\mu = \mu^*$. The third item is a non-degeneracy condition that ensures that x^* is an isolated hyperbolic equilibrium to both vector fields F_1 and F_2. The final stipulation is a nondegeneracy condition with respect to the parameter, that admissible branches of equilibria, say $x^+(\mu)$ and $x^-(\mu)$, of vector fields F_1 and F_2, respectively, cross through the bifurcation point at $\mu = \mu^*$. The condition is derived from the requirement that the total derivative $\frac{dH}{d\mu}(x^\pm(\mu), \mu)$ is non-zero at (x^*, μ^*) . It is worth mentioning here that a boundary equilibrium bifurcation is completely analogous to a border-collision with the admissible equilibrium playing the role of the admissible fixed point of the map.

The existence of different types of dynamics following a BEB was discussed in the book by Bautin & Leontovich [25] as well as the recent work by Freire *et. al.* [108] and Liene and co-workers [176, 177]. In these references such bifurcations are also illustrated through various one-dimensional and two-dimensional examples. Rather than be completely general here, we shall seek to understand the simplest part of the bifurcation, namely what happens to equilibria, in some generality in n dimensions. For that purpose, we will use a modification of Feigin's strategy introduced in Chapter 3 for border-collisions. We shall then deal only briefly with other attractors that may be created or destroyed in the bifurcation, as it is clear that a complete classification in n-dimensions is practically impossible.

5.1.1 Classification of simplest BEB scenarios

Without loss of generality let's assume in what follows that $x = 0$ is a boundary equilibrium for $\mu = 0$; i.e. $F_1(0,0) = F_2(0,0) = 0, H(0,0) = 0$. We shall seek to unfold the bifurcation scenarios that can occur when μ is perturbed away from zero. In complete analogy with what happens to period-one fixed points at border-collisions, we have the following generic possibilities for unfolding the equilibrium behavior at a boundary equilibrium bifurcation. Specifically upon varying μ through zero, we see one of the following:

Persistence (or border-crossing): At the bifurcation point, an admissible equilibrium lying in region S_1 becomes a boundary equilibrium and turns into a virtual equilibrium. Simultaneously, a virtual equilibrium lying in region S_2 becomes admissible. Thus there is one admissible equilibrium on either side of the bifurcation, which is why this is termed persistence.

non-smooth fold: At the bifurcation point, the collision of two branches of admissible equilibria is observed at the boundary equilibrium, before turning into two branches of virtual equilibria past the bifurcation point.

The following extension of Feigin's classification strategy is able to decide which of the above occurs. See was di Bernardo *et. al.* [83]. for more details.

We start by giving more precise definitions of the persistence and non-smooth fold scenarios.

Definition 5.4. *We say that (5.1) exhibits a* **persistence** *(or border-crossing) for $\mu = 0$ if, upon variation of μ through zero (for example, as μ is increased), one branch of admissible equilibria and a branch of virtual equilibria cross at the boundary equilibrium point, $x = 0$. In so doing the virtual equilibrium* *becomes admissible, and* vice versa. *Namely, we assume there exist smooth branches of equilibria $x^+(\mu)$ and $x^-(\mu)$ such that $x^+(0) = x^-(0)$ and*

1. $F_1(x^-,\mu) = 0, H(x^-,\mu) < 0$ and $F_2(x^+,\mu) = 0, H(x^+,\mu) < 0$ for $\mu < 0$
2. $F_1(x^-,\mu) = 0, H(x^-,\mu) > 0$ and $F_2(x^+,\mu) = 0, H(x^+,\mu) > 0$ for $\mu > 0$

or vice versa.

Definition 5.5. *We say, instead, that the BEB at $\mu = 0$ is a* **non-smooth fold** *if, upon variation of μ through zero, two branches of admissible equilibria collide at the boundary equilibrium point $x = 0$ and are both turned into two branches of virtual equilibria past the bifurcation point. Namely, there exist smooth branches of equilibria $x^+(\mu)$ and $x^-(\mu)$ such that, $x^+(0) = x^-(0)$ and*

1. $F_1(x^-,\mu) = 0, H(x^-,\mu) < 0$ and $F_2(x^+,\mu) = 0, H(x^+,\mu) > 0$ for $\mu < 0$
2. $F_1(x^-,\mu) = 0, H(x^-,\mu) > 0$ and $F_2(x^+,\mu) = 0, H(x^+,\mu) < 0$ for $\mu > 0$

Remarks

1. Note that there is no analogy of the period-doubling case for border-collisions of fixed points.
2. We can infer nothing in general about the stability of the admissible equilibria from these definitions. That depends on the precise eigenvalues of $F_{i,x}$ for $i = 1, 2$.
3. In the case of a non-smooth fold, there is no admissible equilibrium beyond the bifurcation point. Under conditions we delineate in the next section, we can anticipate that there must be a more complicated attractor existing after the two admissible equilibria are destroyed.
4. The non-degeneracy condition that $F_{i,x}$ be invertible will be crucial in what follows. If this were not the case then one of the equilibria would be non-hyperbolic and hence would undergo a local bifurcation (with respect to one of the vector fields F_1 or F_2) precisely on the discontinuity boundary. Such a scenario would be of codimension-two and is beyond the scope of this book, although we shall see some preliminary remarks about codimension-two discontinuity-induced bifurcations in Chapter 9.

We will now derive conditions to distinguish between these two fundamental cases in arbitrary n-dimensional systems. Firstly, in order for x^- to be an admissible equilibrium, we must have

$$\begin{aligned} F_1(x^-,\mu) &= 0, \\ H(x^-,\mu) &:= \lambda^- < 0. \end{aligned} \tag{5.3}$$

Similarly, for x^+ to be admissible using (5.2),

$$\begin{aligned} F_2(x^+,\mu) &= F_1(x^+,\mu) + J(x^+,\mu)H(x^+,\mu), \\ H(x^+,\mu) &:= \lambda^+ > 0\,. \end{aligned} \tag{5.4}$$

Then, linearizing about the boundary equilibrium point, $x = 0, \mu = 0$, we have

$$N_1 x^- + M\mu = 0\,, \tag{5.5}$$
$$C^T x^- + D\mu = \lambda^-\,, \tag{5.6}$$

and

$$N_2 x^+ + M_2\mu = N_1 x^+ + M\mu + E\lambda^+ = 0\,, \tag{5.7}$$
$$C^T x^+ + D\mu = \lambda^+\,, \tag{5.8}$$

where $N_1 = F_{1,x}, M = F_{1,\mu}, M_2 = F_{2,\mu}, C^T = H_x, D = H_\mu$ and $E = J$ are all evaluated at $x = 0, \mu = 0$. Note that N_1 is invertible by condition 3 of Definition 5.3. Hence we have

$$x^- = -N_1^{-1} M\mu\,,$$

and substituting into (5.6), we obtain

$$\lambda^- = (D - C^T N_1^{-1} M)\mu\,. \tag{5.9}$$

Similarly using (5.7) and (5.8), we have

$$\lambda^+ = \frac{(D - C^T N_1^{-1} M)\mu}{(1 + C^T N_1^{-1} E)} = \frac{\lambda^-}{(1 + C^T N_1^{-1} E)}\,. \tag{5.10}$$

Therefore, we can state the following.

Theorem 5.1 (Equilibrium points branching from a boundary equilibrium). *For the systems of interest, assuming*

$$det(N_1) \neq 0, \tag{5.11}$$
$$D - C^T N_1^{-1} M \neq 0, \tag{5.12}$$
$$1 + C^T N_1^{-1} E \neq 0. \tag{5.13}$$

1. *A* **persistence** *scenario is observed at the boundary equilibrium bifurcation point if*
$$1 + C^T N_1^{-1} E > 0\,. \tag{5.14}$$
2. *A* **non-smooth fold** *is instead observed if*
$$1 + C^T N_1^{-1} E < 0\,. \tag{5.15}$$

This can be easily proven by considering that, from (5.9) and (5.10), that λ^+ and λ^- have the same sign for the same value of μ (persistence) if condition (5.14) is satisfied, whereas they have opposite signs (non-smooth fold) if condition (5.15) is satisfied instead. Also, since the given conditions ensure that the linearized system is non-singular, an application of the Implicit Function Theorem implies that the conclusions are still valid for the full nonlinear system, in some neighborhood of the boundary equilibrium point of interest.

Remarks

1. Note that, using a similar approach to the one followed in Chapter 3, it is possible to recast the conditions derived above in terms of the same notation introduced by Feigin for border-collisions of fixed points in maps [83]. Namely, label by A (a) a stable (unstable) admissible equilibrium of the flow $\dot{x} = F_1(x, \mu)$ and B (b) a stable (unstable) admissible equilibrium of $\dot{x} = F_2(x, \mu)$. Then, we have that at the boundary equilibrium bifurcation point:
   a) **Persistence** is observed if
$$\sigma_1^- + \sigma_2^- \text{ is even,} \tag{5.16}$$
   e.g. $A \to B$, $A \to b$ etc.
   b) **A non-smooth fold** is observed if
$$\sigma_1^- + \sigma_2^- \text{ is odd,} \tag{5.17}$$
   e.g. $A, b \to \varnothing$, etc.
   where σ_i^+, σ_i^- are the number of positive and negative real eigenvalues of the Jacobians of $F_{i,x}, i = 1, 2$, respectively.
2. Equation (5.17) implies that at a non-smooth fold, a branch of stable admissible equilibria always collides on the boundary equilibrium with a branch of unstable admissible equilibria at the bifurcation point. This can be easily proven by considering that the total number of negative real eigenvalues of the Jacobians, $\sigma_1^- + \sigma_2^-$, can be odd if and only if at least one of them changes sign across the boundary, causing the branches to exchange their stability properties. In the case of all eigenvalues being strictly complex on one side, the system must be even dimensional. Then in order for the sum of eigenvalues less than zero to be odd (to guarantee a non-smooth fold), the number of real negative eigenvalues must be odd. Since n is even this means that at least one real eigenvalue must by positive. Hence one of the two colliding equilibria must be unstable.
3. Without loss of generality, as in Chapter 3, we shall assume (by making a local co-ordinate transformation) that $D = 0$.

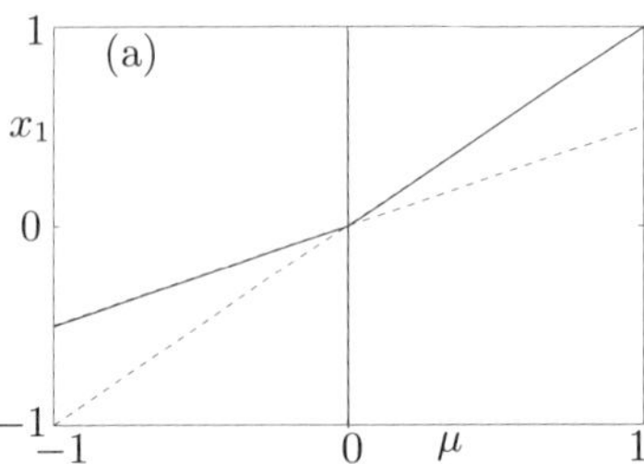

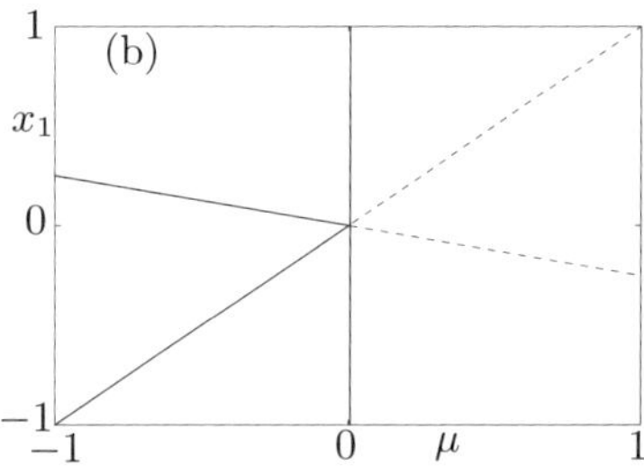

Fig. 5.1. Bifurcation diagrams of (5.18) showing (a) a persistence scenario when $\varepsilon = 1$ and (b) a non-smooth saddle-node when $\varepsilon = 10$, both associated with a BEB.

Example 5.1. Consider the three-dimensional piecewise-linear continuous system of the form:

$$\dot{x} = \begin{cases} N_1 x + M\mu, & \text{if } C^T x < 0, \\ N_2 x + M\mu & \text{if } C^T x > 0, \end{cases} \tag{5.18}$$

with

$$N_1 = \begin{pmatrix} -1 & 1 & 0 \\ -3 & 0 & 1 \\ -2 & 0 & 0 \end{pmatrix}, \qquad N_2 = \begin{pmatrix} -1 & 1 & 0 \\ -3 & 0 & 1 \\ -2+\varepsilon & 0 & 0 \end{pmatrix}$$

and

$$M = [0\,0\,1]^T, \qquad C^T = [1\,0\,0]\,.$$

In this case, we have $E = (\,0 \quad 0 \quad \varepsilon\,)$, thus according to the theory developed in this section we get $1 + C^T N_1^{-1} E = 1/(1-\varepsilon/2)$ and we should get persistence if $\varepsilon < 2$ and a non-smooth saddle node if $\varepsilon > 2$. Fig. 5.1 show the effects of a BEB for $\varepsilon = 1$ and $\varepsilon = 10$. We can clearly see the two different scenarios with branches of admissible and virtual equilibria collapsing onto a boundary equilibrium when $\mu = 0$.

5.1.2 Existence of other attractors

Several calculations in the literature point out that invariant sets other than equilibria can be created or destroyed in a BEB (see e.g. [176, 175, 177] and references therein). It is worth mentioning here that, currently, there is no general theory accounting for all the possible scenarios associated to a boundary equilibrium transition in generic piecewise-smooth systems, nor does it seem that there ever will be. This unlikelihood is because the dynamics close to a nondegenerate BEB is governed by a piecewise-linear flow. Even in three dimensions it is possible to construct piecewise-linear flows with remarkably complex dynamics; see for example the system analyzed in [241], which exhibits Shil'nikov homoclinic chaos.

Here, we focus only on one of the possible scenarios to illustrate the type of analysis needed to classify the simplest possible cases. Specifically, we look at cases where we can definitely say that perturbing a boundary equilibrium causes the trajectory to tend to some attractor (other than an equilibrium) whose amplitude scales with $|\mu|$. We start with the case of planar piecewise-smooth continuous systems where the only such attractors are limit cycles, making this the closest discontinuity-induced equivalent of a Hopf bifurcation for smooth systems. This has sometimes therefore been referred to as a *discontinuity-induced Hopf bifurcation*, although we should stress there is no sense in which eigenvalues cross the imaginary axis. More complex, often non-generic scenarios, which have been discussed in the literature, are also sketched in what follows.

5.1.3 Planar piecewise-smooth continuous systems

In the planar case the only other generic type of invariant set that can be involved in a border-equilibrium bifurcation is a limit cycle. Such a limit cycle must enclose an equilibrium. The cycle will be stable if it encloses an unstable equilibrium, or unstable if it encloses a stable one. limit cycles may branch off a boundary equilibrium bifurcation point either in a persistence scenario or a non-smooth fold. Note that a limit cycle always encircles an equilibrium of 'focus' type. Also there is a fixed area inside the cycle, which means that there must be a balance between area production and destruction inside the cycle.

In what follows we will state the results derived by Freire *et. al.* [108] for piecewise-smooth continuous systems that give conditions for a branch of limit cycles to exist in a BEB. The conditions below are derived in terms of the linearized systems introduced above. But the conditions also guarantee that the piecewise-linear systems are unchanged by small nonlinear perturbations (structural stability). Thus they accurately describe the dynamics of a full nonlinear piecewise-smooth, continuous system close enough to a BEB point. For simplicity, all examples in this section are all in linearized form.

It is possible to summarize the results in the following theorem.

Theorem 5.2 ([108]). *Consider a planar piecewise-linear system of the form*

$$\dot{x} = \begin{cases} N_1 x + M\mu, & \text{if } C^T x < 0, \\ N_2 x + M\mu & \text{if } C^T x > 0, \end{cases} \tag{5.19}$$

with $x \in \mathbb{R}^2$, $\mu \in \mathbb{R}$ *and*

$$N_{2,x} = N_{1,x} + EC^T x.$$

Assume that $\det(N_1) \neq 0$ *and* $1 + C^T N_1^{-1} E \neq 0$, $\operatorname{trace}(N_1) \neq 0$, $\det(N_1) \neq \operatorname{trace}(N_1)^2/4$, $\det(N_1 + EC^T) \neq \operatorname{trace}(N_1 + EC^T)^2/4$.

1. *If* $\operatorname{trace}(N_1)\operatorname{trace}(N_1 + EC^T) > 0$ *then no limit cycle is involved in the bifurcation (because of the area restriction).*
2. *If* $\operatorname{trace}(N_1)\operatorname{trace}(N_1 + EC^T) < 0$ *then:*
   a) *If we have a BEB with* $1 + C^T N_1^{-1} E > 0$ (**persistence**) *and there is at least one focus involved, then*
      i. *If the transition from a focus to a node is observed then the cycle exists and is stable if* $\operatorname{trace}(J) < 0$ *(the node is stable) and unstable if* $\operatorname{trace}(J) > 0$*, where* J *is the Jacobian obtained by linearizing the system about the node.*
      ii. *If instead the transition from a focus to a focus is present, then assuming* $\alpha_i \pm j\omega_i$, $i = 1, 2$ *are the eigenvalues of the two foci with* $\omega_i > 0$*, and* $e^{\frac{\alpha_1}{\omega_1}\pi} e^{\frac{\alpha_2}{\omega_2}\pi} \neq 1$*, then a cycle exists which is stable if* $e^{\frac{\alpha_1}{\omega_1}\pi} e^{\frac{\alpha_2}{\omega_2}\pi} < 1$ *and unstable if* $e^{\frac{\alpha_1}{\omega_1}\pi} e^{\frac{\alpha_2}{\omega_2}\pi} > 1$.

   iii. If a transition of type node/node or saddle/saddle is observed, then no cycle exists.

b) If we have a BEB with $1 + C^T N_1^{-1} E < 0$ (**non-smooth fold**) *then a limit cycle can only surround a focus.*

   i. If the equilibria are node/saddle then no limit cycle exists.

   ii. If we have a saddle/focus bifurcation, then:

      A. If the focus is unstable and the unstable manifold of the saddle point curls inside *the stable one, then a stable limit cycle exists.*

      B. If the focus is stable and the stable manifold of the saddle point curls inside *the unstable one, then an unstable limit cycle exists.*

      C. If the respective manifolds curl outside *instead, then no limit cycle exists.*

To illustrate the theorem, we now look at a set of simple representative examples. In all examples we use

$$C^T = (1 \quad 0), \qquad M = \begin{pmatrix} 0 \\ -1 \end{pmatrix}, \qquad D = 0. \tag{5.20}$$

Example 5.2. Fig. 5.2 shows the bifurcation diagram of a planar system fulfilling the conditions stated above with

$$N_1 = \begin{pmatrix} -1 & 1 \\ -1 & 0 \end{pmatrix}, \quad E = \begin{pmatrix} 3 \\ -4 \end{pmatrix}, \quad N_2 = N_1 + EC^T = \begin{pmatrix} 2 & 1 \\ -5 & 0 \end{pmatrix}.$$

Here, as expected, a stable focus hits the boundary and becomes unstable. We observe that, when this occurs, a limit cycle is indeed generated at the border-collision point and that the amplitude of the limit cycle scales linearly with the parameter.

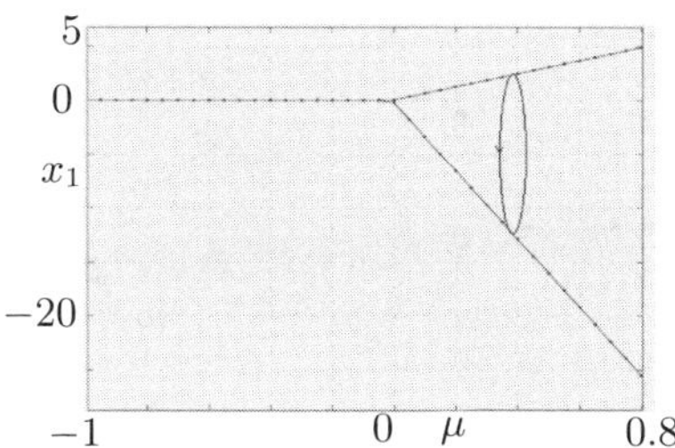

Fig. 5.2. Bifurcation diagram of Example 5.2. showing the occurrence of a Hopf-like transition at $\mu = 0$.

Example 5.3. Suppose

$$N_1 = \begin{pmatrix} -2.1 & 1 \\ -1 & 0 \end{pmatrix}, \qquad E = \begin{pmatrix} 2.6 \\ -4 \end{pmatrix}. \tag{5.21}$$

Then $\text{trace}(N_1) = -2.1 < 0$, $\text{trace}(N_1 + EC^T) = 0.5 > 0$ and $1 + C^T N_1^{-1} E = 5 > 0$ and we have persistence with a stable node for $\mu < 0$. For $\mu > 0$ we have an unstable focus surrounded by a stable limit cycle. See Fig. 5.3.

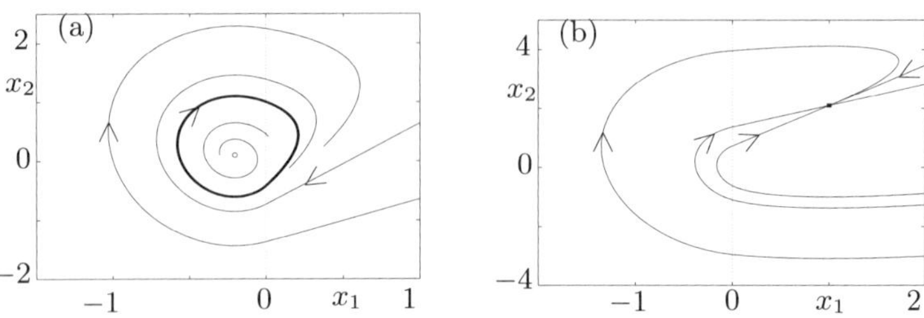

Fig. 5.3. Phase portraits for Example 5.3. (a) Unstable focus with stable limit cycle ($\mu = 1$), (b) stable node ($\mu = -1$).

Example 5.4. Suppose

$$N_1 = \begin{pmatrix} -1 & 1 \\ -1 & 0 \end{pmatrix}, \qquad E = \begin{pmatrix} 1.5 \\ -4 \end{pmatrix}. \tag{5.22}$$

Then $\text{trace}(N_1) = -1 < 0$, $\text{trace}(N_1 + EC^T) = 0.5 > 0$, $1 + C^T N_1^{-1} E = 5 > 0$ and $e^{\frac{\alpha_1}{\omega_1}\pi} e^{\frac{\alpha_2}{\omega_2}\pi} < 1$, and we have persistence with a stable focus for $\mu < 0$. For $\mu > 0$ we have an unstable focus surrounded by a stable limit cycle. See Fig. 5.4.

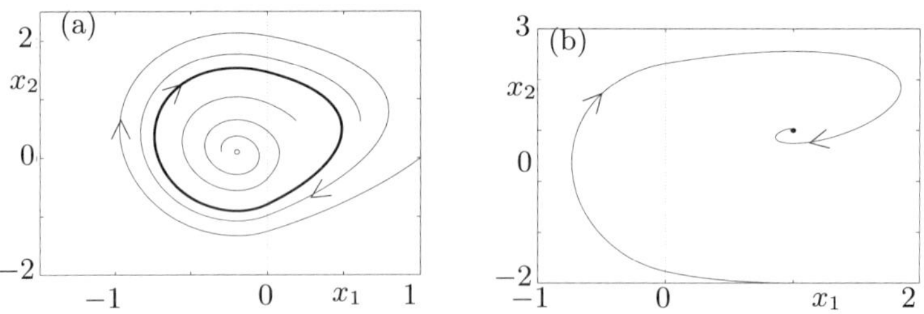

Fig. 5.4. Phase potraits for Example 5.4. (a) An unstable focus with a stable limit cycle ($\mu = 1$), (b) a stable focus ($\mu = -1$).

Example 5.5. Suppose

$$N_1 = \begin{pmatrix} -1 & 1 \\ 1 & 0 \end{pmatrix}, \qquad E = \begin{pmatrix} 1.5 \\ -6 \end{pmatrix}. \tag{5.23}$$

Then $\text{trace}(N_1) = -1 < 0$, $\text{trace}(N_1 + EC^T) = 0.5 > 0$ and $1 + C^T N_1^{-1} E = -5 < 0$, and we have a non-smooth fold with a saddle point, and an unstable focus surrounded by a stable limit cycle for $\mu > 0$, as illustrated in Fig. 5.5.

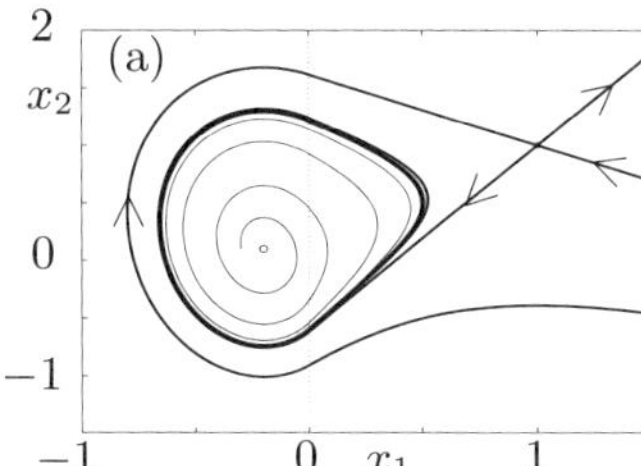

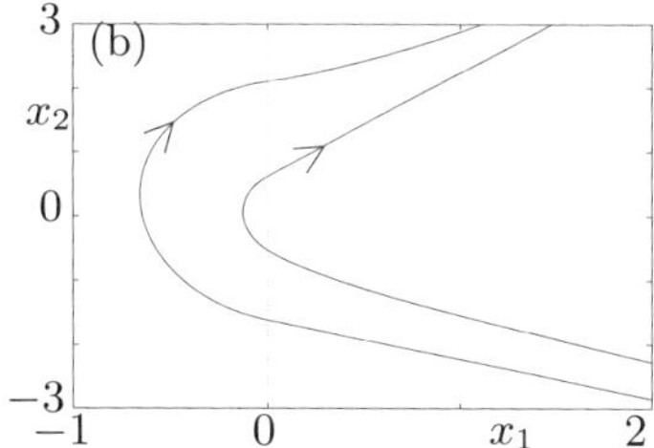

Fig. 5.5. Phase potraits for Example 5.5. (a) Unstable focus with stable limit cycle and saddle point ($\mu = 1$), (b) no limit sets ($\mu = -1$).

5.1.4 Higher-dimensional systems

Classifying the existence of other attractors in higher-dimensional systems is a cumbersome task. In particular, the lack of general results to characterize the existence of limit cycles or chaos in higher-dimensional piecewise-smooth systems makes it impossible to obtain a complete classification.

In n-dimensions it is possible to characterize with some degree of generality the existence of other attractors branching off a boundary equilibrium bifurcation point under some specific circumstances. For example, it is possible to prove the following theorem.

Theorem 5.3. *[87] If (x^*, μ^*) is a boundary equilibrium point that is asymptotically stable (Lyapunov stable and attracting) in a piecewise smooth, continuous system, then for all μ close to μ^*, there is at least one attractor close to x^*. The amplitude of such an attractor scales linearly with μ to leading order.*

Because the boundary equilibrium is assumed to be asymptotically stable for $\mu = 0$, then by continuity with respect to parameter variation, for $\mu \neq 0$ a local neighborhood, say $\mathcal{B}(0)$, of the boundary equilibrium must exist into which trajectories continue to be attracted. As a corollary, if an asymptotically stable equilibrium point is missing on one or both sides of the bifurcation, some other attractor must exist.

The fact that this attracting set scales locally linearly with μ comes from the fact that the linearization $F_{i,x}$ is non-degenerate and hence locally the piecewise-smooth continuous system can be approximated by its linearization

$$\dot{x} = \begin{cases} N_1 x + M\mu, & \text{if } C^T x > 0, \\ N_2 x + M\mu = N_1 x + M\mu + EC^T x, & \text{if } C^T x < 0. \end{cases} \tag{5.24}$$

However this linear system can be made scale-invariant for a fixed sign of μ by substituting $y = x/\mu$ and dividing by μ. Hence any attractor of (5.24) must scale precisely linearly with μ. This then must be the leading-order scaling of any piecewise-smooth continuous system that has (5.24) as its linearization at a BEB.

From the classification viewpoint, in the case of two-dimensional systems, we have seen that only a single attractor is possible in the generic case of a structurally stable linearized system, and that it must either be an equilibrium point or a limit cycle. As we shall see, in higher dimensions, both multiple and more complicated attractors are also possible.

In general, it is a cumbersome task to assess the asymptotic stability of higher dimensional piecewise-smooth systems. For instance, an interesting phenomenon that occurs in higher dimensional systems is that, even though both N_1 and N_2 are Hurwitz matrices (i.e. matrices for which all the eigenvalues lie in the open left half plane), the overall system can exhibit instability. Such an example can be constructed as follows [52]

Example 5.6.

$$N_1 = \begin{pmatrix} -1 & -1 & 0 \\ 1.28 & 0 & -1 \\ -0.624 & 0 & 0 \end{pmatrix} \quad N_2 = \begin{pmatrix} -3.2 & -1 & 0 \\ 25.61 & 0 & -1 \\ -75.03 & 0 & 0 \end{pmatrix}$$

and

$$C^T = (\,1 \quad 0 \quad 0\,)\,.$$

Carmona *et al.* [52] studied the stability of the origin for $n = 3$. With the help of the notion of invariant cones, they reached the following result.

Proposition 5.1 ([52]). *Consider the system (5.24) with $n = 3$. Assume that the pair (C^T, N_1) is observable. Let $N_{1,2}$ and C^T be given by*

$$N_{1,2} = \begin{pmatrix} t_{1,2} & 1 & 0 \\ m_{1,2} & 0 & 1 \\ d_{1,2} & 0 & 0 \end{pmatrix}, \quad C^T = \begin{pmatrix} 1 \\ 0 \\ 0 \end{pmatrix}, \tag{5.25}$$

which are in canonical form. Suppose that the eigenvalues of the matrices $N_{1,2}$ are $\lambda_{1,2} \in \mathbb{R}$ and $\sigma_{1,2} \pm i\omega_{1,2}$, where $\omega_{1,2} > 0$. Also suppose that

$$(\sigma_1 - \lambda_1)(\sigma_2 - \lambda_2) < 0 \tag{5.26}$$

and

$$(t_1 - t_2)(\sigma_2 - \lambda_2) \leq 0. \tag{5.27}$$

Then, the origin is an asymptotically stable equilibrium point if, and only if, λ_1 and λ_2 are both negative.

Example 5.7. Using this result, we can construct a three-dimensional piecewise-linear example of the form (5.24) where, under parameter variation, the boundary equilibrium at the origin becomes unstable, giving rise to a family of stable limit cycles whose amplitude scales linearly with the parameter perturbation. Specifically, Fig. 5.6 shows the bifurcation diagram of a three-dimensional system of the form (5.24) where

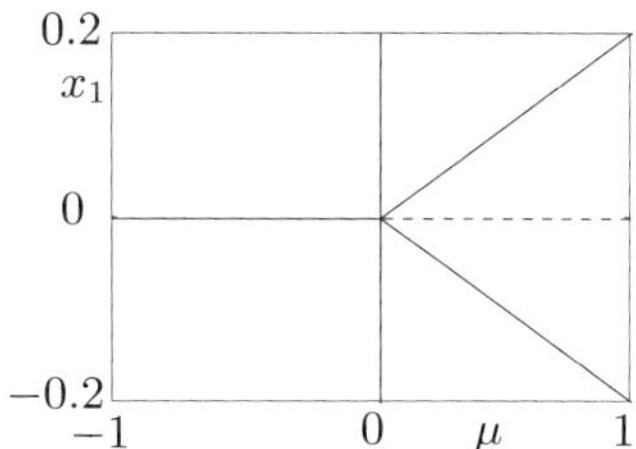

Fig. 5.6. Bifurcation diagram for Example 5.7. The lines for $\mu > 0$ represent the intersections of the limit cycle with a Poincaré section.

$$N_1 = \begin{pmatrix} -5 & 1 & 0 \\ -9 & 0 & 1 \\ -5 & 0 & 0 \end{pmatrix}, \quad N_2 = \begin{pmatrix} -5 & 1 & 0 \\ -12 & 0 & 1 \\ -14 & 0 & 0 \end{pmatrix}, \tag{5.28}$$

and

$$M = \begin{pmatrix} 0 \\ 0 \\ 1 \end{pmatrix}, \qquad C^T = \begin{pmatrix} 1 & 0 & 0 \end{pmatrix}. \tag{5.29}$$

As predicted by Proposition 5.1, since $\lambda_1 = -1$, $\sigma_1 \pm i\omega_1 = -2 \pm i$ and $\lambda_2 = -7$, $\sigma_2 \pm j\omega_2 = 1 \pm i$, in this case the boundary equilibrium at the origin is asymptotically stable when $\mu = 0$. Moreover, we have

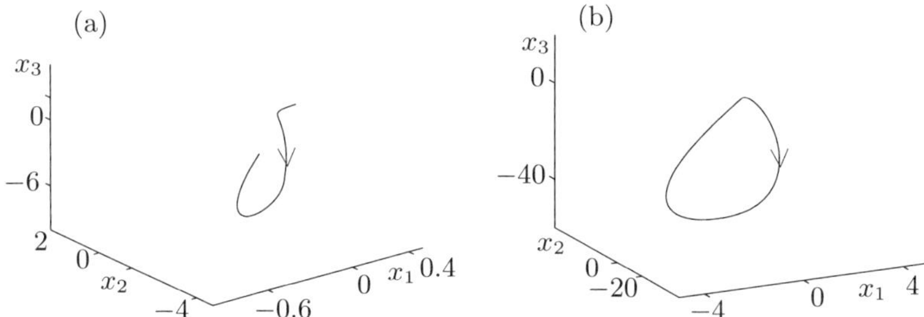

Fig. 5.7. Phase potraits corresponding to a non-smooth Hopf bifurcation for Example 5.7. (a) Stable equilibrium point for $\mu = -1$, and (b) stable limit cycle for $\mu = 1$.

$$1 + C^T (N_1)^{-1} E = 0.3571 > 0.$$

Hence, according to Theorem 5.1, under variations of μ, we expect the branch of stable equilibria for $\mu < 0$ to turn into a branch of unstable equilibria for $\mu > 0$, and as expected from Theorem 5.3, a family of stable limit cycles to be observed locally to the boundary equilibrium transition. See Figs. 5.6 and 5.7.

A yet more striking example, that can be found in higher-dimensional systems, is the sudden transition at a BEB from a stable admissible equilibrium to a stable chaotic attractor whose amplitude scales linearly with μ as predicted by Theorem 5.3. Currently, there are no general analytical tools to account for such a transition so we illustrate this case by means of the following representative example.

Example 5.8. Consider a PWL system of the form (5.24) where

$$N_1 = \begin{pmatrix} -0.8 & 1 & 0 \\ -0.57 & 0 & 1 \\ -0.09 & 0 & 0 \end{pmatrix}, \quad N_2 = \begin{pmatrix} -0.1 & 1 & 0 \\ -0.2 & 0 & 1 \\ -60 & 0 & 0 \end{pmatrix}, \tag{5.30}$$

and

$$M = \begin{pmatrix} 0 \\ 0 \\ 1 \end{pmatrix}, \qquad C^T = \begin{pmatrix} 1 & 0 & 0 \end{pmatrix}. \tag{5.31}$$

This is again a case of persistence, and in this case the transition through a boundary equilibrium causes there to be a chaotic attractor as a result of the bifurcation at $\mu = 0$. See Fig. 5.8.

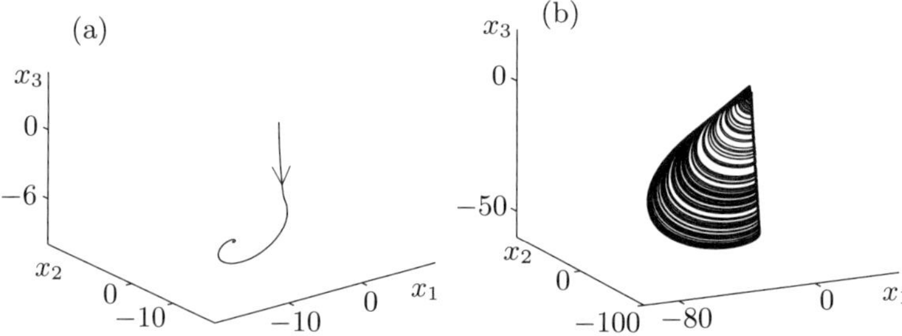

Fig. 5.8. Phase potraits corresponding to a non-smooth bifurcation for Example 5.8. In (a) for $\mu = -1$ stable equilibrium point, and in (b) for $\mu = 1$ stable chaotic trajectory are observed.

5.1.5 Global phenomena for persistent boundary equilibria

Depending on the value of the parameters, admissible equilibria can collide with Σ giving rise to one of the local transitions discussed above. In addition, for some degenerate cases, non-smooth global bifurcations are also possible involving interactions of the invariant manifolds of the equilibria with the discontinuity manifold. Indeed, it has been shown that global phenomena like single- or double-saddle connections (homoclinic or heteroclinic loops) can occur when some parameters of the system are varied. To illustrate the occurrence of such global non-smooth phenomena, we refer to results presented in [108]. Specifically it can be proved that the so-called Lum–Chua conjecture is true; namely, that a continuous piecewise-linear vector field with one boundary condition has at most one limit cycle of period one which visits each region of phase space precisely once. Moreover, it exists, the limit cycle is either attracting or repelling. Also, under some additional conditions, the existence of homoclinic loops can be proved.

Example 5.1 continued. For example, Fig. 5.9(a) shows a bifurcation diagram where a continuum of homoclinic loops (shaded region) is born at the bifurcation point. At this point, the global attractor at the origin changes stability. A phase potrait corresponding to $\mu = 1.75$ is shown in Fig. 5.9(b), where the homoclinic loops are clearly seen.

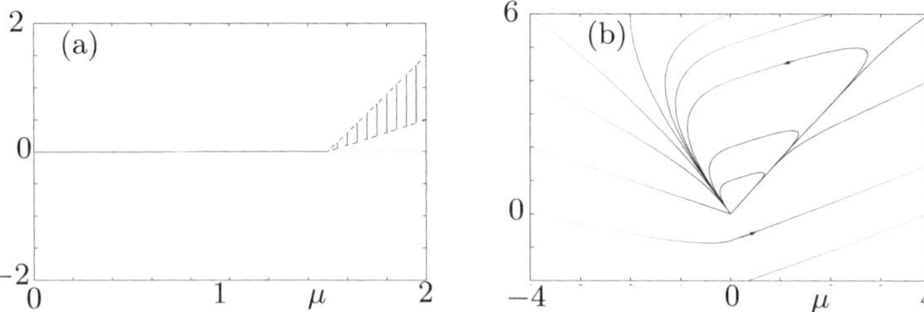

Fig. 5.9. (a) Bifurcation diagram showing the occurrence of a continuum of homoclinic orbits; (b) phase potrait corresponding to $\mu = 1.75$.

5.2 Filippov flows

In this section, we study bifurcations of equilibria in Filippov systems, that is, systems which, local to a single switching manifold, have uniform smoothness of degree one (a jump in the value of F). Thus, generically we will not see codimension-one bifurcations where both vector fields vanish at a boundary equilibrium as in the previous section. However, in Filippov systems there is the possibility of sliding motion and, in particular, the presence of equilibria of the sliding flow (called *pseudo-equilibria* in what follows). We shall therefore find a new boundary equilibrium bifurcation where a pseudo-equilibrium turns into an admissible equilibrium of one of the vector fields away from the boundary. In what follows we shall give a complete unfolding of equilibrium behavior in the neighborhood of such a transition. However, once again a complete description of the dynamics in n-dimensions is not known. Instead we shall focus on the special case of planar Filippov systems, following the results of Kuznetsov *et al.* [169]. A classification of generic bifurcations in sliding vector field was carried out in the earlier work of Teixera [248]. Other equilibrium bifurcations in non-generic classes of Filippov systems were studied by Küpper and co-workers in [167, 283, 284, 286, 285] where, for example, a form of *generalized Hopf bifurcation* was shown to occur as a focus located on the switching surface is perturbed. The transition to sliding cycles (a limit cycle with a segment of sliding motion) in a different class of planar Filippov systems was studied in Giannakopoulos & Pliete [117].

We consider Filippov systems that can be written locally to some region $\mathcal{D}$ in the form

$$\dot{x} = \begin{cases} F_1(x,\mu), & \text{if } H(x,\mu) < 0, \\ F_2(x,\mu), & \text{if } H(x,\mu) > 0. \end{cases} \tag{5.32}$$

As in the case of piecewise-smooth continuous vector fields, it is possible to identify different types of equilibria in a Filippov system. We give the following definitions.

Definition 5.6. *We say that a point $x \in S$ is an* **admissible equilibrium** *of (5.32) if*

$$\begin{aligned} F_1(x,\mu) &= 0, \\ H(x,\mu) &:= \lambda_1 < 0, \end{aligned} \tag{5.33}$$

or

$$\begin{aligned} F_2(x,\mu) &= 0, \\ H(x,\mu) &:= \lambda_2 > 0. \end{aligned}$$

Definition 5.7. *We call a point $\tilde{x}$ a* **pseudo-equilibrium** *if it is an equilibrium of the sliding flow, i.e. for some scalar α,*

$$\begin{aligned} F_1(\tilde{x},\mu) + \alpha(F_2 - F_1) &= 0, \\ H(\tilde{x},\mu) &= 0. \end{aligned} \tag{5.34}$$

Again, we have to check that α lies in the allowed range.

Definition 5.8. *We call a pseudo-equilibrium* **admissible** *if*

$$0 < \alpha < 1.$$

Alternatively, we say that a pseudo-equilibrium is **virtual** *if*

$$\alpha < 0 \quad or \quad \alpha > 1.$$

Note that, typically, pseudo-equilibria are equilibria of neither F_1 nor F_2.

Definition 5.9. *A point $\widehat{x}$ is termed a* **boundary equilibrium** *of (5.32) if*

$$\begin{aligned} &F_1(\widehat{x},\mu) = 0 \text{ or } F_2(\widehat{x},\mu) = 0, \\ &H(\widehat{x},\mu) = 0. \end{aligned}$$

Note that a boundary equilibrium is always located on the boundary of the sliding region where one of the vector field vanishes.

As shown in Sec. 5.1 for non-smooth continuous systems, the appearance of a boundary equilibrium in a Filippov systems represents a codimension-one discontinuity-induced bifurcation. In similarity with Definition 5.3, we have:

Definition 5.10. *The PWS Filippov system (5.1) undergoes a* **boundary equilibrium bifurcation** *at $\mu = \mu^*$ with respect to vector field F_i, $i = 1, 2$, if there exist a point x^* such that*

1. $F_i(x^*, \mu^*) = 0$, *but* $F_j(x^*, \mu^*) \neq 0$.
2. $H(x^*, \mu^*) = 0$.
3. $F_{i,x}(x^*, \mu^*)$ *is invertible (or equivalently* $det(F_{i,x}) \neq 0$) *for* $i = 1$ *and* 2.
4. $H_\mu(x^*, \mu^*) - H_x(x^*, \mu^*) \left[F_{i,x}^{-1} F_{i,\mu}\right] (x^*, \mu^*) \neq 0$.

5.2.1 Classification of the possible cases

Without loss of generality, let us assume that $x = 0$ is a boundary equilibrium with respect to F_1 for $\mu = 0$. We shall now seek conditions to classify the equilibrium behavior in an unfolding of such BEBs. We will show that as μ is varied, scenarios similar to those presented for non-smooth continuous systems are possible. That is, we can observe *persistence* where a branch of admissible equilibria turns into a branch of pseudo-equilibria or, alternatively, *non-smooth fold* where a branch of admissible equilibria disappears after colliding with a branch of pseudo-equilibria on the boundary. The assumption about invertibility of $F_{1,x}$ and $F_{2,x}$ means that we can study what happens by using the linearization of both vector fields. So, let x be an admissible equilibrium of (5.32) and $\tilde{x}$ a pseudo-equilibrium. Then, linearizing (5.33) and (5.34) about the boundary equilibrium point at the origin, we have

$$\begin{aligned} Nx + M\mu &= 0, \\ C^T x + D\mu &= \lambda_1 < 0, \end{aligned} \tag{5.35}$$

and

$$\begin{aligned} N\tilde{x} + M\mu + E\alpha &= 0, \\ C^T \tilde{x} + D\mu &= 0, \\ \alpha &> 0, \end{aligned} \tag{5.36}$$

where $N = F_{1,x}, M = F_{1,\mu}, C^T = H_x, D = H_\mu$ and $E = F_2 - F_1$ all evaluated at $x = 0, \mu = 0$.

Now, from (5.35) we have $x = -N^{-1}E\mu$ and

$$\lambda_1 = (D - C^T N^{-1} M)\mu. \tag{5.37}$$

Moreover, from (5.37), $\tilde{x} = -N^{-1}M\mu - N^{-1}E\alpha$. Hence, we find

$$\alpha = \frac{(D - C^T N^{-1} M)\mu}{C^T N^{-1} E} \tag{5.38}$$

or, equivalently,

$$\alpha = \frac{\lambda_1}{C^T N^{-1} E}. \tag{5.39}$$

In order for x and $\tilde{x}$ to exist for the same value of μ, both λ_1 and α must share the same sign. However, they will exist for opposite values of μ if λ_1 and α have opposite sign. Therefore, using (5.39), we can state the following theorem.

Theorem 5.4 (Equilibrium points branching from a boundary equilibrium). *For the systems of interest, assuming*

$$det(N) \neq 0, \tag{5.40}$$

$$D - C^T N^{-1} M \neq 0, \tag{5.41}$$

$$C^T N^{-1} E \neq 0. \tag{5.42}$$

1. **Persistence** *is observed at the boundary equilibrium bifurcation point if*

$$C^T N^{-1} E < 0.$$

2. a **A non-smooth fold** *is instead observed if*

$$C^T N^{-1} E > 0.$$

Remarks

1. Note that the conditions delineating the two kinds of behavior the found are different from those in Theorem 5.1, as we should expect.
2. As with piecewise-smooth continuous systems, a full unfolding of the dynamics near boundary equilibrium bifurcations in Filippov systems is virtually impossible, as the possibilities in n-dimensions seem almost endless. In the next section we show that even in the planar case there are more cases that one has to consider than for continuous PWS systems.
3. Under certain special conditions, Filippov systems can also exhibit sets of equilibria (see [39]) that can be attracting or repelling. Models of certain systems involving friction naturally give rise to these kinds of dynamics. Discontinuity-induced bifurcations of sets of equilibria, which are bound to occur under parameter variation, remain an important open challenge for further study.

5.2.2 Planar Filippov systems

We present here a summary of the results of Kuznetsov *et al.* [169] which illustrate all the generic topologically distinct phase potraits near a boundary-equilibrium bifurcation. In two dimensions, a sliding region becomes a sliding line, and the boundary of such a region is a point T. Such a point in [169] is called a *tangent point*, and is such that the vectors $F_i(T), i = 1, 2$ are nonzero but at least one of them is tangent to Σ. Suppose that a tangent point $T \in \widehat{\Sigma}$ is such that $H_x(T)F_1(T) = 0$, see Fig.5.10. We say that this tangent point is *visible* if the orbit of $\dot{x} = F_1(x, \mu)$ starting at T belongs to $\mathcal{D}_1$ for all sufficiently small $|t| \neq 0$. We say that it is *invisible* if the orbit belongs to $\mathcal{D}_2$. Similar definitions hold for tangent point with respect to F_2.

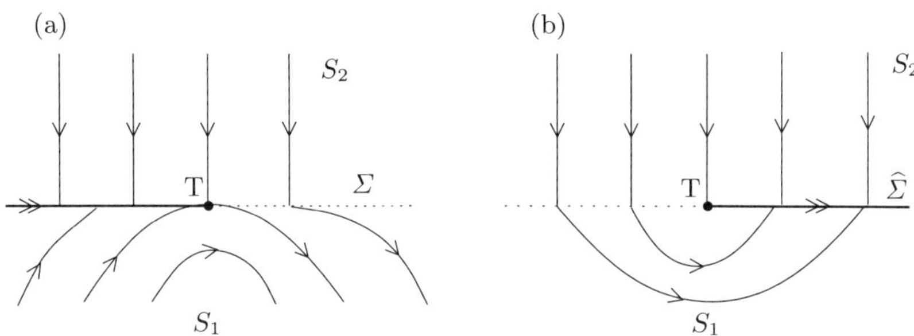

Fig. 5.10. Visible (a) and invisible (b) tangent points of a planar Filippov system. Here the solid portion of the boundary Σ, represents the sliding region $\widehat{\Sigma}$.

We will consider only transitions that involve sliding on the discontinuity boundary. Actually, the appearance or disappearance of a sliding segment is already a discontinuity-induced bifurcation in the topological sense introduced in Chapter 2. To meet all generic one-parameter transitions involving the discontinuity boundary Σ we use the following criterion: for a given parameter value μ, we consider the sliding set $\widehat{\Sigma}$ and find all the pseudo-equilibria and tangent points in it. These points are finite in number but can collide as μ varies, leading to local codimension-one DIBs. Another discontinuity-induced bifurcation can occur when a standard hyperbolic equilibrium in S_1 or S_2 collides with Σ, i.e. a boundary equilibrium bifurcation.

Specifically, we can distinguish three main cases involving the collision of equilibria with the boundary.

Boundary focus: There are five generic critical cases. In all of them there is a visible tangent point for $\mu < 0$ and an invisible tangent point for $\mu > 0$. The cases are distinguished by the relative position of the zero-isoclines of the focus and the behavior of the orbit departing from the visible tangent point into S_1, as well as by the direction of the motion

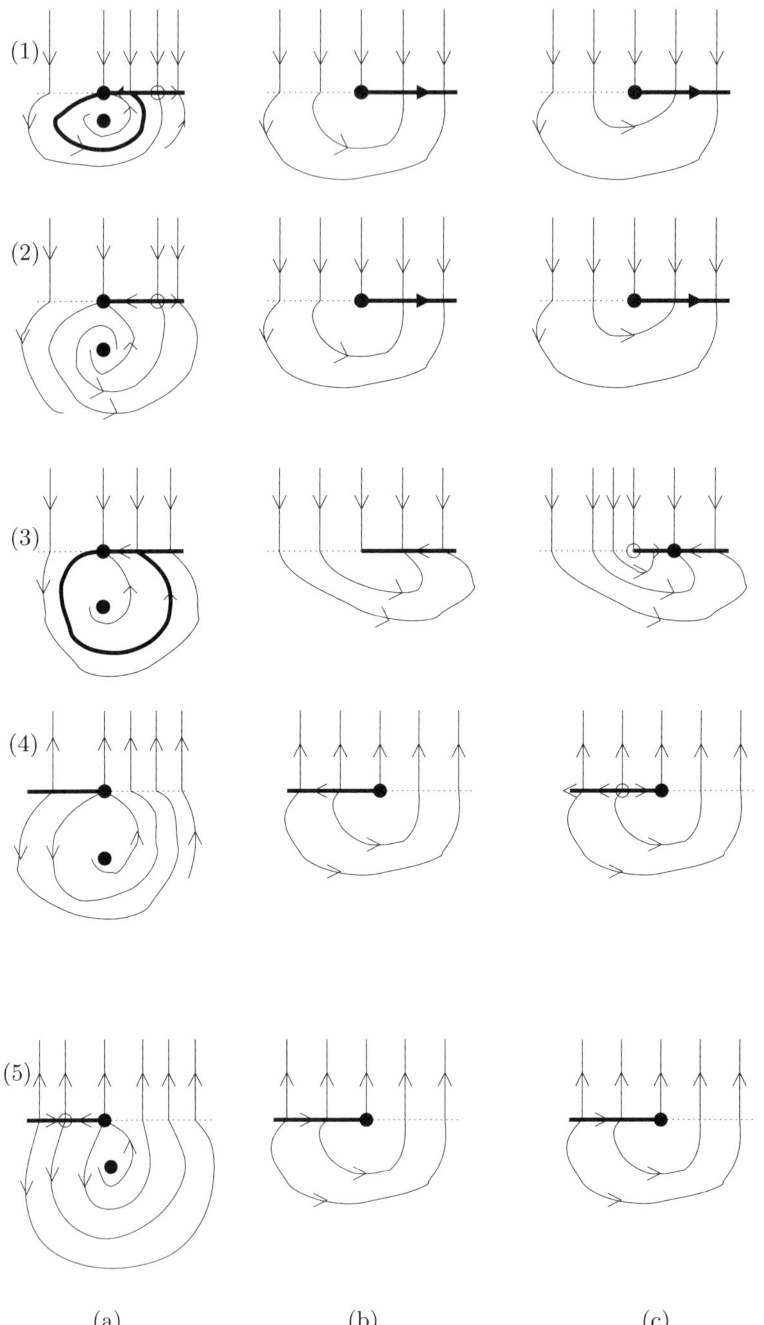

Fig. 5.11. Boundary focus transitions. (a) $\mu < 0$, (b) $\mu = 0$, (c) $\mu > 0$. Cases (1), (2) and (5) are non-smooth fold bifurcations, whereas (3) and (4) correspond to persistence.

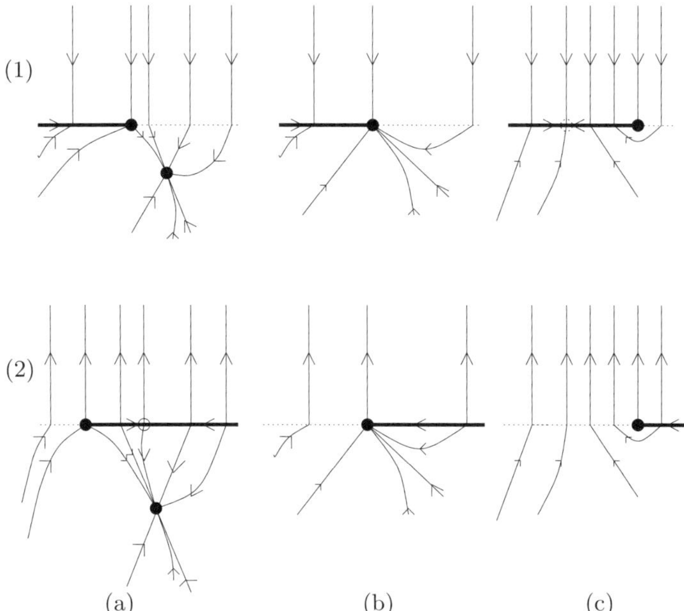

Fig. 5.12. Boundary node transitions. (a) $\mu < 0$, (b) $\mu = 0$, (c) $\mu > 0$. Case (1) is a persistence bifurcation, whereas (2) corresponds to a fold.

in S_2. If we assume that the colliding focus is unstable and has counter-clockwise rotation nearby, we can distinguish all five cases in Fig. 5.11. Cases (1), (2) and (5) are non-smooth fold bifurcations, whereas (3) and (4) correspond to persistence.

Boundary node: Depending on the direction of motion in S_2, there are two generic critical cases, which are shown in Fig. 5.12. Case (1) is a persistence bifurcation whereas (2) corresponds to a fold.

Boundary saddle: When the colliding equilibrium is a saddle, the three generic cases are determined by the slope of the zero-isoclines of the saddle, as shown in Fig. 5.13. In all cases, there is an invisible tangent point for $\mu < 0$ and a visible tangent point for $\mu > 0$. These points delimit the sliding segments on the discontinuity boundary. Cases (1) and (2) are fold bifurcations, whereas (3) corresponds to persistence.

Note that, when μ varies, two pseudo-equilibria can collide and disappear via a standard fold bifurcation, which in this case we will call a pseudo-fold transition. Figure 5.14 shows this transition in the case of a stable sliding segment.

As for piecewise-smooth continuous systems, it is possible to go further in the case of planar Filippov systems and find conditions for the existence of other attractors in a neighborhood of a BEB. In particular, in the two-dimensional case a limit cycle will always contain part of the sliding set, and

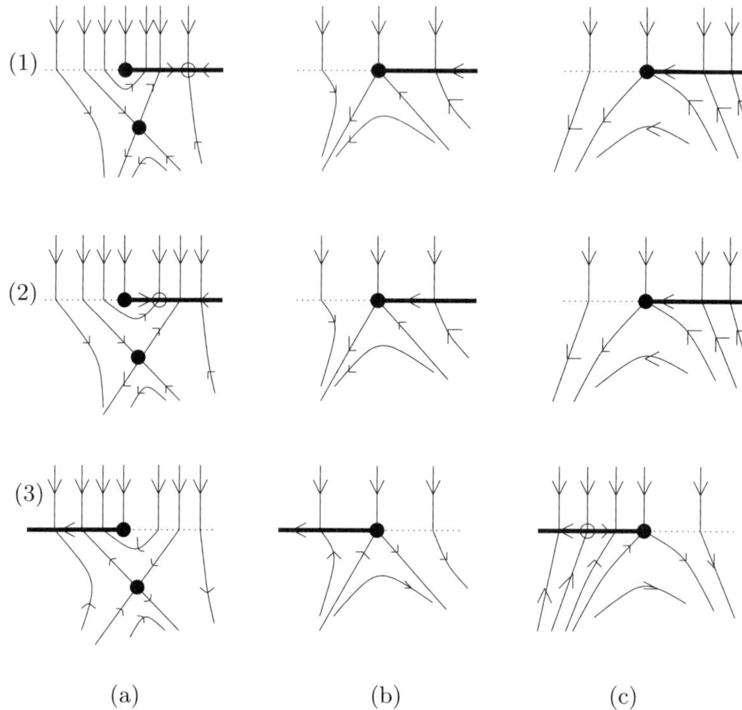

Fig. 5.13. Boundary saddle transitions. (a) $\mu < 0$, (b) $\mu = 0$, (c) $\mu > 0$. Cases (1) and (2) are fold bifurcations, whereas (3) corresponds to persistence.

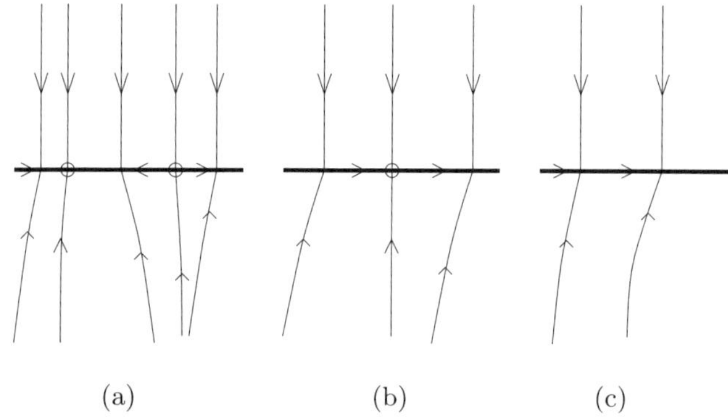

Fig. 5.14. Pseudo-fold transition. (a) $\mu < 0$, (b) $\mu = 0$, (c) $\mu > 0$.

will always encircle a (real) focus. The limit cycle will be stable if the sliding region is attracting ($C^T B > 0$), and unstable if it is repelling ($C^T B < 0$). (Note that trajectories are no longer unique in forward time if the region is repelling).

Theorem 5.5 ([169]). *Assume* $\det(N) \neq 0$, $C^T N^{-1} E \neq 0$, $\mathrm{trace}(N) \neq 0$, $\det(N) \neq \mathrm{trace}(N)^2/4$ *and* $C^T E \neq 0$

1. *If* $\mathrm{trace}(N) C^T E < 0$, *then no limit cycle is involved in the bifurcation (because of the area restriction).*
2. *If* $\mathrm{trace}(N) C^T E > 0$, *then:*
   a) *If we have a BEB with* $C^T N^{-1} E < 0$ *(***persistence***), then a limit cycle will surround a focus:*
      i. *If a transition from a pseudo-node to a focus is observed then the cycle exists and is stable if* $\mathrm{trace}(N) > 0$ *(the focus is unstable) and unstable if* $\mathrm{trace}(N) < 0$.
      ii. *If a transition of type pseudo-node/node or pseudo-saddle/saddle is observed then no the cycle exists.*
   b) *If we have a BEB with* $C^T N^{-1} E > 0$ *(***non-smooth fold***), then a limit cycle can only surround a focus.*
      i. *If the equilibria are pseudo-node/saddle or pseudo-saddle/node, then no limit cycle exists.*
      ii. *If we have a pseudo-saddle/focus bifurcation, then:*
         A. *If the focus is unstable and the unstable manifold of the pseudo-saddle point curls* inside *the stable one, then a stable limit cycle exists.*
         B. *If the focus is stable and the stable manifold of the pseudo-saddle point curls* inside *the unstable one, then an unstable limit cycle exists.*
         C. *If the respective manifolds curl* outside *instead, then no limit cycle exist.*

To illustrate the conclusions of the theorem above we now look at two representative examples. In both examples we use

$$C^T = (1 \quad 0), \qquad M = \begin{pmatrix} 0 \\ -1 \end{pmatrix}, \qquad D = 0. \tag{5.43}$$

Example 5.9. Suppose

$$N = \begin{pmatrix} 0.5 & 1 \\ -1 & 0 \end{pmatrix}, \qquad E = \begin{pmatrix} 5 \\ 3 \end{pmatrix}. \tag{5.44}$$

Then $\mathrm{trace}(N) = 0.5 > 0$, $C^T E = 5 > 0$ and $C^T N^{-1} E = -3 < 0$, and we have persistence with a stable pseudo-node for $\mu > 0$. For $\mu < 0$ we have an unstable focus surrounded by a stable limit cycle. See Fig. 5.15.

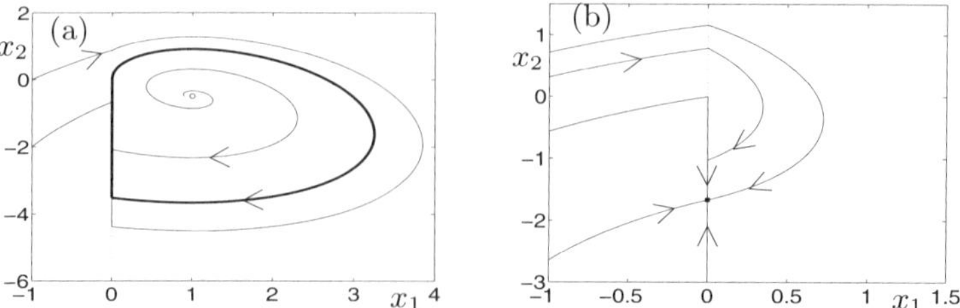

Fig. 5.15. Phase potraits for Example 5.9. (a) Unstable focus with stable limit cycle ($\mu = -1$), (b) stable pseudo-node ($\mu = 1$).

Example 5.10. Suppose

$$N = \begin{pmatrix} 0.5 & 1 \\ -1 & 0 \end{pmatrix}, \qquad E = \begin{pmatrix} 10 \\ -2.6 \end{pmatrix}. \tag{5.45}$$

Then trace$(N) = 0.5 > 0$, $C^T E = 10 > 0$, $C^T N^{-1} E = 2.6 > 0$, and we have a non-smooth fold with a pseudo-saddle point, and an unstable focus surrounded by a stable limit cycle for $\mu > 0$. See figure 5.16.

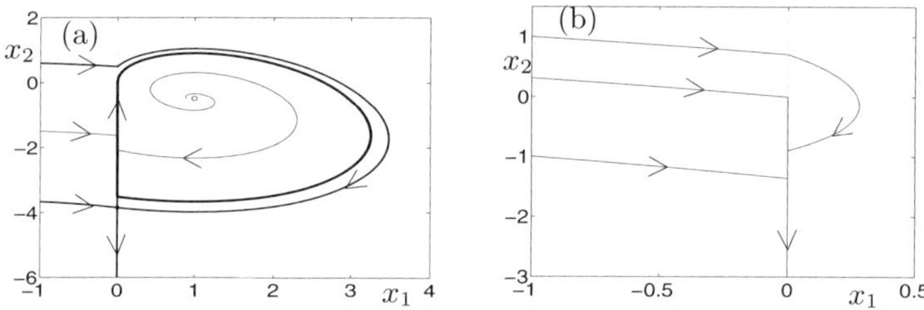

Fig. 5.16. Phase potraits for Example 5.10. (a) Unstable focus with stable limit cycle and pseudo-saddle point ($\mu = -1$), and (b) no limit sets ($\mu = 1$).

5.2.3 Some global and non-generic phenomena

In addition to this classification, global phenomena such as those depicted in Fig. 5.17 are also possible and were studied in [169]. For example, a pseudo-equilibrium $\tilde{x}(\mu)$ can have a sliding orbit that starts and returns back to it for $\mu = 0$. This is possible if $\tilde{x}(0)$ is either a pseudo-fold or a pseudo-saddle. Moreover, a standard saddle x_μ can have a homoclinic orbit containing a sliding segment at $\mu = 0$. Thus we have the cases, which are shown in Fig. 5.17.

Other phenomena concerning equilibria in Filippov systems have been reported in some non-generic cases which are of interest to analyze relay control systems. For example, it has been observed that a branch of limit cycles can appear after a focus changes its stability on the boundary. Specifically, Küpper

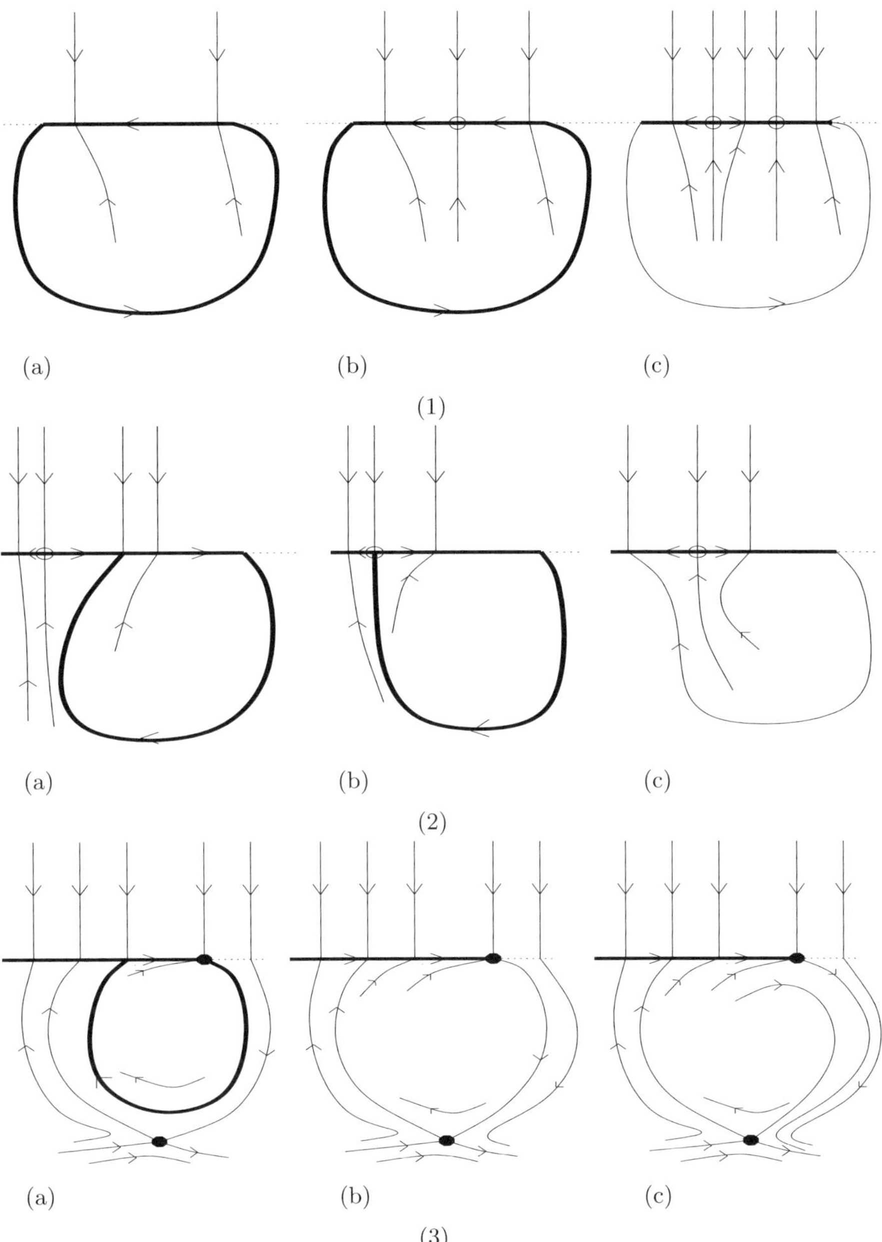

Fig. 5.17. Global phenomena: (1) Sliding homoclinic orbit to a pseudo-fold, (2) sliding homoclinic orbit to a pseudo-saddle, (3) sliding homoclinic orbit to a saddle. (a) $\mu < 0$, (b) $\mu = 0$, (c) $\mu > 0$.

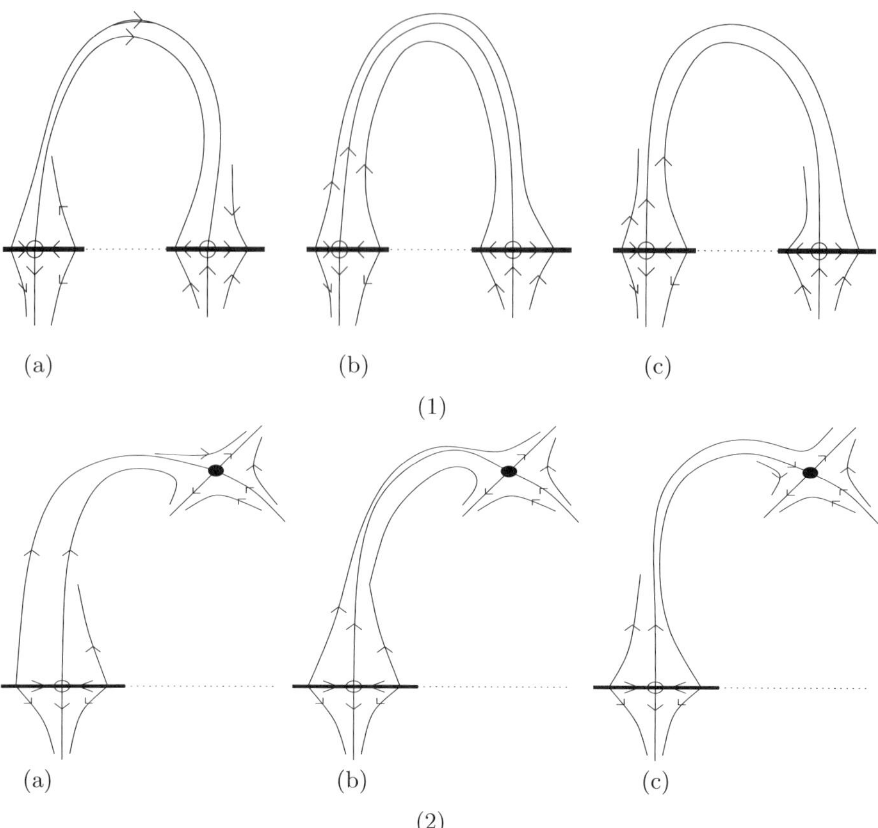

Fig. 5.18. Global phenomena: (1) Heteroclinic connection between two pseudo-saddles, (2) heteroclinic connection between a pseudo-saddle and a saddle. (a) $\mu < 0$, (b) $\mu = 0$, (c) $\mu > 0$.

& Moritz [167] study parameter-dependent Filippov dynamical systems where the focus is always in the origin. Then, it is possible to give conditions for a continuous isolated branch of periodic orbits to bifurcate from the boundary equilibrium at the origin. Another special case is described by Zou *et. al.* [283, 284, 285], where the existence is studied of periodic orbits bifurcating from a corner-like manifold in a planar Filippov dynamical system. There, the creation of a branch of cycles is determined by interactions between the geometrical structure of the corner and the eigenstructure of each smooth subsystem.

A further non-generic Filippov system (with symmetry) modeling a relay system is studied by Giannakopoulos and Pliete in [117]. Specifically, a piecewise-linear system is considered of the form

$$\dot{u} = Au + \text{sign}(w^T u)v,$$

where A is a 2×2 real matrix; u, v, w are two-dimensional real vectors and $\mathrm{sign}(\psi)$ denotes the sign of $\psi \in \mathbb{R}$. The theory of point transformation is applied to obtain conditions for the existence and stability of periodic solutions with and without sliding motion.

5.3 Equilibria of impacting hybrid systems

We study now the case of impacting hybrid systems, i.e. systems with zero degree of smoothness. In particular, we restrict our attention to systems of the form (2.35)–(2.36) given by

$$\dot{x} = F(x) \quad \text{if } H(x) > 0, \tag{5.46}$$

(where for the time being we suppress parameter dependence) with impact at the surface Σ defined by $\Sigma = \{x : H(x) = 0\}$, and where the impact law $R : \Sigma \to \Sigma$ takes the form

$$x^+ = R(x^-) = x^- + W(x^-)H_x F(x^-). \tag{5.47}$$

For convenience, we will also term the velocity and acceleration (of the vector field F relative to H) as

$$\begin{aligned} v(x) &= H_x F(x), \\ a(x) &= (H_x F)_x F(x). \end{aligned}$$

In Chapter 6, we will study the possible dynamical behavior of these systems in detail, motivating their geometry and their connection to mechanical systems. Here we concentrate on the fact that these systems have the possibility of *sticking* motion on the boundary Σ, which is the analogy of sliding motion in Filippov systems. Sticking points satisfy the conditions

$$H(x) = 0 \text{ and } v(x) = 0.$$

where the impact mapping is the identity. To maintain a sticking motion the sticking vector field must take the form

$$F_s(x) = F(x) - \lambda(x)W(x), \tag{5.48}$$

where the value $\lambda(x)$ is chosen to enforce the constraints $H(x) = 0$, $v(x) = 0$. This is possible for typical mechanical impacting systems in which W is parallel to the impact surface. Furthermore, defining

$$b(x) = (H_x F)_x W(x) \tag{5.49}$$

we have for the typical system that $b \leq -1$ at sticking points, since an impact with a negative incoming velocity is mapped to a point with a positive outgoing velocity. It follows that

$$\lambda(x) = a(x)/b(x), \tag{5.50}$$

and since $\lambda(x) > 0$ this is equivalent to the condition $a(x) < 0$, so that the acceleration is directed towards the boundary. Thus the sticking set is determined by the conditions $H(x) = 0$,$v(x) = 0$ and $a(x) < 0$.

We can now distinguish some different types of equilibria.

Definition 5.11. *We say that a point x^* is an* **admissible** *equilibrium of (5.46) if*

$$F(x^*) = 0,$$
$$H(x^*) > 0.$$

We say instead that x^* is a **pseudo-equilibrium** *point of (5.46) if it is an equilibrium of the sticking vector field defined by (5.48); i.e.*

$$\begin{aligned} F(x^*) - \lambda^* W(x^*) &= 0, \\ H(x^*) &= 0, \\ \lambda^* &> 0, \end{aligned}$$

where, for convenience, λ^* is regarded as an independent variable.

Now assume that the system depends on a single parameter μ. Then we can also give the following definition.

Definition 5.12. *A point $x = \bar{x}$, $\mu = \bar{\mu}$ is said to be a* **boundary equilibrium point** *of (5.46) if*

$$F(\bar{x}, \bar{\mu}) = 0,$$
$$H(\bar{x}, \bar{\mu}) = 0.$$

As for non-smooth continuous and Filippov systems, when the parameter μ is changed, admissible and/or pseudo-equilibrium points may branch off the boundary equilibrium. We can easily extend the definition of a boundary equilibrium transition, given in the previous section, to the case of impacting systems. Also, we can again classify the simplest possible cases as further detailed below.

5.3.1 Classification of the simplest BEB scenarios

Let us assume, without loss of generality, that a BEB occurs when $\bar{x} = \bar{\mu} = 0$. Moreover, suppose x^* is an admissible equilibrium of (5.46) whereas $\tilde{x}$ is a pseudo-equilibrium. Then, linearizing the system about $(0, 0)$, we have the conditions

$$
\begin{aligned}
Nx^* + M\mu &= 0, \\
C^T x^* + D\mu &> 0,
\end{aligned}
$$

for the admissible equilibrium, and

$$
\begin{aligned}
N\tilde{x} + M\mu + E\lambda^* &= 0, \\
C^T \tilde{x} + D\mu &= 0, \\
\alpha &> 0
\end{aligned}
$$

for the boundary equilibrium, where $N = F_x(0,0)$, $M = F_\mu(0,0)$, $C^T = H_x(0,0)$, $D = H_\mu(0,0)$ and $E = -W(0,0)$.

If the linear systems are not degenerate, they will be representative of what happens locally in the nonlinear system of interest. Specifically, following derivations similar to those presented in Sec. 5.1.1 and Sec. 5.2.1, we find

Theorem 5.6 (Equilibrium points branching from a boundary equilibrium). *For systems in this class, and assuming*

$$
\begin{aligned}
\det(N) &\neq 0, \\
D - C^T N^{-1} E &\neq 0, \\
C^T N^{-1} E &\neq 0.
\end{aligned}
$$

1. **Persistence** *is observed at the boundary equilibrium bifurcation point if*

$$C^T N^{-1} E < 0.$$

2. A **non-smooth fold** *is instead observed if*

$$C^T N^{-1} E > 0.$$

Remarks

1. The local stability of an admissible equilibrium point is determined by the eigenvalues of the matrix N. The question of stability of a pseudo-equilibrium point can be split into stability of the sticking set and stability of the sticking vector field when restricted to the sticking set, respectively. The stability of the sticking set is guaranteed if

$$-2 < b(\bar{x}) \leq -1. \tag{5.51}$$

(The expression $-(1+b)$ acts like a "coefficient of restitution".) If this is fulfilled, a small disturbance in the initial conditions will decay towards the sticking set through an infinite number of impacts in finite time, so-called *chattering* (see Chapter 6 for further details).

2. The linearization of the sticking vector field at $x = \bar{x}$ is

$$N_s = \left(I - \frac{MC^T N}{C^T NM} \right) N, \tag{5.52}$$

and we see that there is a 2×2 Jordan block corresponding to eigenvalue 0 with left eigenvector $C^T N$ and left generalized eigenvector C^T. This of course corresponds to the invariance of the codimension-two sticking set. The rest of the eigenvalues of N_s correspond to the dynamics within the sticking set, and if all have negative real part, the pseudo-equilibrium is stable within the sticking set.

Example 5.11. Let us consider an unforced single-degree-of-freedom mechanical system, with position x_1 and velocity x_2 characterized by a spring force with spring constant k, negative damping (scaled to unity), and an impact coefficient of restitution r. Such a system can be modeled by equations of the form (5.46)–(5.47) with

$$\begin{aligned} F(x,\mu) &= \begin{pmatrix} x_2 \\ \mu - kx_1 + x_2 \end{pmatrix}, \\ H(x) &= x_1, \\ W &= -(1+r) \begin{pmatrix} 0 \\ 1 \end{pmatrix}. \end{aligned} \tag{5.53}$$

In this case, we have

$$v(x) = x_2, \qquad a(x) = \mu - kx_1 + x_2, \qquad b(x) = -(1+r),$$

and from (5.48) we find

$$F_s(x,\mu) = \begin{pmatrix} x_2 \\ 0 \end{pmatrix}.$$

Clearly, the system undergoes a boundary equilibrium transition at $\bar{x} = 0$, $\bar{\mu} = 0$. Thus, linearizing about the BEB point, we get

$$\begin{aligned} N &= \begin{pmatrix} 0 & 1 \\ -k & 1 \end{pmatrix}, & M &= \begin{pmatrix} 0 \\ 1 \end{pmatrix}, \\ C^T &= (1 \quad 0), & D &= 0, \\ E &= (1+r) \begin{pmatrix} 0 \\ 1 \end{pmatrix}, & A_s &= \begin{pmatrix} 0 & 1 \\ 0 & 0 \end{pmatrix}. \end{aligned}$$

Moreover we have

$$D - C^T N^{-1} M = 1/k \quad \text{and} \quad C^T N^{-1} E = -(1+r)/k.$$

This is consistent with the explicit solution for the admissible equilibrium given by

$$x^* = \begin{pmatrix} \mu^*/k \\ 0 \end{pmatrix}, \qquad \mu^*/k > 0,$$

and the pseudo-equilibrium

$$\tilde{x} = \begin{pmatrix} 0 \\ 0 \end{pmatrix}, \qquad \alpha = -\mu^*/(1+r), \qquad \mu/(1+r) < 0.$$

Using Theorem 5.6, if $k > 0$ the BEB is associated with a persistence scenario. Hence an admissible equilibrium branch exists for $\mu > 0$ and a branch of pseudo-equilibria exists for $\mu < 0$. If, instead, $k < 0$, then no equilibria exist for $\mu > 0$, whereas two exist for $\mu < 0$, because the BEB is associated with a non-smooth fold. Moreover, it is easy to show that the admissible equilibrium point is unstable (a saddle point if $k < 0$) and that the pseudo-equilibrium point is stable if $0 \leq r < 1$ (by proving the stability of the sticking set).

5.3.2 The existence of other invariant sets

Little is known about the existence of invariant sets besides equilibrium points which arise when perturbing a boundary equilibrium in impacting hybrid systems. The type of analysis required clearly has a strong resemblance to what would be needed in the corresponding cases for Filippov and non-smooth continuous systems, where planar systems are fully understood, but only relatively weak results apply in three and higher dimensions. In an impacting system, a limit cycle will always contain one impact. Also, a pseudo-equilibrium will always have focus character, stable if the coefficient of restitution r is less than unity and unstable if $r > 1$. As shown in [87], the following theorem can be proven.

Theorem 5.7 ([87]). *Assume* $\det(N) \neq 0$, $C^T N^{-1} M \neq 0$, $\mathrm{trace}(N) \neq 0$, $\det(N) \neq \mathrm{trace}(N)^2/4$, $r = C^T N M - 1 \geq 0$ *and* $r \neq 1$.

1. *If* $\mathrm{trace}(N)(r-1) > 0$, *then no limit cycle is involved in the bifurcation (because of the area restriction).*
2. *If* $\mathrm{trace}(N)(r-1) < 0$, *then:*
   a) *If we have a BEB with* $C^T N^{-1} M < 0$ *(***persistence***), then*
      i. *If a transition from a pseudo-focus to a node is observed, then the cycle exists and is stable if* $\mathrm{trace}(N) < 0$ *(the node is stable) and unstable if* $\mathrm{trace}(N) > 0$.
      ii. *If instead a transition from a pseudo-focus to a focus is present, then assuming* $\alpha \pm i\omega$ *are the eigenvalues of the focus, with* $\omega > 0$, *and* $re^{\frac{\alpha}{\omega}\pi} \neq 1$, *then a limit cycle exists. It is stable if* $re^{\frac{\alpha}{\omega}\pi} < 1$ *and unstable if* $re^{\frac{\alpha}{\omega}\pi} > 1$.
   b) *If we have a BEB with* $C^T N^{-1} M > 0$ *(a* **non-smooth fold***) then we have a pseudo-focus and a saddle point. Both the stable and the unstable half manifold of the saddle will intersect the boundary* Σ. *Let* $\lambda_1 > 0$ *be the unstable eigenvalue of the saddle point, and assume* $\mathrm{trace}(A) \neq \lambda_1(1-r)$.

i. If trace$(N) < \lambda_1(1-r) < 0$, *the pseudo-focus is unstable and the unstable manifold of the saddle point curls* inside *the stable one, and a stable limit cycle exists.*

ii. If trace$(N) > \lambda_1(1-r) > 0$, *the pseudo-focus is stable and the stable manifold of the saddle point curls* inside *the unstable one, and an unstable limit cycle exists.*

iii. If $\lambda_1(1-r)/\text{trace}(N) > 1$, *the respective manifolds curl* outside *instead, and no limit cycle exists.*

Rather than giving the lengthy proof of the above theorem, we will illustrate the conditions presented by means of some simple examples. In all examples we use

$$C^T = (1 \quad 0), \qquad M = \begin{pmatrix} 0 \\ -1 \end{pmatrix}, \qquad D = 0. \tag{5.54}$$

Example 5.12. Suppose

$$N = \begin{pmatrix} -2.1 & 1 \\ -1 & 0 \end{pmatrix}, \qquad E = \begin{pmatrix} 0 \\ 2.5 \end{pmatrix}. \tag{5.55}$$

Then trace$(N) = -2.1 < 0$, $r = 1.5 > 1$, $C^T N^{-1} E = -2.5 < 0$ and we have persistence with a stable node for $\mu < 0$ and for $\mu > 0$ we have an unstable pseudo-focus surrounded by a stable limit cycle. See figure 5.19.

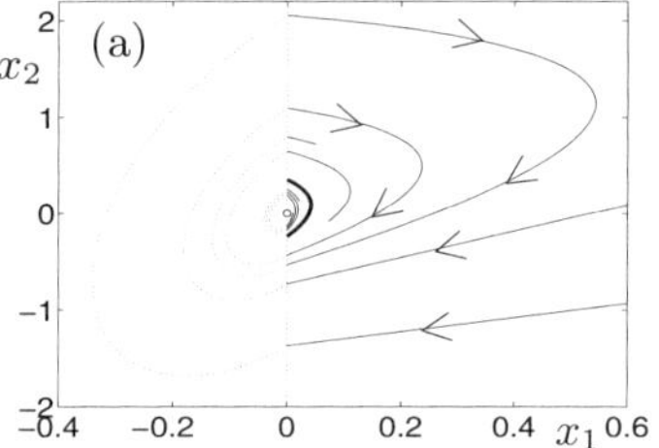

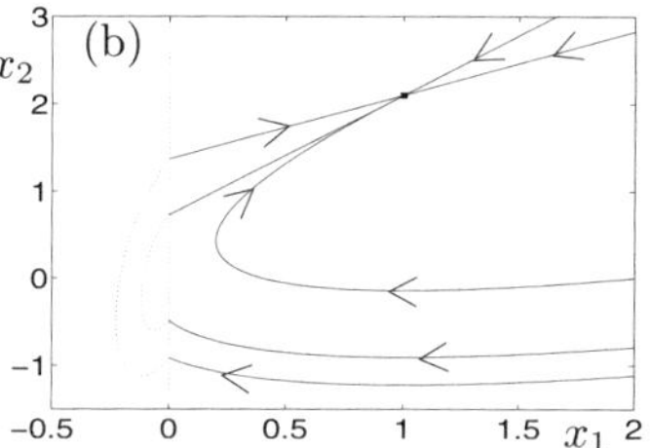

Fig. 5.19. Phase potraits for Example 5.12. (a) Unstable pseudo-focus with stable limit cycle ($\mu = 1$) and (b) stable node ($\mu = -1$). (Dotted curves to the left of $x_1 = 0$ are not part of the trajectories; they merely indicate how the trajectories connect during impact.)

Example 5.13. Suppose

$$N = \begin{pmatrix} -1 & 1 \\ -1 & 0 \end{pmatrix}, \qquad E = \begin{pmatrix} 0 \\ 2.5 \end{pmatrix}. \tag{5.56}$$

Then trace$(N) = -1 < 0$, $r = 1.5 > 1$, $C^T N^{-1} E = -2.5 < 0$, $re^{\frac{\alpha}{\omega}\pi} < 1$, and we have persistence with a stable focus for $\mu < 0$. For $\mu > 0$ we have an unstable pseudo-focus surrounded by a stable limit cycle. See Fig. 5.20.

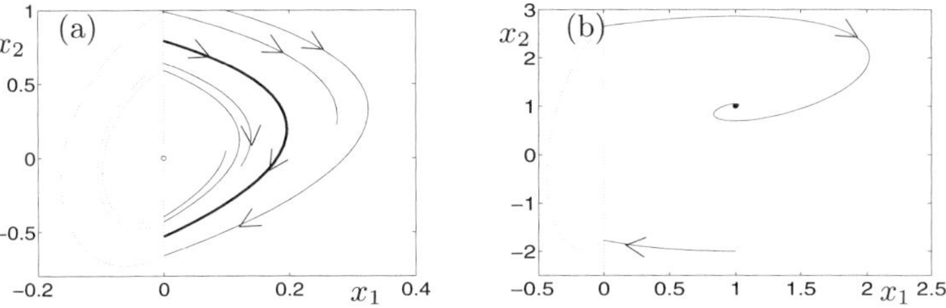

Fig. 5.20. Phase potraits for example 5.13. (a) Unstable pseudo-focus with stable limit cycle ($\mu = 1$) and (b) stable focus ($\mu = -1$).

Example 5.14. Suppose

$$N = \begin{pmatrix} 0.5 & 1 \\ -1 & 0 \end{pmatrix}, \qquad E = \begin{pmatrix} 0 \\ 1.5 \end{pmatrix}. \tag{5.57}$$

Then trace(N) $= 0.5 > 0$, $r = 0.5 < 1$, $C^T N^{-1} E = -1.5 < 0$, $r e^{\frac{\alpha}{\omega}\pi} < 1$, and we have persistence with a stable pseudo-focus for $\mu > 0$. For $\mu < 0$ we have an unstable focus surrounded by a stable limit cycle. See Fig. 5.21.

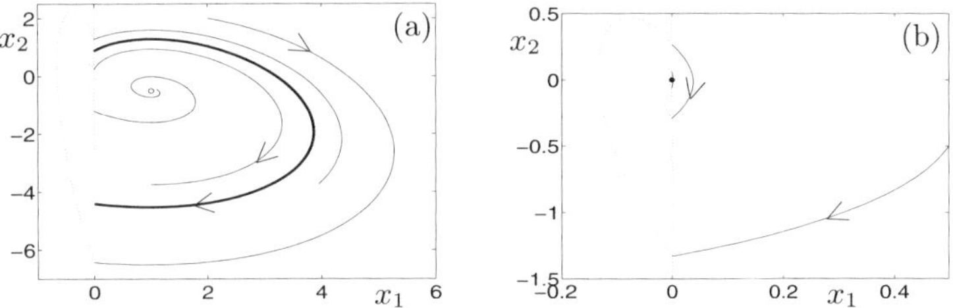

Fig. 5.21. Phase potraits for Example 5.14. (a) Unstable focus with stable limit cycle ($\mu = -1$) and (b) stable pseudo-focus ($\mu = 1$).

Example 5.15. Suppose

$$N = \begin{pmatrix} -1 & 1 \\ 1 & 0 \end{pmatrix}, \qquad E = \begin{pmatrix} 0 \\ 3 \end{pmatrix}. \tag{5.58}$$

Then trace(N) $= -1 < 0$, $r = 2 > 1$, $C^T N^{-1} E = 3 > 0$, trace(N) $< \lambda_1(1 - r) < 0$, and we have a non-smooth fold with a saddle point, and an unstable pseudo-focus surrounded by a stable limit cycle for $\mu > 0$. See Fig. 5.22.

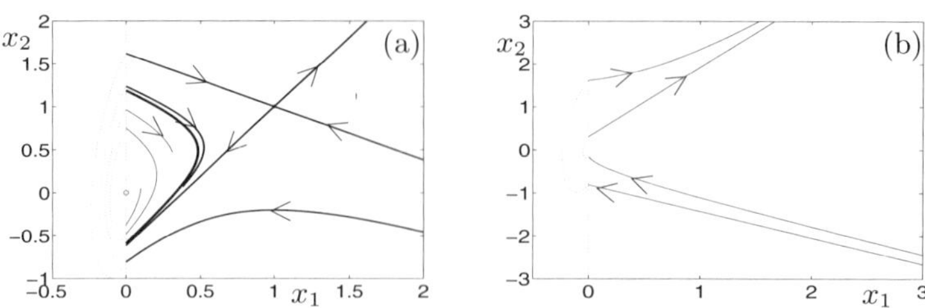

Fig. 5.22. Phase potraits for Example 5.15. (a) Unstable pseudo-focus with stable limit cycle and saddle point ($\mu = 1$), and (b) no limit sets ($\mu = -0.5$).

6

Limit cycle bifurcations in impacting systems

This chapter begins by motivating the class of impacting hybrid systems that were introduced in Chapter 2, introducing several practical examples of impact-oscillator systems. We shall then see how such systems naturally lead to differing kinds of Poincaré maps on suitably defined surfaces. Next, in Sec. 6.2, we show in detail how to calculate the *discontinuity mapping* to unfold the dynamics close to a *grazing* impact. We then show in Sec. 6.3 how to analyze the discontinuity-induced bifurcation (DIB) associated with a grazing limit cycle, by reducing locally to Poincaré maps with square-root singularities. The hybrid system can then be analyzed using the methods developed in Chapter 4. In Sec. 6.4 we then explore more global issues — including chattering, chaos and domains of attraction — through seeking to explain the dynamics observed in Chapter 1 on the single degree-of-freedom impact oscillator. The chapter ends in Sec. 6.5 with a brief discussion of hybrid systems with more than one impact surface, motivated by multi-body impacting systems for that a triple collision corresponds to the crossing of an intersection between two discontinuity surfaces

6.1 The impacting class of hybrid systems

Recall Definition 2.25 of impacting hybrid systems from Chapter 2. Mostly in this chapter we shall be treating phenomena that occur with respect to a single impact surface, in that case we shall use the reduced form of such systems:

$$\dot{x} = F(x) \quad \text{for } x \in S^+ = \{x \in \mathcal{D} \subset \mathbb{R}^n : H(x) > 0\} \tag{6.1}$$

with impact at the surface $\Sigma := \{x \in \mathcal{D} : H(x) = 0\}$, where the impact law $R : \Sigma \to \Sigma$ takes the form

$$x^+ = R(x^-) := x^- + W(x^-)v(x^-). \tag{6.2}$$

Here, W is a smooth function such that $x + W(x)v(x)$ is also smooth, and

$$v(x) = H_x F(x), \qquad a(x) = (H_{xx} + H_x F_x) F(x) \tag{6.3}$$

are the velocity and acceleration, respectively, of the flow relative to H. Often we shall be interested in the case that W is a constant vector, in that case the requirement that W maps Σ to itself means that

$$H_x W = 0. \tag{6.4}$$

Note that the expressions for a and v have a particularly convenient form when expressed in the Lie derivative notation introduced in Chapter 2. Specifically $\mathcal{L}_F(H)(x)$ represents the time derivative of $H(x)$ along a trajectory generated by the vector field $F(x)$. That is, if $\Phi(x,t)$ is the flow generated by F, we have

$$\mathcal{L}_F H(x) = \frac{\partial}{\partial t} H(\Phi(x,t)) = H_x \frac{\partial}{\partial t}\Phi(x,t) = H_x F(x).$$

Hence, this scalar quantity depends only on the value of the vector field F at the point x. Similarly,

$$\mathcal{L}_F^m H(x) = \left.\frac{\partial^m H(\Phi(x,t))}{\partial t^m}\right|_{t=0} = (\ldots((H_x F)_x F)_x F \ldots)_x F(x)$$

is a scalar quantity that represents the mth derivative of H along the flow. Thus, we have

$$v(x) = \mathcal{L}_F H(x), \quad a(x) = \mathcal{L}_F^2 H(x) \text{ and } R(x) = x + W(x)\mathcal{L}_F(H)(x). \tag{6.5}$$

Also, it is useful to divide the impacting surface into three separate regions

$$\Sigma^- = \{x \in \Sigma : v(x) < 0\}, \quad \Sigma^+ = \{x \in \Sigma : v(x) > 0\}$$

$$\text{and } G = \{x \in \Sigma : v(x) = 0\}.$$

In general, the impact rule R is defined so that if $x^- \in \Sigma^-$, then $x^+ \in \Sigma^+$. Thus flow in S^+ that intersects Σ^- is mapped to Σ^+ and then continues in S^+ again. The set G is called the *grazing region*. It is also useful to define the virtual region

$$S^- = \{x \in \mathcal{D} : H(x) < 0\},$$

in that the vector field F and its corresponding flow $\Phi(x,t)$ is well defined, despite not being part of the hybrid system.

In Chapter 5 we analyzed simple equilibrium bifurcations in systems of the form (6.1)–(6.5). Before proceeding to more detailed analysis of DIBs associated with limit cycles in such systems, let us consider some motivating examples, of increasing complexity.

6.1.1 Examples

Example 6.1 (The single degree-of-freedom (1DoF) impact oscillator). The simple, forced impact oscillator introduced as case study I is defined by the three-dimensional autonomous system:

$$du/dt = v, \quad dv/dt = -u - 2\zeta v + w(s), \quad ds/dt = 1, \quad \text{if} \quad u > \sigma, \tag{6.6}$$

so that

$$x = \begin{pmatrix} u \\ v \\ s \end{pmatrix}, \quad F(x) = \begin{pmatrix} v \\ -u - 2\zeta v + w(s) \\ 1 \end{pmatrix}.$$

As already explained in Chapter 2, this system can be written in the form (6.1)–(6.3) with

$$H(x) = u - \sigma, \qquad v(x) = v \quad \text{and} \quad a(x) = -2\zeta v + w(s).$$

On $\Sigma := \{x : u > \sigma\}$ the Newtonian impact law $R : v^+ = -rv^-$ is of the form (6.2) with $W(x) = (0, -(1 + r), 0)$, where r is the coefficient of restitution.

In case study I in Chapter 1, we saw transitions in the dynamics that result when limit cycles *graze* with the impact surface Σ. Much of this chapter is devoted to an analysis of this discontinuity-induced bifurcation in some generality. We also saw in Chapter 1 how fingered strange attractors can occur close to chattering sequences and how complex basins of attraction can arise when there are competing attractors. A complete analysis of such phenomena is beyond the scope of this book, but section 6.4 below is devoted to an explanation of them in the context of this example system.

Example 6.2 (Two impacting masses). The so-called Newton's cradle toy consists of suspended masses swinging independently when not in contact, and interacting through impact with each other. The simplest such device, with two masses, is illustrated in Fig. 6.1. Such devices also have industrial application, for example to test the behavior of the material in the impacting part of the pendulum under the effects of impact induced wear. See for example the work of Blackmore *et al.* [32] for testing the tolerance of pills to impact.

For small angles of swing, we can model this system in terms of two independent oscillating masses at locations $p < q$ of respective mass M and m, each undergoing simple harmonic motion

$$\frac{d^2p}{dt^2} + \omega_1^2 p = w_1(s), \quad \frac{d^2q}{dt^2} + \omega_2^2 q = w_2(s), \quad ds/dt = 1, \tag{6.7}$$

with natural frequencies ω_1 and ω_2 respectively. Note that ordinarily, for a forced pendulum system there will be a gross qualitative error involved with linearizing the sine nonlinearity inherent in the angular DoF of the motion. However, for small angles of swing, the impact law will provide a far stronger

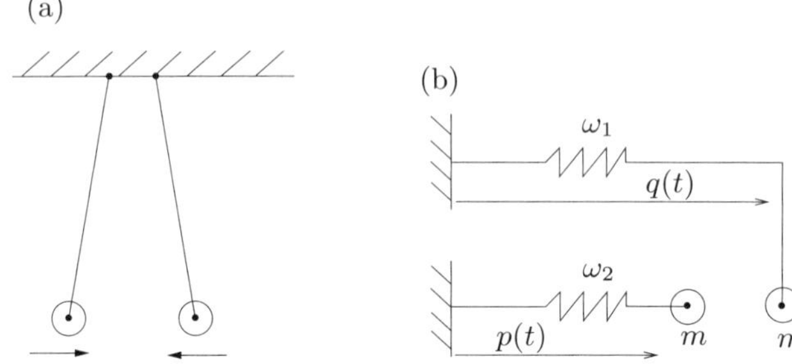

Fig. 6.1. (a) Schematic illustration of simplest Newton's cradle toy, and (b) its mathematical model for small angles of swing.

form of nonlinearity than the geometric nonlinearity. Roughly speaking then we can linearize the sine function without making qualitative error. The Newtonian rule for impact of two masses conserves momentum, reverses the direction of relative velocity between the masses while reducing this velocity by a coefficient of restitution $0 \le r \le 1$:

$$p^+ = p^-, \quad q^+ = q^-, \quad M\dot{p}^+ + m\dot{q}^+ = M\dot{p}^- + m\dot{q}^- \quad \dot{q}^+ - \dot{p}^+ = -r(\dot{q}^- - \dot{p}^-). \tag{6.8}$$

Now we have a five-dimensional autonomous system that can be written in the form (6.1)–(6.3) with

$$x = (p, \dot{p}, q, \dot{q}, s)^T, \quad F = (\dot{p}, f_1 - \omega_1^2 p, \dot{q}, f_2 - \omega_2^2 q, 1)^T,$$

where the smooth flow applies in region $S^+ = \{x : q - p > 0\}$ and the impacting surface is given by

$$\Sigma = \{x : q - p = 0\}, \quad \text{so that} \quad H(x) = q - p, \quad v = \dot{q} - \dot{p}.$$

The impact rule (6.8) can be put into the canonical form (6.2) with

$$W(p, \dot{p}, q, \dot{q}, s) = \left(0, \frac{(1+r)m}{M+m}, 0, -\frac{(1+r)M}{M+m}, 0\right)^T$$

Sticking flow in this system occurs when $p = q, \dot{p} = \dot{q}$ and

$$a(x) := d^2q/dt^2 - d^2p/dt^2 = f_2 - f_1 + (\omega_1^2 - \omega_2^2)p \le 0.$$

A particular example arises in the Newton cradle problem when the two masses swing only under the action of gravity, so that $f_1 = f_2 = 0$ and $\omega_1 = \omega_2$. In this case it is possible for the two masses to swing whilst being in continuous contact with $p = q$. If the system is started in a general motion and $M \gg m$, then the observed dynamics will be a periodic motion of p that undergoes a series of impacts, with q resulting in a chattering sequence before the mass at q sticks to, and moves in permanent contact with the mass at p. The time taken for the chattering sequence to converge can be used as a test of the wear on the mass at q [32].

An interesting simplification of this problem arises when p is infinitely massive when compared with q, so that q rebounds away from p at impact without changing the motion of p. We can take $p(t)$ to be a general periodic function, so that the impacting system becomes

$$\frac{d^2q}{dt^2} + \omega^2 q = w(t), \quad q(t) > p(t),$$

with impact law

$$\dot{q}^+ - \dot{p} = -r(\dot{q}^- - \dot{p}).$$

In standard form this system has

$$x = (q, \dot{q}, s)^T, \quad F = (\dot{q}, w(s) - \omega^2 q, 1)^T, \quad H(x) = q - p(s), \quad v = \dot{q} - \dot{p}$$

with

$$W(q, \dot{q}, s) = (0, -(1+r), 0)^T.$$

Note that an example of this case where the angle of swing is large so that the sine nonlinearity must be taken into account is treated in detail as an experimental case study in Chapter 9. Another practical example is that of an internal combustion engine in that valve rods on springs at positions $q(t)$ are driven by a rotating cam, leading to the periodic function $p(t)$ describing the impact surface [3, 210, 211]; a schematic of such a cam-follower system is shown in Fig. 6.2.

Example 6.3 (An impacting cantilever beam). A beam is a structural element that has a finite resistance to bending and so, unlike a rigid rod, can undergo continuous deflection along its length. Such an element may be held clamped at one end and allowed to impact with a rigid stop at a set distance along its length, see Fig. 6.3. Such an apparatus provides a natural extension to the 1DoF impact oscillator, as it can be thought of as a system that evolves in an infinite-dimensional phase space, but with a single impact surface.

Specifically, consider a beam of length L that is clamped at one end, $z = 0$, and free at the other. Suppose the transverse displacement of the beam is given by $U(z, t)$, where τ is time and z is position along the axis of the beam. We suppose that the beam impacts a rigid stop at $U = \sigma$, $z = z^*$, where $0 < z^* < L$. We suppose the beam to be subject to a spatially uniform external acceleration $\widehat{w}(z, t)$ in a direction perpendicular to its axis, and that the motion of the beam is confined to the plane. In the simplest case, assuming linear elasticity and small angles of deflection, the free (non-impacting) motion of such a beam satisfies a fourth-order linear partial differential equation (PDE):

$$U_{\tau\tau} + EIU_{zzzz} = \widehat{w}(z, t), \quad 0 < z < L,$$

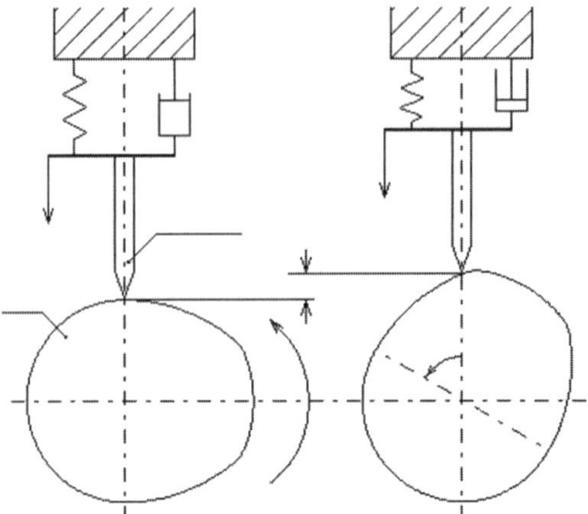

Fig. 6.2. Schematic diagram of an ideal cam-follower system at two different time instants.

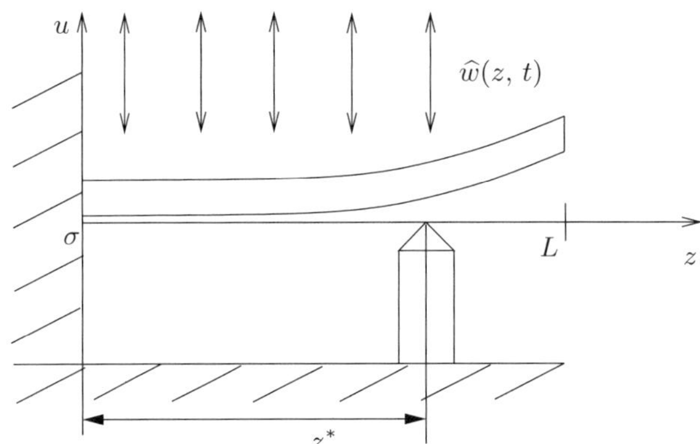

Fig. 6.3. A forced cantilever beam impacting with a rigid stop.

$$\text{with} \quad U(0) = U_z(0) = 0, \quad U_{zz}(L) = U_{zzz}(L) = 0,$$

where E is Young's modulus and I is the second moment of area of the cross-section. Nondimensionalization via

$$t = \frac{L^2}{\sqrt{EI}}\tau, \qquad y = \frac{z}{L},$$

leads to the dimensionless PDE

$$u_{tt} + u_{yyyy} = w(y, t), \quad 0 < y < 1, \tag{6.9}$$

$$\text{with} \quad u(0) = u_y(0) = 0, \quad u_{yy}(1) = u_{yyy}(1) = 0,$$

where $w(t) = \widehat{w}(t)L^4/EI$

By applying the method of separation of variables, the general solution of this system can then be expressed in the modal form

$$u(y,t) = \sum_i x_i(t)\psi_i(y), \tag{6.10}$$

where the $\psi_i(y)$ are eigenfunctions of the fourth-derivative operator on the unit interval subject to the boundary conditions in (6.9). The corresponding modal frequencies of the system $\{\omega_i\}$ are all real, positive and well ordered $\omega_1 < \omega_2 < \omega_3 < \ldots$. Each modal amplitude satisfies an equation of the form

$$\frac{\mathrm{d}^2 x_i}{\mathrm{d}t^2} + \omega_i^2 x_i = w_i(y,t), \tag{6.11}$$

where $w_i(t) = \int_0^1 \psi_i(y)w(y,t)\mathrm{d}y$. The motion can then be expressed in terms of the evolution of the single vector $x = (x_1, x_2, x_3, \ldots, \dot{x}_1, \dot{x}_2, \ldots)^T$ where each component x_i is a solution of a harmonic oscillator equation (6.11). Now, since the impact occurs at the point $y = y^* := z^*/L$, $u = \sigma$, the impact surface is given by

$$\Sigma(x) := \{x : H(x) := \Sigma_i x_i \psi_i(z^*) - \sigma := c^T x - \sigma = 0\}.$$

Thus the cantilever beam has a simple impacting surface. However, a simple impact law

$$u_t(y,t^+) = -r u_t(y,t^-) \tag{6.12}$$

does not uniquely specify the values of $x_i(t^+)$. At impact, there is typically an exchange of energy from low-frequency to high-frequency modes. The overall law can be written as

$$\frac{\mathrm{d}x_i}{\mathrm{d}t}(t^+) = \sum_j R_{ij}\frac{\mathrm{d}x_j}{\mathrm{d}t}(t^-).$$

However, the evaluation of the coefficients R_{ij} are still unclear for any particular system and need a detailed knowledge of the nature of the elastic waves that can move up and down the beam. An example of an experimental, reasonably rigid beam was seen in Fig. 1.5. In this we can see the excitation of higher modes at impact. Typically these are strongly damped, and provided that impacts are not too close together, we can approximate the overall motion of the cantilever beam as that of a 1DoF oscillator with a high coefficient of restitution [31].

Example 6.4 (A one-dimensional lattice with impacts). Consider a system of n particles u_i, coupled together with springs (such as in a Toda lattice [254]) so that

$$\frac{d^2 u_i}{dt^2} = -k_i(u_i - u_{i-1}) + k_{i+1}(u_{i+1} - u_i),$$

but with each particle *individually constrained* by a stop, so that the motion of this particle is *smooth* if $u_i > \sigma_i$ and if $u_i = \sigma_i$ the particle rebounds with coefficient of restitution r_i. In this system, S^+ is a subset of $\mathbb{R}^n$ and it is bounded by the union of the sets $\Sigma_i = \{x : u_i - \sigma_i = 0\}$. This is thus another example of a system with a multiple impact surface. However in this case there is a possibility of a non-trivial intersection of these surfaces over the codimension-two sets

$$\{x : u_i - \sigma_i = 0, u_j - \sigma_j = 0\}.$$

Locally, that is, close to each set Σ_i, the motion is essentially identical to the 1DoF impact oscillator. It is possible for the solution to become stuck to one of these surfaces, for example x may enter Σ_1 through a chattering sequence. In this case the system loses a degree of freedom, with the dynamics confined to the remaining $n-1$ particles, until the acceleration a_1 of $u_1 - \sigma_1$ becomes positive. A detailed description of the dynamics is given in [259].

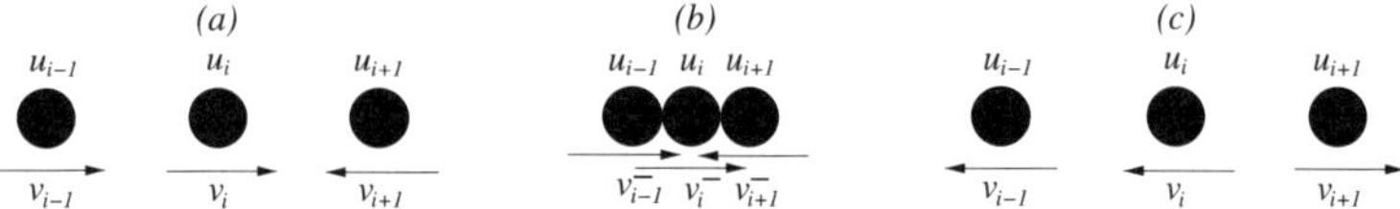

Fig. 6.4. Possible fate of three particles (a) before, (b) during and (c) after a triple collision event.

Example 6.5. Multiple impacting particles. Of course there are far more complex mechanical systems, such as robot arms [38], clock mechanisms, or rattling bundles of heat exchanger tubes [135], that comprise many impacting components, all tightly coupled to each other. These components can move freely and come into contact with each other. As a simple model of such a system we might suppose that the problem is reduced to that of particles constrained to one spatial dimension. The ith particle is assumed to occupy position $u_i \in \mathbb{R}$, subject to a smooth evolution equation

$$\frac{\mathrm{d}^2 u_i}{\mathrm{d}t^2} = w(u_i, \dot{u}_i, t)$$

for $u_{i-1} < u_i < u_{i+1}$, where w is some external forcing. Smooth motion is bounded by the surfaces $\Sigma_i = \{x : u_i = u_{i+1}\}$. Note that there is thus the potential for a multiple collision when $x \in \Sigma_i \cap \Sigma_{i+1}$. This would represent a codimension-one DIB, a full analysis of that is beyond the scope of this book. Nevertheless, Sec. 6.5 below gives some preliminary results on how to analyze such events in a special case.

6.1.2 Poincaré maps related to hybrid systems

As described in Chapter 2, the overall dynamics of an impacting hybrid system of the form (6.1)–(6.3) is an alternating sequence of smooth flows in the set S^+ interrupted by impacts with the bounding set Σ, where the orbits either re-enter S^+ instantaneously or, after a chattering sequence, stick to the set G for a non-zero time. We then have a *hybrid flow map* $\Psi(x,t)$ that describes the evolution of the hybrid system for positive times t to new positions in $S^+ \cup \Sigma$. For a typical initial condition, this map will consist of a composition of smooth flow maps Φ in S^+ and (a possible infinite number of) reset maps at any impacts.

A natural method for studying these systems is to derive discrete-time maps from the flow Ψ and then to apply the theory presented in Chapters 3 and 4 to understand the dynamics of such maps. As outlined in Chapter 2, a common procedure for defining such discrete-time maps is via choice of suitable transverse Poincaré sections Π. For hybrid flows, several choices of the surface Π are possible, leading to subtly different maps.

Suppose, for ease of discussion, that we have a time-periodically forced system, and for the time being, we shall ignore any parameter dependence. As explained in Chapter 2, this in effect means we can consider the phase space to be cylindrical, with time playing the role of an additional phase space variable s that repeats every period T, say, of the forcing. Then, a natural Poincaré surface Π_S is defined by sampling every time T,

$$\Pi := \Pi_S = \{x : s := t \bmod(T) = s_0\} \cap S^+$$

with corresponding *stroboscopic Poincaré map*

$$P_S : \Pi_S \to \Pi_S \quad \text{given by} \quad P_S(x(t_0)) = x(t_0 + T) = \Psi(x_0, T).$$

This map is easy to compute numerically, as all that needs to be done is to evolve the hybrid system forward through time T, allowing for whatever impacts or sticking occurs. Such a stroboscopic map was used to present the experimental work on impact oscillators described in Chapter 1 [208]. For genuinely autonomous systems, one can make an analogous map to P_S by taking a local section Π_S *away* from Σ that intersects trajectories transversally.

Continuity of the flow, and of the impact law, implies that the stroboscopic map is generally continuous as a function of its arguments. Even if grazing impacts occur, the map remains continuous although, as we shall see, the map is not smooth at the intersection of such trajectories with Π_S. Although simple to define and simulate, the stroboscopic map has the analytical disadvantage that, for a given x, the number m of impacts that occur in the time interval $t_0 < t < t_0 + T$ is not known *a priori*, and indeed can change discontinuously as x varies.

Another convenient Poincaré section is the impact surface itself so that

$$\Pi := \Pi_I = \Sigma.$$

The so-called *impact map*, P_I, is then obtained by applying the hybrid flow map Ψ from Σ to itself followed by the reset map R. Some geometric properties of the map P_I have been studied extensively for 1DoF impact oscillators in [264, 263, 265], Budd *et al.* [44] and Chillingworth [53]. The map P_I has the disadvantage that it can only describe orbits that intersect Σ. For this reason the map is not suitable for analyzing grazing bifurcations, because one needs to describe the transition from non-impacting to impacting orbits.

A better conditioned map, that we shall call the *normal* Poincaré mapP_N, arises by choosing Π to intersect Σ transversally at the boundary G between Σ^+ and Σ^-. The simplest example of such a surface is given by taking

$$\Pi := \Pi_N = \{x : v(x) = H_x F = 0\}.$$

Π_N is everywhere transverse to the flow provided that the normal acceleration $a(x)$ does not vanish. The Poincaré section discontinuity mapping (PDM), as introduced in Chapter 2, allows P_N to be defined even for $x \in S^-$; see Sec. 6.2 below for more details.

To show the action of these three maps, in Fig. 6.5 we illustrate a typical trajectory and its intersection with the Poincaré surfaces Π_S, Π_N and $\Pi_I = \Sigma$. With the aid of the figure, consider a trajectory starting from an impact at the point $A \in \Sigma^+$. This point advances to B where it intersects Π_S, and onwards to Σ^- at the point C. An application of the impact law takes the flow to the point D, from where it continues within S^+, intersecting Π_S at the point E at time T later than point B. Now, suppose the impact at C was ignored and the flow was allowed to continue into S^-; it would then intersect Π_N at the point Y. Also, if the flow from A were continued backwards in time within in S^-, it would intersect Π_N at the point X. Similarly, the backwards flow from D intersects Π_N at the point Z. Hence, we can define the action of the various maps

$$P_S : B \to E, \qquad P_I : A \to C \qquad P_N : X \to Z.$$

Note that to define P_N in this case we have had to include (non-physical) parts of the flow in the set S^-.

In order to understand possible dynamical behavior in hybrid systems, let us briefly consider the effect of the stroboscopic map P_S on sets of initial data. Suppose the flow $\Psi(x_0, t)$ of a hybrid system intersects Π_S at points x_0 at time $t = 0$ and x_1 at time $t = T$. Trajectories starting close to x_0 will remain close to $\Psi(x_0, t)$ for all $t \in (0, T)$. Hence, an open neighborhood $\Omega_0 \subset \Pi_S$ of x_0 will evolve, under the action of the flow $\Psi(\cdot, T)$ to an open neighborhood Ω_1 of the point x_1. Thus $P_S : \Omega_0 \to \Omega_1$. Four possible scenarios result, depending on the initial condition x_0.

1. The flow $\Psi(x_0, t), \quad 0 < t < T$ lies wholly in S^+ and does not intersect the discontinuity surface Σ.
2. The flow $\Psi(x_0, t)$ intersects Σ transversally a *finite* number $m \geq 1$ of times at normal velocity $v > 0$. It does not enter a sticking region.

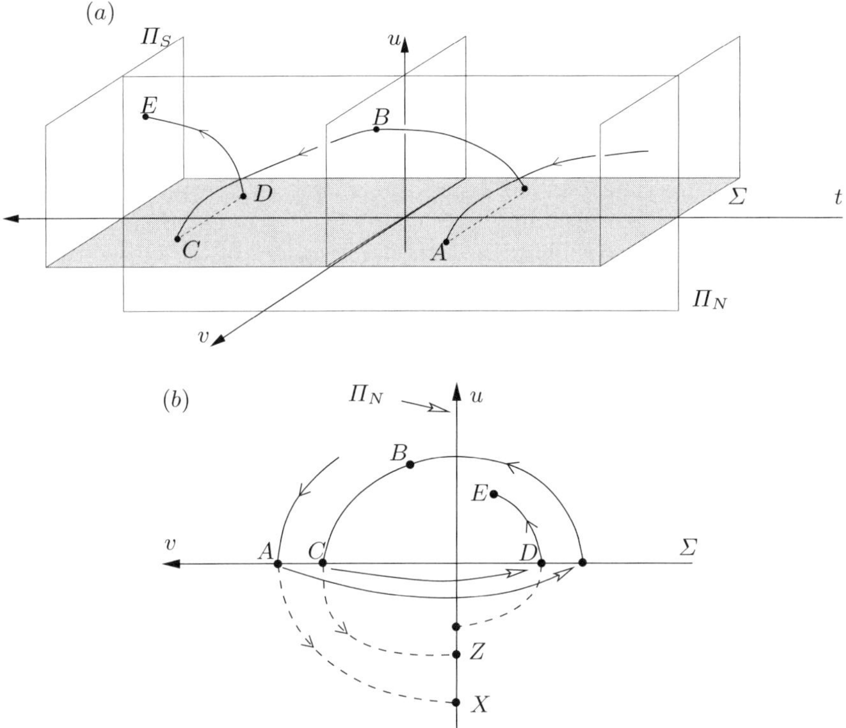

Fig. 6.5. An illustration of an impacting non-periodic flow. In these figures P_S : $B \to E$, $P_I : A \to D$ and $P_N : X \to Z$.

3. The flow $\Psi(x_0, t)$ intersects Σ tangentially (with $v(x) = 0$) at a grazing impact at a point $x = x^* \in G$.
4. The flow $\Psi(x_0, t)$ intersects Σ an *infinite* number of times, culminating at an accumulation point, and enters a sticking region.

Of course, the flow from Π_S back to itself can comprise a hybrid combination of the above. Let us now deal with each possibility in more detail.

Case 1. The evolution of an open neighborhood of x_0 leads to trajectories that remain wholly inside S^+ and do not intersect Σ. The Poincaré map P_S for these is then fully described by the smooth flow map $\Phi(x_0, T)$ of the dynamical system in S^+. This map will be as smooth as the flow. So, if the vector field F is analytic at x_0, this map will have a regular Taylor series expansion;

$$P_S(x) = x_1 + N_T(x - x_0) + O(\|(x - x_0\|^2),$$

where $N_T = \Phi_x(x_0, T)$.

Case 2. The evolution of an open neighborhood Ω_0 of x_0 leads to trajectories that intersect Σ transversally a finite number m times, with m constant for all flows starting from Ω_0. This leads to a well defined map P_S that differs from the smooth flow map Φ due to the impacts. However, as the intersections are transversal, this map is as smooth as R is. The process for calculating this linearization, using the transverse discontinuity mapping Q, was already described in Sec. 2.5.2 of Chapter 2. The derivative of the map is given by a combination of the linearization of the flow Φ between impacts combined with the *saltation matrix* Q_x as defined in (2.78) in Chapter 2, that is the linearization of the discontinuity mapping at impact. Let $Q_{k,x}$ be the saltation matrix at the kth impact with $k = 1 \ldots m$, and N_T be defined as in Case 1. Consequently, if the impacts are at times t_k we have

$$P_S(x) = x_1 + \widehat{N}(x - x_0) + O(\|x - x_0\|^2),$$

where

$$\widehat{N} = N_{T-t_m} Q_{m,x} \ldots Q_{2,x} N_{t_2-t_1} Q_{1,x} N_{t_1}.$$

Having constructed these linearizations it is possible to analyze impacting hybrid flows in much the same manner as we would analyses the smooth flows in Case 1. Indeed, it is straightforward to calculate fixed points of the map, P_S, corresponding to periodic orbits, and to determine their stability.

Case 3. This is the case that will concern us for most of this chapter and differs from Case 1 and Case 2 in that it has dynamics peculiar to the discontinuous nature of the dynamical system. The set Ω_0 is divided into the subset of (i) initial data leading to orbits that do not intersect Σ close to x^* (although they may have other high-velocity impacts with Σ), (ii) initial data leading to orbits that intersect Σ transversally with low-velocity orbits close to x^* and (iii) initial data leading to orbits that graze Σ close to x^*. The latter is a subset of the set of initial data in Π leading to grazing impacts. This set has a complex geometry, illustrated in part in Fig. 6.6. The Poincaré map in Case *(i)* is again the flow map described in Cases 1 or 2. In cases *(ii)* and *(iii)* we must take account of the near grazing incidence that leads to considerable stretching of the phase plane by the Poincaré map. This leads to novel dynamics and a discontinuity-induced bifurcation. We will study this case in detail presently.

Case 4. This case is in general hard to analyze. The resulting map P_S is highly contracting in phase space due to the loss of energy at each of the infinitely many impacts. In fact the set Ω_0 will be mapped into the accumulation point of the chattering sequence, and hence it will enter the codimension-two manifold G. All trajectories will leave G through points at that $a(x) = 0$. This is a codimension-three manifold (a point in the case of the 1DoF impact oscillator example). The forward image of this manifold on Π_S will also be a set of codimension three. Thus there is a significant memory loss; for the case of a 1DoF oscillator where Π_S is a plane, an entire open region of Ω_0 will be mapped to a single point!. In Sec. 6.4 we study the special geometry of set G_Π and its effect on chattering behavior in more detail for 1DoF impact oscillator.

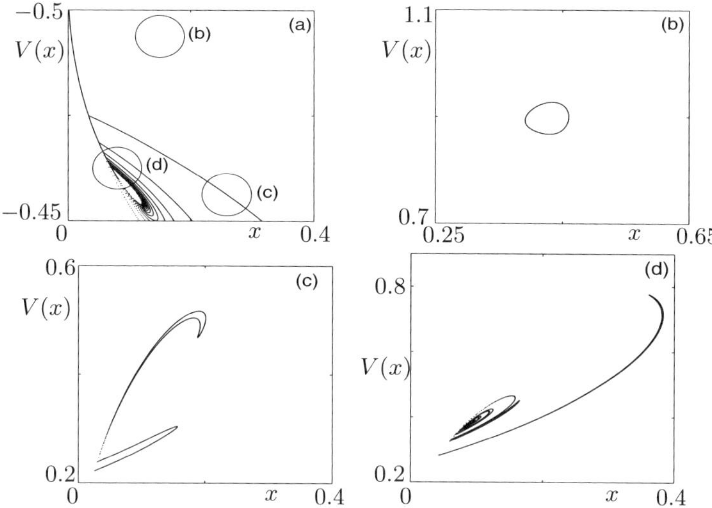

Fig. 6.6. (a) In this figure the branched solid line represents the set of initial data G_Π leading to a grazing impact, and the circles give sets of initial data whose images under P_S are depicted in one of the next three panels with the corresponding label. (b) Orbits in this set have one high-velocity impact per period (c) Grazing between areas with one and two impacts per period. (d) Chatter and incomplete chatter.

We illustrate these four cases in Fig. 6.6 by taking a representative set Ω_0 illustrated by the interior of the solid curve, and evolving it forward a time T to give the curves in panels (b), (c) and (d). In (b) we see case 2 where the effect of the mapping is clearly a regular perturbation of Ω_0. In panel (c) we see the effects of grazing by allowing Ω_0 to intersect a curve G_Π comprising those trajectories that have grazing impacts. Notice how the set of orbits with low-velocity impacts is greatly distorted and is mapped to a long thin set tangent to G_Π, whereas the set of those orbits with high-velocity impacts is much less distorted by the flow. In panel (d) we look at an open neighborhood of a chattering orbit that also includes points with *incomplete chatter* (where there are a large, but finite, number of low velocity impacts).

6.2 Discontinuity mappings near grazing

Let us now look at the dynamics associated with grazing and near impacts, as described in Case 3 above. To do this we assume that there is a grazing trajectory that intersects Σ non-transversally at a point x^* (as well as possibly intersecting it transversally at other distant points) and that close to this trajectory are initial conditions that lead to impact with low normal velocity v, and trajectories that do not impact at all locally. As described in Chapter 2, the effects of the low-velocity impacts can be corrected for, by using a discontinuity mapping; either finding a zero time discontinuity map (ZDM) or a Poincaré discontinuity map (PDM). In this section we shall carefully derive local expressions for the ZDM and PDM, without any assumption that x^* is part of a distinguished trajectory. Section 6.3 below then considers the effect of embedding these expressions within the calculation of Poincaré maps valid around grazing limit cycles.

6.2.1 The geometry near a grazing point

Recall Definitions 2.35 and 2.34 of the ZDM and PDM and consider again the situation illustrated in Fig. 6.7. Here, a near grazing trajectory starting at time $t = 0$ at the point $x = x_0$ intersects the impact surface Σ at the point x_2 that is then mapped to x_3. This trajectory is close to one with a grazing impact at $x = x^*$. As described in Chapter 2, the ZDM is given by the map from x_0 to x_4 and the PDM by the map x_1 to x_5. We will constantly refer to this figure as we derive the analytic forms of the ZDM and PDM.

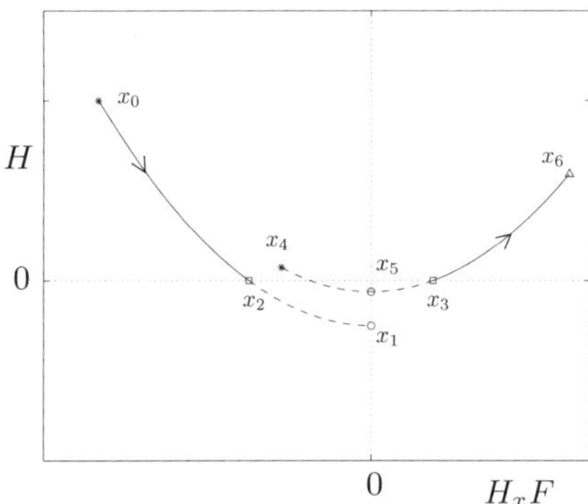

Fig. 6.7. In this figure the solid line represents the actual flow of the hybrid system in the region S^+, and the dashed line the extended flow in the region S^-.

Let us start with some generic hypotheses. Suppose the system in question is written locally near Σ in the form (6.1)–(6.5), where we assume that the scalar function H is well defined at the grazing point x^*:

$$H_x(x^*) \neq 0. \tag{6.13}$$

The grazing point itself is defined by the three conditions

$$H(x^*) = 0, \tag{6.14}$$

$$v(x^*) = \frac{\partial}{\partial t} H(\Phi(x^*, 0)) = \mathcal{L}_F H(x^*) = 0, \tag{6.15}$$

$$a(x^*) = \frac{\partial^2}{\partial t^2} H(\Phi(x^*, 0)) = \mathcal{L}_F^2 H(x^*) := a^* > 0. \tag{6.16}$$

Definition 6.1. *We shall refer to a point x^* satisfying (6.13)–(6.16) as being a* **regular grazing point***.*

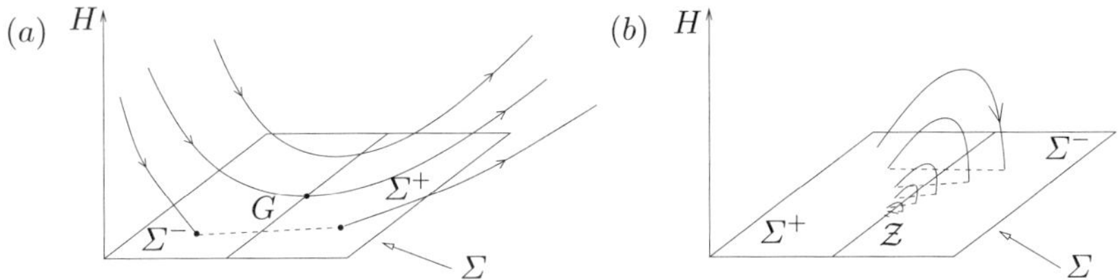

Fig. 6.8. (a) Dynamics close to a regular grazing point x^*. (b) The case when condition (6.16) is replaced by $a^* < 0$.

Note the necessity of the open condition (6.16). This means the grazing trajectory is locally a parabola that points 'downwards' towards Σ, see Fig. 6.8. If the normal acceleration a^* were negative then trajectories would be parabolic in the opposite sense and grazing would not occur, see Fig. 6.8(b). The region of Σ with $a(x) < 0$ is called the *sticking* region $\mathcal{Z}$, the dynamics near that we shall analyze in Sec. 6.4 below in the context of the 1DoF impact oscillator.

A full local neighborhood of x^* (including points in S^- with $H(x) < 0$) can be divided into two regions comprising the set $\mathcal{G}^+$ of points along that the trajectories generated by the flow Φ *do not* impact Σ close to x^*, and the set $\mathcal{G}^-$ of points along that the trajectories *do* impact Σ. These two regions are separated by the set $\mathcal{G}$ of points on trajectories (including the trajectory through x^*) that have a grazing contact, with $G = \mathcal{G} \cap \Sigma$. The *global* geometry of the set $\mathcal{G}$ is complex and will be considered further in Sec. 6.4 below; however, locally it is relatively easy to describe; see Fig. 6.9.

If x_0 is a point close to x^*, we can consider the value of $H(\Phi(x_0, t))$ along this flow starting at x_0. We define

$$H_{\min}(x_0) \equiv \text{the local minimum value of } H(\Phi(x_0, t)) \text{ with the smallest } |t|. \tag{6.17}$$

Note that $H_{\min}$ is well defined for x_0 close to x^* since, by the hypothesis (6.16), x^* is a local minimum H.

Definition 6.2. *Locally, in a neighborhood $\mathcal{D}$ of a regular grazing point x^* of a hybrid flow (6.1)–(6.3), the* **grazing manifold** $\mathcal{G}$*, the* **impacting set** $\mathcal{G}^-$*, and the* **non-impacting set** $\mathcal{G}^+$ *are given by*

$$\mathcal{G} := \{x : H_{\text{min}} = 0\}, \quad \mathcal{G}^- := \{x : H_{\text{min}} < 0\}, \quad \mathcal{G}^+ := \{x : H_{\text{min}} > 0\}.$$

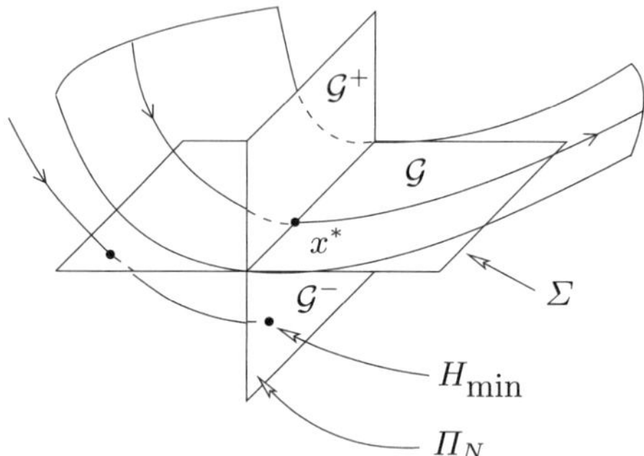

Fig. 6.9. The grazing manifold $\mathcal{G}$, the impacting set $\mathcal{G}^-$, and the **non-impacting set** $\mathcal{G}^+$ close to a regular grazing point x^*.

In a small neighborhood $\mathcal{D}$ of a regular grazing point x^*, the surface $\mathcal{G}$ is given by

$$\mathcal{G} = \left\{x_0 \in \mathcal{D} : H(x_0) - v(x_0)^2 \left(\frac{1}{2a^*} + r(x)\right) = 0\right\}, \tag{6.18}$$

where $r(x)$ is a smooth function that is to leading-order linear in $\|x - x^*\|$. Hence, this surface is locally tangent to the set Σ at x^*.

To see how the expression (6.18) arises, suppose $\|x_0 - x^*\| = \varepsilon \ll 0$. Then, since H and Φ are smooth, we can expand $H(\Phi(x_0, t))$ about x_0 for small t:

$$\begin{aligned} H(\Phi(x_0, t)) &= H(x_0) + v(x_0)t + a(x_0)\frac{t^2}{2} + O(t^3), \\ &= H(x_0) + v(x_0)t + a^*\frac{t^2}{2} + O(t^3, t^2\varepsilon). \end{aligned} \tag{6.19}$$

Now, seeking a local minimum with respect to t of (6.19), we find

$$t = -\frac{v(x_0)}{a^*} + O(\varepsilon^2) \tag{6.20}$$

Moreover if $v(x_0) = 0$, then by definition $x = x^*$, that is a local minimum of H and so $t = 0$. Hence $v(x_0)$ must be a factor of (6.20) and so we can write

$$t = -v(x_0)\left(\frac{1}{a^*} + O(\varepsilon)\right). \tag{6.21}$$

Substitution of (6.21) into (6.19), leads to (6.18).

Using this construction of $\mathcal{G}$ we are now in a position to state the form that the ZDM and PDM take.

Theorem 6.1 (The local ZDM close to a grazing impact). *Suppose that the point x^* is a regular grazing point of an impacting hybrid system written in local form as (6.1)–(6.3). Then, the ZDM, defined for all points x in a sufficiently small neighborhood of x^*, may be written in the form*

$$x \mapsto \begin{cases} x, & \text{if} \quad H_{\min}(x) \geq 0, \\ x - \sqrt{2a^*}W(x)y + O(y^2), & \text{if} \quad H_{\min}(x) < 0, \end{cases} \tag{6.22}$$

where

$$y = \sqrt{-H_{\min}(x)},$$

with $H_{\min}$ defined in (6.17).

The need to find $H_{\min}$ introduces a technical difficulty into this definition, that can be avoided by considering flows starting from a Poincaré section on that $H_{\min} = v = 0$. Such a surface is given by

$$\Pi = \Pi_N = \{x : v(x) = 0\},$$

which also contains the grazing set G.

Theorem 6.2 (The local PDM close to a grazing impact). *Suppose that the point x^* is a regular grazing point of an impacting hybrid system written in local form as (6.1)–(6.3). Then, the PDM, defined for all points $x \in \Pi_N$ in a sufficiently small neighborhood of x^*, may be written in the form*

$$x \mapsto \begin{cases} x, & \text{if} \quad H(x) \geq 0, \\ x + \beta(x,y)y, & \text{if} \quad H(x) < 0, \end{cases} \tag{6.23}$$

where

$$\beta(x,y) = -\sqrt{2a(x)}\left(W(x) - \frac{b(x)}{a(x)}F(x)\right) + O(y),$$

$$y(x) = \sqrt{-H(x)}, \quad b(x) = (H_x F)_x W(x).$$

Remarks

1. Note that the map (6.23) stretches phase space in the direction of the vector β, which is tangent to Σ at the point x^*. To see this, note that $F(x^*)$ and $W(x^*)$ are both in the tangent space of Σ by assumption. Provided that a, b, F and W can be calculated, each of these maps can be readily constructed. We contrast the form taken by these maps with that given by the expression (2.78) in Chapter 2 when v is not close to zero. Here we have a term proportional to y, that gives a square-root singularity. Instead, the discontinuity mapping for transverse intersection is smooth and is linear to leading-order.

2. The proofs of Theorems 6.1 and 6.2 are constructive and use local expressions for the orbit given by the Taylor series expressions and simple calculations of the evolution of H along the flow. In the next three subsections, we give a sketch proof by calculating the leading-order terms of the PDM and ZDM, to give an intuitive motivation of the form that they take. This is followed in Sec. 6.2.5 by a more detailed derivation using the notation of Lie derivatives, that includes a procedure for calculation of the higher-order terms. This latter method is the one that will be adopted in Chapters 7 and 8 for deriving discontinuity mappings in piecewise-smooth flows.

Example 6.6. We illustrate the PDM and ZDM construction by again looking at the forced 1DoF impact oscillator, without dissipation. Consider

$$\ddot{u} + u = \cos(\omega t) - \sigma, \quad \dot{s} = 1, \quad x = (u, v, s)^T$$

with impact at $u = 0$. Thus $H(x) = u$. We look for solutions with initial conditions $x_0 = (u_0, v_0, s_0)$ close to grazing at a point where $u = v = s = 0$. In this case we write

$$H_{\min} := u_{\min} = u_0 - \frac{v_0^2}{2a_0},$$

where

$$a_0 = a(u_0, v_0, s_0) = \gamma \cos(\omega s_0) - \sigma \approx a^* = \gamma - \sigma,$$

where a^* is the value of a at the grazing point. Also we have

$$b(x) = -(1+r), \quad F(x) = (0, a(x), 1)^T, \quad W(x) = (0, -(1+r), 0)^T.$$

Thus, to leading-order

$$\beta = -\sqrt{\frac{2}{a}} \begin{pmatrix} 0 \\ 0 \\ 1+r \end{pmatrix}.$$

On the impact side ($u < 0$), we then have to leading-order

$$\text{ZDM}(x) = x + \sqrt{2a} \begin{pmatrix} 0 \\ 1+r \\ 0 \end{pmatrix} \sqrt{-u_{\min}}\,,$$

$$\text{PDM}(x) = x - \sqrt{\frac{2}{a}} \begin{pmatrix} 0 \\ 0 \\ 1+r \end{pmatrix} \sqrt{-u}\,.$$

Note that in the PDM, the correction is purely in the direction of the time variable to leading-order. In fact, as we shall see in detail in Example 6.7, there are also higher-order corrections proportional to u, that affect the coordinate direction u as well. Significantly, there is a square-root singularity in both the ZDM and the PDM for this example. In the ZDM it is manifest as a change in v proportional to $\sqrt{-u_{\min}}$ and in the PDM a change in phase, again proportional to $\sqrt{-u}$. As the function $\sqrt{-u}$ has an infinite gradient as $u \to 0$, this simple analysis demonstrates the stretching of phase space in the limit of $x \to 0$.

6.2.2 Approximate calculation of the discontinuity mappings

The method of construction of the maps is composed of three steps related to the points shown earlier in Fig. 6.7. See also Fig. 6.10.

1. For a general initial condition $x_0 \in \mathcal{G}^-$ with $\|x_0 - x^*\| \ll 1$, we follow the flow through a time δ_0 until the trajectory intersects Σ, at the point x_2.
2. The reset map R is applied to x_2 to obtain x_3.
3. To construct the local ZDM, we then compute the flow from x_2 *backwards* through time $-\delta_0$ to the point x_4. Thus, the total transformation of the flow from x_0 to x_4 takes *zero time.*
4. Alternatively, to construct the PDM, we assume the initial point $x_1 \in \Pi_N$ and, as above, follow the flow through a time $\delta < 0$ to reach the impact point x_2 and to compute its image $x_3 = R(x_2)$. For the final step, though, we flow backwards through a new time $-\Delta$ until the original Poincaré section Π_N is reached, at the point x_5.

To compute the local forms for these maps, we shall use Taylor series expansion, exploiting the smoothness of the vector field F both above and below Σ. To obtain the leading-order expressions, it is sufficient to assume that the surfaces Σ and G are flat, i.e., linear manifolds. (By choosing appropriate co-ordinate changes, this may be assumed without loss of generality when calculating higher-order approximations too; see [78].) Thus, we take

$$\Sigma = \{x \in \mathcal{D} : H_x x = 0\}, \quad \text{and} \quad G = \{x \in \Sigma | z(x - x^*) = 0\}, \tag{6.24}$$

where $z = (H_x F)_x$ is a vector normal to the grazing set within Σ.

6.2.3 Calculating the PDM

It is most instructive to start with the derivation of the PDM, for that the surface Π_N becomes the linear manifold

$$\Pi_N = \{x \in \mathcal{D} : z(x - x^*) = 0\}. \tag{6.25}$$

To calculate the PDM, we take an initial point x_1 lying in Π_N with $H(x_1) < 0$ and such that $\|x_1 - x^*\| = \varepsilon$, where ε is small. Now, we can expand $H(\Phi(x_1, t))$ as a Taylor series in time, using the fact from the definition of Π_N that

$$H(x_1) := -y^2 < 0, \quad \frac{\partial}{\partial t} H(\Phi(x_1, 0)) = z(x_1 - x^*) = 0$$

and

$$\frac{\partial^2}{\partial t^2} H(\Phi(x_1, 0)) = (H_x F)_x F(x_1) = a^* + O(\varepsilon).$$

Hence, along the flow starting at x_1 we have

$$H(\Phi(x_1, t)) = H(x_1) + \frac{1}{2} a^* t^2 + O(t^3) = -y^2 + \frac{1}{2} a^* t^2 + O(t^3). \tag{6.26}$$

Note that we expanded $H(\Phi(x_1, t))$ in t, and then the coefficients of the Taylor series can be further expanded around $x = x^*$.

Expanding now the flow in t about the grazing point $x = x^*$, we also find that

$$\Phi(x_1, t) = x_1 + F^* t + O(t^2, \varepsilon t, \varepsilon^2), \tag{6.27}$$

where henceforth we shall use a superscript $*$ to mean evaluation at $x = x^*$.

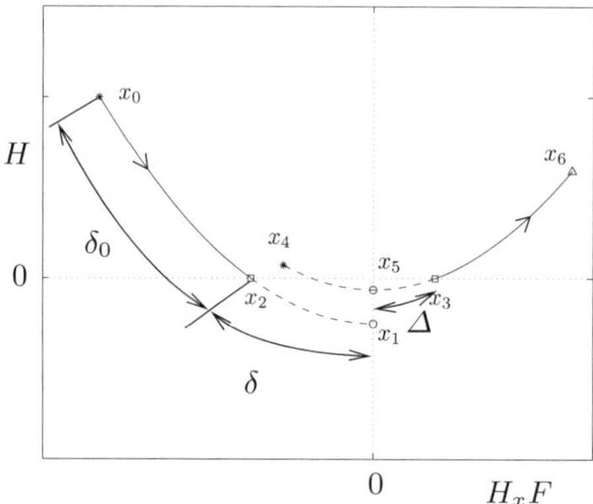

Fig. 6.10. Similar to Fig. 6.7 but with times δ, δ_0, and Δ.

From these expressions we can calculate the negative time δ and point $x_2 = \Phi(x_1, \delta)$ for that $H(x_2) = 0$. From (6.26), we obtain

$$\delta = -\sqrt{\frac{2}{a^*}} y + O(y^2),$$

that, using (6.27), gives

$$x_2 = x_1 - F^* \sqrt{\frac{2}{a^*}} y + O(y^2).$$

It is here that we immediately see the source of the square-root term in the discontinuity map (cf. Fig. 6.10). Due to the locally quadratic nature of the trajectory, the time (of the virtual flow) spent with $H(x) < 0$ scales like y, the square root of the penetration $H(x_1)$.

The normal velocity at x_2 can be expanded in a Taylor series about x_1

$$v(x_2) = H_x F(x_2) = H_x F(x_1) + a^* \delta + O(y^2) = 0 - a^* \sqrt{\frac{2}{a^*} y} + O(y^2).$$

Hence we can apply the reset map (6.2) to obtain

$$\begin{aligned} x_3 &= R(x_2) := x_2 + W(x_2) v(x_2), \\ &= x_2 - W(x_2) \left(a^* \sqrt{\frac{2}{a^*} y} + O(y^2) \right), \\ &= x_1 - F^* \sqrt{\frac{2}{a^*} y} - W^* \sqrt{2a^* y} + O(y^2). \end{aligned} \tag{6.28}$$

As a final step, we take this point and flow backwards from x_3 for a time $-\Delta$ (comparable with δ) to reach Π_N at a point x_5. From Taylor expansion, we obtain

$$x_5 = x_3 - \Delta F^* + O(y^2). \tag{6.29}$$

Now, using (6.28) and the identities

$$z(x_1 - x^*) = 0, \quad zF^* := a^* \quad \text{and} \quad zW^* := b^*,$$

we obtain to leading-order

$$0 = z(x_5 - x^*) = -a^* \sqrt{\frac{2}{a^*} y} - b^* \sqrt{2a^* y} - a^* \Delta.$$

Hence

$$\Delta = (-b^* - 1) \sqrt{\frac{2}{a^*} y} = -(b^* + 1)\delta.$$

Substitution into (6.29) gives finally,

$$x_5 = x_1 - F^* \sqrt{\frac{2}{a^*} y} - W^* \sqrt{2a^* y} + F^*(b^* + 1) \sqrt{\frac{2}{a^*} y} \tag{6.30}$$

$$= x_1 - \sqrt{2a^*} \left(W^* - \frac{b^*}{a^*} F^* \right) y + O(y^2), \tag{6.31}$$

which is the leading-order expression for the PDM given by (6.22).

6.2.4 Approximate calculation of the ZDM

Calculation of the ZDM follows along similar lines. Here we start from a general initial point x_0. There is an additional step that involves the calculation of the time δ_1 to reach the point x_1 at that $H = H_{\min}$, see Fig. 6.10. From a Taylor expansion about $x = x^*$ we have, to leading-order,

$$x_1 = x_0 + \delta_1 F^*$$

for some time δ_1 yet to be determined. Then the derivation follows closely that of the PDM to obtain

$$\delta = \sqrt{\frac{2}{a^*}y},$$

as the leading-order expression for the time to flow from x_1 to x_2. The point $x_3 = R(x_2)$ is again given by

$$x_3 = x_1 - F^*\sqrt{\frac{2}{a^*}y} - W^*\sqrt{2a^*y}.$$

The final step for the ZDM is different though. To reach the final point x_4, we must flow for precisely the time

$$\delta_0 = (\delta - \delta_1).$$

Thus we have

$$\begin{aligned}
x_4 &= x_3 + (\delta - \delta_1)F^* \\
&= x_1 - F^*\sqrt{\frac{2}{a^*}y} - W^*\sqrt{2a^*y} - \delta_1 F^* + F^*\sqrt{\frac{2}{a^*}y} \\
&= (x_0 + \delta_1 F^*) - W^*\sqrt{2a^*y} - \delta_1 F^* \\
&= x_0 - W^*\sqrt{2a^*y},
\end{aligned}$$

that is the leading-order expression for the ZDM given by (6.22). Notice that, due to cancellation, for this leading-order expression, we did not need to compute the unknown time δ_1.

6.2.5 Derivation of the ZDM and PDM using Lie derivatives

The asymptotic expansion process leading to the above leading-order expressions for the ZDM and the PDM in Theorems 6.1 and 6.2 can in principle be continued to compute the $O(y^2)$ and higher terms; however, the algebra soon becomes unwieldy. It is better to use the Lie derivative notation (6.5), since then we can express everything in terms of scalar quantities $\mathcal{L}_F^m(H)(x^*)$ for different orders m. Therefore, we shall use the hypotheses (6.14)–(6.16) written in the form

$$H(x^*) = 0 \quad v(x^*) = \mathcal{L}_F(H)(x^*) = 0 \quad \text{and } a(x^*) = \mathcal{L}_F^2(H)(x^*) > 0.$$

Bearing in mind the construction in Fig. 6.10, note that the ZDM may be written concisely as the following combinations of flows and mappings:

$$\Phi(R(\Phi(x_0, \delta_0)), -\delta_0), \tag{6.32}$$

where δ_0 is the time taken to flow from x_0 to the point x_2 at that impact occurs.

As in the above approximate derivation of the ZDM, we split the time δ_0 into the time $\delta < 0$ taken to flow to x_2 from a point x_1 at that $H = H_{\min}$, and the time δ_1 taken to flow from the initial point x_0 to x_1, that is

$$\delta_0 = \delta + \delta_1.$$

By introducing the scalar variable $y = \sqrt{-H_{\min}}$ for $H_{\min} < 0$, we find a regular expression for the flow combination (6.32) in terms of x_0, y, δ and δ_1. Before evaluating (6.32), we first calculate y, δ and δ_1 in terms of the initial condition x_0. Thus, every step of our derivation can be treated as a separate calculation, and it is thus highly suited to implementation in computer algebra.

Calculation of the time δ. Let us start by supposing that x is a point where, in the absence of the impacting surface, an impacting trajectory would attain its minimum H-value; so that $H(x) = H_{\min} < 0$ and $\mathcal{L}_F(H)(x) = 0$. As before, we define $y > 0$ by

$$y^2 + H_{\min} = 0$$

and seek the time $t = \delta < 0$ so that

$$H(\Phi(x, \delta)) = 0. \tag{6.33}$$

Now, in order to write expressions that remain asymptotically valid, even when $\mathcal{L}_F(H)(x) \neq 0$, and $y \neq \sqrt{-H_{\min}}$, it is convenient to rewrite (6.33) in the form

$$H(\Phi(x, \delta)) - (y^2 + H(x)) - \mathcal{L}_F(H)(x)\delta = 0. \tag{6.34}$$

Note that the two subtracted terms are both zero for the x in question, but for more general x this leads to a regular asymptotic expansion in the two variables x and y, that we can think of as independent. Expanding (6.34) in powers of δ, we get

$$\mathcal{L}_F^2(H)(x)\frac{\delta^2}{2} + \mathcal{L}_F^3(H)(x)\frac{\delta^3}{6} + O(\delta^4) - y^2 = 0. \tag{6.35}$$

In order to solve for δ, we first recast (6.35) as

$$(\sqrt{A} + y)(\sqrt{A} - y) = 0, \tag{6.36}$$

where

$$A = \mathcal{L}_F^2(H)(x)\frac{\delta^2}{2} + \mathcal{L}_F^3(H)(x)\frac{\delta^3}{6} + O(\delta^4).$$

Our assumption that y is positive means that, in order to solve (6.36), we need the second factor to be zero. Moreover, since by construction $\delta < 0$, we have

$$\delta\sqrt{\left(\mathcal{L}_F^2(H)(x)\frac{1}{2}+\mathcal{L}_F^3(H)(x)\frac{\delta}{6}+O(\delta^2)\right)}+y=0. \tag{6.37}$$

Now, since $\mathcal{L}_F^2(H)(x^*)/2$ is non-zero by the hypothesis (6.16), the Implicit Function Theorem guarantees the existence of a smooth function $\delta(x,y)$ in a sufficiently small neighborhood of the grazing point $(x,y)=(x^*,0)$ that solves equation (6.37). Thus, we can invert the power series (6.37) term by term to obtain

$$\delta(x,y)=y\left(-\sqrt{\frac{2}{\mathcal{L}_F^2(H)(x)}}-\frac{1}{3}\frac{\mathcal{L}_F^3(H)(x)}{(\mathcal{L}_F^2(H)(x))^2}y+O(y^3)\right). \tag{6.38}$$

The velocity before the impact. We define x_2 to be the point of first impact, so that

$$x_2(x,y)=\Phi(x,\delta(x,y))$$

and

$$v_2(x,y)=\mathcal{L}_F(H)(x_2(x,y))-\mathcal{L}_F(H)(x). \tag{6.39}$$

Once again, in (6.39) we have subtracted a term, in this case $\mathcal{L}_F(H)(x)$, that is zero for the point x in question but that makes the asymptotics become uniformly valid for any x and y. Expanding v_2 in powers of δ, we have

$$v_2(x,y)=-y\left(\sqrt{2a(x)}-\frac{2}{3}\frac{c(x)}{a(x)}y+O(y^2)\right),$$

where

$$a(x)=\mathcal{L}_F^2(H)(x) \text{ and } c(x)=\mathcal{L}_F^3(H)(x).$$

The velocity after the impact. We define x_3 to be the image of x_2 after impact, $x_3=\widehat{R}(x_2(x,y),v_2(x,y))$, where $\widehat{R}(x,v)$ is the expression for $R(x)$ where x and v are considered as independent co-ordinates, i.e., $\widehat{R}(x,v)=R(x)=x+W(x)v$. Hence, we write

$$x_3(x,v)=x+W(x)v,$$

where $v=\mathcal{L}_F(H)(x)$, and we set

$$v_3(x,v)=\mathcal{L}_F(H)(x_3(x,v))-\mathcal{L}_F(H)(x)+v,$$

where the term $\mathcal{L}_F(H)(x)-v$ that is zero by definition, has been subtracted to make the expression valid for any independent x and v. This gives

$$v_3(x,v)=v(1+\mathcal{L}_W\mathcal{L}_F(H)(x)+O(v)).$$

The expansion for the ZDM. Finally, we define x_4 via the condition that the total time taken is zero, so that

$$x_4(x, \delta, v) = \Phi(\widehat{R}(\Phi(x, \delta), v), -\delta). \tag{6.40}$$

Note that this expression is just the flow combination introduced in (6.32) for $\delta_0 = \delta$. Expanding (6.40) in v and δ, we get

$$x_4(x, \delta, v) = x + W(x)v + W_x F(x)\delta v - F_x W(x)\delta v + vO((\delta, v)^2). \tag{6.41}$$

We are now in a position to derive general expressions for the ZDM and PDM. There are several cases to consider.

The ZDM for a point x such that $\mathcal{L}_F(H)(x) = 0$. For such special points, the ZDM is simply given by

$$ZDM(x, y) = x_4(x, \delta(x, y), v_2(x, y)). \tag{6.42}$$

Using expansion (6.41) we have that

$$ZDM(x, y) = x - \sqrt{2a(x)}W(x)y + O(y^2), \tag{6.43}$$

that is the leading-order term of the ZDM given in (6.22). The next order term is

$$2y^2\left(\frac{c(x)}{3a(x)}W(x) + W_x F(x) - F_x W(x)\right),$$

and, in principle, we can carry out the expansion process to any arbitrary order.

The ZDM for a general point x. To obtain the ZDM for a point x such that $\mathcal{L}_F(H)(x) = v \neq 0$, we must apply a correction to the ZDM presented in (6.43) obtained by employing an additional projection onto the zero velocity surface. To determine this projection we must solve for the time δ_1 taken to reach a point for that the velocity is zero. That is

$$\mathcal{L}_F(H)(\Phi(x, \delta_1)) - \mathcal{L}_F(H)(x) + v = 0, \tag{6.44}$$

where, once again, we have subtracted a zero term $\mathcal{L}_F(H)(x) - v$, that in effect defines v as an independent variable.

Using the previous techniques we find that

$$\delta_1(x, v) = -v\left(\frac{1}{a(x)} + O(v^2)\right); \tag{6.45}$$

and we define

$$x_1(x, v) = \Phi(x, \delta_1(x, v)) \tag{6.46}$$

$$H_{\min}(x, v) = H(x_1(x, v)) + [v - \mathcal{L}_F(H)(x)]\delta_1(x, v). \tag{6.47}$$

Substituting for δ_1 from (6.45) into (6.47) we find

$$H_{\min}(x,v) = H(x) - v^2\left(\frac{1}{a(x)} + O(v)\right). \tag{6.48}$$

The final time of flow from x_3 to x_4 should now be $\delta_0 = \delta + \delta_1$. We can therefore write the ZDM as

$$ZDM(x,y,v) = x_4(x, \delta_1(x,v) + \delta(x_1(x,v),y), y, v_2(x_1(x,v),y)), \tag{6.49}$$

where x_4 is given by (6.41).

Evaluation of (6.49), shows that the leading-order term of this general ZDM is identical to the ZDM (6.42) calculated from the zero-velocity surface, that is

$$ZDM(x,y,v) = x - y\left(\sqrt{2a(x)}W(x) + O(y,v)\right), \tag{6.50}$$

but there are differences in the expressions for the higher-order terms.

The PDM map. To obtain the PDM it is sufficient to consider the ZDM (6.43) projected onto the zero velocity surface. We define the point x_5 by

$$x_5(x, \delta(x,y), \Delta_0, v_2(x,y)) = \Phi(x_4(x, \delta(x,y), v_2(x,y)), \Delta_0), \tag{6.51}$$

where $\Delta_0 = -\Delta + \delta_0$ is the time taken to flow from x_4 to x_5. Expanding (6.51) in the small scalar variables Δ_0 and v_2, we find that

$$x_5(x,t,\Delta_0,v_2) = ZDM + F\Delta_0 + F_x W(x)\Delta_0 v_2 + F_x F(x)\frac{\Delta_0^2}{2} + O((t,\Delta_0,v_2)^3). \tag{6.52}$$

Now we must determine the time Δ_0 by solving the identity

$$\mathcal{L}_F(H)(\Phi(x_4(x,t,v_2),\Delta_0)) - \mathcal{L}_F(H)(x) = 0, \tag{6.53}$$

that can be expanded in powers of Δ_0. The Implicit Function Theorem guarantees the existence of a smooth function $\Delta_0(x, v_2)$ provided that $\mathcal{L}_F^2(x) \neq 0$. This latter condition is guaranteed by the hypothesis (6.16) for x sufficiently close to x^*. Thus, we can invert the power series expansion of (6.53) to obtain

$$\Delta_0 = -\frac{b(x)}{a(x)}v_2 + \left(-\frac{b(x)^2c(x)}{2a(x)^2} - H\right)v_2^2 + O(v_2^3),$$

where

$$H = \frac{\mathcal{L}_F\mathcal{L}_W\mathcal{L}_F(H)(x) - \mathcal{L}_W\mathcal{L}_F^2(H)(x)}{a(x)} - \frac{b(x)}{a(x)},$$

and $b(x) = \mathcal{L}_W\mathcal{L}_F(H)(x)$. We are now ready to define the PDM as

$$\begin{aligned}
PDM(x,y) &= x_5(ZDM(x,y), v_2(x,y)) \\
&= x + y\Bigg[-\sqrt{2a(x)}W(x) + \frac{\sqrt{2a(x)}b(x)}{a(x)}F(x) + \\
&+2y\left(\frac{b(x)^2}{2a(x)}F_xF(x) + W_xF(x) - (1+b(x))F_xW(x) - \frac{c(x)}{3a(x)}W(x) + \right. \\
&\left. -\frac{b(x)c(x)}{3a(x)^2}F(x) + \left(-\frac{b(x)^2c(x)}{2a(x)} - B + C\right)F(x)\right) + O(y^2)\Bigg],
\end{aligned} \tag{6.54}$$

where $B = \mathcal{L}_F\mathcal{L}_W\mathcal{L}_F(H)(x)$ and $C = \mathcal{L}_W\mathcal{L}_F^2(H)(x)$ and the overall error is of $O(y^3)$. This gives not only the leading-order expression (6.23) but the $O(y^2)$ term too.

Remark. Notice that the functional expressions for the ZDMs and the PDM, when applied to a particular system locally around the grazing contact, are functions of single variable x. The other independent variables (y, v) are functions of x. However, the functional expressions for the ZDMs and the PDM as derived are obviously valid when the variables x, y and v are treated as independent quantities [with y and v being of $\mathcal{O}(\varepsilon)$].

6.3 Grazing bifurcations of periodic orbits

Throughout this section we assume that a regular grazing point x^* that locally satisfies (6.13)–(6.16) is part of a limit cycle of an impacting hybrid flow $\Psi(x,t)$, the ODE part of that can be written as $\dot{x} = F(x,\mu)$. Now, since H is a scalar function, a periodic orbit will generically have isolated local minimum values $H_{min}(\mu)$ of H. When one of these minima becomes zero, upon varying a single parameter μ, this corresponds to a *grazing bifurcation* of the periodic orbit. Thus grazing bifurcations are of codimension-one in parameter space.

Definition 6.3. *We say that a* **limit cycle grazing bifurcation** *occurs at a parameter value $\mu = \mu^*$ if there exists a hyperbolic limit cycle $p(t)$ of a hybrid flow $\Psi(x,t)$ that contains a regular grazing point x^*, and that the parameter μ perturbs the local phase potraits near the grazing point and around the limit cycle in a non-degenerate way (so that $H_{min}(\mu^*) = 0$ and $\frac{dH_{min}}{d\mu} \neq 0$).*

Now, we wish to unfold the dynamics in the neighborhood of such a grazing bifurcation point. There are two steps to this process. First, we use the discontinuity mappings calculated in the previous section, valid in the neighborhood of the grazing point. These maps must be composed with a Poincaré mapvalid in the neighborhood of the critical periodic orbit at $\mu = \mu^*$ under the supposition that no impact occurs close to x^*. Thus we obtain a compound Poincaré mapthat is uniformly valid for both kinds of orbit close to the grazing limit cycle; orbits that have low-velocity impact near x^*, and those that do not. The construction of such compound maps forms the subject of Sec. 6.3.1 below. The second step is then to analyze the dynamics of these maps. It should not be a surprise given that the DMs constructed above contain square-root singularities that the compound maps also contain such terms. Thus, as shown in a suite of examples in Sec. 6.3.2 below, we use the analysis of Sec. 4.3 of Chapter 4 to explain the different dynamical scenarios that may result from a limit cycle grazing bifurcation, period-adding cascades, chaos, etc.

6.3.1 Constructing compound Poincaré maps

We wish to construct a map from a Poincaré section Π to itself that is valid in a neighborhood of the grazing limit cycle $p(t)$ in both phase and parameter space. There are two possible approaches to doing this, either using the specific Poincaré section Π_N that contains the grazing point x^* and is by construction *normal* to the flow, or using a remote Poincaré section Π_S away from the grazing point. We shall call the latter a *stroboscopic* Poincaré mapby analogy with periodically forced systems, where Π_S is often chosen to be given by a constant value of the phase variable $s = t \bmod T$, where T is the period of the forcing. The induced map $P_S : \Pi_S \to \Pi_S$ is most useful for numerical investigation, and representation of chaotic attractors and domains of attractions. In contrast, the map $P_N : \Pi_N \to \Pi_N$ is more useful for theoretical analysis as it contains only terms evaluated at the grazing point itself; thus such a map may be termed the *grazing bifurcation normal form map*, by analogy with the normal forms of local bifurcations introduced in Chapter 2.

Let us start by considering the stroboscopic map P_S; see Fig. 6.11. Assume the system has a natural period T, for example, the period of any forcing. Then the map P_S maps x at time t to $\Psi(x, t+T)$. Suppose that grazing occurs on a particular orbit $p(t)$ at the point x^* at time $t = 0$. (In order to be completely general, $p(t)$ may not necessarily be a periodic orbit in what follows.) The orbit may also have further finite-velocity impacts for times not close to zero. For any given $T > s_0 > 0$, we may then construct surfaces Π^-, Π^0 and Π^+ at times $t = -s_0$, $t = 0$ and $t = T - s_0 > 0$ so that the map P_S acts from Π^- to Π^+. Now we have $x^* = p(t) \cap \Pi_0$, and let us define $x^- = p(t) \cap \Pi^-$, $x^+ = p(t) \cap \Pi^+$. If $p(t)$ is a periodic orbit, then we necessarily have $\Pi^- = \Pi^+$ and $x^- = x^+$. We can then define natural flow maps P_1 from Π^- to Π^0 and P_2 from Π^0 to Π^+, by the evolution of the hybrid flow operators $\Psi(\cdot, s_0)$ and $\Psi(\cdot, T - s_0)$, respectively, allowing for any transverse impacts that are distant from x^*, but ignoring any impact close to x^*. Thus, we can construct the compound Poincaré mapvia

$$P_S = P_2 \circ \mathrm{ZDM} \circ P_1.$$

where ZDM is the zero-time discontinuity map constructed in the previous section. Note, by definition of the ZDM, this map defines the evolution through time T as required.

The methods described in Sec. 2.5 of Chapter 2 allow us (by using a saltation matrix where necessary in the case of distant transverse impacts) to linearize each of the maps P_1 and P_2 about $p(t)$. Let us call such linearizations

$$N_1 := \left.\frac{d}{dx} P_1\right|_{x=x^-} \qquad \text{and} \quad N_2 := \left.\frac{d}{dx} P_2\right|_{x=x^*}.$$

Now, in Π_0 the set $\mathcal{G}$ separates orbits that impact close to x^* from those that do not. Let us define the pre-image G_π of the set $\mathcal{G}$ in Π^-, and its image G_2

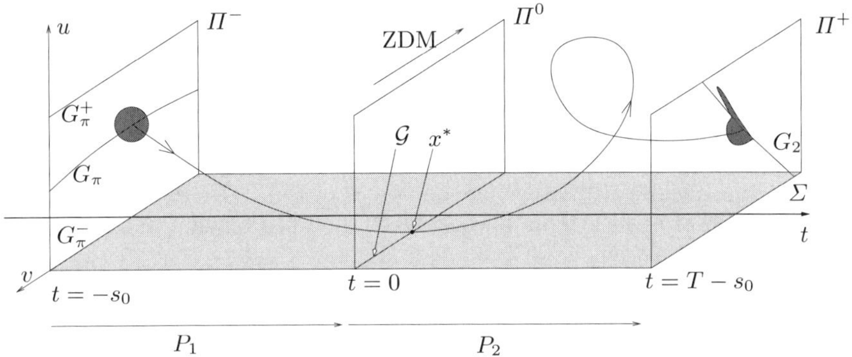

Fig. 6.11. An illustration of the stroboscopic Poincaré map. See text for details.

in Π^+ via

$$\mathcal{G} = P_1(G_\pi) \quad \text{and} \quad G_2 = P_2(\mathcal{G}).$$

The set G_π can be called a *discontinuity set* as it separates initial conditions, $x \in G_\pi^-$, in P_1 that have an impact near x^* from those, $x \in G_\pi^+$, that do not. The following result then follows immediately from the leading-order form of the discontinuity map (6.22).

Theorem 6.3 (The linearized stroboscopic grazing map). *Calculating the map P_S for a point x close to x^- by taking P_1, applying the ZDM and then taking P_2 we have to leading-order:*

$$P_S(x) = \begin{cases} N_2 N_1(x - x^-), & \text{if} \quad x \in G_\pi^+, \\ N_2 N_1(x - x^-) + \sqrt{2a}\sqrt{-H_{\min}} N_2 W(x^*), & \text{if} \quad x \in G_\pi^-, \end{cases}$$

where the error terms are $O\left(\|x - x^-\|^2\right)$. Linearizing $H_{\min}$ around x^* gives

$$H_{\min} = H_x N_1(x - x^-) + O\left(\|x - x^-\|^2, v^2\right),$$

where locally around the grazing contact $v = O\left(\|x - x^-\|\right)$. Thus, the effect of grazing is to stretch the phase space adjacent to the point x^- in the direction of the vector $N_2 W$.

In order to unfold the grazing bifurcation, one needs to include parameters in the derivation of the compound Poincaré map. We do this next, in the context of maps derived using the normal Poincaré section.

Consider the Poincaré section $\Pi_N = \{x : H_x F(x) = v = 0\}$, that we assume intersects a grazing periodic orbit $p(t; \mu)$ at a regular grazing point $x = x^*$ at parameter value $\mu = \mu^*$. We wish to define a normal form that can be used to unfold the grazing bifurcation. To do this, we need to calculate the full Poincaré map P_N from Π_N to itself, with the aid of the PDM to correct for points with $v < 0$ that are in the non-physical region S^-. We begin by constructing the *natural Poincaré map* $\tilde{P}_N$ from Π_N to itself, which is obtained by computing the flow Ψ ignoring *low-velocity* impacts close to Π_N, but allowing for possible transverse impacts distant from Π. Such a map is computed by ignoring the presence of Σ close to x^* and thus allowing points to have $v < 0$ while still following the flow Φ of the ODE without correction. By definition of the PDM in Theorem 6.2, the full Poincaré mapfrom Π_N to Π_N is then defined by

$$P_N(x) = PDM \circ \tilde{P}_N(x); \tag{6.55}$$

see Fig. 6.12. Note there is a subtle point here. We could equally well have chosen P_N to be $\tilde{P}_N \circ PDM(x)$. These two maps are topologically equivalent, and it is essentially a matter of taste whether one chooses to flow first and correct later as in (6.55), or vice versa.

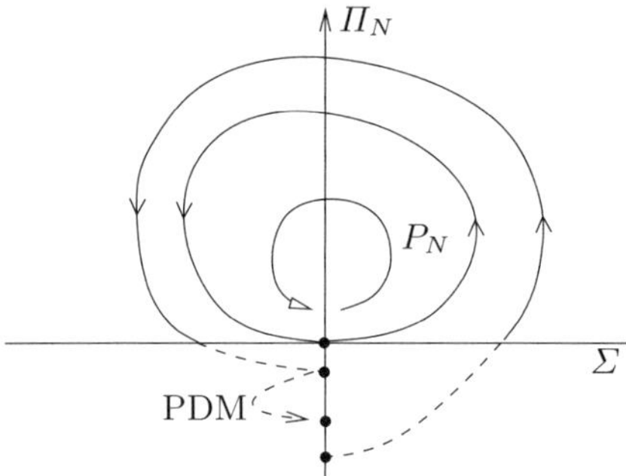

Fig. 6.12. Construction of the normal form close to a grazing bifurcation using the normal Poincaré section Π_N.

Let us now derive the leading-order expression in $x - x^*$ and $\mu - \mu^*$ for the normal form map (6.55). The assumption that all impacts of $p(t;\mu^*)$ other than at x^* are transverse means that, using the transverse discontinuity mapping as presented in Sec. 2.5.2 at all such impacts, the natural Poincaré mapwill be smooth in both x and μ close to $(x,\mu) = (x^*,\mu^*)$. Hence, we can write

$$\tilde{P}_N(x,\mu) = N(x - x^*) + M(\mu - \mu^*) + O\left(\|x - x^*\|^2, (\mu - \mu^*)^2\right),$$

where

$$N := \frac{\partial}{\partial x}\tilde{P}_N\bigg|_{x=x^*,\mu=\mu*} \quad \text{and} \quad M := \frac{\partial}{\partial \mu}\tilde{P}_N\bigg|_{x=x^*,\mu=\mu*}.$$

Similarly, there is a row vector $C^T = H_x$ and a scalar $D = H_\mu$ such that to leading-order

$$H(x, \mu) = C^T(x - x^*) + D(\mu - \mu^*).$$

We are now in position to combine this local form of $\tilde{P}_N$ with the PDM in (6.22) to write down the leading-order form of the compound map P_N.

Theorem 6.4 (The normal form map at a grazing bifurcation). *Suppose a periodic orbit $p(t; \mu)$ has a regular grazing at $(x, \mu) = (x^*, \mu^*)$ of an impacting hybrid system that is written in local co-ordinates in the form (6.1)–(6.3). Let $\widehat{x} = x - x^*$, $\widehat{\mu} = \mu - \mu^*$. Then the Poincaré map P_N from Π_N to itself defined by (6.55) can be written to leading-order in the form*

$$P_N(\widehat{x}, \widehat{\mu}) = \begin{cases} N\widehat{x} + M\widehat{\mu}, & \text{if } C^T N\widehat{x} + (C^T M + D)\mu \geq 0, \\ N\widehat{x} + M\widehat{\mu} + Ey, & \text{if } C^T N\widehat{x} + (C^T M + D)\widehat{\mu} < 0, \end{cases} \tag{6.56}$$

where

$$y = \sqrt{(-C^T N\widehat{x} - (C^T M + D)\widehat{\mu})} + O(x, \mu)$$

and

$$E = \beta(C^T N\widehat{x}, 0)\big|_{\widehat{x}=0},$$

with β as defined in (6.23). The error term is $O(\widehat{x}^2, y^2, \widehat{\mu}^2)$ for both signs of $C^T N\widehat{x} + (C^T M + D)\widehat{\mu}$.

Remarks

1. In the absence of the PDM correction, the periodic orbit $p(t; \mu)$ intersects Π_N at a point x_1 that satisfies the leading-order equation
$$x_1 - x^* = (I - N)^{-1} M(\mu - \mu*),$$
so that to leading-order
$$\begin{aligned} H(x_1, \mu) &= C^T(x_1 - x^*) + D(\mu - \mu^*) = [C^T(I - N)^{-1}M + D](\mu - \mu^*) \\ &:= e(\mu - \mu^*) \end{aligned} \tag{6.57}$$
Thus the periodic orbit is in $S^+ \cap \Pi_N$ if $e(\mu - \mu^*) > 0$. So, the condition for the existence of a simple periodic orbit is that $e\widehat{\mu} > 0$. The dynamics for $e\widehat{\mu} < 0$ are likely to be much more subtle and may involve the existence of period-adding cascades, chaotic attractors, and so on, as we now motivate.
2. As with the stroboscopic map, the main feature of the map P_N, for $e\widehat{\mu} < 0$ is the stretching associated with the square-root in the direction of the vector E. Note, though, that by the assumption (6.4) on W in the case that H and W are linear, we must have in this case that $C^T E = 0$. That is, the stretching is in a direction tangent to Σ.

6.3.2 Unfolding the dynamics of the map

Note that the normal form map (6.56) we have just constructed is identical in structure to the square-root map (4.29) analyzed in Chapter 4.

We can thus completely describe the dynamics close to the grazing point at $\mu = \mu^*$ by using this earlier theory, The nature of this dynamics depends on the matrices N, E, C^T, M and D and in particular on the eigenvalues of N. These in turn depend upon the nature of the flow Ψ around the periodic orbit, but in principle the whole range of eigenvalue space can be looked at by appropriate choices of the parameters of the flow. We will present examples that show how different behaviors of impacting systems at grazing bifurcations can be explained simply by finding the eigenvalues of the associated matrix N. First though, let us briefly recall some key results from Chapter 4. (See Chapter 4 for the derivation of these results.)

Theorem 6.5 (Existence of a period-one orbit). *If*

$$\alpha = C^T N(I - N)^{-1} E \neq 0, \quad \det(I - N) \neq 0 \ \textit{and} \ e \neq 0,$$

where e was defined in (6.57), then for small $\mu - \mu^*$ there exists a unique impacting period-one point branching off from the grazing orbit. If $\alpha < 0$ then this orbit coexists with the non-impacting orbit (when $\mu < \mu^*$) and if $\alpha > 0$, the impacting periodic orbit exists for $\mu > \mu^*$.

Recall also that periodic orbits of so-called *maximal* type are most likely to be observed for μ close to μ^* when N has *complex* eigenvalues.

In contrast, if N has *real* eigenvalues and satisfies the condition

$$C^T N^n E > 0 \quad \text{for all} \quad n > 0,$$

then we see attracting periodic or chaotic behavior as μ increases beyond μ^*, as follows:

1. If λ_1 is the principal eigenvalue of N and if $2/3 < \lambda_1 < 1$, then there is a chaotic attractor close to the origin for *all* small negative values of $e(\mu - \mu^*)$.
2. If $1/4 < \lambda_1 < 2/3$, then for *all* small negative values of $e(\mu - \mu^{*)}$ there is an alternating series of chaotic and stable periodic motions, accumulating in a period-adding cascade as $\mu \to \mu^*$.
3. If $0 < \lambda_1 < 1/4$, then the chaotic motion disappears and is replaced by periodic bands that overlap and increase in period as $\mu \to \mu^*$.

It is also helpful to recall Fig. 4.14 and the accompanying text in the special case of two-dimensional maps. Here differing behavior can be classified according to the different eigenvalue configurations of the matrix N. In particular, if N has complex eigenvalues, then we can see a multiplicity of different behaviors depending on the ratio of the real to the imaginary parts.

6.3.3 Examples

We shall now apply the above theory to the 1DoF impact oscillator studied in case study I, at various parameter values where different unfoldings of grazing bifurcations occur. In particular we look at the oscillator

$$\ddot{u} + 2\zeta\dot{u} + u = w(s), \quad \dot{s} = 1, \quad u > 0, \quad \dot{u} \to -r\dot{u} \quad \text{when} \quad u = 0, \tag{6.58}$$

where the forcing $w(s)$ is either constant or periodic. For convenience, and ease of notation, we will consistently set $v = du/dt$ and $x = (u, v, s)$.

Before embarking on various examples, it would be helpful to construct general expressions for the Jacobian N of the Poincaré map$\tilde{P}_N$ of the flow. We do this in three stages:

Time-T map of flows without impact. Suppose $(u(t), v(t) = du/dt)$ is a solution to the equation $\ddot{u}+2\zeta\dot{u}+u = w(t)$, and consider the perturbed flow $(u(t) + \delta u(t), v(t) + \delta v(t))$. The function δu satisfies the *variational equation*

$$\delta\ddot{u} + 2\zeta\delta\dot{u} + \delta u = 0, \quad (\delta u(t_0), \delta v(t_0)) = (\delta u_0, \delta v_0),$$

so that

$$\frac{d}{dt}\begin{pmatrix} \delta u \\ \delta v \end{pmatrix} = L\begin{pmatrix} \delta u \\ \delta v \end{pmatrix} := \begin{pmatrix} 0 & 1 \\ -1 & -2\zeta \end{pmatrix}\begin{pmatrix} \delta u \\ \delta v \end{pmatrix}.$$

Thus

$$\begin{pmatrix} \delta u_T \\ \delta v_T \end{pmatrix} := \begin{pmatrix} \delta u(t_0 + T) \\ \delta v(t_0 + T) \end{pmatrix} = e^{LT}\begin{pmatrix} \delta u(t_0) \\ \delta v(t_0) \end{pmatrix}.$$

The Jacobian matrix of the flow associated with passage through time T is therefore given by

$$N_T = e^{LT} = e^{-\zeta T}\begin{pmatrix} C_T & S_T \\ -\zeta C_T - \omega_0 S_T & \omega_0 C_T - \zeta S_T \end{pmatrix}, \tag{6.59}$$

where

$$\omega_0 = \sqrt{1 - \zeta^2}, \qquad C_T = \cos(\omega_0 T), \qquad S_T = \sin(\omega_0 T).$$

Observe that

$$\det(N_T) = e^{-2\zeta T},$$

so that the flow is dissipative if $\zeta > 0$.

Time-T map for flows with impact. Suppose now that in the time interval T the orbit (u, v) has a single transverse impact at time $t = t_0 + t_I$ with impact velocity $v_I < 0$ (just before impact) and impact acceleration a_I^- just before impact and a_I^+ just after impact. The flow is modified by the impact and the effect of this, as described in Chapter 2 (2.78), can be estimated by calculating the saltation matrix Q_x associated with the impact. In particular, we have

$$\begin{pmatrix} \delta u_T \\ \delta v_T \end{pmatrix} = \tilde{N}_T \begin{pmatrix} \delta u_0 \\ \delta v_0 \end{pmatrix},$$

where the $\tilde{N}_T$ is the Jacobian derivative of the hybrid flow Ψ in the presence of a single impact, and it is given by

$$\tilde{N}_T = N_{T-t_I} Q_x N_{t_I}, \tag{6.60}$$

with N_T defined by (6.59). From Chapter 2 the saltation matrix is given by

$$Q_x = \begin{pmatrix} -r & 0 \\ \frac{ra_I^- + a_I^+}{v_I} & -r \end{pmatrix},$$

therefore, in this case

$$\det(\tilde{N}_T) = e^{-2\zeta T} r^2.$$

Clearly we may extend this calculation to the case of m impacts in the period T, and in this case,

$$\det(\tilde{N}_T) = e^{-2\zeta T} r^{2m}. \tag{6.61}$$

Observe that the flow is therefore dissipative if $r < 1$ provided that $\zeta \geq 0$.

Linearization of the return map P_N. Assume next that the flow $u(t)$ is periodic, with $v(t_0) = v(t_0 + T) = 0$ and with acceleration $a_0 := a(t_0) \neq 0$. We now compute the linearization N of the flow map P_N from the surface $\Pi_N = \{(u,t) : v = 0\}$ to itself. To do this we consider making a small change to the initial conditions and find the evolution of a flow starting at time $t_0 + \delta t_0$ with $(u, v) = (u + \delta u_0, 0)$, and finishing at a time $t_0 + T + \delta T$ with $v(t_0 + T + \delta T) = 0$, and $u(t_0 + T + \delta T) = u(t_0 + T) + \delta u_T$. For small δt_0 and δu_0 this map can be calculated directly in terms of the linearization of the evolution map P_T by considering a projection at times close to t_0 and $t_0 + T$ from the surface Π_N to Π and *vice versa*. Noting that if $v(t_0 + \delta t_0) = 0$ and $u(t_0 + \delta t_0) = u + \delta u_0$ we have (to leading-order in δt_0) that $v(t_0) = -a_0 \delta t_0$ and $u(t_0) = u + \delta u_0$. A similar identity applies at time T. We therefore have

$$\begin{pmatrix} \delta u_T \\ \delta T \end{pmatrix} = N \begin{pmatrix} \delta u_0 \\ \delta t_0 \end{pmatrix} := \begin{pmatrix} 1 & 0 \\ 0 & -1/a_0 \end{pmatrix} \tilde{N}_T \begin{pmatrix} 1 & 0 \\ 0 & -a_0 \end{pmatrix} \begin{pmatrix} \delta u_0 \\ \delta t_0 \end{pmatrix}. \tag{6.62}$$

Note that

$$\det(N) = \det(\tilde{N}_T).$$

We are now in a position to consider various examples of grazing bifurcations.

Example 6.7 (Constant Forcing). We consider first an undamped impact oscillator with constant forcing

$$\ddot{u} + u = 1, \quad u > 0, \quad R : v \to -rv, \text{ at } u = 0 \text{ where } v = \dot{u}$$

for $0 < r < 1$. The simplicity of this example makes it possible to construct the maps involved in the preceding theory more or less explicitly, with $x = (u, v)^T$. In particular we can construct closed form expressions for both stroboscopic map P_S and the grazing set G_Π. There is no natural time period for the forcing in this example; however, all non-impacting periodic orbits have period $T = 2\pi$, so we take this as the natural period to describe the stroboscopic map. We illustrate the calculation in Fig. 6.13.

Orbits without impact correspond to points (u, v) moving on circles so that

$$(u-1)^2 + v^2 = \rho^2,$$

with phase

$$\theta = \tan^{-1}(v/(u-1)).$$

If $\rho < 1$, then $u > 0$ for all t, and no impacts occur. If $\rho = 1$, we have a *periodic grazing orbit.* In particular, if we set $t = 0$, then the set G_Π of initial conditions leading to a grazing impact is simply the circle of radius one given by

$$G_\Pi = \{(u, v) : (u-1)^2 + v^2 = 1\}.$$

Observe that when u and v are small and we express u as a power series in v, we have

$$G_\Pi = \{(u, v) : u = v^2/2 + \mathcal{O}(v^4)\},$$

that agrees with earlier local analysis [see (6.18)]. Now, consider a set of initial conditions with $\rho > 1$. This will lead to an impact when

$$u = 0, \quad v = -\sqrt{\rho^2 - 1},$$

that is then mapped to the point

$$u = 0, \quad v = r\sqrt{\rho^2 - 1}.$$

After impact, the orbit is again a circle satisfying

$$(u^+ - 1)^2 + v^{+2} = 1 + r^2(\rho^2 - 1) := \widehat{\rho}^2$$

However, the phase of the point on this circle is instantaneously changed by the impact, from θ^- to θ^+, where

$$\theta^- = -\tan^{-1}(\sqrt{\rho^2 - 1}) \quad \text{and} \quad \theta^+ = \tan^{-1}(r\sqrt{\rho^2 - 1}).$$

Consequently, there is a jump $\Delta\theta$ in the phase given by

$$\Delta\theta = \tan^{-1}(\sqrt{\rho^2 - 1}) + \tan^{-1}(r\sqrt{\rho^2 - 1}).$$

From this analysis, we can easily construct the stroboscopic Poincaré map. Suppose that at $t = 0$ we have $(u, v) = (1 + \rho\cos(\theta), \rho\sin(\theta))$; then if $\rho > 1$, we have

$$(\rho,\theta) \mapsto (\widehat{\rho},\theta+\Delta\theta) := (\sqrt{1+r^2(\rho^2-1)}, \tan^{-1}(\sqrt{\rho^2-1})+\tan^{-1}(r\sqrt{\rho^2-1}). \tag{6.63}$$

Similarly, if $\rho < 1$, then

$$(\rho,\theta) \mapsto (\rho,\theta).$$

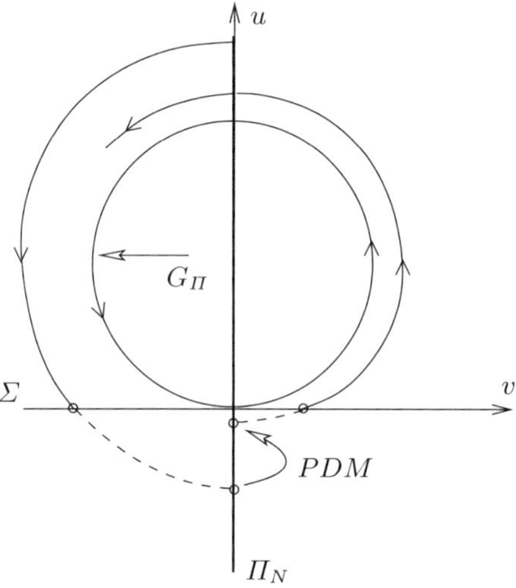

Fig. 6.13. The effect of impact on an orbit with $\rho = \sqrt{2}$ in that we can see the effect of the PDM.

Of specific interest is the case when $\rho = 1+\varepsilon$, corresponding to a near grazing orbit when $\varepsilon \ll 1$, with $H_{\min} = -\varepsilon$. In this case (6.63) becomes to leading-order,

$$(1+\varepsilon,\theta) \mapsto (1+r^2\varepsilon, \theta + (1+r)\sqrt{2\varepsilon}).$$

These results are in perfect agreement with the ZDM, (6.22), because when u is close to zero, the velocity is changed by

$$\sqrt{2a}(1+r)\sqrt{\varepsilon} := -\sqrt{2a}[-(1+r)]\sqrt{-H_{\min}}\ ,$$

where $-(1+r)$ is the v component of $W(x)$.

We can also calculate the PDM. On the surface $\Pi_N \equiv \{v=0\}$, an orbit of radius $\rho^- = 1+\varepsilon$ is instantaneously mapped to one of radius $\rho^+ = 1+r^2\varepsilon$, and v is unaltered. Hence on Π_N, when $u = 1-\rho$, we have

$$(u,v) \to (r^2u, v) = (u,v) + (r^2-1, 0)u. \tag{6.64}$$

Note that u is changed by a linear term only.

This exact expression (6.64) can be replicated using the expression for the PDM given in Theorem 6.2. To see this, we rewrite the system in the form

$$F = \begin{pmatrix} v \\ -u+1 \end{pmatrix}, \quad F_x = \begin{pmatrix} 0 & 1 \\ -1 & 0 \end{pmatrix}, \quad W = -(1+r)\begin{pmatrix} 0 \\ 1 \end{pmatrix}, \quad H_x = \begin{pmatrix} 1 \\ 0 \end{pmatrix}^T,$$

and we calculate the PDM on the zero-velocity surface $\{v = 0\}$. We find, in the notation of (6.22), that $b(x) = -(1+r)$, $a(x) = -(u+1)$, $c(x) = v = 0$ and

$$W_x F = (0,0)^T$$

and that E and C are both zero. We can now use the formula (6.23) to compute the terms of order y, where $y = -u^2$. We find

$$-\sqrt{2(-u+1)}\begin{pmatrix} 0 \\ -(1+r) \end{pmatrix} + \frac{-\sqrt{2(-u+1)}(1+r)}{-u+1}\begin{pmatrix} v \\ -u+1 \end{pmatrix} = \begin{pmatrix} 0 \\ 0 \end{pmatrix}$$

as $v = 0$ by assumption. We therefore need to find the next order term. We have

$$2y^2\left[\frac{(1+r)^2}{2(-u+1)}\begin{pmatrix} -u+1 \\ 0 \end{pmatrix} - (1-(1+r))\begin{pmatrix} -(1+r) \\ 0 \end{pmatrix}\right] =$$
$$2y^2\left(\frac{(1+r)^2}{2} - r^2 - r\right) = y^2(1-r^2) = u(r^2-1).$$

Hence, we can write the PDM as

$$(u,0) \mapsto (u,0) + (r^2-1,0)u,$$

that in our particular case is the exact formula for the PDM (there are no terms of higher order).

Using these results, we can look at the effect of the map on a small square of side length 2ε in (ρ,θ) co-ordinates, so that $\rho \in (1-\varepsilon, 1+\varepsilon)$, $\theta \in (-\varepsilon,\varepsilon)$. For $\rho < 1$ the part of the square is unchanged. If $\rho > 1$, then the sides of the square are mapped to a parabola parallel to $\rho = 1+\theta^2$. The square is thus stretched to a set that is tangent to the line $\rho = 1$, as illustrated in Fig. 6.14. Note that the area of the stretched portion of the square is reduced by a factor of r^2.

Example 6.8 (periodic orbits with a single low-velocity impact). Consider the example of a periodically forced undamped harmonic oscillator

$$d^2u/dt^2 + u = \cos(\omega t) - \mu, \quad u > 0, \quad \text{impact at} \quad u = 0,$$

with $\omega \approx 2$. As we saw in case study I in Chapter 1, if ω is fixed and $\mu < \mu^* = 1/(1-\omega^2)$, this system has a non-impacting periodic orbit, with period $T = 2\pi/\omega$ given by

$$u(t) = \cos(\omega t)/(1-\omega^2) - \mu$$

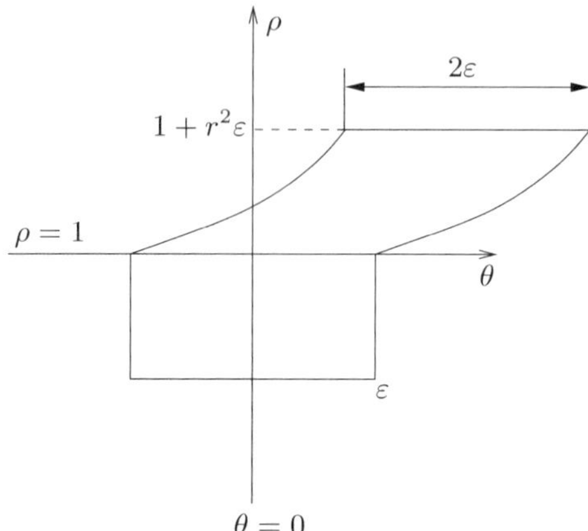

Fig. 6.14. The stretching of a square of initial data either side of G_Π.

that grazes with the surface $\Sigma = \{u = 0\}$ when $\mu = \mu^*$. Consider the Poincaré map from the surface Π_N to itself close to this periodic orbit when μ is close to μ^*. The intersection of the periodic orbit with this surface is given by

$$(u, t) = (\mu^* - \mu, \pi/\omega).$$

The corresponding matrices and vectors in Theorem 6.4 can readily be calculated to be

$$N = \begin{pmatrix} 1 & 0 \\ 0 & -1/a \end{pmatrix} \begin{pmatrix} \cos(T) & \sin(T) \\ -\sin(T) & \cos(T) \end{pmatrix} \begin{pmatrix} 1 & 0 \\ 0 & -a \end{pmatrix}, \quad E = -\sqrt{\frac{2}{a}} \left(\begin{pmatrix} 0 \\ (1+r) \end{pmatrix} \right)$$

and

$$C = (1, 0)^T, \quad M = \begin{pmatrix} 1 & 0 \\ 0 & -1/a \end{pmatrix} \begin{pmatrix} \cos(2\pi/\omega) - 1 \\ -\sin(2\pi/\omega) \end{pmatrix}, \quad D = 0,$$

where a is the acceleration when $v = du/dt = 0$. A graze occurs when $u = v = 0$ so that at this point $a = \cos(2\pi\omega) - \mu$. Hence

$$N = \begin{pmatrix} S_T & -aS_T \\ S_T/a & C_T \end{pmatrix}, \quad M = \begin{pmatrix} C_T - 1 \\ S_T/a \end{pmatrix},$$

where $S_T = \sin(2\pi/\omega)$, $C_T = \cos(2\pi/\omega)$. Hence, in the notation of Theorem 6.5,

$$(I - N)^{-1} M = (-1 \; 0)^T, \text{ thus } e = -1,$$

and

$$\alpha = C^T N (I - N)^{-1} E = (1 + r)\sqrt{2a}s/2(1 - c).$$

Now, at the grazing point, we have $a > 0$ and $1 - C_T > 0$; thus, the sign of α is the same as the sign of $S_T = \sin(2\pi/\omega)$. We conclude that if $\sin(2\pi/\omega) < 0$

(locally if $\omega < 2$), then the impacting orbit coexists with the non-impacting one, and if $\sin(2\pi/\omega) > 0$ (locally $\omega > 2$), then the impacting orbit replaces the non-impacting one. This is precisely the behavior observed in case study I corresponding, respectively, to the sub-resonant and super-resonant ellipses described in Fig. 1.8.

Example 6.9 (Chaotic behavior for orbits with a single low-velocity impact). Consider next the same impacting system but with non-zero damping:

$$\ddot{u} + 2\zeta\dot{u} + u = \cos(2t) - \mu, \quad u > 0,$$

for $\zeta > 0$, and the same impact law $v \to -rv$ when $u = 0$. We study the change in the dynamics as μ varies through the value

$$\mu^* = -1/\sqrt{9 + 16\zeta^2}.$$

At $\mu = \mu^*$, the system has a period-π grazing orbit

$$u(t) = \frac{3\cos(2t) + 4\zeta\sin(2t)}{9 + 16\zeta^2} - \mu^*,$$

with a grazing impact at $t = t^*$ where $\tan(2t^*) = 4\zeta/3$ and no other impacts. There is a positive acceleration $a = \cos(2t) - \mu^*$ at the grazing point. The linearization of the Poincaré map P_N about this orbit is now given by

$$N = \begin{pmatrix} 1 & 0 \\ 0 & -1/a \end{pmatrix} N_\pi \begin{pmatrix} 1 & 0 \\ 0 & -a \end{pmatrix},$$

where

$$N_\pi = e^{-\pi\zeta} \begin{pmatrix} C_T + \frac{S_T\zeta}{\omega_0} & \frac{S_T}{\omega_0} \\ -\omega S_T - \frac{\zeta^2 S_T}{\omega_0} & C_T - \frac{\zeta S_T}{\omega_0} \end{pmatrix},$$

$$C_T = \cos(\omega_0\pi), \quad S_T = \sin(\omega_0\pi), \quad \omega_0^2 = 1 - \zeta^2.$$

The matrix N is similar to, and therefore has the same eigenvalues, as N_π. These eigenvalues are complex and given by

$$\lambda_{1,2} = e^{-\pi\zeta}[\cos(\sqrt{1-\zeta^2}\pi) \pm i\sin(\sqrt{1-\zeta^2}\pi)]$$

if $0 < \zeta < 1$. In contrast if $\zeta \geq 1$, the eigenvalues are both real:

$$\lambda_{1,2} = e^{-\pi\zeta}e^{\pm\sqrt{\zeta^2-1}\pi}.$$

Note the special case of $\zeta = \sqrt{3}/2 = 0.867$, where the eigenvalues are pure imaginary.

Fig. 6.15 shows the results of computing Monte Carlo bifurcation diagrams for this example in four illustrative cases. The diagram in panel (a) is for

$\zeta = 0.26$; in which case, the eigenvalues of N_π are $-0.776 \pm 0.083i$. Thus, with reference to Fig. 4.14 in Chapter 4, we are in Region 3 close to the right-hand parabola-shaped boundary with Region 2. Hence, according to the theory presented in Chapter 4, we should expect to see a jump to orbits with a high period or aperiodic. Indeed, the numerics suggest a jump to broad-band chaos at the grazing bifurcation value $\mu^* = -0.3149$.

Fig. 6.15(b) shows the case $\zeta = 0.867$, where the eigenvalues $\lambda_{1,2}$ are pure imaginary. Hence we are in the center of Region 3 of Fig. 4.14 where we should expect to see a maximal periodic orbit of period-two. This is precisely what the numerical bifurcation diagram shows to occur for $\mu > \mu^*$. Panels (c) and (d) show the cases $\zeta = 1.2$ and 2, respectively, where the eigenvalues are real; with dominant eigenvalue $\lambda_1 = 0.184$ for $\zeta = 1.2$ and 0.4301 for $\zeta = 2$. Thus, according to the results of Chapter 4, we should have an overlapping period-adding cascade for $\zeta = 1.2$, and a period-adding cascade with intermediate regions of chaotic motion. This is again what is observed in the numerical simulations.

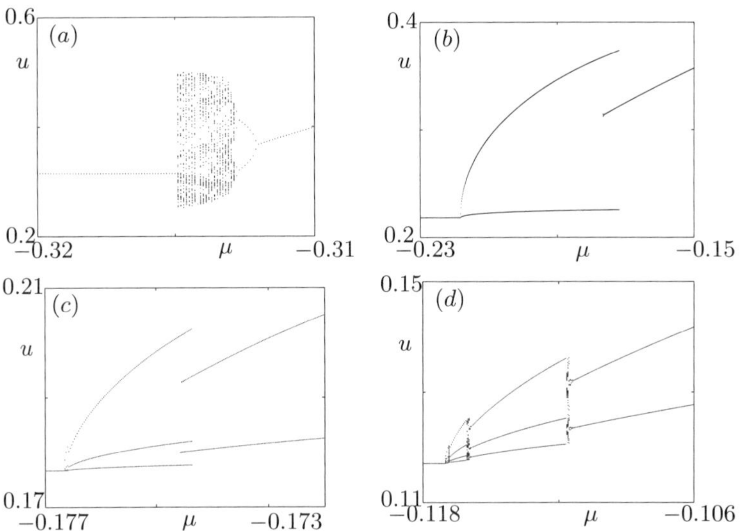

Fig. 6.15. Bifurcation diagrams for the cases (a) $\zeta = 0.26$, (b) $\zeta = \frac{\sqrt{3}}{2}$, (c) $\zeta = 1.2$ and (d) $\zeta = 2$. The corresponding grazing bifurcations happen at (a) $\mu = -0.3149$, (b) $\mu = -0.2182$, (c) $\mu = -0.1766$ and (d) $\mu = -0.1170$. In figures (c) and (d) we see hysteresis and a period-adding cascade, whereas (a) shows an immediate jump to chaos.

Example 6.10 (Chaotic behavior with high-velocity impact in the non-dissipative limit). We modify the previous example to consider the grazing bifurcation that occurs close to a periodic orbit that also contains a transverse impact. For simplicity of calculation we set the damping constant ζ to zero, that is justified because dissipation is introduced by the restitution law at the non-grazing impact. Thus we take the periodically forced impact oscillator given by

$$\ddot{u} + u = \cos(2t) - \mu, \quad u > 0, \quad v \to -rv \quad \text{when} \quad u = 0.$$

Taking the forcing frequency equal to 2 also greatly simplifies the resulting calculations. Again, we consider the effect on the dynamics of varying μ, this time for $\mu > 0$. We shall show that there is a critical penetration $\mu = \mu^*$ at that the system has a periodic solution $u(t)$ with period $T = \pi$ and a single high-velocity impact at time $t = t_I$, velocity $v_I < 0$ immediately before impact, and velocity $-rv_I > 0$ immediately after. This orbit is given by

$$u(t) = \left[-rv_I - \frac{2}{3}\sin(2t_I)\right]\sin(t - t_I) - \frac{1}{3}\cos(2t) - \mu,$$

with

$$\cos(2t_I) = -3\mu, \quad v_I = \frac{4}{3(1-r)}\sin(2t_I).$$

Note that v_I and μ lie on the ellipse

$$\frac{9}{16}(1-r)^2 v_I^2 + 9\mu^2 = 1.$$

When $\mu = 0$ there is a large amplitude π-periodic orbit with $v_I = -4/(1-r)$ and $u \geq 0$ for all t . When $\mu = 1/3$ there is a grazing orbit with $v_I = 0$, for that $u \leq 0$ for all t. By continuity, it follows that there is a point $0 < \mu^* < 1/3$ at that the orbit $u(t) \geq 0$ for all t and that has a grazing impact, between the high-velocity impacts. This is illustrated in Fig. 6.16(a).

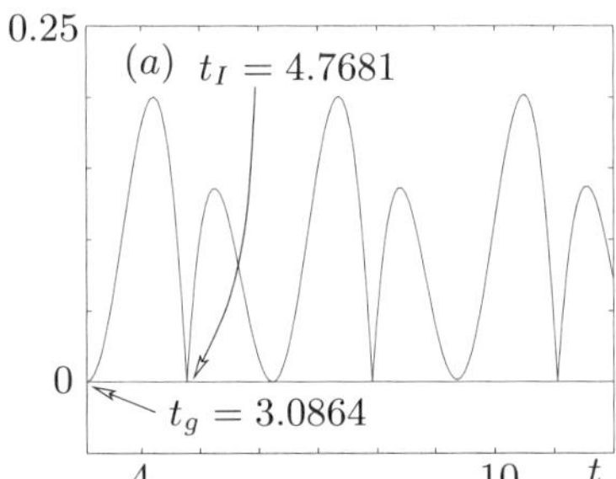

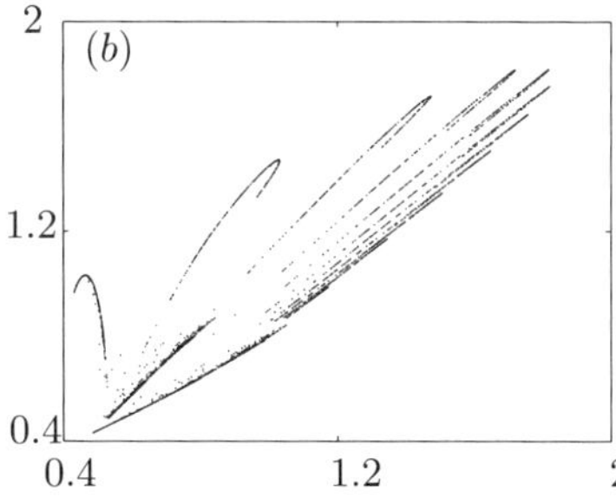

Fig. 6.16. (a) Periodic orbit with high-velocity impact and grazing when $\mu = 0.3312$ and $r = 0.8$ indicating the grazing impact at time t^* and the high velocity impact at time t_I. (b) Strange attractor for $r = 0.8$ at $\mu = 0.3459$.

Suppose that at $\mu = \mu^*$ the grazing impact occurs at time t^* with acceleration $a > 0$ and the high-velocity impact at time t_I with $t^* < t_I < t^* + \pi$.

Let a_I be the acceleration at time t_I (Note that, as the dissipation is zero, we have $a_I^- = a_I^+ = a_I$.) From (6.62), it follows that the linearization of the map $\tilde{P}_N$ about the grazing orbit at μ^* is given by

$$N = \begin{pmatrix} 1 & 0 \\ 0 & -1/a \end{pmatrix} N_{\pi+t^*-t_I} Q_x N_{t_I-t^*} \begin{pmatrix} 1 & 0 \\ 0 & -a \end{pmatrix},$$

where the saltation matrix is given by

$$Q_x = \begin{pmatrix} -r & 0 \\ (1+r)\frac{v_I}{a_I} & -r \end{pmatrix}.$$

Now N is similar to the matrix $N_{\pi+t^*-t_I} N_{t_I-t^*}$, which is in turn similar to the matrix

$$\widehat{N} = N_{t_I-t^*} N_{\pi+t^*-t_I} Q_x N_{t_I-t^*} N_{t^*-t_I} = N_\pi Q_x.$$

But, for a system without damping between impacts, we have simply that

$$N_\pi = -I,$$

so that the eigenvalues of N are the same as those of Q_x but with *opposite sign.* We conclude that the eigenvalues of N are simply r and r and, consequently, the bifurcation diagram is determined by the value of r. Hence we expect to see the periodic orbit evolving immediately into a chaotic orbit if $r > 2/3$, a period-adding cascade if $1/4 < r < 2/3$ and hysteresis if $r < 1/4$ provided that $C^T N^n E > 0$.

For example, taking $r = 0.8$ we have

$$\mu^* = 0.331269,\ t^* = 3.0865,\ a = 0.6629,$$
$$t_I = 4.7681,\ v_I = -0.7408,\ a_I = -1.3251.$$

Thus, from the above

$$N = \begin{pmatrix} 0.446225 & -2.10926 \\ 0.059337 & 1.15378 \end{pmatrix}$$

with (degenerate) eigenvalues $0.8, 0.8$. For this example we also have

$$E = \sqrt{2/a^*}(0,-1)^T, \quad C = (1,0)^T,$$

and it is easy to see that $C^T N^n E$ is positive for all n.

In contrast, when $r = 0.3$ we have

$$\mu^* = 0.2808, t^* = 2.8551, a = 0.5593, t_I = 4.9966, v_I = -1.0255, a_I = -1.1247,$$

and thus

$$N = \begin{pmatrix} -0.346321 & -0.561721 \\ 0.743657 & 0.946313 \end{pmatrix}.$$

The vectors C^T and E are as before, and again $C^T N^n E > 0$ for all n.

A plot of the bifurcation diagrams in these two cases is presented in Fig. 6.17. In panel (a) we see (as expected) that when $r = 0.8$ there is an immediate transition into a region of chaos, that terminates in a maximal period-21 orbit, followed by a period-subtracting cascade. In contrast, when $r = 0.3$, there is a period-adding cascade separated by regions of chaos. Fig. 6.16(b) presents the strange attractor (as iterates of the map P_S) for the case of $r = 0.8, \mu = 0.3459$ in which its fingered structure is clearly visible.

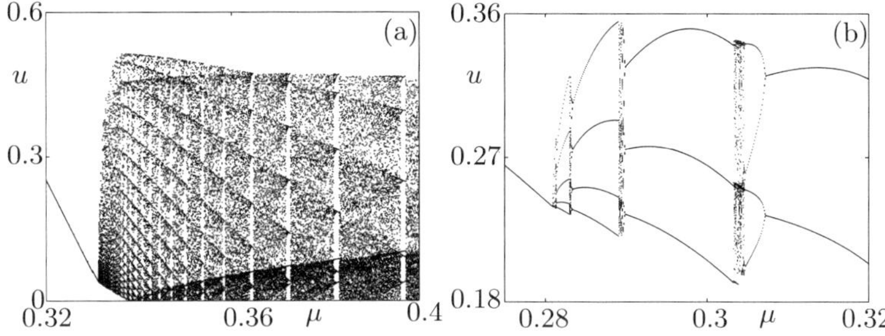

Fig. 6.17. Bifurcation diagram for (a) $r = 0.8$, (b) $r = 0.3$ showing, respectively, a square-root growth of a chaotic attractor and a period-adding cascade with chaotic windows.

6.4 Chattering and the geometry of the grazing manifold

We now study the *global* nature of the flow of impacting hybrid systems, to help explain phenomena such as complex domains of attraction and the effects of chattering observed in case study I in Chapter 1. Our main tool of analysis will be to examine the geometry of the grazing manifold $\mathcal{G}$ via its intersection with a stroboscopic Poincaré section. By doing this we will be able to see the influence of grazing on the dynamics, motivating the form of the domains of attraction of certain periodic orbits and the shape of certain strange attractors. We can also understand how to estimate Lyapunov exponents, that determine the average rate of expansion or compression in an attractor.

The results of this section are discussed solely in the context of the 1DoF impact oscillator (6.58). General theory, in more phase space dimensions, is the subject of ongoing research and is beyond the scope of this book.

6.4.1 Geometry of the stroboscopic map

The matrix $\tilde{N}_T$ calculated in (6.60) is simply the Jacobian of the stroboscopic map P_S away from any grazing point. We observe from (6.61) that if T is held fixed, then this matrix has constant determinant over any set Ω_α for that the number of impacts m of any trajectory starting in Ω_α is *constant.* Consequently, if we set $\Omega_\beta = P_S(\Omega_\alpha)$ and let $A_{\alpha,\beta}$ represent the areas of the sets $\Omega_{\alpha,\beta}$, respectively; then we have

$$A_\beta = e^{-2\zeta T} r^{2m} A_\alpha. \tag{6.65}$$

Now, suppose we take k iterations of the stroboscopic map P_S defined over a time T [that we assume to be the period $T = 2\pi/\omega$ of the forcing function $w(s)$]. Suppose that the flow representing the jth iteration of the map has m_j transverse impacts with $\Sigma := \{(u, v, s) : u = 0\}$. Since each impact is transverse, it will survive under small perturbation. Let Ω_0 be the set of initial conditions for that the number of impacts in each iteration remains the same, and each impact remains transverse. Let Ω_k be the kth iterate of Ω_0 under the stroboscopic map. Then, applying the result (6.65) repeatedly we deduce that the corresponding areas of these two sets are related by

$$A_k = e^{-2k\zeta T} r^{2(m_0+m_1+\ldots+m_{k-1})} A_0.$$

As $k \to \infty$ the *average* rate of contraction of the phase space between successive actions of the map P_S defined by $(A_k/A_0)^{1/k}$ is given $e^{-2\zeta T} r^{2z}$, where the *winding number* of the flow is defined by

$$z = \lim_{k\to\infty} \left(\frac{1}{k} \sum_{j=0}^{k-1} m_j \right).$$

Using z we can estimate the Lyapunov exponents of the flow directly. If the Lyapunov exponents are Λ_1 and Λ_2, then the average rate of contraction of the area of a set over the time interval T is given by $e^{(\Lambda_1+\Lambda_2)T}$. However, we have already shown that this rate of contraction is given by $e^{-2\zeta T} r^{2z}$. Equating, we have the following estimate for the sum of the Lyapunov exponents

$$\Lambda_1 + \Lambda_2 = 2(z \log(r)/T - \zeta).$$

The winding number z helps to distinguish between different types of flow, in particular z will be rational for a periodic orbit and may not be so for a chaotic orbit. For an orbit with no impacts at all, z is zero and for a *chattering* orbit [42] in that one of the m_j is infinite, z is also infinite, and there is a consequent infinite contraction of the phase space because $\Lambda_1 + \Lambda_2 = -\infty$.

6.4.2 Global behavior of the grazing manifold $\mathcal{G}$.

It is now natural to consider the boundaries of regions Ω of trajectories that have the same event history. Such boundaries represent initial conditions for that the corresponding flow has a non-transverse impact with Σ. Thus, these initial conditions must form part of the grazing manifold $\mathcal{G}$; see Definition 6.2.

Let us now define $G_\Pi = \mathcal{G} \cap \Pi$ to be the set of initial data in a Poincaré section Π leading to an orbit with a grazing impact at some future time. Specifically throughout this discussion, we shall take Π to be the stroboscopic Poincaré section $t = t_0 \bmod T$ where T is the period of the forcing function $w(s)$. We shall look at the geometry of G_Π, that was first described by Whiston [264]. We also refer to the careful topological study of Chillingworth [53] for more precise information, including rigorous proofs.

The set G_Π is made up of the union of the sets $G_\Pi^{(k)}$ that comprise initial data that lead to a grazing impact together with $k-1$ transversal impacts before intersecting Π again. The set G_Π is *locally* a one-dimensional submanifold of Π. However, $\mathcal{G}_\Pi$ can self-intersect at points in Π corresponding to the initial data of orbits with more than one grazing impact (we already saw this in Fig. 6.6(a)). As calculation of G_Π is difficult in general, we confine attention here to calculating it for the periodically forced, undamped, impact oscillator given by

$$\ddot{u} + u = \cos(\omega s) - \sigma, \quad u > 0, \quad v \to -rv, \quad \text{at} \quad u = 0,$$

with $s = t \bmod T = 2\pi/\omega$, and taking $\Pi = \Pi_S = \{(x, v, s) : s = 0\}$

To determine the form taken by G_Π, consider a grazing impact at $u = v = 0$ occurring at time $t = s_0 > 0$. Close to the grazing point we have

$$u = (\sigma - \gamma\cos(\omega s_0))C_0 + \omega\gamma\sin(\omega s)S_0 + \gamma\cos(\omega t) - \sigma, \tag{6.66}$$

$$v = -(\sigma - \gamma\cos(\omega s))S_0 + \omega\gamma\sin(\omega s)C_0 - \omega\gamma\sin(\omega t), \tag{6.67}$$

$$\gamma = \frac{1}{1-\omega^2}, \quad S_0 = \sin(t - s_0), \quad C_0 = \cos(t - s_0). \tag{6.68}$$

This orbit is only a solution of the hybrid system if $u > 0$, which requires that the local acceleration a and penetration σ both be positive. That is, we are not in the *sticking region* $\mathcal{Z} \subset G \subset \Sigma$ defined by

$$\mathcal{Z} = \{(u, v, s) : u = v = 0, \ a := \cos(\omega s) - \sigma < 0\}.$$

Observe that if $\sigma < -1$, then $\mathcal{Z}$ is the empty set and if $\sigma > 1$ then $\mathcal{Z}$ is the whole interval $s \in [0, T]$. For $-1 < \sigma < 1$, $\mathcal{Z}$ is a line, a subinterval of the grazing set G, centered on the point $s = \pi/\omega$. Namely, we have

$$\mathcal{Z} = \{(u, v, s) : u = v = 0, t_\alpha < s < t_\beta\},$$

where t_α and t_β are defined by

$$\cos(\omega t_{\alpha,\beta}) - \sigma = 0, \quad t_\alpha + t_\beta = \frac{2\pi}{\omega}.$$

Provided $(0, 0, s_0)$ is not in the sticking set, we may continue the orbit (6.66)-(6.68) *backwards in time* allowing for any further impacts as necessary. The intersection of this orbit with the manifold Π at $t = 0$ is then a point on

G_Π. If *no impacts* occur in the interval $[0, s_0]$, then the corresponding part of G_Π is given by

$$G^{(0)} = \{(u, v) :$$
$$u = (\sigma - \gamma \cos(\omega s_0) \cos(s_0) - \omega\gamma \sin(\omega s_0) \sin(s_0) + \gamma - \sigma,$$
$$v = (\sigma - \gamma \cos(\omega s_0)) \sin(s_0) + \omega\gamma \sin(\omega s_0) \cos(s_0)\}.$$

If $s_0 \ll 1$ and $\sigma < 1$, then this set is given locally by the half parabola

$$\{(u, v) : u = (1 - \sigma)s_0^2/2, v = (\sigma - 1)s_0\}.$$

Now, as s_0 increases from zero, we have

$$\frac{du}{ds_0} = \sin(s_0)(\cos(\omega s_0) - \sigma), \quad \frac{dv}{ds_0} = \cos(s_0)(\sigma - \cos(\omega s_0)).$$

Thus, the curve $G^{(0)}$ has a stationary point when $(\sigma - \cos(\omega s_0)) = 0$, which is precisely when s_0 lies on the boundary of the sticking region $\mathcal{Z}$, at $s_0 = t_\alpha$. We call this point G_α.

If $s_0 > t_\beta$, then we may again calculate the pre-image on Π of the grazing impact. In general this pre-image will lie on an orbit that has several transverse impacts in the interval $[0, s_0]$. Suppose that there are k such impacts at times $t_i, i = 1 \dots k$. As s_0 increases it is likely that further impacts will occur between, or before, these k impacts. The s_0-values at that the number k changes will correspond to the occurrence of an additional grazing impact. Suppose, for example, that this grazing impact occurs before the first transverse impact at a time $t = t_0 < t_1$. The pre-image on Π of the orbit starting from this s_0-value is then also the pre-image on Π of an grazing orbit starting from $s_0 = t_0$, that defines a point in the set $G^{(0)}$. Hence we see an intersection point between two parts of G_Π, between $G^{(0)}$ and $G^{(k)}$. Also for higher s_0-values, the set $G^{(k)}$ will now become $G^{(k+1)}$.

The effect of this scenario is to increase the number of impacts on the orbit starting from $(0, 0, s_0)$ by one. Thus $G^{(k)}$ spawns a new segment $G^{(k+1)}$ within Π_S. For further increase in s_0, $G^{(k+1)}$ will likely cross $G^{(0)}$ again. In fact this sequence of transitions typically continues making curve segments $G^{(n)}$ for arbitrarily large n. It was shown in [42] that these curves accumulate on a limiting curve $G^{(\infty)}$ comprising orbits that impact an infinite number of times before a final grazing impact; that is on orbits that undergo a complete chattering sequence. This scenario is illustrated in the section Π_S in Fig. 6.18. In Fig. 6.19, we show the various orbits corresponding to the initial data labeled in the previous figure. Observe the sticking orbit for initial condition $E = G_\alpha$ and an orbit with complete chatter starting from the point labeled F. Note that as s_0 varies we see that while each curve $G^{(k)}$ intersects $G^{(0)}$ transversally, the new segment $G^{(k+1)}$ leaves tangentially (with high speed as s_0 varies). Note also from Fig. 6.18 that the curve $G^{(\infty)}$ terminates at the point G_α, denoted by E in the figure. Notice the complex geometry of the

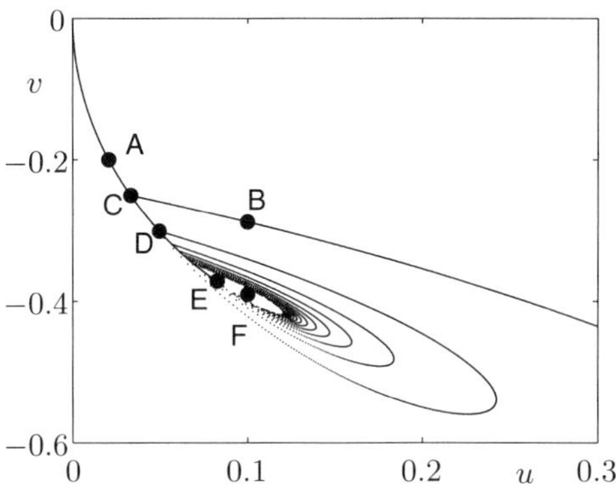

Fig. 6.18. The curve G_Π close to the point $E = G_\alpha$ when $\omega = 2.6, \sigma = 0$ and $r = 0.8$. In this figure the labeled points correspond to initial data at 'A' ($u = 0.020718, v = -0.2$),'B' ($u = 0.1, v = -.2881056$), 'C' ($u = 0.033268, v = -0.250478$), 'D' ($u = 0.049762, v = -0.301258$), 'E' ($u = 0.08282, v = -0.371567$), 'F' ($u = 0.1, v = -0.3905$).

set G_Π close to the G_α, where the many leaves $G^{(k)}$ accumulate on the curve $G^{(\infty)}$.

For more general calculations it is useful not only to consider the set G_Π but also its various pre-images under the action of the map P_S. To illustrate this calculation we show in Fig. 6.20 the full curve G_Π plus 5 pre-images calculated for the case of $\sigma = 0, \omega = 2.6$ and $r = 0.8$.

For many values of ω and σ the impact oscillator has several competing attractors. Each of these has a domain of attraction. The enormous stretching of the phase space close to the set G_Π and its pre-images under P_S means that the domain of attraction of these attractors will be greatly influenced by this set. This process is described in detail in [42]. In particular, G_Π typically acts as a set separating the domains of attraction of the attracting states. As the geometry of G_Π is typically complex, it leads to great complexity in the domains of attraction. We can see some of this behavior in Fig. 6.21. Here we see the domains of attraction of a period-one and a period-six orbit when $\omega = 2.6, r = 0.8$ and $\sigma = 0$. This figure should be compared with the plot of G_Π, and its pre-images, for the same parameter values given in Fig. 6.20

6.4.3 Chattering and the set $G^{(\infty)}$

Note that there is a region close to G_α of initial data on Π for that *all* orbits (in forward time) intersect Σ an infinite number of times before sticking in $\mathcal{Z}$. The region bounded by $G^{(\infty)}$ is thus part of the pre-image of the sticking region $\mathcal{Z}$. We shall now investigate this sticking behavior, and the chattering dynamics as sticking is approached.

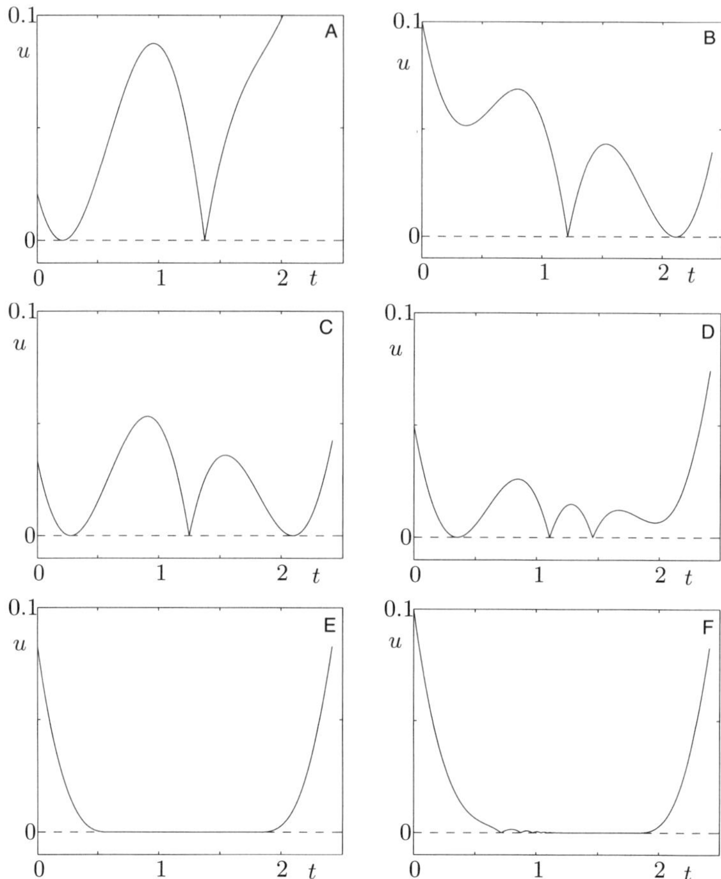

Fig. 6.19. The orbits labeled on the figure of G_{Π}. Observe the double grazing orbit in C, the sticking orbit in E and the chattering orbit in F.

Orbits with initial data close to G_α will impact in the sticking region with low-velocity when the acceleration is towards the impact surface Σ. In this case, we expect to see *complete chatter*. That is, there is a sequence of impact times t_k accumulating at a time t_∞ lying inside the sticking region $\mathcal{Z}$.

Suppose that a sequence of impacts occur at times t_k in the sticking region with velocity v_k *immediately after impact* and with acceleration $a_k < 0$. If the times t_k are close, and we are not close to the boundary of the sticking region (where a_k is close to zero), then we may approximate the acceleration by the constant $-a < 0$. Suppose that an impact occurs at time t_0 with (low) velocity $v_0 > 0$, then on the assumption of constant a, we have

$$v_{k+1} = rv_k > 0 \quad \text{and} \quad \Delta_k := t_{k+1} - t_k = \frac{2v_k}{a}.$$

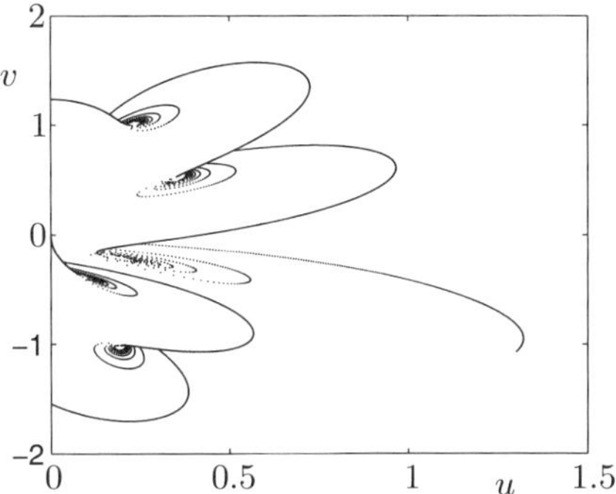

Fig. 6.20. The curve G_Π together with 5 pre-images when $\omega = 2.6, \sigma = 0$ and $r = 0.8$..

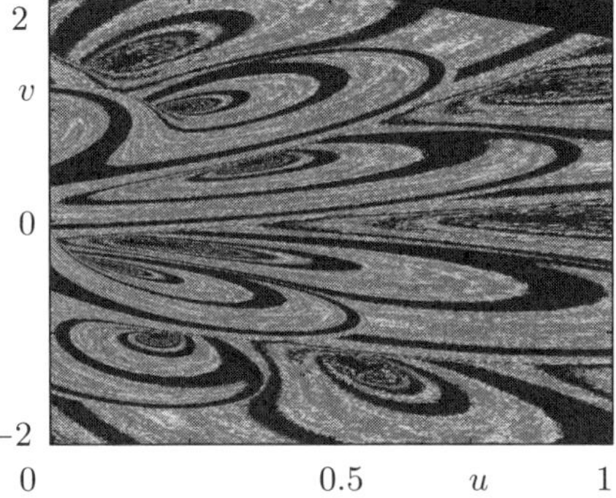

Fig. 6.21. Domain of attraction when $\sigma = 0, r = 0.8$ and $\omega = 2.6$. In this figure the dark regions are attracted to a period-one periodic orbit, and the light regions to a period-six orbit.

If u_k is the maximum value of u in the interval $[t_k, t_{k+1}]$, then from the parabolic nature of the trajectory under constant acceleration, it follows that $u_k = \frac{v_k^2}{2a}$. It follows that there are an infinite number of impacts (a chattering sequence) occurring in the time interval $[t_0, t_\infty)$ where

$$t_\infty = t_0 + \sum_{i=0}^{\infty} \Delta_k = t_0 + \frac{2v_0}{a} \sum r^k = t_0 + \frac{2v_0}{a(1-r)}.$$

Note that

$$t_k = t_\infty - \frac{2v_k}{a(1-r)} = t_\infty - \sqrt{\frac{8u_k}{a(1-r)^2}}. \tag{6.69}$$

Given a sequence of impact times and velocities (or displacements), either of the two expressions in (6.69) can be used to estimate t_∞. The chattering sequence will terminate with the solution at rest ($u = v = 0$) inside the sticking region $\mathcal{Z}$. We term such sequence *complete chatter*. The chatter set $\mathcal{C}$ (that is the pre-image of $\mathcal{Z}$) is defined to be the set of orbits that have a complete chattering sequence.

In contrast, *incomplete chatter* occurs when the system has a (large) finite number of impacts in the sticking region, but it does not come to rest. Typically this arises when the acceleration a changes sign during the flight time between impacts, and there is an impact close to, but outside, the sticking region. Incomplete chatter is closely linked with the calculation of the curve G^{∞}. Detailed results may be found in the work of Budd & Dux [42]. Loosely speaking, the curves $G^{(k)}$ bound incomplete chatter regions that have k impacts before leaving the local region of analysis. These curves accumulate on G^{∞} in a universal manner that can be analyzed recursively. The set G^{∞} then can be shown to bound the complete chatter region $\mathcal{C}$, that is the pre-image of the sticking region $\mathcal{Z}$. Preliminary results show that transitions from complete to incomplete chattering, under parameter variations, can lead to dramatic changes in the system behavior. These phenomena, that were first proposed by Nordmark in [201], have been recently observed in cam-follower mechanical systems [211, 3].

6.5 Multiple collision bifurcation

We have seen how the problem of a single forced particle impacting with a surface can be modeled as a hybrid system and then analyzed by considering the action of square-root and related maps. However, in many applications, such as in granular materials, there are many such particles that interact through impact. Whilst the interaction of two such particles through impact can be readily analyzed by using the preceding analysis, we may be faced with the much more difficult problem of determining the dynamics following a multiple collision between several particles at once. Whilst this may be thought of as a rather rare event, as parameters in a system vary, then it is almost certain to occur, and the resulting dynamics is significantly affected by it. In fact, in a multiple impact surface system such as Example 6.5, a triple collision represents a trajectory hitting the intersection point between two different impacting surfaces, Σ_{ij} and Σ_{ik}. Thus, a scenario where a limit cycle crosses such an intersection point would represent a codimension-one bifurcation in the system.

In this section, we briefly consider the possible richness of the dynamics that results from the triple collision among three objects. To do this we look at the case of a massive wall, that moves smoothly and periodically, and interacts through impact with two particles that are constrained to move in a single coordinate direction. The positions of the wall, and the two particles are given, respectively, by $w(t)$, $z(t)$ and $u(t)$, where we assume that $w(t) \leq z(t) \leq u(t)$.

The dynamics following a simultaneous collision with $w = z = u$ is not well understood, but we show through the numerical studies that if the particle at z is massive compared with the particle at u, that periodic samples of $u(t)$ behave in a similar manner to the iterations of the discontinuous, piecewise-smooth maps described in Chapter 4.

We suppose that the massive wall is at the position

$$w(t) = \kappa + \sin(\omega_0 t), \tag{6.70}$$

where $\kappa, \omega_0 \in \mathbb{R}$. We will also suppose that the motion of the particles at $z(t)$ and $u(t)$ are governed by the differential equations

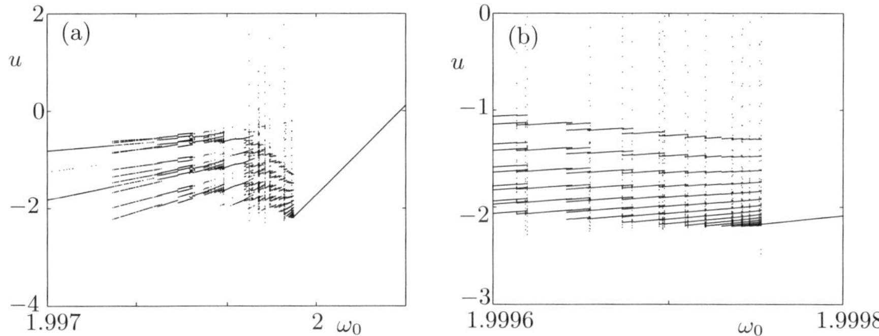

Fig. 6.22. Bifurcation diagrams under ω_0 variations, where (b) is a zoom-in of (a) close to a corner bifurcation. Here $\omega_1 = 1.001$, $\omega_2 = 1$, $\kappa = 0$ and $r_1 = r_2 = 0.8$.

$$\frac{d^2 z}{dt^2} + \omega_1^2 z = 0 \quad \text{and} \quad \frac{d^2 u}{dt^2} + \omega_2^2 u = 0, \quad \text{for} \quad w < z < u. \tag{6.71}$$

Furthermore, let the mass of the first particle be M and the mass of second particle m, and let the mass ratio between the two particles be

$$\mu = \frac{M}{m}.$$

An impact between the two free particles occurs at a time t if $z(t) = u(t)$. Let the positions of the particles before and after the impact be $z^{\pm}$ and $u^{\pm}$ and the velocities $y^{\pm}$ and $v^{\pm}$, respectively. A reasonable model for the impact is that the combined momentum of the system is conserved but that the relative velocity is reversed and reduced by a factor r so that

$$u^+ = u^- = z^+ = z^-, \quad My^- + mv^- = My^+ + mv^+,$$

and

$$\left(v^{+}-y^{+}\right)=-r\left(v^{-}-y^{-}\right).$$

Solving this system we have

$$\begin{pmatrix} z^+ \\ u^+ \\ y^+ \\ v^+ \end{pmatrix} = \begin{pmatrix} 1 & 0 & 0 & 0 \\ 0 & 1 & 0 & 0 \\ 0 & 0 & \frac{\mu-r}{1+\mu} & \frac{1+r}{1+\mu} \\ 0 & 0 & \frac{\mu(1+r)}{1+\mu} & \frac{1-r\mu}{1+\mu} \end{pmatrix} \begin{pmatrix} z^- \\ u^- \\ y^- \\ v^- \end{pmatrix}. \tag{6.72}$$

In the limit of $\mu \to \infty$ this reduces to the usual restitution impact law.

Interesting dynamics occurs when, as a parameter varies, an impact of u with z coincides at a time $t = t_k$ with an impact between z and w (with a *triple impact*). If $\omega_1 = \omega_2 = 1$, then when z is very massive compared to u, such a triple impact occurs when $\omega_0 = 2$ and $\kappa = 0$. We show in Fig. 6.22 the dynamics observed at time intervals of $T = 2\pi/\omega_0$ when $\mu = 1000$, $r = 0.8$, $\omega_1 = 1.001$, $\omega_2 = 1$, keeping $\kappa = 0$ fixed and varying ω_0. It is clear from this figure that the dynamics is relatively simple if $\omega_0 > 1.99975$ but becomes much more complex if $\omega_0 < 1.99975$. Indeed, we see the creation of a series of periodic orbits, with a period-adding structure. Thus, the existence of a triple impact implies the creation of a large number of new periodic orbits. This calculation clearly demonstrates the rich complexities of the dynamics likely to be observed when studying *systems* of impact oscillators. Note that the structure of the bifurcation diagram is different from that observed following a grazing bifurcation. In fact it is remarkably similar to that observed in Fig. 4.4 for the piecewise-linear map with a jump discontinuity described in Chapter 4. In the study [46] by Budd and Piiroinen the link between such triple impacts and piecewise-linear discontinuous maps is made more precise.

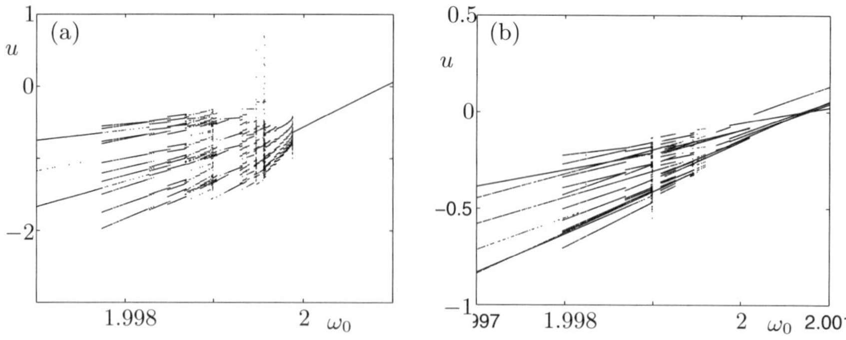

Fig. 6.23. Bifurcation diagrams under ω_0 variations, for (a) $\mu = 100$ (b) $\mu = 10$. In both cases $\omega_1 = 1.001$, $\omega_2 = 1$, $\kappa = 0$ and $r_1 = r_2 = 0.8$.

In comparison, we plot in Fig. 6.23 the same scenario as in Fig. 6.22 but now with $\mu = 100$ and $\mu = 10$, respectively. In the first case ($\mu = 100$)

we find that the bifurcation scenario has not changed significantly from $\mu = 1000$. However, in the second case ($\mu = 10$) the bifurcation diagram is clearly different and the structure that was visible before cannot be located. The number of attractors is higher for $\omega_0 > 2$ than before that could explain why the sudden jump to high-periodic orbits cannot be seen. In conclusion, this brief discussion shows the additional complexity that the increase in the number of impacting objects gives to the dynamical behavior, and how this depends on the mass ratio of the particles as well as many other factors.

The next chapter considers a different generalization of the analysis in this chapter: not in regard to multiple impacts, but to the case where the impact law is replaced by smooth dynamic evolution. That is, trajectories are allowed to *flow* on the far side S^- of a discontinuity manifold, Σ.

7

Limit cycle bifurcations in piecewise-smooth flows

This chapter concerns the analysis of the discontinuity-induced bifurcations (DIBs) that arise in piecewise-smooth flows in the case where no sliding occurs. Two specific DIBs are treated: in Sec. 7.2 the dynamics that arise when a trajectory (typically a limit cycle) becomes tangent to (grazes) a single discontinuity boundary Σ; and, in Sec. 7.3, when a trajectory passes transversally through an intersection between two boundaries Σ_1 and Σ_2. In each case, local discontinuity maps, valid close to such degenerate trajectories, are derived using similar methods to those used in the previous chapter. The resulting maps can be analyzed using the theory presented in Chapters 3 and 4 in order to classify the ensuing dynamics. In each case, the theory is applied to both explicitly constructed examples and those arising in applications; including the bilinear oscillator, and DC–DC converter case studies from Chapter 1, and a more complicated stick-slip friction oscillator model.

7.1 Definitions and examples

Recall Definition 2.20 of a general n-dimensional piecewise-smooth flow from Chapter 2. This consists of a set of autonomous smooth systems of ODEs

$$\dot{x} = F_i(x, \mu), \quad \text{if} \quad x \in S_i,$$

where $\cup_i S_i = \mathcal{D} \subset \mathbb{R}^n$ and each S_i has a non-empty interior. The flow corresponding to each F_i is written as $\Phi_i(x, t)$, is assumed to be smooth, and is defined in a full neighborhood of each boundary $\Sigma_{ij} = \overline{S}_i \cap \overline{S}_j$.

In this chapter we shall be interested in DIBs that involve at most two discontinuity boundaries. So, for ease of notation, we shall exclusively consider systems defined locally in a neighborhood of the appropriate number of boundaries. Specifically, in Sec. 7.2 we shall consider systems with a single discontinuity boundary $\Sigma := \{x \in \mathcal{D} : H(x) = 0\}$ for some smooth function H, for which the complete system can be written in the form

$$\dot{x} = f(x,\mu) = \begin{cases} F_1(x,\mu), & \text{if } H(x) > 0, \\ F_2(x,\mu), & \text{if } H(x) < 0. \end{cases} \tag{7.1}$$

Note that it is without loss of generality that we assume that H is independent of the parameter μ, since a change of co-ordinates can incorporate parameter variation into the functions F_1 and F_2. In contrast, Sec. 7.3 concerns a neighborhood of two transversally intersecting boundaries $\Sigma_1 := \{x \in \mathcal{D} : H_1(x) = 0\}$ and $\Sigma_2 := \{x \in \mathcal{D} : H_2(x) = 0\}$, for separate smooth functions H_1 and H_2, for which the system can be written in the form

$$\dot{x} = f(x,\mu) = \begin{cases} F_1(x,\mu), & \text{if } H_1(x) < 0 \text{ and } H_2(x) < 0, \\ F_2(x,\mu), & \text{if } H_1(x) < 0 \text{ and } H_2(x) > 0, \\ F_3(x,\mu), & \text{if } H_1(x) > 0 \text{ and } H_2(x) < 0, \\ F_4(x,\mu), & \text{if } H_1(x) > 0 \text{ and } H_2(x) > 0. \end{cases} \tag{7.2}$$

A key feature of these systems is that the form of the derived (Poincaré) maps close to grazing bifurcations is strongly influenced by the *degree of smoothness* m at a point x in the discontinuity boundary. Recall from Definition 2.21 that this degree is the order of the first non-zero partial derivative with respect to t of the difference between the flows $\Phi_i(x,t) - \Phi_j(x,t)$ evaluated at the time t at which the flow intersects Σ. As we might expect, the higher the degree of smoothness, the smoother the derived map associated with grazing at such a point x. If the degree of smoothness is independent of x, or at least is constant in a neighborhood of x, then the boundary Σ is *uniformly discontinuous* (Definition 2.22). Recall that in such a case [written in terms of the local description (7.1)] the vector field on one side of the boundary, F_2 say, can be written in terms of the vector field on the far side via

$$F_2(x,\mu) = F_1(x,\mu) + J(x,\mu)H(x)^{m-1}, \tag{7.3}$$

where J, F_1 and F_2 are all sufficiently smooth in a neighborhood of x, and m is the degree of smoothness

We shall assume in this chapter that no *sliding flow* takes place close to the points considered, so that trajectories do not evolve within the set Σ. In particular, we will be interested in systems that do not have sliding trajectories close to points where a flow grazes the discontinuity boundary Σ. One assumption that ensures this is that the degree of smoothness across Σ is two or more so that if $x^* \in \Sigma_{ij}$, then $F_i(x^*) = F_j(x^*)$ and any grazing of the flow Φ_i with Σ implies simultaneous grazing of Φ_j. Generally speaking, the onset of grazing in systems with $F_i(x^*) \neq F_j(x^*)$ can lead to sliding nearby, because grazing of the flow Φ_i does not necessarily imply grazing of the flow Φ_j. Tangency with Σ and sliding in such systems will form the topic of the next chapter. However, in certain examples a special structure of the system can mean that sliding can be prevented in piecewise-smooth discontinuous systems (with degree of smoothness one).

Let us further motivate the analysis given in this chapter by considering some typical examples of piecewise-smooth flows.

Example 7.1 (The bilinear oscillator). In case study II in the Introduction we introduced the bilinear oscillator, which is a simple canonical example of a piecewise-smooth flow. To continue with our study of such systems, we include the possibility of a variable stiffness k_i, damping coefficient ζ_i, offset a_i and forcing amplitude b_i, to give the second-order oscillator

$$\frac{d^2u}{dt^2} + 2\zeta_i \frac{du}{dt} + k_i^2 u = a_i + b_i \cos(\omega t), \tag{7.4}$$

where $i = 1, 2$ with $i = 1$ if $u > 0$ and $i = 2$ if $u < 0$. This can be written as the first-order system

$$\begin{aligned} \dot{x}_1 &= x_2, \\ \dot{x}_2 &= -2\zeta_i x_2 - k_i^2 x_1 + a_i + b_i \cos(\omega x_3), \\ \dot{x}_3 &= 1, \end{aligned}$$

in which $x = (x_1, x_2, x_3) = (u, du/dt, t)$. In the context of the above definitions $S_1 = \{x \in \mathbb{R}^3 : x_1 > 0\}$, $S_2 = \{x \in \mathbb{R}^3 : x_1 < 0\}$ and Σ is the set $\{x \in \mathbb{R}^3 : H(x) = 0\}$, where $H(x) = x_1$. The set of grazing points where flows intersect Σ tangentially is given by the line $G = \{x \in \mathbb{R}^3 : x_1 = x_2 = 0\}$.

As in Chapter 6, we are interested in the behavior of the flows of piecewise-smooth systems close to a grazing trajectory. Suppose such a trajectory grazes with Σ at the point $x^* = (0, 0, x_3^*)$. Then locally to this point we can write to leading-order that

$$\dot{x} = \begin{cases} A_1 x + B_1, & \text{if } H(x) = C^T x > 0, \\ A_2 x + B_2, & \text{if } H(x) = C^T x < 0, \end{cases} \tag{7.5}$$

where

$$\begin{aligned} A_1 &= \begin{pmatrix} 0 & 1 & 0 \\ -k_1^2 & -2\zeta_1 & 0 \\ 0 & 0 & 0 \end{pmatrix}, \quad B_1 = \begin{pmatrix} 0 \\ a_1 + b_1 \cos(\omega x_3^*) \\ 1 \end{pmatrix}, \\ A_2 &= \begin{pmatrix} 0 & 1 & 0 \\ -k_2^2 & -2\zeta_2 & 0 \\ 0 & 0 & 0 \end{pmatrix}, \quad B_2 = \begin{pmatrix} 0 \\ a_2 + b_2 \cos(\omega x_3^*) \\ 1 \end{pmatrix}, \\ C^T &= \begin{pmatrix} 1 & 0 & 0 \end{pmatrix}. \end{aligned} \tag{7.6}$$

In general, the degree of smoothness is *one* across Σ since $B_1 \neq B_2$. However, depending on which coefficients are equal across Σ we can have differing degrees of smoothness. Also under quite general conditions we do not get sliding motion, as we now show.

As a first example, let $k_1 = k_2 = k$, $a_1 = a_2 = a$, and $b_1 = b_2 = b$ and suppose only the damping coefficient varies across the boundary. At a grazing point $x = (0, 0, x_3^*)$, we then have

$$\frac{dF_i^*}{dx} = \begin{pmatrix} 0 & 1 & 0 \\ -k^2 & -2\zeta_i & -b\omega\sin(\omega x_3^*) \\ 0 & 0 & 0 \end{pmatrix} \qquad F_i^* = \begin{pmatrix} 0 \\ a + b\cos(\omega(x_3^*)) \\ 1 \end{pmatrix},$$

where a superscript $*$ is always used, as in the previous chapter, to represent a quantity evaluated at a grazing point. Hence along the grazing set $G \subset \Sigma$ the degree of smoothness is *two* at the point, since the lowest-order jump is in the first derivative of the vector field. Thus, there is no sliding in this case, because the flows Φ_1 and Φ_2 always cross Σ in the same sense of direction. However, for a general point $x \in \Sigma$ with $x_2 \neq 0$ the degree of smoothness is one, since

$$F_i\big|_{x_1=0} = \begin{pmatrix} x_2 \\ -2\zeta_i x_2 + a + b\cos(\omega x_3) \\ 1 \end{pmatrix},$$

This example of non-uniform discontinuity is illustrated in Fig. 7.1(a).

Alternatively, consider the case where $\widehat{b}_1 \neq \widehat{b}_2$, where $\widehat{b}_i = a_i + b_i\cos(x_3^*)$. Here, in addition to being discontinuous at all other points in Σ, the vector field itself is also discontinuous at the grazing point, since

$$F_i^* = \begin{pmatrix} 0 \\ \widehat{b}_i \\ 1 \end{pmatrix}.$$

So, the degree of smoothness is uniformly one in this case. However, even in this instance, there is no sliding close to the grazing point. This is because the grazing line of the two vector fields F_1 and F_2 coincide, and for both flows $G := \{x \in R^3 : x_1 = x_2 = 0\}$. Moreover if $\widehat{b}_1$ and $\widehat{b}_2$ are both positive, then the acceleration

$$\frac{\partial^2 H}{\partial t^2}(\Phi_i(x^*, t)) := \mathcal{L}_{F_i}^2 H(x^*) = \widehat{b}_i$$

has the same sign of direction for both vector fields. Hence, in addition to both flows sharing the same grazing set on Σ, trajectories of both flows cross Σ in the same sense of direction near the grazing set; see Fig. 7.1(b).

Finally, consider the case where $\widehat{b}_1 = \widehat{b}_2$, $\zeta_1 = \zeta_2$ but $k_1 \neq k_2$; see Fig. 7.1(c). Here the degree of smoothness is uniformly two, since for any point x_0 in Σ, $F_1(x_0) = F_2(x_0)$ but

$$\frac{dF_2}{dx} - \frac{dF_2}{dx} = \begin{pmatrix} 0 & 0 & 0 \\ -k_2^2 - k_1^2 & 0 & 0 \\ 0 & 0 & 0 \end{pmatrix}.$$

Note in this case that we can write the system in the form (7.3)

$$F_2(x) = F_1(x) + J(x)H(x) \quad \text{with} \quad J(x) = (0, k_2^2 - k_1^2, 0)^T.$$

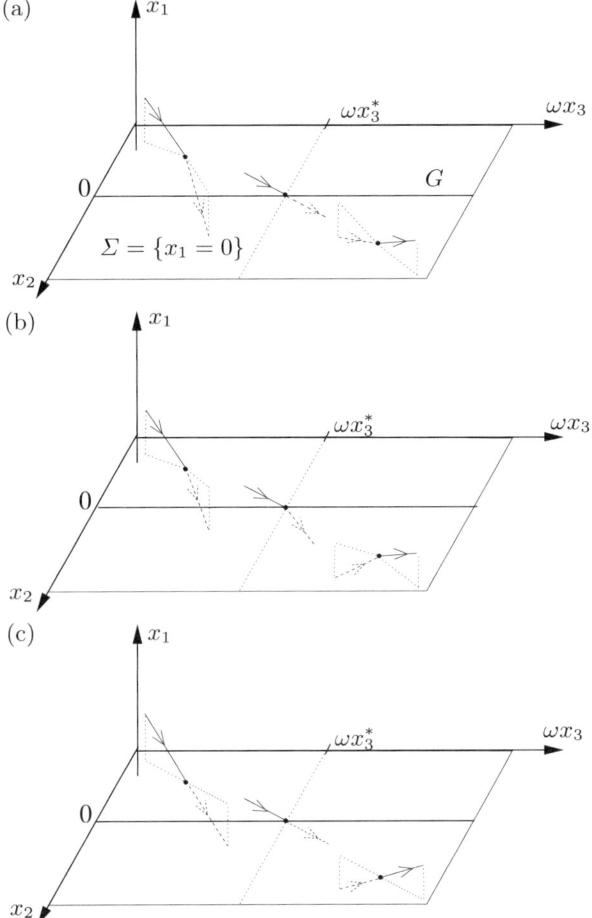

Fig. 7.1. Illustrating the continuity properties of the vector fields F_1 (solid line, above Σ) and F_2 (dashed line, below Σ) for the bilinear oscillator in the three cases: (a) $\widehat{b}_1 = \widehat{b}_2$, $\zeta_1 \neq \zeta_2$; (b) $\widehat{b}_1 \neq \widehat{b}_2$; (c) $\widehat{b}_1 = \widehat{b}_2$, $\zeta_1 = \zeta_2$, $k_1 \neq k_2$. Here $x_3^* = t^*$ is the time at which grazing occurs. Dotted lines depict the vertical projection of the vector fields onto Σ.

Generalizing from this example, we require a general sufficient condition that can apply to non-uniformly discontinuous systems, which ensures that no sliding takes place close to a grazing point. Specifically, upon writing the system in the local form (7.1), we require that under the flow the boundary $\{H = 0\}$ should never be simultaneously attracting (or repelling) from both sides. This is given by the condition

$$\mathcal{L}_{F_1} H(x) \mathcal{L}_{F_2} H(x) \geq 0 \quad \forall x \in \Sigma. \tag{7.7}$$

In particular, such a condition implies that if $\mathcal{L}_{F_1} H(x)$ smoothly goes through zero as we vary x along some curve in Σ, then $\mathcal{L}_{F_2} H(x)$ must also go through zero. Hence the grazing sets $G_i = \{x \in \Sigma : \mathcal{L}_{F_i} H(x) = 0\}$ for the two flows must coincide and also the sets $\Pi_N^{(i)} = \{x \in \mathcal{D} : \mathcal{L}_{F_i} H(x) = 0\}$.

Example 7.2 (The rocking block). An important problem related to the behavior of buildings in an earthquake is that of the motion of a block rocking on an oscillating table. The block takes the form illustrated in Fig. 7.2 and

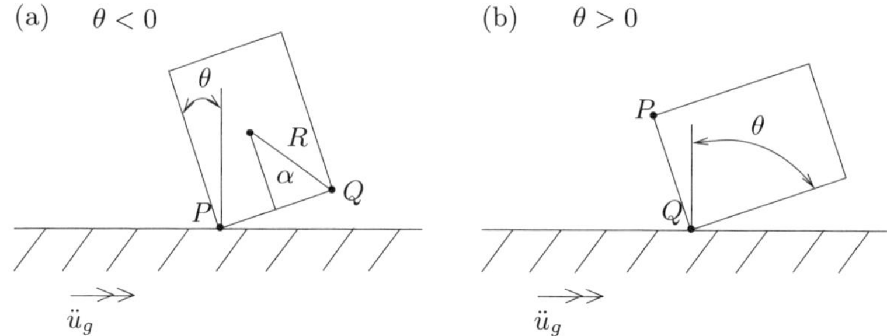

Fig. 7.2. Schematic diagram of the dynamics of a rocking block.

typically can either teeter on a single corner, or wobble on both of its corners alternately. Modeling the transition from rocking on the corner P to rocking on the corner Q as an instantaneous change leads to a piecewise-smooth model for this system. In particular, if θ gives the angle of the block to the vertical position, then it rocks on corner P if $\theta < 0$ and on corner Q if $\theta > 0$. Let α be the (fixed) angle between the side of the block and the center of mass and R the length of the line from the corner to the center of mass, m the mass and I the moment of intertia. If we assume that the system is undamped and has a horizontal ground acceleration of $\ddot{u}_g$, the equations of motion take the following form [12, 133]:

$$I\ddot{\theta} + mgR\sin(-\alpha - \theta) = -m\ddot{u}_g R\cos(-\alpha - \theta), \quad \text{if} \quad \theta < 0,$$

$$I\ddot{\theta} + mgR\sin(\alpha - \theta) = -m\ddot{u}_g R\cos(\alpha - \theta), \quad \text{if} \quad \theta > 0.$$

In terms of the usual notation, this system has $S_1 = \{\theta < 0\}$, $S_2 = \{\theta > 0\}$ and $\Sigma = \{\theta = 0\}$. The rocking block system that has received the most attention in the literature is that for which α and θ are relatively small, so that we can use linear approximations for the various trigonometric functions. By using various appropriate rescalings and taking a sinusoidal ground acceleration with u_g proportional to $\sin(\omega t + \phi)$, the linearized system becomes

$$\ddot{\theta} - \theta = \beta\sin(\omega t + \phi) + \alpha \quad \text{if} \quad \theta < 0,$$

$$\ddot{\theta} - \theta = \beta \sin(\omega t + \phi) - \alpha \quad \text{if} \quad \theta > 0.$$

This system is thus an example of a bilinear oscillator with a forcing term that is discontinuous across Σ; that is, it has degree of smoothness equal to one. The dynamics of the rocking block described for example in the work of Hogan [133, 134] include a wide variety of periodic motions, characterized by differing numbers of rocks with $\theta > 0$ and with $\theta < 0$, together with chaotic motions.

Example 7.3 (The DC–DC buck converter). We return to case study V from the Introduction. Recall that the equations of state of the converter can be written in terms of the output voltage $V(t)$ and a corresponding current $I(t)$, as

$$\dot{V} = -\frac{1}{RC}V + \frac{I}{C} \tag{7.8}$$

$$\dot{I} = -\frac{V}{L} + \begin{cases} 0 & \text{if } V \geq V_r(t) \\ E/L & \text{if } V < V_r(t) \end{cases}, \tag{7.9}$$

where V_r is the *ramp* signal

$$V_r(t) = \gamma + \eta t (\bmod T) . \tag{7.10}$$

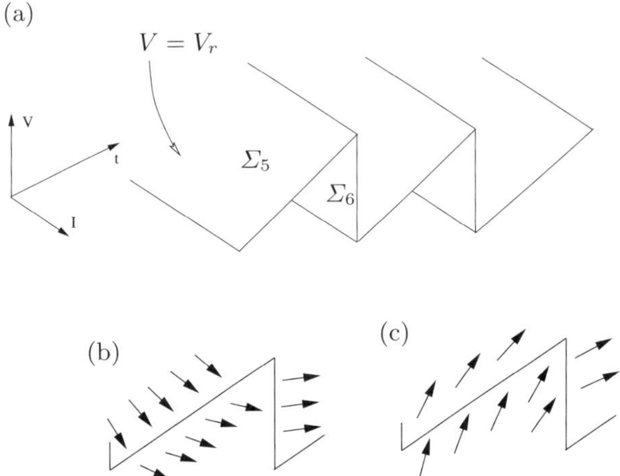

Fig. 7.3. (a) Qualitative figure illustrating the discontinuity surfaces Σ_5 and Σ_6 of the DC–DC buck converter (7.8)-(7.10) and (b), (c) the changes in the dynamics that arises when the flow crosses them in two different ways for $I < C\eta + V_r/R$ and $I > C\eta + V_r/R$, respectively.

Now, if we include time t as a third dynamic variable, then we can consider the model (7.8)–(7.10) as evolving in a three-dimensional phase space, with two discontinuity surfaces

$$\Sigma_5 := \{V = V_r(t), \quad t \neq 0 \bmod T\}, \quad \Sigma_6 := \{t = 0 \bmod T\},$$

which meet at a corner; see Fig. 7.3(a). Note that the vector field is discontinuous across Σ_5 since a constant term E/L is added to or subtracted from $\dot{I}$ as the surface is crossed in the direction of decreasing or increasing V [see Fig. 7.3(b) and Fig. 7.3(c) respectively]. Hence the degree of smoothness is one. However, note that it is not possible for the flow to be tangent to Σ_5 for the parameter values chosen in this case study

$$\begin{aligned} &R = 22\Omega,\ C = 4.7\mu F,\ L = 20mH,\ T = 400\mu s, \\ &\gamma = 11.75238V,\ \eta = 1309.524Vs^{-1}, \quad E \in [20, 40], \end{aligned} \tag{7.11}$$

except possibly for $t = 0 \bmod T$. In order to see this, note that the parameters (7.11) satisfy the inequality

$$ER > L\eta(\gamma + \eta T)R. \tag{7.12}$$

Let a subscript 1 represent flow for $V \geq V_r(t)$ and a subscript 2 represent flow for $V < V_r(t)$. Now suppose that we have a flow that is tangent to the ramp signal Σ_5 for a time $t^* \neq 0 \bmod T$. Then, $V = V_r$ and $\dot{V} = \eta$. Substitution of these values into (7.10), using the inequality (7.12), shows that $\ddot{V}_1(t^*) > 0$, whereas $\ddot{V}_2(t^*) < 0$. Thus these accelerations are of the wrong sign for a trajectory to reach the ramp and to become tangent to it.

The only possibility for motion tangent to Σ_5 is if a trajectory approaches $V = V_r = \gamma$ at $t = 0 \bmod T$. In order to be tangent to Σ_5, we thus require $t = 0$, $V = \gamma$ and $\dot{V} = \eta$, which implies $I = C\eta + V/R$, which is a unique point in the three-dimensional phase space of the system. Note that the trajectory through this point then undergoes sliding motion along the line $V = V_r(t)$, $I = C\eta + V/R$, until $t = T$. For a distinguished trajectory, such as a limit cycle, the existence of such sliding motion will represent a bifurcation event of codimension-two, because two conditions must be imposed, one on V and one on I when $t = 0$; see Fig. 7.4(a).

Across Σ_6 the degree of smoothness is also one, provided V is in the range $[\gamma, \gamma + \eta T]$. Outside of this range, there is no change in the dynamics as we cross Σ_6. Note, though, that flow always crosses Σ_6 transversally, since time increases along trajectories. Hence we have for this model that a codimension-one grazing bifurcation of a periodic orbit with either of the discontinuity sets is not possible. However, a different kind of codimension-one DIB is possible in this system, which occurs if a distinguished trajectory crosses the point of intersection between Σ_5 and Σ_6; see Fig. 7.4(b); specifically if $t = T$ and $V(T) = V_r(T) = \gamma + \eta T$. Note that this is a codimension-one condition since only the value of V need be specified at $t = T$, whereas $I(T)$ can be

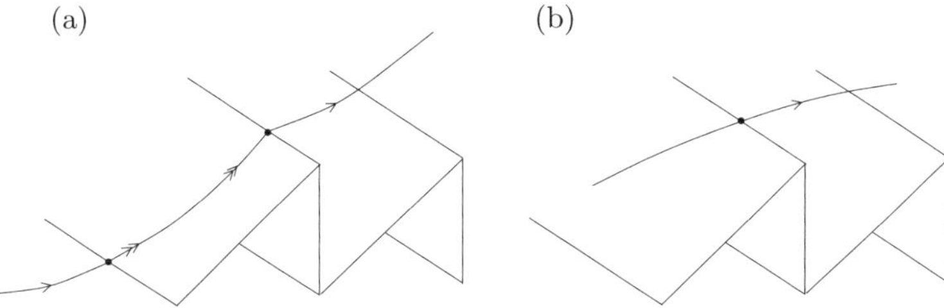

Fig. 7.4. (a) Schematic representation of a trajectory undergoing a codimension-two sliding bifurcation for the buck converter (7.8)–(7.10). The bold line represents the sliding portion of the trajectory. (b) A codimension-one corner-collision bifurcation.

arbitrary. We shall see in Sec. 7.3 below that complex dynamics can result from such a *boundary intersection crossing* bifurcation in the case where the distinguished trajectory is a limit cycle. This DIB can also be called a *corner-collision bifurcation* in this case, since we could view the intersection of Σ_5 and Σ_6 as providing a sharp corner in the boundary between two regions $S_1 : \{V > V_r(t)\}$ and $S_2 : \{V \leq V_r(t)\}$ in which smooth dynamics apply.

Example 7.4 (The Chua circuit). Another piecewise-smooth electronic system, which has historical importance in the development of dynamical systems theory, is the Chua electronic circuit, depicted in Fig. 7.5 and extensively documented in the book [185]. This electronic circuit has proved to be a major stim-

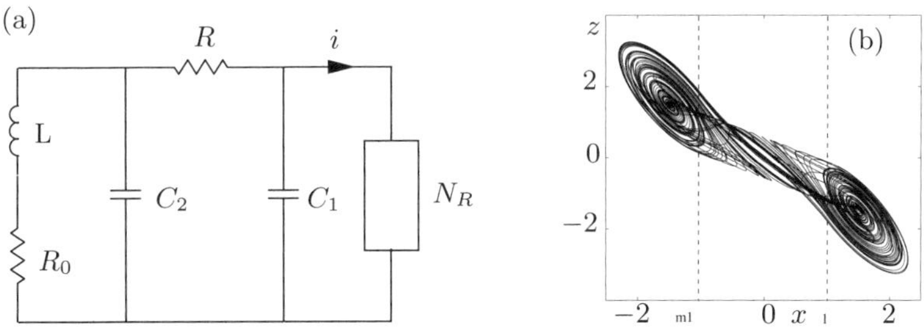

Fig. 7.5. The Chua circuit: (a) a schematic circuit diagram, and (b) the 'double scroll' chaotic attractor at parameter values $\alpha = 9.85$, $\beta = 14.3$, $m_0 = -1/7$, and $m_2 = 2/7$. Here C_1 and C_2 are capacitors, R_0 and R are resistors and L is an inductor. The component N_R represents non linear resistance of the circuit, and it can be written in dimensionless form via (7.13) below.

ulus for the development of the theory of certain types of dynamical systems. A Chua circuit comprises two capacitors, an inductor and a nonlinear resistor for which the current-voltage response closely approximates a piecewise-linear function. The Chua circuit is known to exhibit chaotic behavior under certain operating conditions, arising from a period-doubling cascade. The dynamics of the Chua circuit can be shown to be governed by the following dimensionless system of equations:

$$\begin{aligned}\dot{x} &= \alpha(y - h(x)),\\ \dot{y} &= x - y + z,\\ \dot{z} &= -\beta y.\end{aligned}$$

Here α, and β are positive parameters, x and y measure the nondimensionalized voltage of the capacitors C_1, and C_2 respectively, and z is the current through the inductance L. Finally,

$$h(x) = m_1 x + \frac{m_0 - m_1}{2}(|x+1| - |x-1|) \tag{7.13}$$

represents the characteristics of the nonlinear resistor. Hence, the Chua circuit closely resembles a bilinear oscillator and has a degree of smoothness equal to two across either of the two non-intersecting discontinuity boundaries $\Sigma_1 = \{x = 1\}$ and $\Sigma_2 = \{x = -1\}$. As we shall see from the analysis in Sec. 7.2 below, any grazing bifurcation with Σ_1 and Σ_2 will not lead to an immediate change in the attractor. Indeed, the occurrence of chaotic dynamics in this example can be attributed to a sequence of smooth bifurcations, although the shape of the chaotic attractor depicted in Fig. 7.5(b) is clearly influenced by the presence of the discontinuity boundaries; the attractor appears to fold about the planes $\{x = \pm 1\}$. Also, since Σ_1 and Σ_2 do not intersect, there is no possibility of the boundary intersection crossing bifurcation that we will analyze in Sec. 7.3.

Example 7.5 (A stick-slip oscillator). We conclude our examples with an autonomous friction oscillator model. Rather than modeling dry friction with discontinuous Coulomb type laws as we did in case study IV (a study we shall return to in Chapter 8), such models take account of the micro-dynamics of friction. Specifically, Dankowicz & Nordmark [64] studied the following five-dimensional model of a friction oscillator:

$$\begin{aligned}\dot{y}_1 &= y_2,\\ \dot{y}_2 &= -1 + \left[1 - \gamma U|1 - y_4|y_2 + \beta U^2(1 - y_4)^2\sqrt{K}(y_1)\right]e^{y_1 - d},\\ \dot{y}_3 &= y_4,\\ \dot{y}_4 &= -sy_3 + \frac{\sqrt{g\sigma}}{U}e^{-d}\left[\mu(y_5 e^{-y_1} - 1) + \alpha U^2 S(y_1, y_4)\right],\\ \dot{y}_5 &= \frac{1}{\tau}[(1 - y_4) - |1 - y_4|y_5],\end{aligned} \tag{7.14}$$

where

$$K(y_1) = 1 - \frac{y_1 - d}{\Delta}, \qquad S(y_1, y_4) = (1 - y_4)|1 - y_4|K(y_1)e^{-y_1} - 1 + \frac{d}{\Delta}. \quad (7.15)$$

This dimensionless model is based on the derivation in [63], which aims to explain experimentally observed stick-slip motion using more realistic laws than simple Coulomb friction. Here the variable y_1 is a vertical and y_3 a horizontal degree of freedom of a mass m being pulled across a horizontal surface by a spring of stiffness k whose other end moves at constant speed U. The corresponding velocities are y_2 and y_4. The extra co-ordinate $y_5 \in [-1, 1]$ is an internal variable measuring the shear deformation between the surface and the mass. Here g is the acceleration due to gravity, μ is the equivalent to a static friction coefficient, σ measures the roughness of the surface and s, τ are dimensionless constants given by

$$s = \frac{k\sigma}{mg}, \qquad \tau = \frac{\mu\sigma}{U}\sqrt{\frac{g}{\sigma}}. \quad (7.16)$$

The constants α, β, γ and d are parameters that describe the dynamic friction law, with d representing the horizontal displacement beyond which there is no longer a normal component to the friction force; see [63, 64] for more details. Finally, Δ defines the equilibrium vertical displacement, which is given by the formula

$$\beta U^2 = \frac{e^{\Delta} - 1}{\sqrt{1 - \frac{\Delta}{d}}}. \quad (7.17)$$

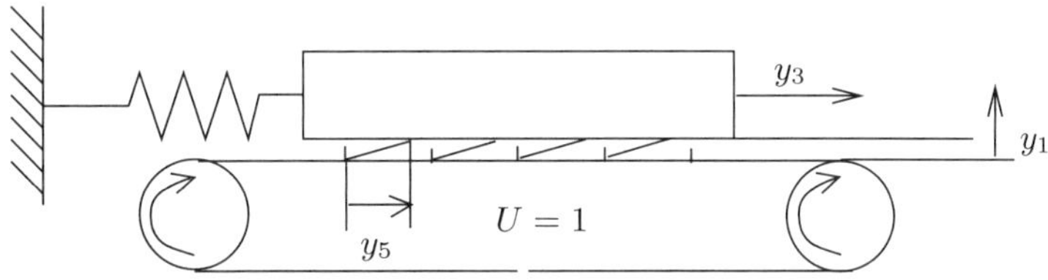

Fig. 7.6. Schematic diagram of the autonomous friction oscillator model (7.14).

The dynamics is such that the mass is subject to a constant downwards gravitational force (scaled to be -1 in these dimensionless co-ordinates) and a vertical repulsion/adhesion force that depends on the horizontal velocity y_4, which keeps the mass at a small fixed vertical height d from the friction surface. The horizontal velocity variable y_4 is scaled by U and is measured in a frame that moves with the moving end of the spring so that $y_4 = 1$ corresponds to the sticking motion where the mass is stationary. If $y_4 < 1$, then slipping takes place; that is, the mass is being dragged across the surface. Thus, the discontinuity in this system occurs as the dynamics crosses $\{y_4 = 1\}$.

Now let

$$\Sigma = \{x \in \mathbb{R}^5 : H(x) = 0\}, \quad \text{where} \quad x = (y_1, \dots y_5)^T, \quad H(x) = 1 - y_4$$

and the two regions of smooth dynamics be given by

$$S_1 = \{x \in \mathbb{R}^5 : H(x) > 0\}, \quad S_2 = \{x \in \mathbb{R}^5 : H(x) < 0\}.$$

Let F_1 represent the vector field (7.14) in region S_1 and F_2 the vector field in region S_2. Then we have

$$F_2(x) - F_1(x) = H(x) \begin{pmatrix} 0 \\ -2\gamma U y_2 e^{y_1 - d} \\ 0 \\ -2\frac{\sqrt{g\sigma}}{U} e^{-y_1 - d} \alpha U^2 H(x) K(y_1) \\ -\frac{1}{\tau} y_5 \end{pmatrix}, \tag{7.18}$$

from which we see that $F_2 - F_1$ vanishes when $H(x) = 0$. Thus the degree of smoothness is more than one. In fact, it is easy to see that this system has degree of smoothness two at all points on Σ except when $y_2 = y_5 = 0$, where the degree of smoothness is at least three. Specifically we can write the system in the form (7.3) with $m = 2$, and $F_2(x) = F_1(x) + J(x)H(x)$, where $J(x)$ is the smooth vector function appearing in the bracket in (7.18).

Having finished with these motivating examples, we are now ready to analyze the two discontinuity-induced bifurcations that form the heart of this chapter.

7.2 Grazing with a smooth boundary

Motivated by the analysis of grazing bifurcations for impacting hybrid systems presented in Chapter 6, we now ask the question of what maps are associated with the event of a limit cycle of a piecewise-smooth system grazing with a smooth boundary Σ. In Chapter 6 we saw that the analogous maps for impacting hybrid systems had square-root type singularities. The situation for piecewise-smooth systems is richer. It depends crucially upon the degree of smoothness of the system and whether or not the discontinuity is uniform or not. We start by looking at the local neighborhood of a grazing point, then state the form that the discontinuity mappings take. At this point we should recall the discussion in Sec. 2.5.3 about the differences between the zero-time and Poincaré-section discontinuity mappings (ZDM and PDM) and the experience from the previous chapter about how they can be used in practice.

The derivation of the ZDM and PDM is technical for piecewise-smooth systems and so a detailed explanation is relegated to Sec. 7.2.5. Before doing so, in Sec. 7.2.3 we derive the form that the compound Poincaré maps take in a neighborhood of a limit cycle that grazes and Sec. 7.2.4 shows how to compute the discontinuity mappings in examples: several forms of a bilinear oscillator and the above stick-friction model.

7.2.1 Geometry near a grazing point

As with grazing in impacting systems, we start with some generic hypotheses. Suppose the system in question is written locally in the form (7.1) in the neighborhood $\mathcal{D}$ of a single discontinuity boundary

$$\Sigma := \{x \in \mathcal{D} : H(x) = 0\}.$$

To begin with we shall suppress any parameter dependence and consider systems of the form

$$\dot{x} = \begin{cases} F_1(x), & \text{if } H(x) > 0, \\ F_2(x), & \text{if } H(x) < 0. \end{cases} \tag{7.19}$$

We assume that the scalar function H is well defined at the grazing point x^*:

$$H_x(x^*) \neq 0. \tag{7.20}$$

We suppose, without loss of generality, that the grazing trajectory $\Phi_1(x^*, t)$ approaches from the side $S_1 : \{x \in \mathbb{R}^3 : H(x) > 0\}$. Thus, grazing occurs at a point where the vector field F_1 is tangent to Σ. Hence we require the three conditions (see Fig. 7.7)

$$H(x^*) = 0, \tag{7.21}$$

$$v_1(x^*) = \frac{\partial}{\partial t} H(\Phi_1(x^*, 0)) = \mathcal{L}_{F_1} H(x^*) = 0, \tag{7.22}$$

$$a_1(x^*) = \frac{\partial^2}{\partial t^2} H(\Phi_1(x^*, 0)) = \mathcal{L}_{F_1}^2 H(x^*) := a_1^* > 0. \tag{7.23}$$

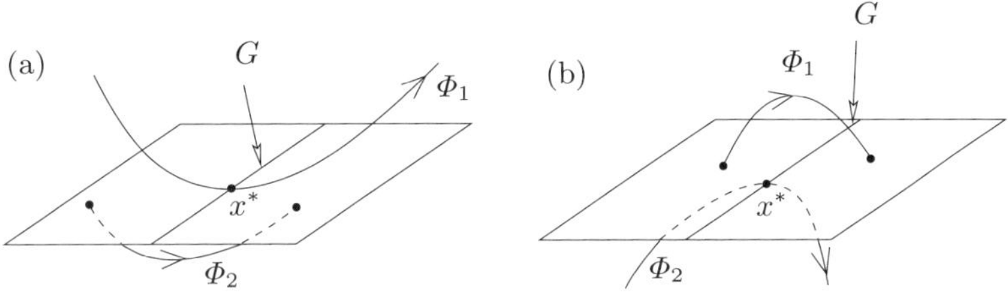

Fig. 7.7. (a) The geometry near a grazing point satisfying (7.20)–(7.23) in a system with uniform discontinuity of degree two. (b) The geometry if (7.23) is replaced by $a_1(x^*) < 0$. Here grazing occurs from the side S_2.

The first condition (7.21) states that $x^* \in \Sigma$, whereas equation (7.22) is the defining equation that the flow is tangent to Σ at x^*. Note the necessity of the sign condition (7.23) on the acceleration. The opposite sign of a^* leads to trajectories that curve toward the discontinuity surface Σ and would not correspond to a trajectory that grazes from the side S_1; see Fig. 7.7(b).

Note that if the degree of smoothness is two or more across Σ in a neighborhood of the grazing point, then (7.20)–(7.23) are sufficient to assume that $v_2(x^*) = \mathcal{L}_{F_2}H(x^*) = 0$. Also, by continuity, we find that $a_2(x^*) = \mathcal{L}^2_{F_2}H(x^*) > 0$ must be satisfied in order for trajectories of flows Φ_1 and Φ_2 to cross Σ in the same sense, see Fig. 7.7(a).

In the case where the degree of smoothness is one at some points of Σ, we shall require a condition that ensures that no sliding motion is possible near x^* (cases where sliding is possible form the subject of Chapter 8). A sufficient condition to avoid sliding was given by (7.7). This condition has several consequences. First, as already mentioned, the grazing sets $G_i = \{x \in \Sigma : \mathcal{L}_{F_i}H(x)\}$ for the two flows Φ_1 and Φ_2 must coincide (see Fig. 7.8). Moreover, if x^* satisfies (7.22), then (7.7) implicitly implies that

$$v_2(x^*) = \mathcal{L}_{F_2}H(x^*) = 0, \tag{7.24}$$

$$\text{sign}(a_2(x^*)) = \text{sign}(\mathcal{L}^2_{F_2}H(x^*)) = \text{sign}(\mathcal{L}^2_{F_2}H(x^*)) > 0. \tag{7.25}$$

Definition 7.1. *We shall refer to a point x^* satisfying (7.7), (7.20)–(7.23) as being a* **regular grazing point** *of a piecewise-smooth continuous system written in the local form (7.1).*

Note that condition (7.7) is *automatically satisfied* for systems with uniform discontinuity with smoothness of degree two or more. Grazing for such systems is therefore a true codimension-one bifurcation, with the single defining condition for the bifurcation being (7.22). The other conditions (7.21) and (7.23) are non-degeneracy conditions that ensure the bifurcation unfolds in a regular way. In systems with degree of smoothness one, condition (7.7) might be seen as defining a higher-codimension grazing bifurcation, since we require an additional equality condition to hold. Rather, we think in this chapter that (7.7) defines a class of systems we are interested in this case, as motivated by the bilinear oscillator example, in which no sliding motion can occur. It is also possible to relax the assumptions slightly and still avoid sliding solutions. For example, one case would be if the grazing sets G_1 and G_2 were not coincident, but under variation of a bifurcation parameter, the periodic orbit undergoing the grazing does not enter the sliding regions depicted in Fig. 7.8(a). We shall not pursue such more general conditions here, as our motivation is to explain the behavior observed in more general examples.

We are now in a position to state normal form results for grazing bifurcations, by constructing discontinuity maps in terms of these local co-ordinates.

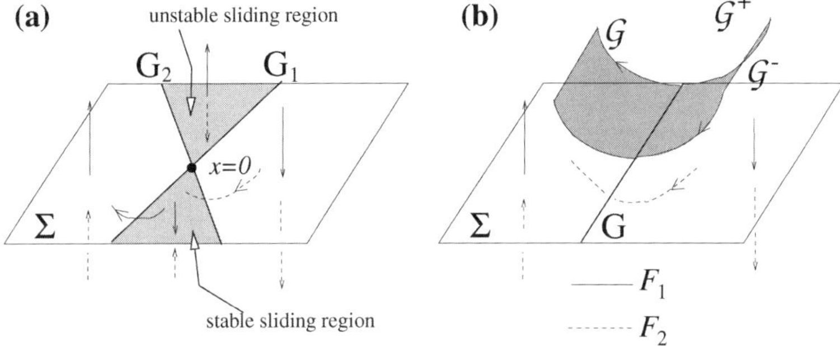

Fig. 7.8. (a) Under weaker hypotheses the two grazing sets G_1 and G_2 do not necessarily coincide and there can be a region of sliding (shown as the shaded portion of Σ). (b) The geometry near a grazing point x^* defined by (7.20)–(7.23) in systems with degree of smoothness one under the assumption (7.7), which implies that the grazing sets coincide.

7.2.2 Discontinuity mappings at grazing

To motivate, and help to define, the discontinuity maps associated with a piecewise-smooth flow without sliding, we consider the trajectories shown in Fig. 7.9, which is an adaptation of Fig. 2.30 in Chapter 2. Consider a flow in S_1 close to a trajectory that contains a regular grazing point x^*. Consider a general initial condition x_0 and the flow Φ_1 passing through x_0, which first intersects Σ at the point x_2 after a time $t = \delta_0$ (this time may be positive or negative depending on whether $x_0 \in S_1$ or S_2). The piecewise-smooth system continues in S_2 using flow Φ_2 until the trajectory hits Σ again at the point x_3, after a further time $t = \delta_2$. Now, if we continue backwards from x_3 using flow Φ_1 for a time $-(\delta_0 + \delta_2)$, we reach the point x_4. The map $x_0 \to x_4$ represents the zero-time discontinuity mapping (ZDM). Thus, we can write

$$\mathrm{ZDM} = \Phi_1(\Phi_2(\Phi_1(x_0, \delta_0), \delta_2), -(\delta_0 + \delta_2)) \tag{7.26}$$

If instead we had continued to flow forward using Φ_1 from the point x_2, we would reach the normal Poincaré surface

$$\Pi_N = \{x \in D : \mathcal{L}_{F_1} H(x) = 0\} \tag{7.27}$$

at the point x_1, where the time to go from x_1 to x_2 is $\delta < 0$. Thus, in order to define the Poincaré-section discontinuity map (PDM) we flow Φ_1 backward in time from x_1, forwards in time using Φ_2 through a time δ_2 until we hit Σ for a second time at x_3, and then finally backwards through a time $-\delta_3$ using Φ_1 until we reach Π_N again, at the point x_5. The PDM is then the map $x_1 \to x_5$:

$$\mathrm{PDM} = \Phi_1(\Phi_2(\Phi_1(x_1, \delta), \delta_2), -\delta_3). \tag{7.28}$$

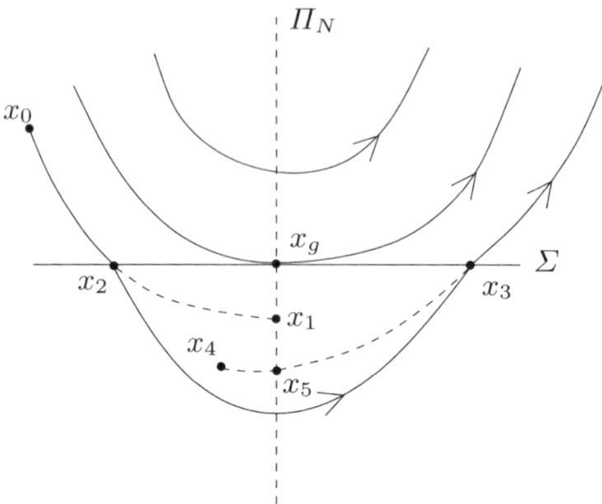

Fig. 7.9. Illustrating the construction of the ZDM and the PDM for a piecewise-smooth flow. Here, the ZDM maps x_0 to x_4 and the PDM maps x_1 to x_5. Solid lines represent the actual trajectories where the trajectory changes from being the flow Φ_1 to the flow Φ_2, and the dashed lines to the smooth extensions of the trajectories in S_1 into the region S_2 under the action of the flow Φ_1 alone.

The detailed calculation of the leading-order expressions for the ZDM and PDM can be rather cumbersome in the general case. However, the leading-order expression for the ZDM takes a particularly simple form in the case of uniform discontinuity.

Theorem 7.1 (The ZDM for surfaces with a uniform discontinuity [200]). *Let x^* be a regular grazing point of a piecewise-smooth system (7.3), (7.19) with a uniform degree of smoothness $m \geq 2$. Let x be a general point near x^* and $H_{\min}(x)$ be the minimum value of $H(x_0)$ attained along the flow $\Phi_1(x_0, t)$. Then, the zero-time discontinuity mapping for small $\|x_0 - x^*\|$ is given by*

$$x_0 \mapsto \begin{cases} x_0, & \text{if } \; H_{\min} \geq 0, \\ x_0 + e(x, y)y^{2m-1}, & \text{if } \; H_{\min} < 0, \end{cases} \tag{7.29}$$

where

$$y = \sqrt{-H_{\min}} \quad \text{for} \quad H_{\min} < 0$$

and e is a sufficiently smooth function of its arguments within $\mathcal{D}$ whose lowest order term is given by

$$e(0, 0) = 2(-1)^{m+1} I(m) J(0) \sqrt{\frac{2}{(H_x F_1)_x F_1(0)}},$$

with

$$I(m) = \int_0^1 (1 - \xi^2)^{m-1} d\xi; \qquad I(2) = \frac{2}{3}, \quad I(3) = \frac{8}{15}, \quad I(4) = \frac{16}{35}, \ldots.$$

Remarks

1. The proof of this theorem was given by Nordmark in [200], using a combination of Lie derivatives, the Implicit Function Theorem and a Picard iteration scheme.
2. The key feature of this ZDM is the existence of *fractional powers* 1/2, 3/2, 5/2, etc. of $y^2 = H_{\min}$ that arise in the correction term provided by the ZDM. As was the case for the impacting systems considered in the last chapter, these fractions arise because the grazing trajectory is locally parabolic close to Σ. Thus the time spent within S_2, scales like the square root of the penetration $-H_{\min}$.

Table 7.1. The relationship between the degree of smoothness of the system at the grazing point and the local form of the corresponding map.

degree of smoothness	**System at grazing point**	**Map singularity**	
	jump in	Uniform Case	Non-uniform
1	F	-	1/2
2	F_x	3/2	3/2
3	F_{xx}	5/2	3/2

3. Intuitively, the fact that decreasing the degree of smoothness m by 1 leads to a decrease in the fractional power by 1 can be motivated by considering the derivative of the flow. That is, if flows Φ_1 and Φ_2 lead to a ZDM with a correction term proportional to y^{2m-1}, then flows $\frac{\partial}{\partial t}\Phi_1$ and $\frac{\partial}{\partial t}\Phi_2$ have degree of smoothness one less, and will lead to a ZDM correction proportional to $\frac{d}{dx}y^{2m-1} \propto y^{2m-3}$.
4. The basic order of singularity of the leading-order term of the ZDM (and hence the PDM) are summarized in Table 7.1. This also includes the more general results for the ZDM where we do not assume uniform discontinuity, as given in the next theorem, first derived in [78].

Theorem 7.2 (The ZDM at a general grazing bifurcation). *Let x^* be a regular grazing point of a piecewise-smooth system (7.19). Then, the ZDM describing trajectories in a neighborhood of the grazing trajectory has:*

(i) a square-root singularity if $F_1^* \neq F_2^*$;
(ii) a 3/2-type singularity at the grazing point in the case where $F_1^* = F_2^*$ while $F_{1,x}^* \neq F_{2,x}^*$ or $F_{1,xx}^* \neq F_{2,xx}^*$,

where an asterisk represent terms that are evaluated at the grazing point $x = x^*$. Specific formulae for these maps are given in the two cases as follows.

(i) If the vector field is discontinuous *at the grazing point, i.e., $F_1(x^*) \neq F_2(x^*)$ we have:*

$$x \mapsto \text{ZDM} = \begin{cases} x, & \text{if } H_{\min} \geq 0, \\ v\sqrt{-H_{\min}} + O(x), & \text{if } H_{\min} < 0, \end{cases} \tag{7.30}$$

where

$$v = 2\sqrt{2}\frac{(H_x F_2)_x F_1}{(H_x F_2)_x F_2((H_x F_1)_x F_1)^{\frac{1}{2}}}(F_2 - F_1).$$

Or, in Lie derivative notation

$$v = 2\sqrt{2}\frac{\mathcal{L}_{F_1}\mathcal{L}_{F_2}(H)(x)}{\mathcal{L}^2_{F_2}(H)(x)\sqrt{\mathcal{L}^2_{F_1}(H)(x)}}(F_2 - F_1).$$

(ii) *If the vector field is* continuous *at the grazing point, i.e.,* $F_1(x^*) = F_2(x^*) := F$, *but has a discontinuous first or second derivative, then:*

$$x \mapsto \text{ZDM}(x) = \begin{cases} x, & \text{if } H_{\min} > 0, \\ x + 2\sqrt{H_{\min}}\sqrt{\dfrac{2}{(H_x F_1)_x F}}v(x) + O(|x|^2), & \text{if } H_{\min}(x) < 0, \end{cases} \tag{7.31}$$

where $v(x) = v_1 + v_2 + v_3$ *with* $v_1, v_2, v_3 \in \mathbb{R}^n$ *being each proportional to* x *and given by*

$$v_1 = -\left\{-\frac{\left((H_x F_2)_x \left(F_1 - \frac{2}{3}F_2\right)\right)_x F}{(H_x F_2)_x F}(F_2 - F_1)_x F + \left(F_{1,x}F_2 - \frac{1}{3}F_{1,x}F_1 - \frac{2}{3}F_{2,x}F_2\right)_x F\right\}\frac{H_x x}{(H_x F_1)_x F}, \tag{7.32}$$

$$v_2 = (F_2 - F_1)_x x, \tag{7.33}$$

$$v_3 = -(F_2 - F_1)_x F\frac{(H_x F_2)_x x}{(H_x F_2)_x F}, \tag{7.34}$$

$$H_{\min}(x) = H_x x + O\left(|x|^2\right). \tag{7.35}$$

The proof of Theorem 7.2 is given in Sec. 7.2.5 below.

In the general, non-uniform case it is possible to derive equivalent expressions for the PDM applied at the local normal Poincaré section Π_N that contains the grazing point. As will become apparent in the proof, note that this map may be expressed as a smooth projection, S of the ZDM map

$$\text{PDM} = S(\text{ZDM})$$

for points x that satisfy $x \in \Pi_N$, where S is the *smooth projection operator* that takes trajectories along flow lines until they hit Π_N. Note that the smooth projection that converts from the ZDM to the PDM does not change the order of the leading singularity of the map in general. A specific form for the PDM map is given by the next Theorem.

Theorem 7.3 (The PDM for a general grazing bifurcation). *Let x be a point in Π_N, the Poincaré section given by (7.27). Then, sufficiently close to a regular grazing point of a system (7.19) satisfying (7.7)–(7.23), the Poincaré-section discontinuity mapping can be written as follows:*

(i) If, at the grazing point, $F_2^* \neq F_1^*$

$$x \mapsto \begin{cases} x & \text{if } H(x) > 0, \\ v_z\sqrt{-H(x)} + O(x,\mu) & \text{if } H(x) < 0, \end{cases} \tag{7.36}$$

where

$$v_z = 2\sqrt{2}(F_2 - \frac{F_{2z}}{F_{1z}}F_1)\frac{H_xF_{2,x}F_1}{H_xF_{2,x}F_2((H_xF_1)_xF_1)^{\frac{1}{2}}}. \tag{7.37}$$

A subscript z denotes the projection along $z = (\mathcal{L}_{F_1}(H)(x))_x$, the normal vector to Π_N, so that $F_{2z} = (H_xF_1)_xF_2$, $F_{1z} = (H_xF_1)_xF_1$, etc.

(ii) If the vector field is continuous at the grazing point, i.e., $F_1^* = F_2^* := F$,

$$x \mapsto \begin{cases} x, & \text{if } H(x) \geq 0, \\ x + v_{1z}(-H(x))^{\frac{3}{2}} + v_{2z}x(-H(x))^{\frac{1}{2}} & \\ +v_{3z}H_xF_{2,x}x(-H(x))^{\frac{1}{2}} + O(x^2), & \text{if } H(x) < 0, \end{cases} \tag{7.38}$$

$$\begin{aligned} v_{1z} &= \frac{2}{(H_xF_{1,x}F)^{3/2}}\Big\{\frac{1}{3}(F_{2,xx} - F_{1,xx})F^2 + F_{2,x}F_{1,x}F - \frac{1}{3}\big[F_{1,x}^2 + 2F_{2,x}\big]F \\ &\quad -\frac{1}{H_xF_{2,x}F}(F_{2,x} - F_{1,x})F\big[\frac{1}{3}H_xF_{2,xx}F^2 + H_xF_{2,x}F_{1,x}F \\ &\quad -\frac{2}{3}H_xF_{2,xx}F^2\big]\Big\}, \\ v_{2z} &= \frac{2\sqrt{2}}{\sqrt{H_xF_{1,x}F}}(F_{2,x} - F_{1,x}), \\ v_{3z} &= \frac{2\sqrt{2}}{H_xF_{2,x}F\sqrt{H_xF_{1,x}F}}(F_{2,x} - F_{1,x})F. \end{aligned}$$

Here, v_1, v_2 and v_3 are given by (7.33) and a subscript z has the same meaning as in case (i) above.

7.2.3 Grazing bifurcations of periodic orbits

The above theory was all posed local to a grazing point, without reference to parameter variation. In order to describe a discontinuity induced *bifurcation* event, we must suppose that the grazing point x^* is part of a distinguished trajectory of a parameter dependent system with $x \in \mathbb{R}^n$ and $\mu \in \mathbb{R}$, at a certain parameter value μ^*. In particular we suppose that local co-ordinates can be chosen close to the grazing point that puts the system in the form (7.1):

$$\dot{x} = \begin{cases} F_1(x,\mu), & \text{if } H(x) > 0, \\ F_2(x,\mu), & \text{if } H(x) < 0. \end{cases}$$

Suppose we take this distinguished trajectory to be a limit cycle $p(t;\mu^*)$, which may or may not intersect other discontinuity boundaries other than $\Sigma = \{x \in \mathbb{R}^n : H(x) = 0\}$. We simply stipulate that any such intersections are transversal for parameter values μ close μ^*, except at $x = x^*$, which is a regular grazing point when $\mu = \mu^*$; see Fig. 7.10. As in Chapter 6, for grazing bifurcations in impacting systems, we shall then use the PDM map to define a *grazing bifurcation normal form*, the leading-order terms of which will capture all the recurrent dynamics in a neighborhood of the grazing orbit, for sufficiently small $\|x - x^*\|$ and $\|\mu - \mu^*\|$.

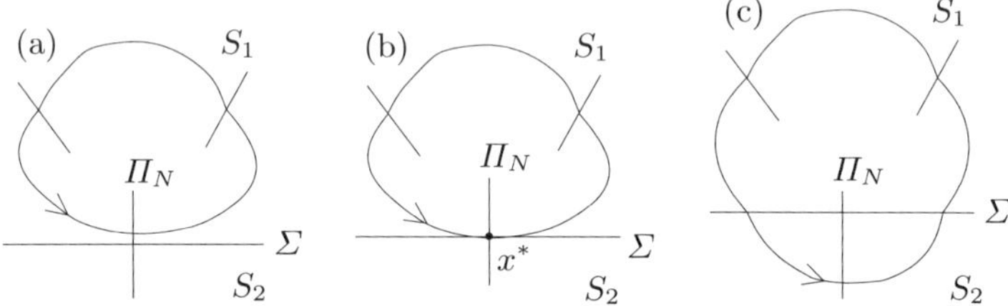

Fig. 7.10. A periodic orbit undergoing a grazing bifurcation (a) for $\mu - \mu^* < 0$ before/after the bifurcation, (b) for $\mu - \mu^* = 0$ at the bifurcation and (c) for $\mu - \mu^* > 0$ after/before the bifurcation.

The assumption of transverse boundary crossing away from the grazing point means that we can use transverse discontinuity mapping theory from Chapter 2 to define a natural Poincaré map $\tilde{P}_N$ from Π_N to itself that ignores the presence of Σ near x^*; see Fig. 7.10. That is, local to x^*, the flow map used to construct $\tilde{P}_N$ is assumed to be governed by vector field F_1 alone. That is, close to x^*, we map points using trajectories of flow Φ_1. Such a Poincaré mapwill be smooth and have a well-defined linearization

$$\tilde{P}_N(x,\mu) = N(x - x^*) + M(\mu - \mu^*) + O\left(\|x - x^*\|^2, (\mu - \mu^*)^2\right)$$

for an $n \times n$ matrix N and $1 \times n$ matrix M satisfying

$$N := \frac{\partial}{\partial x}\tilde{P}_N\bigg|_{x=x^*,\mu=\mu*} \quad \text{and} \quad M := \frac{\partial}{\partial \mu}\tilde{P}_N\bigg|_{x=x^*,\mu=\mu*}.$$

Similarly, there is a row vector $C^T = H_x$ and a scalar $D = H_\mu$ such that to leading-order

$$H(x,\mu) = C^T(x - x^*) + D(\mu - \mu^*).$$

Then we have the following description of the compound Poincaré mapthat describes the unfolding of the grazing bifurcation.

Theorem 7.4 (The normal form map at a grazing bifurcation). *Suppose a periodic orbit $p(t;\mu)$ of a piecewise-smooth system that is written in local co-ordinates in the form (7.1) has a regular grazing at $(x,\mu)=(x^*,\mu^*)$. Let $\widehat{x}=x-x^*$, $\widehat{\mu}=\mu-\mu^*$. Then the Poincaré map P_N from Π_N defined by (7.27) to itself can be written as*

$$P_N(x,\mu) = \mathrm{PDM}(\tilde{P}_N(x,\mu)), \tag{7.39}$$

In the case of degree of smoothness 1, this map takes the form

$$P_N(x,\mu) = \begin{cases} N\widehat{x} + M\widehat{\mu}, & \text{if} \quad C^T N\widehat{x} + (C^T M + D)\mu \geq 0, \\ Nq + M\widehat{\mu}, & \text{if} \quad C^T N\widehat{x} + (C^T M + D)\widehat{\mu} < 0, \end{cases}$$

where $q = v_z y$ with

$$y = \sqrt{(-C^T N\widehat{x} - (C^T M + D)\widehat{\mu})} + O(x,\mu)$$

and v_z given by (7.37) in the case of degree of smoothness one and $q = x + v_{1z} y^{\frac{3}{2}} + v_{2z} x y^{\frac{1}{2}} + v_{3z} H_x F_{2,x} x y^{\frac{1}{2}} + O(x^2)$ given by (7.38) in the case of degree of smoothness two or more.

7.2.4 Examples

We now look at some examples that illustrate the above theory.

Example 7.6 (Example 7.1 continued: grazing in the undamped, constantly forced bilinear oscillator). To illustrate the above theory, we take a special case of the bilinear oscillator, for which calculations can be carried out explicitly from first principles. In particular, we show the effect that the degree of smoothness of the flow has on the form of the discontinuity map. Specifically, consider the undamped bilinear oscillator

$$\frac{d^2u}{dt^2} + k_i^2 u = a_i, \quad v = \frac{du}{dt}, \tag{7.40}$$

with $\Sigma = \{x : u = 0\}$ and $i = 1$ if $u > 0$, $i = 2$ for $u < 0$. In the absence of damping, the motion is conservative, such that along trajectories, the energy E

$$E_i = \frac{1}{2}\left(\frac{du}{dt}\right)^2 + \frac{1}{2}k_i u_i^2 - a_i u$$

is conserved. Note that E is conserved along all trajectories, even if they intersect Σ (that is $E = E_1$ each time $x \in S_1$ and $E = E_2$ each time $x \in S_2$).

If we take the initial condition $v = du/dt = 0$ at $t = 0$, then (7.40) has the exact solution

$$u(t) = \frac{a_i}{k_i^2} + \left(u(0) - \frac{a_i}{k_i^2}\right)\cos(k_i t). \tag{7.41}$$

Now, consider the PDM for a trajectory starting on the (normal) Poincaré surface $\Pi_N = \{(u,v,t) : v = 0\}$ at the point $x_1 = (u_0, 0, 0)$ with $u_0 := -y_1^2 < 0$.

To construct the PDM we follow the procedure outlined in Fig. 7.9. That is we follow the trajectory (7.41) for a time interval $\delta < 0$ using the ODE (7.40) with $i = 1$ until it intersects Σ at $x_2 = (0, v_2, \delta)$. Next, we flow forward using the ODE with $i = 2$ through a time δ_2 until we reach the point $x_3 = (0, v_3, \delta_2 + \delta)$ in Σ. Finally, we flow back through a time δ_3 using the ODE with $i = 1$ to reach Π_N again at the point $x_5 = (u_5, 0, \delta_2 + \delta - \delta_3)$. As E is conserved throughout the motion, it is immediate that $\delta = -\delta_3$ and $u_1 = u_5$. However, the map takes a non-zero time $\Delta = \delta_2 + 2\delta$. To calculate this time we observe (from symmetry and the energy conservation) that when the trajectory moves in S_2 from x_2, then it intersects Π_N after a time $\delta_2/2$. Let $(u, v, t) = (-y_2^2, 0, \delta_2/2 + \delta)$ be the point of intersection. By applying the conservation law and also considering the exact solution, one can obtain the following identities:

$$\frac{1}{2}k_1 y_1^4 + a_1 y_1^2 = \frac{1}{2}k_2 y_2^4 + a_2 y_2^2,$$

$$0 = \frac{a_1}{k_1^2} - \left(y_1^2 + \frac{a_1}{k_1^2}\right)\cos(k_1\delta) = \frac{a_2}{k_2^2} - \left(y_2^2 + \frac{a_2}{k_2^2}\right)\cos(k_2\delta_2/2).$$

If y_1 and y_2 are both small, then so are δ and δ_2. We can then express δ and δ_2 as a power series in y_1 and y_2 respectively, and use the Taylor series expansion of $\cos(k_1\delta)$ and $\cos(k_2\delta_2/2)$ about $\delta = 0$ and $\delta_2 = 0$. After inverting the power series, it then follows that

$$\begin{aligned} \delta &= -y_1\sqrt{\frac{2}{a_1}}\left(1 - \frac{5}{12}\frac{y_1^2 k_1^2}{a_1} + O(y_1^4)\right), \\ \delta_2 &= 2y_2\sqrt{\frac{2}{a_2}}\left(1 - \frac{5}{12}\frac{y_2^2 k_2^2}{a_2} + O(y_2^4)\right), \end{aligned} \tag{7.42}$$

and

$$y_2^2 = \frac{a_1}{a_2}y_1^2 + \frac{1}{2a_2^3}y_1^4(k_1 a_2^2 - k_2 a_1^2) + O(y_1^6). \tag{7.43}$$

By substituting the expression (7.43) for y_2 into (7.42) we obtain an expression for the time $\Delta = \delta_2 + 2\delta$ that is given by:

$$\Delta = y_1 \frac{2\sqrt{2}}{a_2\sqrt{a_1}}(a_1 - a_2) + \mathcal{O}(y_1^3). \tag{7.44}$$

This then is the PDM correction in the time direction if $a_1 \neq a_2$ is proportional to $y_1 = |u_1|^{1/2}$.

Now, the equation (7.40) can be written in first-order form with $x = (u, v, t)$ as

$$\begin{aligned}\dot{x}_1 &= x_2,\\ \dot{x}_2 &= -k_i^2 x_1 + a_i,\\ \dot{x}_3 &= 1.\end{aligned}$$

It is easy to check that a grazing point with $v = u = 0$ satisfies the assumptions of the PDM Theorem (7.3) with $H = x_1$ and, when evaluated at a grazing point, $(H_x F_i)_x F_1 = a_1$ and $(H_x F_i) F_2 = a_2$, for $i = 1$ or 2. We can now apply the formula (7.36) for the PDM for which we obtain

$$v_z = \frac{2\sqrt{2}a_1}{a_2\sqrt{a_1}} \frac{(a_1 - a_2)}{a_1} \begin{pmatrix} 0 \\ 0 \\ 1 \end{pmatrix} = \frac{2\sqrt{2}}{a_2\sqrt{a_1}} (a_1 - a_2) \begin{pmatrix} 0 \\ 0 \\ 1 \end{pmatrix}. \tag{7.45}$$

Hence the PDM correction to the time variable in the case that $a_1 \neq a_2$ is

$$y_1 \frac{2\sqrt{2}}{a_2\sqrt{a_1}} (a_1 - a_2),$$

which agrees with the leading-order term of Δ calculated explicitly.

Note that when $a_1 = a_2$, the vector field is continuous at the grazing point. Then, as expected from the theory, we observe in this case that the leading-order term given by (7.45) is annihilated and that the leading-order correction term of the PDM is then proportional to y_1^3 or equivalently $|u_1|^{3/2}$.

Example 7.7 (A third-order oscillator). We next consider a third-order oscillator used to describe the dynamics of a relay feedback controller [255], which is similar to the one we considered in case study III in Chapter 1, as an example of a system that can undergo sliding motion. For simplicity, we consider a model that can be written in the form

$$\frac{d^3 u}{dt^3} = -a_{3i} \frac{d^2 u}{dt^2} - a_{2i} \frac{du}{dt} - a_{1i} u + b, \quad i = 1, 2, \tag{7.46}$$

where $i = 1$ corresponds to $u > 0$ and $i = 2$ to $u < 0$. For simplicity we will take the forcing term b to be constant in time. Equation (7.46) can be recast as a system of three first-order differential equations in the form

$$\dot{x} = \begin{cases} A_1 x + B, & \text{if } H(x) = C^T x > 0, \\ A_2 x + B, & \text{if } H(x) = C^T x < 0, \end{cases} \tag{7.47}$$

where

$$A_i = \begin{pmatrix} 0 & 1 & 0 \\ 0 & 0 & 1 \\ -a_{1i} & -a_{2i} & -a_{3i} \end{pmatrix}, \quad B = \begin{pmatrix} 0 \\ 0 \\ b \end{pmatrix}, \quad C^T = \begin{pmatrix} 1 \\ 0 \\ 0 \end{pmatrix}^T.$$

First, we notice that, when $x = 0$, system (7.47) is such that:

1. $H(x) = C^T x = 0$ for $x = 0$
2. $H_x(0,0) = C^T \neq 0$
3. $\mathcal{L}_{F_i}(H(0)) = H_x F_i(0,0) = C^T B = 0$
4. $\mathcal{L}^2_{F_i} H(0) = (H_x F_{i,x}) F_i = C^T A_i B = 0$.

Hence, the system does not satisfy all the conditions required for a regular grazing to occur at $x = 0$; specifically it violates the condition on the curvature of the vector fields (7.23).

In fact, at a grazing point we have $x_1 = u = 0$, $x_2 = \dot{u} = 0$, and solving (7.46) we get:

$$x_3 = \ddot{u} = -b/a_{3i}.$$

Thus, for the oscillator described by (7.46), the grazing point is located at $x^* = (0, 0, -\beta_1/a_{31})$. In order to apply the local theory presented in this chapter, we need therefore to consider an appropriate change of co-ordinates to shift the grazing point from $x = x^*$ to $x = 0$ as required. Specifically, let $w = x - x^*$ so that system (7.47) becomes

$$\dot{w} = \begin{cases} A_1 w + B_1, & \text{if } C^T w > 0, \\ A_2 w + B_2, & \text{if } C^T w < 0, \end{cases} \tag{7.48}$$

where, in this case:

$$A_i = \begin{pmatrix} 0 & 1 & 0 \\ 0 & 0 & 1 \\ -a_{1i} & -a_{2i} & -a_{3i} \end{pmatrix}, \tag{7.49}$$

$$B_1 = \begin{pmatrix} 0 \\ \lambda \\ a_{31}\lambda + b \end{pmatrix}, B_2 = \begin{pmatrix} 0 \\ \lambda \\ a_{32}\lambda + b \end{pmatrix}, C^T = \begin{pmatrix} 1 \\ 0 \\ 0 \end{pmatrix}^T, \tag{7.50}$$

with $\lambda = -b/a_{31}$.

Note that system (7.48) now satisfies all the properties required for a regular grazing point to occur at $w = 0$. Now consider an initial point $w = \varepsilon(x_1, x_2, x_3)$ with $\varepsilon \ll 1$. According to Theorem 7.2, we will have terms of the form $\varepsilon^{1/2}$ in the map if the vector field is discontinuous, i.e., if

$$B_1 \neq B_2.$$

From (7.49), (7.50), we can deduce that this occurs if and only if $a_{31} \neq a_{32}$, i.e., the coefficient of $\ddot{y}$ varies across Σ. For all other cases, we have $B_1 = B_2$, and terms in the map proportional to $\varepsilon^{3/2}$ arise instead.

In fact, assuming $a_{31} \neq a_{32}$ while $a_{11} = a_{12}$,$a_{21} = a_{22}$ in (7.49), (7.50) the Taylor expansion of the ZDM in w, around $w^* = 0$, gives

$$\text{ZDM} = \begin{pmatrix} 0 \\ 0 \\ 2\lambda(a_{32} - a_{31})\gamma_1 \end{pmatrix} \varepsilon^{\frac{1}{2}} + \ldots,$$

for $x_1 < 0$, where

$$\gamma_1 = \sqrt{\frac{|x_1|}{\lambda}} := \sqrt{-a_{31}\frac{x_1}{b}}.$$

Thus, the third component of the state perturbation exhibits $\varepsilon^{\frac{1}{2}}$ behavior as expected.

If, instead, we assume $a_{31} = a_{32} = \alpha$, but suppose that $a_{11} \neq a_{12}$ and $a_{21} \neq a_{22}$, then careful substitution of all the relevant terms in (7.31) yields

$$\text{ZDM} = \varepsilon w + \begin{pmatrix} 0 \\ \frac{2}{3}(a_{21} - a_{22})\gamma_1^3 \\ \frac{1}{3}\left[a_{11} - a_{12} + 2\alpha(a_{21} - a_{22})\right]\lambda\gamma_1^3 + 2(a_{11} - a_{12})x_1\gamma_1 \end{pmatrix} \varepsilon^{\frac{3}{2}} + \dots$$

We should comment here that we implicitly assumed that $H_{\min}(w) = H(w)$ to leading-order for the initial points. This in turn implies that we restrict the set of initial points to some subset of the phase space where the ZDM correction applies. Note that this assumption is true for all points that lie on the zero-velocity manifold given by $\mathcal{L}_{F_1}(H)(w) = 0$.

Example 7.8 (Example 7.5 continued, the stick-slip oscillator). We conclude our examples by looking again at the autonomous friction oscillator model introduced in Example 7.5, which is a piecewise-smooth system with degree of smoothness equal to 2.

Figure 7.11 shows a bifurcation diagram for this system in which the bifurcation parameter is the spring stiffness k and the other parameters are held at

$$m = 0.01, \quad \sigma = 10^{-6}, \quad \alpha = 4000, \quad \gamma = 20000, \tag{7.51}$$

$$g = 9.82, \quad \mu = 0.4, \quad \beta = 20000, \quad d = \ln 2, \quad \Delta = 0.1, \tag{7.52}$$

with the horizontal velocity U calculated from (7.17). When $k = 214.2529$, an unstable limit cycle grazes with Σ. This causes the onset, upon decreasing k of a *stick-slip* motion that makes repeated tiny penetrations into the region with $y_4 > 1$. This motion can be quite involved and features chaotic dynamics and period-doubling bifurcations.

The onset of this rich dynamics, observed when k is decreased through the grazing value, can be largely explained by the theory treated here. Specifically, an evaluation of the relevant terms in the normal form (7.31) in Theorem 7.2. gives the local ZDM

$$x \longrightarrow \begin{cases} x & \text{for } y_4 \geq 1, \\ x + 38.58432719\, J(x^*)(1 - y_4)^{3/2} & \text{for } y_4 < 1 \end{cases} \tag{7.53}$$

where $x = (y_1, \dots, y_5)^T$ and $x^* = (-0.099724, 0.000393, 12.258254, 1, 1)^T$ $J(x^*) = (0, 0.037457, 0, 0, 3.80423)^T$.

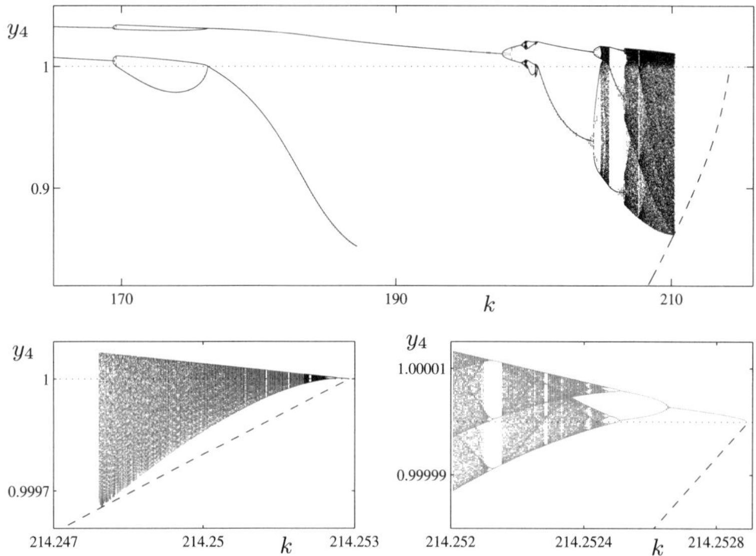

Fig. 7.11. Successive enlargements of a Monte Carlo bifurcation diagram of (7.14)–(7.17) for varying spring stiffness k with other parameters given by (7.51)-(7.52). The vertical axis depicts local maxima of y_4. The dotted line corresponds to the discontinuity set $\Sigma = y_4 = 1$ and the dashed line to a branch of unstable limit cycles that are born in a sub-critical Hopf bifurcation for a lower value of k. (Reprinted from [64] with permission from Elsevier.)

Recall the discussion at the end of Chapter 4 on the location of fold bifurcations of a map with a leading-order nonlinear term proportional to $x^{3/2}$ close to the grazing bifurcation in the case that the coefficient of the $O(3/2)$-term is large. Note from (7.53) that the coefficient in question, that of the term in the y_4 direction is equal to 146.784. Also, computing the Poincaré maparound the grazing orbit, we find that the Jacobian matrix of this map has eigenvalues (corresponding to Floquet multipliers of the flow) approximately equal to 1.2, 1 (the trivial Floquet multiplier), and (approximately) 0, 0 and 0. However, the non-zero elements of the Jacobian matrix are approximately of size 10^2. Hence, a simple sensitivity argument shows that a perturbation caused by the discontinuity mapping of size 10^{-2} has the potential to change these multipliers by an $O(1)$ amount. However, since the coefficient multiplies the $\frac{3}{2}$ term (η in the notation of Chapter 4) is actually $O(10^3)$, clearly the DM is likely to have a massive influence on the dynamics. This observation is borne out in the numerics in Fig. 7.11 where a non-smooth fold bifurcation appears to occur at precisely the parameter value of the grazing bifurcation, at least to an accuracy of four decimal places in the parameter. However, we know in theory that the discontinuity mapping is actually smooth at the grazing bifurcation point, with a singularity only in the $O(3/2)$ terms. We conclude that there must be a smooth fold bifurcation approximately $O(10^{-5})$ or closer away from the grazing bifurcation.

These observations are confirmed by the results in Fig. 7.12, which reproduce a computation from [64] that compares the results of iteration of the compound discontinuity map over one whole period with the results of the numerics. Note over this small scale the close agreement between the mapping and the simulations. The variables in this figure have been rescaled so that the grazing bifurcation occurs at $\tilde{\nu} = 0$. However, even at the scale depicted it is hard to see the existence of a fold bifurcation for small $\tilde{\nu}$. In fact, further zooming shows the fold occurs at a $\tilde{\nu}$-value within 10^{-3} of the grazing point. Returning to the physical co-ordinates, this implies a fold for k within 10^{-7} of the grazing point!

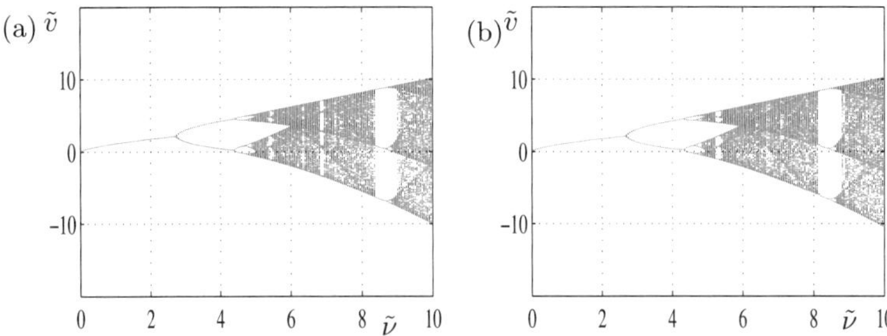

Fig. 7.12. Comparison between the numerical simulations (left panel) and the discontinuity mapping (right panel) local to the grazing bifurcation at $k = 214.2528$. Here $\tilde{v}$ is a rescaling of y_4 and $\tilde{\nu}$ is a rescaling of $-k$ [cf. Fig. 7.11(c)] such that the grazing bifurcation occurs at $\tilde{\nu} = 0$. (Reprinted from [64] with permission from Elsevier.)

This example serves to illustrate a key point about grazing bifurcations where the degree of smoothness is 2 or more. The local analysis of the normal form of the map, as given earlier, shows that it (and its first derivative) is continuous at the grazing point and that it has a 3/2-type discontinuity there. At the grazing point, there should not be a change in the tangent to the branch of fixed points and we might conclude that this would rule out any complex dynamics emerging from such transitions. But, if no instantaneous transition occurs, grazing in piecewise-smooth systems with degree of smoothness 2 can cause a rapid change in the curvature of a bifurcation branch giving rise to a nearby fold, followed in this example by a period-doubling cascade and many nearby classical bifurcations.

It is worthwhile to point out some more features of the bifurcation diagram in Fig. 7.11. First, there is another grazing bifurcation at $k \approx 176.3429$. This bifurcation again induces a smooth bifurcation at a nearby parameter value. In this case it is a period-doubling bifurcation, as evidenced by the obvious period-two attractor in the figure for k larger than this value. Second, notice what appears to be a period-adding cascade, interspersed by regions of chaotic dynamics for approximately $k \in (200, 210)$. This occurs below a k value for which there is no immediate attractor in the dynamics. What actually happens here is that the dynamics created as k is decreased through 214.2529 (the value of the grazing bifurcation we analyzed above) rapidly disappears at around $k = 214.2485$ in a boundary crisis involving the unstable limit cycle (see the second zoom of Fig. 7.11). The attracting behavior then reappears in another boundary crises at around $k = 210$. Recall, these results were all for $\Delta = 0.1$. Now, other results presented in [64] show a similar grazing bifurcation for the case of $\Delta = 0.01$ [and all other parameters held fixed at their values given in (7.51)–(7.52)]. Then there is no equivalent pair of boundary crises, and the period-adding behavior exists over a wide range of parameter values, right up to that of the grazing. Thus, the period-adding would appear to be symptomatic of the global bifurcation diagram of maps with an $O(3/2)$ singularity, as described in Chapter 4.

7.2.5 Detailed derivation of the discontinuity mappings

We now present the necessary analytic steps required to compute the leading-order expressions of the ZDM and PDM given in Theorems 7.2 and 7.3. The derivation follows in a similar manner to that of the equivalent maps for impacting hybrid systems described in Chapter 6, using Lie derivative notation. Guided by Fig. 7.9 the derivation of the local discontinuity mappings is divided into three different stages:

1. Calculating the trajectory starting from x under the action of the flow Φ_1, until it intersects (if it does so) the manifold Σ at the point x_2 after an elapsed time $t = \delta_0$;
2. Calculating the trajectory starting from x_2 under the action of the flow Φ_2 until the second crossing of Σ at the point x_3 after an elapsed time δ_2;
3. Calculating the trajectory starting from x_3 under the action of the flow Φ_1 *backwards in time* either until either it reaches the point x_4 after a zero total time has elapsed (the ZDM case) or until the trajectory intersects the Poincaré section Π_N at the point x_5.

In this calculation Steps 1 and 3 are identical to the equivalent steps described in Chapter 6. The difference here comes in the new step 2, which requires following flow Φ_2 rather than applying a restitution map R. Thus, referring to Fig. 7.9, we define the ZDM as (7.26)

$$\Phi_1(\Phi_2(\Phi_1(x_0, \delta_0), \delta_2) - (\delta_0 + \delta_2))$$

and the PDM as (7.28)

$$\Phi_1(\Phi_2(\Phi_1(x_1, \delta), \delta_2), -\delta_3).$$

Before calculating these expressions, we calculate separately each of the times δ, δ_0, δ_1, δ_2 and δ_3 through Taylor expansion, appealing to the Implicit Function Theorem where necessary. As in Chapter 6, we shall decompose the calculation of δ_0 into the calculation of times δ_1 to flow from the initial point to the point x_1 at which $\mathcal{L}_{F_1}H(x) = 0$ and the time $\delta < 0$ to flow from such a point to Σ, so that

$$\delta_0 = \delta_1 + \delta.$$

Step 1. To find δ and hence the point x_2, we solve the problem

$$H(x_2) = H(\Phi_1(x_1, \delta)) = 0$$

for δ as a function of y, where

$$\mathcal{L}_{F_1}H(x_1) = 0 \quad \text{and} \quad H_{\min} = H(x_1), \quad y^2 = -H(x_1).$$

This step is identical to the equivalent step for impacting systems presented in Chapter 6 but with F replaced by F_1. There, application of the Implicit Function Theorem gave the existence of a unique, smooth function (6.38) given by

$$\delta(x_1, y) = y\left(-\sqrt{\frac{2}{\mathcal{L}^2_{F_1}H(x_1)}} - \frac{1}{3}\frac{\mathcal{L}^3_{F_1}H(x_1)}{(\mathcal{L}^2_{F_1}H(x_1))^2}y + O(y^3)\right), \tag{7.54}$$

where $\mathcal{L}^2_{F_1}H(x_1) > 0$ is the normal acceleration at the point x.

Note that x_1 determines the value of y, hence the left-hand side is a function of x_1 only. However, (7.54) is a valid expression when x_1 and y are treated as independent variables.

Step 2. We next need to find the time δ_2. To do this we expand the expression $H(\Phi_2(x_2, \delta_2)) = 0$ in powers of δ_2. As $H(x_2) = 0$ this gives

$$H(\Phi_2(x_2, \delta_2)) = \mathcal{L}_{F_2}H(x_2)\delta_2 + \frac{1}{2}\mathcal{L}^2_{F_2}H(x_2)\delta_2^2 + O(\delta_2^3). \tag{7.55}$$

Note that $\mathcal{L}_{F_2}H(x_2)$ is a small quantity. To be able to find an expression for δ_2 as a power series in y, we first expand in the Taylor series $\mathcal{L}_{F_2}H(x_2)$ and $\mathcal{L}^2_{F_2}H(x_2)$ in x_2 about x_1. Using properties of Lie derivatives, (7.55) becomes

$$H(\Phi_2(x_2, \delta_2)) = \mathcal{L}_{F_2}H(x_1)\delta_2 - C(x_1)\mathcal{L}_{F_1}\mathcal{L}_{F_2}H(x_1)y\delta_2 + \frac{1}{2}\mathcal{L}^2_{F_2}H(x_1)\delta_2^2 + O(\delta_2^3), \tag{7.56}$$

where

$$C(x_1) = \sqrt{\frac{2}{\mathcal{L}^2_{F_1}H(x_1)}}\,.$$

Now, we can divide the right-hand side of (7.56) by δ_2 since we seek a time $\delta_2 > 0$. Also, since the term $\frac{1}{2}\mathcal{L}^2_{F_2}H(x_1)$ is non-zero and of $O(1)$, by assumption (7.23) and the fact that x_1 is sufficiently close to x^*, we can appeal to the Implicit Function Theorem to guarantee the existence of a smooth function $\delta_2(x_1, y)$ for which $H(\Phi_2(x_2, \delta_2)) = 0$. Moreover, the term $\mathcal{L}_{F_2}H(x_1)$ is zero because $\mathcal{L}_{F_1}H(x_1) = 0$ by definition, and the assumption (7.7) that means the sets

$$\{x \in \mathcal{D} : \mathcal{L}_{F_1}H(x_1) = 0\} \quad \text{and} \quad \{x \in \mathcal{D} : \mathcal{L}_{F_2}H(x_1) = 0\}$$

coincide. After inverting the power series (7.56) to leading-order in y, we have

$$\delta_2(x_1, y) = 2\frac{\mathcal{L}_{F_1}\mathcal{L}_{F_2}H(x_1)}{\mathcal{L}^2_{F_2}H(x_1)}C(x_1)y + O(y^2). \tag{7.57}$$

Step 3. We can now complete this calculation to find the ZDM for an initial point $x \in \Pi_N$ close to a grazing point. To do this we define

$$\mathrm{ZDM}(x, y) = x_4(x, \delta(x, y), \delta_2(x, y)) = \Phi_1(\Phi_2(\Phi_1(x, \delta), \delta_2), -(\delta + \delta_2))$$

A Taylor expansion of this expression gives

$$x_4 = x + \delta_2(F_2 - F_1) + O((\delta, \delta_2)^2)).$$

After substituting for δ_2 from (7.57), we can write the leading-order ZDM for a point $x \in \Pi_N$ as

$$\mathrm{ZDM}(x, y) = x + \begin{cases} 0, & \text{if } H(x) \geq 0, \\ 2C\frac{\mathcal{L}_{F_1}\mathcal{L}_{F_2}H(x)}{\mathcal{L}^2_{F_2}H(x)}(F_2(x) - F_1(x))y, & \text{if } H(x) \leq 0, \end{cases} \tag{7.58}$$

where the error term is $O(y^2)$. Finally we can expand (7.58) in x around the grazing point at $x = x^*$ giving to leading-order

$$\mathrm{ZDM}(x) = x + \begin{cases} 0, & \text{if } H(x) > 0, \\ C^*D^*(F_2(x^*) - F_1(x^*))y + O(H(x)), & \text{if } H(x) < 0, \end{cases} \tag{7.59}$$

with $C^* = C(x^*)$, $D^* = D(x^*)$ where

$$D(x) = 2\frac{(H_xF_2)_xF_1}{(H_xF_2)_xF_2}(x).$$

Each calculation (for the time δ, δ_1, the point x_2 etc.) should be treated as a separate calculation, and only when the final expression for the ZDM map is introduced we make use of the expressions found earlier. Now we are in a position to start stating the forms that the PDM and ZDM take for general initial conditions in a neighborhood of the point x^*.

The ZDM for a general initial point. The above expressions for the ZDM were derived assuming that the starting point x was in Π_N; that is $\mathcal{L}_{F_1}H(x) = 0$. In order to apply these expressions for a general initial point x, we must first advance the flow using vector field F_1 for a time δ_1 to a point where $\mathcal{L}_{F_1}H(\Phi_1(x,\delta_1)) = 0$. Hence, we consider the following flow combination

$$\Phi_1(\mathrm{ZDM}(\Phi_1(x,\delta_1),y),-\delta_1), \tag{7.60}$$

where ZDM denotes the zero time correction (7.59), where we now define

$$y = \sqrt{-H(\Phi_1(x,\delta_1))} = \sqrt{-H_{\min}(x)},$$

since $\Phi_1(x,\delta_1)$ is the point where H has its local minimum along a trajectory through x. Let us also set

$$v = \mathcal{L}_{F_1}H(x)$$

and solve for δ_1 in the expression

$$\mathcal{L}_{F_1}H(\Phi_1(x,\delta_1)) - \mathcal{L}_{F_1}H(x) + v = 0. \tag{7.61}$$

where $|x - x^*|$ and v are small. Note that the introduction of the additional variable v is used to make this a regular expression at all points in a neighborhood of x^*.

Now, by Taylor expansion of the flow we have

$$\mathcal{L}_{F_1}H(\Phi_1(x,\delta_1)) = v + \mathcal{L}^2_{F_1}H(x)\delta_1 + O\left(\delta_1\right)^2 = 0,$$

and since $\mathcal{L}^2_{F_1}H(x) \neq 0$ by (7.23), we can appeal to the Implicit Function Theorem to guarantee the existence of a smooth function $\delta_1(x,v)$, which solves (7.61). After inverting the power series we get

$$\delta_1(x,v) = -\frac{v}{\mathcal{L}^2_{F_1}H(x)} + O\left(v^2\right). \tag{7.62}$$

Now, we can write $H_{\min}$ as

$$\begin{aligned} H_{\min}(x,v) &= \mathcal{L}_{F_1}H(\Phi(x,\delta_1)) + (v - \mathcal{L}_{F_1}H(x))\delta_1(x,v) \\ &= H(x) + v\delta_1(x,v) + O\left(\delta_1(x,v)^2\right). \end{aligned} \tag{7.63}$$

After substituting for δ_1 (7.62) into (7.63), we get

$$H_{\min}(x,v) = H(x) - v^2 r(x,v),$$

where $r(x,v)$ contains the remaining terms. Note that, similarly to grazing bifurcations in impacting systems, the correction that accounts for the fact that we start at different initial point does not affect the leading-order approximation for the ZDM obtained for x such that $\mathcal{L}_{F_1}H(x) = 0$. This follows from the fact that if we expand (7.60) in δ_1, there is no correction to this flow combination for the terms of order y.

Thus, we can write the final form of the ZDM as

$$\mathrm{ZDM}(x,y,v) = x + \begin{cases} 0, & \text{if } H_{\min}(x) \geq 0, \\ 2C^* D^* F_d^* y + \mathcal{O}[(y,v)^2], & \text{if } H_{\min}(x) \leq 0, \end{cases} \tag{7.64}$$

where

$$C(x)D(x) = \sqrt{\frac{2}{\mathcal{L}_{F_1}^2 H(x)} \frac{\mathcal{L}_{F_1}\mathcal{L}_{F_2} H(x)}{\mathcal{L}_{F_2}^2 H(x)}} \qquad \text{and} \qquad F_d(x) = F_2(x) - F_1(x),$$

and an asterisk denotes evaluation at the grazing point $x = x^*$. Note that to leading-order, (7.64) is exactly the same as (7.59), and indeed provides the proof of the first part of Theorem 7.2. However, the ZDM for a general point is a function not just of x, y but also of an additional small variable v that measures the closeness of the initial point to Π_N. Therefore the expansion of (7.64) to higher order will lead in general to more complex expressions than the expansion of (7.59).

The ZDM derivation in the case when $F_1 = F_2 = F$. Note that the leading-order (square root) term of (7.64) vanishes if $F_1(x^*) = F_2(x^*)$. To obtain the next-order expression for the map, we first need to establish the order of the next non-vanishing term in the expansion. Roughly speaking, given degree of smoothness 2, the value of $F_2 - F_1$ in (7.58) is proportional to x, which will lead to terms in (7.59) proportional to $|x - x^*|^{3/2}$. The precise derivation of these terms follows exactly the same steps as above, but carrying out the computation of the times δ, δ_1, and δ_2 to higher-order. Since these expressions were derived using the Implicit Function Theorem, we know that each of them has a regular Taylor series expansion. In particular, we find that if $F_2(x^*) = F_1(x^*)$, then

$$\delta_2 = -2\delta - 2\frac{(H_x F_2)_x x}{(H_x F_2)_x F_2} + O(|x|^{3/2}).$$

The precise expressions that lead to the coefficient of the $O(3/2)$ term presented in the second part of Theorem 7.3 are rather lengthy and are best produced with the aid of computer algebra. Instead we merely motivate here how each of the terms v_1, v_2 and v_3 arise.

Note first that the term $F_2 - F_1$ expanded in x around $x = x^*$ is to leading-order given by $(F_2 - F_1)_x(x - x^*)$. This gives the term v_2 of (7.31) and is $O(|x - x^*|^{3/2})$. To find the terms v_1 and v_3, we need to expand the flow combination (7.60) to third order. After some algebraic manipulations, we can show that the remaining terms that contribute to v_3 are of the form $(F_2 - F_1)_x F \delta \delta_2$, with δ_2 containing only terms of order $|(H_x F_2)_x (x - x^*)|$. We can also obtain the expression

$$\frac{1}{2}\left((F_2 - F_1)_x F\right)\delta_2^2$$

that contributes to the first part of the v_1 term. The remaining part of the v_1 term comes from the contribution of the third-order terms of the flow expansion (7.55) with δ and δ_2 being considered to leading-order only, in which case $\delta_2 = -2\delta$. Finally we find that in the case where $F_2 - F_1$ vanishes at x^*, consideration of the time δ_1 does not change the coefficient of the leading $O(3/2)$ term.

The PDM for a grazing bifurcation. Since the ZDM above was derived assuming that the starting point $x \in \Pi_N$ satisfying $\mathcal{L}_{F_1}H(x) = 0$ to obtain the PDM from Π_N to itself, it is the most convenient to compose the ZDM with a projection map that maps its image onto Π_N. Hence, we follow the procedure outlined in Chapter 6 and calculate the PDM by considering the following combination of flows:

$$\Phi_1(\mathrm{ZDM}(x,y),\widehat{\delta}) = \mathrm{ZDM}(x,y) + \widehat{\delta}F_1(x_4) + O\left((y,\widehat{\delta})^2\right) \tag{7.65}$$

where $\mathrm{ZDM}(x,y)$ denotes the expression (7.58) that was derived under the assumption that $\mathcal{L}_{F_1}H(x) = 0$, and

$$\widehat{\delta} = \delta_3 - (\delta_0 + \delta_2)$$

is the time of flow under Φ_1 from the final point x_4 of the ZDM construction to intersect Π_N at the point x_5. We know that $\mathcal{L}_{F_1}H(\Phi_1(x_4,\widehat{\delta})) = 0$. Expanding this expression in powers of $\widehat{\delta}$, we get

$$\mathcal{L}_{F_1}H(\Phi_1(x_4,\widehat{\delta})) = \mathcal{L}_{F_1}H(x_4) + \widehat{\delta}\mathcal{L}_{F_1}^2H(x_4) + O\left(\widehat{\delta}^2\right) = 0. \tag{7.66}$$

We now solve (7.66) for $\widehat{\delta}$ as a power series in y yielding

$$\widehat{\delta}(x_4,y) = -\frac{\mathcal{L}_{F_1}H(x_4)}{\mathcal{L}_{F_1}^2H(x_4)}y + O\left(y^2\right). \tag{7.67}$$

From the properties of Lie derivatives we have

$$\widehat{\delta}(x,y) = \left(\frac{\mathcal{L}_{F_2}\mathcal{L}_{F_1}H(x)C(x)D(x)}{\mathcal{L}_{F_1}^2H(x)} - C(x)D(x)\right)y + O\left(y^2\right).$$

From expression (7.65), we can then obtain the leading-order expression of the PDM given by

$$\mathrm{PDM}(x,y) = x + \begin{cases} 0 & H(x) \geq 0 \\ Z(x^*)y + O\left(y^2\right) & H(x) < 0 \end{cases}, \tag{7.68}$$

with

$$Z(x) = 2\left(F_2 - \frac{\mathcal{L}_{F_2}\mathcal{L}_{F_1}H(x)}{\mathcal{L}_{F_1}^2H(x)}F_1\right)\sqrt{\frac{2}{\mathcal{L}_{F_1}^2H(x)}\frac{\mathcal{L}_{F_1}\mathcal{L}_{F_2}H(x)}{\mathcal{L}_{F_2}^2H(x)}}$$

evaluated at the grazing point, and $H_{\min}$ replaced by $H(x)$ since Π_N has been chosen so that the initial point x lies at a local minimum of H. Alternatively we can express Z as

$$Z(x) = 2\left(F_2 - \frac{(H_x F_1)_x F_2}{(H_x F_1)_x F_1} F_1\right)\sqrt{\frac{2}{(H_x F_1)_x F_1}\frac{(H_x F_2)_x F_1}{(H_x F_2)_x F_2}},$$

which is the term v_z given in the Theorem 7.3.

For simplicity we have only derived here the leading-order expression for the PDM for the case of $F_1(x^*) \neq F_2(x^*)$ for which the PDM has square-root form. The 3/2 form of the PDM arises when $F_1 = F_2$ at the point of grazing, follows after lengthy calculation of the next-order terms in the expansion (7.68). We omit the details here. A complete derivation, using asymptotic expansion methods rather than Lie derivatives is given in [78].

7.3 Boundary-intersection crossing bifurcations

A basic hypothesis, assumed so far in this and previous chapters, is that the discontinuity boundary Σ should be a smooth subset of phase space. However, in many applications, for example the DC–DC converter circuit in case study V, this is not the case since the switching manifold itself is a sawtooth function. In general, in many control systems and electronic switching devices, switching conditions may be governed by several overlapping inequalities. A generic feature of such examples is that the discontinuity boundary will have corner-type singularities formed by the transverse intersection, along a set $\mathcal{C}$, of two smooth codimension-one surfaces Σ_i and Σ_j, which locally divide the phase space into four regions with differing vector fields $F_i, i = 1 \ldots 4$.. If the system has a parameterized periodic orbit $p(t, \mu)$, then interesting behavior occurs when, as μ is varied, this orbit intersects the set $\mathcal{C}$. We will now show that such an interaction is characterized by continuous maps that are locally piecewise-linear. Thus the behavior can be understood in terms of the theory developed in Chapter 3.

The locus of *boundary intersection points* $\mathcal{C}$ will in general be a $(n-2)$-dimensional subset of the phase space $\mathbb{R}^n$. The passage of a trajectory through a point in $c \in \mathcal{C}$ is a DIB in the sense of a topological change of a piece of phase potrait introduced in Chapter 2 because in a neighborhood of the boundary intersection, there are distinct trajectories that do not behave similarly with respect to phase-space regions on either side of $\Sigma = \Sigma_i \cup \Sigma_j$. This situation is illustrated in Fig. 7.13(a).

The special case illustrated in Fig. 7.13(b) that corresponds to three vector fields in 7.13(a) being identical, arises in many applications (such as the DC–DC converter) and has previously been called a *corner-collision bifurcation* [76]. In this case we consider two discontinuity surfaces Σ_5 and Σ_6 that meet

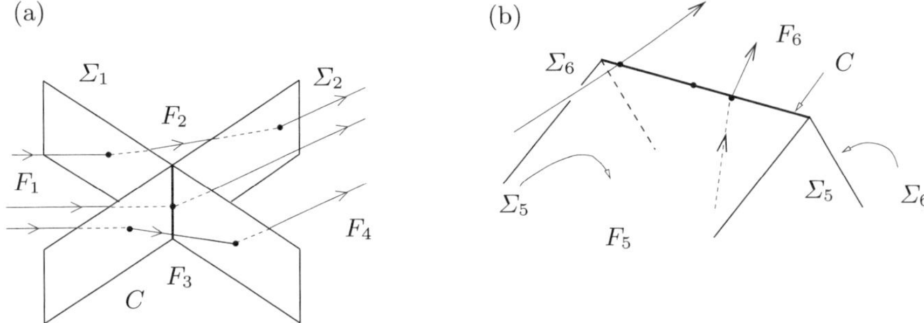

Fig. 7.13. (a) A boundary-intersection crossing trajectory that intersects the crossing manifold C between two discontinuity surfaces Σ_1 and Σ_2, and two nearby trajectories each starting in the region of phase space for which the vector field is given by F_1. Here it is assumed that a different smooth vector field F_i, $i = 1...4$ applies in each of the four local phase space regions. (b) The special case of a corner crossing where only two different vector fields, F_5 and F_6 apply, and the crossing manifold can be described as the corner in a single discontinuity surface made up of two smooth pieces Σ_5 and Σ_6. Two distinct kinds of corner-intersecting trajectories are depicted, *external* (trajectory starting in S_6) and *internal* (trajectory starting in S_5) corner-collisions.

along a corner $\mathcal{C}$. We call the region *inside* the corner S_5 and that *outside* S_6, with corresponding vector fields F_5 and F_6. There are two generic kinds of such intersections between a trajectory and $\mathcal{C}$, leading to differing bifurcations that we shall refer to either as *external* or *internal* corner-collision. These are both illustrated in Fig. 7.13(b). A sufficient condition for a boundary-intersection crossing to occur is that the periodic orbit $p(t, \mu)$ intersects the codimension-two corner manifold $\mathcal{C}$. Since we can choose the phase of $p(t, \mu)$ for which this happens, a boundary-intersection crossing is thus a codimension-one bifurcation.

7.3.1 The discontinuity mapping in the general case

In this chapter, we shall consider only the case where the overall vector field is discontinuous across each of Σ_i and Σ_j, under conditions that no sliding occurs, and shall show that to lowest order this leads to a piecewise-linear normal form. The case where the vector field is continuous can be similarly shown to lead to a discontinuity mapping with a jump at quadratic order. In general, degree of smoothness m leads to maps with a jump in the mth derivative.

Consider first the general case depicted in Fig. 7.13(a) where the four vector fields all differ and the trajectory is assumed to start in the region of phase space where the vector field is given by F_1 and the discontinuity surfaces are Σ_1 and Σ_2. Suppose this trajectory is part of a periodic orbit

$p(t,\mu)$, which is parameterized by μ. We set up local co-ordinates such that the point of intersection of the periodic orbit with $\Sigma_1 \cap \Sigma_2$ occurs when $\mu = 0$ at the point $x = 0$. Let the boundaries Σ_1 and Σ_2 be given by the zero sets of smooth functions $H_1(x)$ and $H_2(x)$ respectively, which for simplicity we take to be linear; $\Sigma_1 = \{H_1 = 0\}$ and $\Sigma_2 = \{H_2 = 0\}$, and the sense of their normal vectors is as depicted in Fig. 7.15.

Now, it will follow that the linear approximation to the flow and to the boundaries is sufficient to determine the leading-order expression for the discontinuity mapping in a neighborhood of $(x,\mu) = (0,0)$. Thus the vector field $F_i(x,\mu)$ can be replaced by $F_i(0,\mu)$ and we suppose for simplicity that the local situation near the point $x = 0$ is unchanged by the variation of μ, so that $F_i(x,\mu) \approx F_i(0,0) := F_i$. Let the flow associated with the vector field F_i be Φ_i. Also let a final subscript indicate a component in the direction perpendicular to each surface given by $\frac{dH_j}{dx} := H_{j,x}$, so that $F_{ij} = H_{j,x}F_i(0)$ and $x_j = H_{j,x}x$, for $j = 1, 2$.

We make the further assumption that there is no sliding or grazing in the neighborhood of $x = 0$, so that all four vector fields cross both Σ_1 and Σ_2 transversely and in the same sense. That is,

$$F_{ij} > 0 \quad \text{for} \quad i = 1,\ldots,4, \quad j = 1,2. \tag{7.69}$$

Because of this assumption, it is possible and indeed convenient, to take one of the surfaces Σ_i to be a Poincaré section. Without loss of generality we take this section to be

$$\Pi := \Sigma_1 = \{x : H_1(x) = 0\},$$

as in Fig. 7.15. We construct the resulting Poincaré map associated with this section. The global form of this map is illustrated in Fig. 7.14 in two cases, one where the trajectory passes through the region S_2 close to the point of intersection, and the second where it passes through the region S_3 close to the point of intersection. In the first case the Poincaré map from Σ_1 to itself follows from a trajectory that starts from the point A, intersects Σ_2 at the point B and is then mapped (under the action of the vector field F_2) to the point C. If the flow from B is continued under the action of the vector field F_1, then it will intersect Σ_1 at the point D. This map is a composition of the return map from Σ_1 to itself (mapping A to D and ignoring the impact with Σ_2) *followed* by the PDM(x) which maps D to C. Similarly, in the second case, the Poincaré map can also follow from a trajectory that maps the point P to the point S, so that the trajectory moves from P to $Q \in \Sigma_2$ under the action of the vector field F_3 and the trajectory continued backwards from Q under the action of the vector field F_4 intersects Σ_1 at the point R. In this case the Poincaré map from P to S is a composition of *first* taking the PDM mapping P to R followed by the return map from Σ_1 to itself, mapping R to S and again ignoring the impact with Σ_2.

To construct the global Poincaré map we must first calculate the PDM described above. The calculation for this is illustrated in Fig. 7.15. After some algebra we arrive at the following

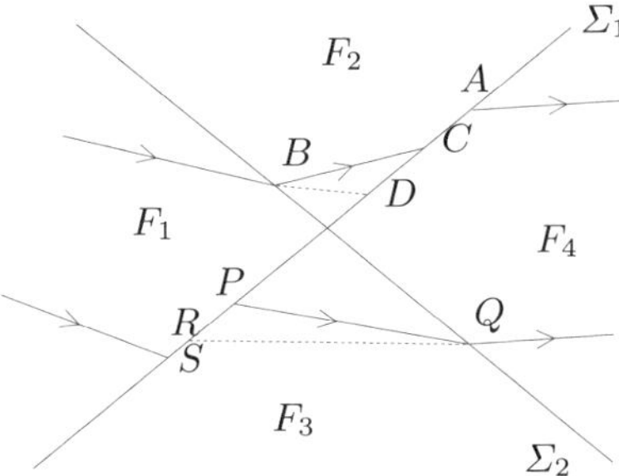

Fig. 7.14. A schematic representation of the action of the global Poincaré map from Σ_1 to itself. In this case either A is mapped to C or P is mapped to S. The solid line shows the actual trajectory, and the dashed line the continuation of the trajectory using different vector fields.

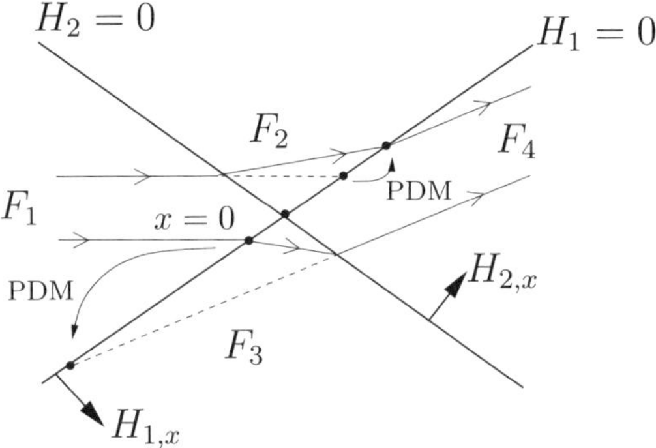

Fig. 7.15. A more detailed representation of the construction of the local PDM in a neighborhood of a boundary-crossing intersecting trajectory.

Theorem 7.5 (The local PDM at a boundary crossing point intersection). *Under the above assumptions, the local PDM based on the Poincaré section Σ_1 is given by*

$$\mathrm{PDM}(x) = \begin{cases} x + \frac{x_2}{F_{12}}\left(F_2\frac{F_{11}}{F_{21}} - F_1\right) + \mathcal{O}\left(|x|^2\right), & \textit{if} \quad x_2 > 0, \\ x + \frac{x_2}{F_{32}}\left(F_4\frac{F_{31}}{F_{41}} - F_3\right) + O\left(|x|^2\right), & \textit{if} \quad x_2 < 0. \end{cases} \tag{7.70}$$

Here the correction is made to a trajectory which is assumed to evolve in region S_1 before hitting the intersection point of the manifolds Σ_1 and Σ_2 and finally evolving in region S_4.

Proof. Taking the Poincaré section to be Σ_1 the PDM can be written as the following flow composition:

$$\Phi_2(\Phi_1(x, \delta_1), \delta_2), \quad \text{when} \quad H_{2,x}x > 0,$$

where δ_1 is the (negative) time required to move from Σ_1 to Σ_2 following the flow Φ_1 associated with the vector field F_1 and δ_2 is the (positive) time required to move from Σ_2 to Σ_1 following the flow Φ_2 associated with the vector field F_2.

On the other hand, when $H_{2,x}x < 0$, the PDM can be expressed as the flow composition

$$\Phi_4(\Phi_3(x, \delta_3), \delta_4).$$

This time the flow directions are reversed so that δ_3 is positive and δ_4 is negative.

Following the earlier approach we expand $\Phi_2(\Phi_1(x, \delta_1), \delta_2)$ in terms of the scalar variables δ_1 and δ_2 to give

$$\Phi_2(\Phi_1(x, \delta_1), \delta_2) = x + F_1\delta_1 + F_2\delta_2 + O\left((\delta_1, \delta_2)^2\right). \tag{7.71}$$

To find the sought time δ_1 we must determine when this trajectory intersects the surface Σ_2. To do this we expand the identity

$$H_2(\Phi_1(x, \delta_1)) = 0$$

in δ_1. This gives

$$H_2(\Phi_1(x, \delta_1)) = H_2(x) + H_{2,x}F_1\delta_1 + O\left(\delta_1^2\right) = 0. \tag{7.72}$$

We now define the auxiliary variable $y = H_2(x)$ and solve (7.72) for δ_1 as a power series in y. Provided that $H_2F_1 \neq 0$, the Implicit Function Theorem guarantees the existence of a solution $\delta_1(x, y)$ with $\delta_1(0, 0) = 0$. After inverting the power series, it is straightforward to show that

$$\delta_1(x, y) = -\frac{y}{H_{2,x}F_1} + O\left(y^2\right). \tag{7.73}$$

To find the time δ_2 when the trajectory intersects Σ_1, we solve the identity

$$H_1(\Phi_2(\Phi_1(x, \delta_1), \delta_2)) = 0.$$

After expanding the above expression in δ_1 and δ_2, we have

$$H_1(x) + H_{1,x}F_1\delta_1 + H_{1,x}F_2\delta_2 + O\left((\delta_1, \delta_2)^2\right) = 0. \tag{7.74}$$

Note that by the choice of Poincaré section $H_1(x) = 0$. Substituting for δ_1 (7.73) into (7.74) gives

$$-\frac{y}{H_{2,x}F_1}H_{1,x}F_1 + H_{1,x}F_2\delta_2 + O\left((\delta_1, \delta_2)^2\right) = 0. \tag{7.75}$$

A further application of the Implicit Function Theorem implies the existence of a smooth function $\delta_2(x, y)$ with $\delta_2(0, 0) = 0$ that is a solution of (7.75). After inverting the power series, we have

$$\delta_2(x, y) = H_{1,x}F_1\frac{y}{H_{2,x}F_1}\frac{1}{H_{1,x}F_2} + O\left(y^2\right). \tag{7.76}$$

We now define a final point $x_f = M(x)$ by $x_f(x,y) = \Phi_2(\Phi_1(x,\delta_1(x,y)), \delta_2(x,y))$.

Substituting for δ_1 (7.73) and for δ_2 (7.76) into (7.71) yields

$$x_f(x,y) = x - F_1 \frac{y}{H_{2,x}F_1} + F_2 H_{1,x} F_1 \frac{y}{H_{2,x}F_1} \frac{1}{H_{1,x}F_2} + O\left(y^2\right). \qquad (7.77)$$

To obtain the PDM when $H_{2,x}x < 0$, we can follow the same steps. We give the final expression for times δ_3 and δ_4.

$$\delta_3(x,y) = -\frac{y}{H_{2,x}F_3} + O\left(y^2\right).$$

$$\delta_4(x,y) = H_{1,x}F_3 \frac{y}{H_{2,x}F_3} \frac{1}{H_{1,x}F_4} + O\left(y^2\right).$$

Define $x_f(x,y) = \Phi_4(\Phi_3(x,\delta_3(x,y)), \delta_4(x,y))$. Thus, we have

$$x_f(x,y) = x - F_1 \frac{y}{H_{2,x}F_3} + F_2 H_{1,x} F_3 \frac{y}{H_{2,x}F_3} \frac{1}{H_{1,x}F_4} + O\left(y^2\right). \qquad (7.78)$$

Expanding (7.77) and (7.78) in x gives the PDM expressed by (7.70), which, when expressing these results in the appropriate notation, proves the theorem. □

Remarks

1. To leading-order, the PDM takes the form of a piecewise-linear map of the form described in Chapter 3 such that each of the maps for $x_2 > 0$ and $x_2 < 0$ is a rank-one update of the identity. Consequently, when studying boundary-crossing events we expect to see the full range of the dynamics introduced in Chapter 3. This will include such behavior as non-smooth fold, Hopf and period-adding bifurcations and also transitions to robust chaos.
2. The ZDM can be constructed in a similar manner and again takes the form of a piecewise-linear map.

It is now straightforward to construct the global Poincaré map close to the (parameterized) periodic orbit $p(t,\mu)$ that passes through the boundary crossing point at $x = 0$, $\mu = 0$, crossing Σ_1 transversally, passing from the region with vector field F_1 to that with vector field F_4. Consider a trajectory, close to $p(t)$ starting from a point x close to the origin, and evolve this ignoring any intersections with Σ_2 close to the point $x = 0$. If $|x|$ and μ are both small, then this will intersect Σ_2 at a point $Q(x)$, which to leading-order is given by

$$Q(x) = Nx + M\mu,$$

where N is the linearization of the Poincaré around the periodic orbit, as derived in Chapter 2. The global Poincaré map $P(x)$ is then given by

$$P(x) = \mathrm{PDM} \circ Q(x) \quad \text{if} \quad x_2 > 0, \qquad P(x) = Q \circ \mathrm{PDM}(x) \quad \text{if} \quad x_2 < 0.$$

This is, of course, a piecewise-linear map.

7.3.2 Derivation of the discontinuity mapping in the corner-collision case

The corner-collision illustrated in Fig. 7.16 is a special case of the above more general event. With reference to Fig. 7.13 an *external corner-collision* arises if we set $F_3 = F_5 \neq F_6 = F_1 = F_2 = F_4$ and we set $\Sigma_1 = \Sigma_5$ and $\Sigma_2 = \Sigma_6$. Similarly an *internal corner-collision* arises if we set $F_1 = F_5 \neq F_6 = F_2 = F_3 = F_4$ and we set $\Sigma_2 = \Sigma_5$ and $\Sigma_1 = \Sigma_6$.

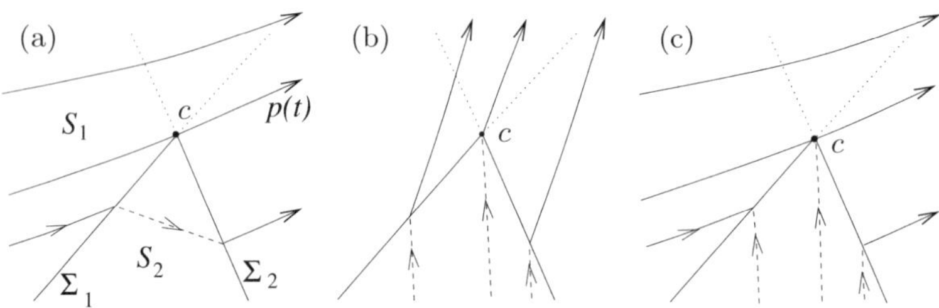

Fig. 7.16. Illustrating, in a general two-dimensional slice, trajectories in the neighborhood of the three types of interaction with the corner depicted in Fig. 7.13(b): (a) external corner-collision; (b) internal; and (c) which is not a corner-collision. Note the topological difference between cases (a) and (b). In(b), all trajectories enter both regions S_5 and S_6, whereas in (a), some trajectories remain locally in S_6. Case (c) fails the hypotheses because the left-hand portion of the boundary, $\Sigma_1 = 0$, is attracting from both sides and hence sliding solutions would occur.

To derive the local form of the PDM for trajectories close to the corner we apply Theorem 7.5 using the relabeled vector fields described above, and make the assumption that the Poincaré surface is always the set corresponding to Σ_1, so that it is Σ_5 for an external corner-collision and Σ_6 for an internal one. Doing this, the following theorem follows immediately.

Theorem 7.6. *(PDM at a corner-collision) The PDM for the external corner-collision is*

$$\mathrm{PDM}_{\mathrm{ext}}(x) = \begin{cases} x, & \text{if } x_6 > 0, \\ x + \frac{x_6}{F_{56}}\left(F_6\frac{F_{55}}{F_{65}} - F_5\right) + O(|x|^2), & \text{if } x_6 < 0. \end{cases} \tag{7.79}$$

Similarly, the PDM for the internal corner-collision is given by

$$\mathrm{PDM}_{\mathrm{int}}(x) = \begin{cases} x + \frac{x_5}{F_{55}}\left(F_6\frac{F_{56}}{F_{66}} - F_5\right) + O(|x|^2), & \text{if } x_5 > 0, \\ x, & \text{if } x_5 < 0. \end{cases}$$

Note that the choice of Poincaré section here is rather arbitrary and we could have used either of the two surfaces Σ_5 or Σ_6. A similar calculation allows us to determine the ZDM at a corner-collision and the global Poincaré map. Further details are given in [76].

7.3.3 Examples

To motivate this theory we consider two examples of a corner-collision.

Example 7.9 (An explicitly calculable corner-collision).

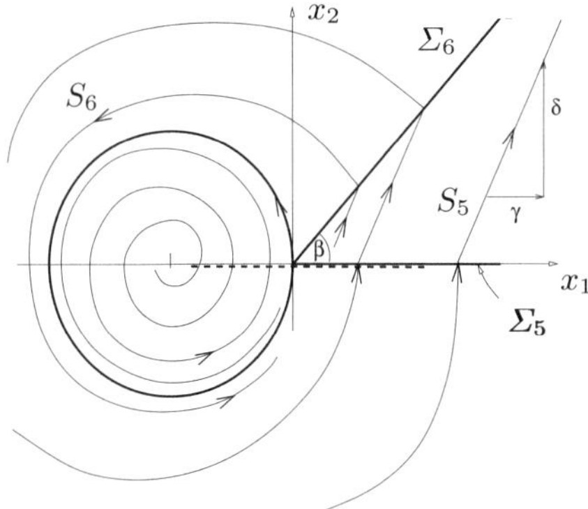

Fig. 7.17. Sketch of the phase potrait of (7.80), (7.81) with $a = 1$.

Consider first an example where a hyperbolic limit cycle grazes with a corner in an autonomous, piecewise-smooth vector field that is soluble in closed from. Specifically we take a system

$$\begin{array}{l} \dot{x}_1 = \gamma \\ \dot{x}_2 = \delta \end{array}, \quad \text{for } x_1 > 0,\ x_2 > 0,\ x_2 < x_1 \tan(\beta) \quad (\text{REGION } S_5), \tag{7.80}$$

$$\begin{array}{l} \dot{r} = \varepsilon r(a - r) \\ \dot{\theta} = 1 \end{array}, \quad \text{otherwise } (\text{REGION } S_6). \tag{7.81}$$

Here

$$x_1 + 1 = r\cos(\theta), \quad x_2 = r\sin(\theta),$$

and γ, δ, β, ε and a are real constants satisfying the constraints

$$0 < \beta < \pi/2, \quad \delta > \gamma \tan(\beta). \tag{7.82}$$

See Fig. 7.17(a). Consider the system (7.81); for $a > 0$ there is a limit cycle that is stable if $\varepsilon > 0$; at $a = 1$ this limit cycle collides with the boundary of region S_5 in an external corner-collision bifurcation. Specifically we take

$$H_5(x_1, x_2) = -x_2, \quad H_6(x_1, x_2) = x_2\cos(\beta) - x_1\sin(\beta).$$

The constraints (7.82) ensure that the inequality constraints (7.69) are satisfied along $H_5 = 0$ and $H_6 = 0$.

Since the systems in regions S_5 and S_6 are solvable in closed form we can explicitly construct the Poincaré map associated with the Poincaré section $\{x_2 = 0, x_1 > -1\}$ which is a portion of Σ_5. In region S_6 the general solution takes the form

$$r(t) = \frac{a r_0 \exp(\varepsilon a t)}{r_0 \exp(\varepsilon a t) + a - r_0}, \qquad \theta = \theta_0 + t. \tag{7.83}$$

For $x_1(0) < 0$ ($r_0 < 1$), this equation defines the Poincaré map after setting $r_0 = x_1(0) + 1$, $\theta_0 = 0$ and $t = 2\pi$. For $x_1 > 0$, however, we must first solve (7.80) until the time

$$\widehat{t} = \frac{x_1(0)\tan(\beta)}{\delta - \gamma \tan(\beta)} \tag{7.84}$$

at which $x_2 = x_1 \tan(\beta)$. Taking the (x_1, x_2)-values at this point, converting to polar co-ordinates $(\widehat{r}, \widehat{\theta})$ where

$$\widehat{r}\cos(\widehat{\theta}) = x_1(0) + \gamma\,\widehat{t} + 1, \quad \widehat{r}\sin(\widehat{\theta}) = \delta\,\widehat{t}, \tag{7.85}$$

substituting these values as r_0 and θ_0 in (7.83) and evaluating at $t = 2\pi - \widehat{\theta}$ then gives us an analytic expression for the Poincaré mapfor $x_1(0) > 0$. Thus we obtain

$$x_1(0) \mapsto \frac{a\widehat{r}\exp[\varepsilon a(2\pi - \widehat{\theta})]}{\widehat{r}\exp[\varepsilon a(2\pi - \widehat{\theta})] + a - \widehat{r}}, \qquad x_1(0) > 0, \tag{7.86}$$

where $\widehat{\theta}$ and $\widehat{r}$ are related to $x_1(0)$ via (7.84) and (7.85).

As shown, in [77], in the case where $\varepsilon = 0.1, \beta = \pi/4, \gamma = 3/8$ and $\delta = 0.5$, the slopes of the map are such that the corner-collision has the effect of destroying the limit cycle. Namely, for these parameter values, before the bifurcation, i.e., for $a < 1$, there is a stable limit cycle lying solely in region S_6, but this coexists with an unstable limit cycle that passes through region S_5. At $a = 1$ these two periodic solutions coalesce and for $a > 1$ they have disappeared. Note, finally, that unlike a saddle-node bifurcation for a smooth system, the Floquet multipliers of the two periodic orbits (the slopes of the two portions of the map) do not approach 1 as $a \to 1^-$ (see [77] for further details).

Example 7.10 (Example 7.3 continued; corner-collision in the buck converter).

We conclude the examples, and the chapter, by returning to the DC–DC buck converter, which is described in case study V, of Chapter 1. Here we shall limit ourselves to an analytical explanation of a phenomenon observed numerically in [75, 104, 81], namely that corner-collision of a periodic orbit is associated with a piecewise-linear map. Moreover the dynamics of this map are such that it causes a fold (actually a sharp corner) in the bifurcation diagram of a branch of periodic orbits. Specifically a sequence of such folds was found for certain $3T$ and $5T$-periodic orbits, as part of a bigger picture of a spiraling bifurcation diagram that can be explained in terms of a local analysis close to a (codimension-two) sliding periodic orbit. What we shall show here, using the preceding analysis, is that by calculating a few features of the single trajectory undergoing the corner-collision, we can calculate precisely the angle of the fold in the bifurcation diagram and determine the stability of orbits.

Consider the model (7.9). Suppose that at some $E = E_0$, an nT-periodic orbit for some $n > 1$ collides with the upper corner of the function $V_r(t)$ at

$$t = t_0 = 0 \bmod T, \quad V = \gamma + \eta T, \quad I = I_0,$$

for some I_0 representing the value of current as the periodic trajectory crosses the corner. Moreover (see Fig. 7.18), we can have both internal and external corner-collision. We shall treat here only the external corner-collision case. Internal corner-collisions are similarly treated in [77].

As a first step, define local co-ordinates

$$x_1 = V - (\gamma + \eta T), \quad x_2 = I - I_0, \quad x_3 = t - t_0 \tag{7.87}$$

and rewrite the equation (7.9) in the form

$$\begin{aligned} \dot{x_1} &= -a_1 + b_1 x_1 - c_1 x_2, \\ \dot{x_2} &= -a_2 - c_2 x_2 + d\Theta(\sigma(x_3) - x_1), \\ \dot{x_3} &= 1, \end{aligned}$$

in which

$$a_1 = \frac{\gamma + \eta T - RI_0}{RC}, \; b_1 = \frac{1}{C}, \; c_1 = \frac{1}{RC}, \; a_2 = \frac{\gamma + \eta T}{L}, \; c_2 = \frac{1}{L}, \; d = \frac{E}{L},$$

Θ is the Heaviside step function and

$$\sigma(x_3) = \eta[(x_3 \bmod T) - T].$$

For this system we have

$$\Sigma_5 := \{H_5 = 0\} = \{x_1 = \sigma(x_3)\}, \quad \Sigma_6 := \{H_6 = 0\} = \{x_3 = 0\},$$

and $\mathcal{C} = \{x_1 = 0, x_3 = 0\}$.

The corner-collision happens at $x_1 = x_2 = x_3 = 0$, and it can be checked that the conditions of the preceding theory are met there.

It is convenient to take a Poincaré section $\Pi = \{(x, y, z) : z = 0\}$ in order to calculate the local effect of trajectories that are close to the corner-collision (the PDM and ZDM are then equivalent). The derivatives of the functions defining the discontinuity surfaces at the origin are then given by

$$H_{5,x} = (-1, 0, \eta), \quad H_{6,x} = (0, 0, 1),$$

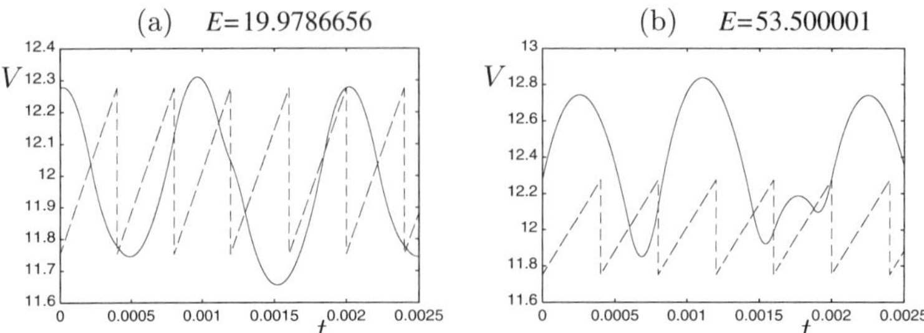

Fig. 7.18. Periodic orbits of the DC/DC buck converter with period $5T$ undergoing (a) an external and (b) an internal corner collision.

with the vector fields given by

$$F_5 = (-a_1, -a_2, 1), \quad F_6 = (-a_1, -a_2 + d, 1).$$

Using (7.79), the discontinuity map for an *external* corner-collision associated with the Poincaré surface $z = 0$ then takes the form

$$P_{\text{PDM}} : \begin{pmatrix} x_1 \\ x_2 \end{pmatrix} \mapsto \begin{pmatrix} x_1 \\ x_2 + k_1(E)x_1 \end{pmatrix} + \text{ h.o.t}, \tag{7.88}$$

where

$$\begin{aligned} k_1(E) &= \frac{-d\eta}{a_1 + \eta} = -\frac{E\,RC}{L(\gamma + \eta T - RI_0 + \eta RC)}, \\ k_2(E) &= \frac{-da_1}{a_1 + \eta} = -\frac{E(\gamma + \eta T - RI_0)}{L(\gamma + \eta T - RI_0 + \eta RC)}. \end{aligned} \tag{7.89}$$

Note that this map depends on the bifurcation parameter E.

We must next compose the map P_{PDM} with a global return map associated with the flow close to the periodic orbit at $E = E_0$ ignoring the effects of the corner. To leading-order this return map takes the form

$$P_\Pi : \begin{pmatrix} x_1 \\ x_2 \end{pmatrix} \mapsto N \begin{pmatrix} x_1 \\ x_2 \end{pmatrix} + M(E - E_0), \tag{7.90}$$

where the coefficients of the matrix N and vector M must in general be calculated numerically for the particular periodic orbit undergoing the corner-collision.

For trajectories that do not cross the ramp signal $x = \sigma(z)$ close to the corner the global Poincaré mapis P_Π. But for trajectories that do cross, it takes the form

$$\begin{aligned} \begin{pmatrix} x_1 \\ x_2 \end{pmatrix} &\mapsto \begin{bmatrix} a_{11} & a_{12} \\ a_{21} + a_{11}k_1(E_0) & a_{22} + a_{12}k_1(E_0) \end{bmatrix} \begin{pmatrix} x_1 \\ x_2 \end{pmatrix} \\ &+ \begin{pmatrix} b_{11} \\ b_{21} + k_1(E_0)b_{11} \end{pmatrix} (E - E_0), \end{aligned} \tag{7.91}$$

where

$$N = \begin{bmatrix} a_{11} & a_{12} \\ a_{21} & a_{22} \end{bmatrix}, \quad M = \begin{pmatrix} b_{11} \\ b_{21} \end{pmatrix}.$$

As an example, an external corner colliding $5T$-periodic orbit depicted in Fig. 7.18(a) occurs for the values

$$E = E_0 = 19.9786656, \quad I_0 = 0.56929860.$$

From these data we calculate from (7.89) that

$$k_1(E_0) = -0.934\,.$$

By computing this trajectory over five periods and using numerical differencing to calculate derivatives, we find that

$$N = \begin{bmatrix} -3.96 & -44.0 \\ -2.03 & -23.0 \end{bmatrix}, \quad M = \begin{pmatrix} 0.31 \\ 0.15 \end{pmatrix}. \tag{7.92}$$

Figure 7.19 shows the result of substituting these numerical values into the analytical Poincaré map (7.90), (7.91) and its comparison with a map calculated from straightforward numerical integration of trajectories. We have chosen to illustrate just a one-dimensional approximation to this two-dimensional map, by only displaying the effect of changes in initial current y. The results in Fig. 7.19(a) and (b) show good quantitative and qualitative agreement between the local theory and the numerical calculations at $E = E_0$. They also illustrate the extent of the region of validity for the local analysis; for $-0.006 < y(0) < -0.0035$ at $E = E_0$, the local map is qualitatively correct, but outside of this region the numerical map shows extra corners. This is due to other corner-collisions taking place at $t = nT$ for some $n \leq 5$. Note from panel (b) in particular that there is no corner in the x-component of the numerically computed map — this component of the map is smooth — which is in complete agreement with the analytical result (7.91) (there is no change in the x-component).

Panels (c) and (d) show the effect of variation of E, with the existence of a fixed point on such a graph of $y(5T)$ against $y(0)$ being indicative only of a fixed point of the full two-dimensional map. Here again there is good agreement between theory and numerics on how the map is perturbed as E varies and that two fixed points (corresponding to unstable periodic orbits of the ODEs) are created at $E = E_0$ and coexist for $E > E_0$.

We have, thus, shown that the Poincaré map close to the corner-collision has the form of a continuous piecewise-linear map. We can now use the results of the analysis of border-collision bifurcations given in Chapter 3 to classify the dynamic behavior of the DC–DC converter. At $E = E_0$ the map takes the form

$$\begin{pmatrix} x_1 \\ x_2 \end{pmatrix} \mapsto \begin{cases} N_1 \begin{pmatrix} x_1 \\ x_2 \end{pmatrix}, \text{ when } & -3.96x_1 - 44x_2 < 0, \\ N_2 \begin{pmatrix} x_1 \\ x_2 \end{pmatrix}, \text{ when } & -3.96x_1 - 44x_2 \geq 0, \end{cases} \tag{7.93}$$

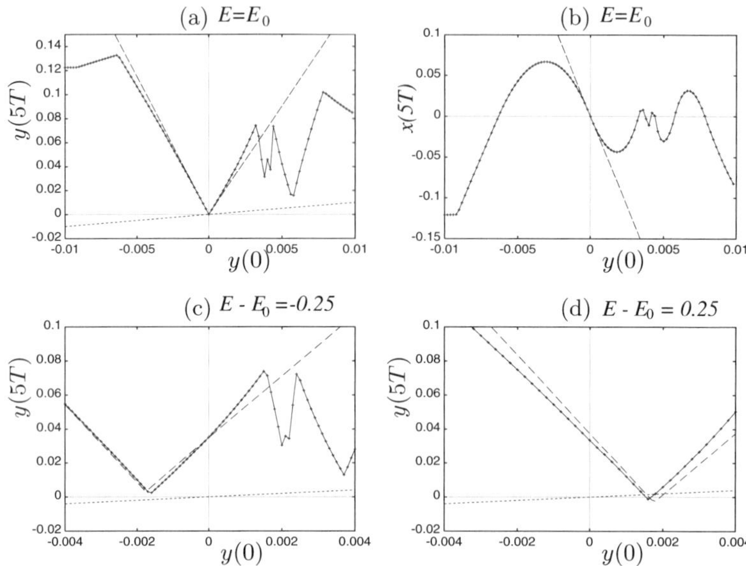

Fig. 7.19. The Poincaré mapfor a $5T$-periodic external corner colliding orbit at $E_0 = 19.9786656$, computed numerically (solid line with crosses) and via the corner-collision analysis (dashed line). A one-dimensional slice of the map is taken considering the effect of varying only the initial current $y(0)$. (a) and (b) depict the final current and voltage, respectively, for $E = E_0$; (c) and (d) show the effect on the final current of variation of the bifurcation parameter E. In the final current versus initial current figures, the $45°$ line is depicted as dotted; viewing the graphs as approximations of one-dimensional maps, intersections with this line are indicative of nearby fixed points of the two-dimensional map.

with $N_1 = \begin{bmatrix} -3.96 & -44 \\ 1.67 & 18.10 \end{bmatrix}$ and $N_2 = \begin{bmatrix} -3.96 & -44 \\ -2.03 & -23 \end{bmatrix}$. Following the methodology introduced in Chapter 3 we now calculate eigenvalues of N_1 and N_2. We get the following four eigenvalues: $\lambda_{l1} = 0.126$, $\lambda_{l2} = 14.010$, $\lambda_{n1} = -0.065$ and $\lambda_{n2} = -26.895$. Subscript 'l' denotes the eigenvalues of matrix N_1 and 'n' the eigenvalues of N_2. The number of the eigenvalues of N_1 and N_2, with real part greater than 1, is odd (λ_{l2} is the only eigenvalue with the real part greater than 1). Therefore, according to our discussion in Chapter 3 we expect a non-smooth fold bifurcation. This agrees with our discussion on the character of the map for neighboring values of parameter E. However, using the analysis given in Chapter 3 we can also make prediction on the possible existence of period-two points that are involved in the bifurcations. We note that there is an odd number of the eigenvalues of N_1 and N_2 that have the real part less than -1. Therefore, a period-two point is involved in the border-collision bifurcation. The stability and whether the period-two point collides and vanishes together with the fixed points, or whether it is born in the collision, can be determined by calculating the eigenvalues of the matrix compositions N_1N_2 and N_1N_1. We find, using a similar notation, the following set of the eigenvalues $\lambda_{ln1} = -0.008$, $\lambda_{ln2} = -384.618$, $\lambda_{nn1} = 0.004$ and $\lambda_{nn2} = 723.317$. The eigenvalues of N_1N_2 determine the stability of the period-two orbit. Clearly the orbit is unstable because λ_{ln2} is less than -1. Moreover, because the number of the eigenvalues of the N_1N_2 and N_2N_2 that are greater than 1 is odd, the period-two orbit collides and vanishes with the fixed points at the border-collision.

This chapter has considered grazing in piecewise-smooth systems specifically in the absence of sliding motion when discontinuity boundaries are simultaneously attracting from both sides. The next chapter shall consider specifically the case of Filippov systems and the DIBs associated with periodic orbits doing structurally unstable things with respect to the boundary of the sliding region of a discontinuity boundary.

8

Sliding bifurcations in Filippov systems

We have already shown in Section 1.3.2 that the onset of sliding motion in Filippov piecewise-smooth systems can lead to intricate dynamics. The current chapter focuses on the classification of discontinuity-induced bifurcations (DIBs) caused by the interaction of a trajectory with the boundary of a sliding region. First, in Section 8.1, we classify the four principal codimension-one cases — *crossing-sliding*, *grazing-sliding*, *switching-sliding* and *adding-sliding* — and give expressions for the local discontinuity mappings (DMs) close to such points. Section 8.2 then gives a detailed motivating example, before going on in Section 8.3 to an explanation of how these DMs were calculated. Section 8.4 discusses the composition of these DMs with a Poincaré mapin the case that the trajectory undergoing the sliding bifurcation is part of a limit cycle. We show in the *grazing-sliding* case that the dynamics local to the bifurcation point are governed by a piecewise-linear map that is singular in one region, as studied in Section 3.6. We then explicitly calculate this normal form map for the example of a friction oscillator (case study IV in Chapter 1) in order to explain its dynamics close to grazing-sliding. Finally, Section 8.6 presents briefly other possible bifurcations involving limit cycles and sliding.

8.1 Four possible cases

As in the previous chapter we consider n-dimensional piecewise-smooth continuous autonomous systems of ODEs, which, local to some discontinuity boundary $\Sigma = \{x \in \mathcal{D} : H(x) = 0\}$, can be written in the form

$$\dot{x} = \begin{cases} F_1(x), & \text{if } H(x) > 0, \\ F_2(x), & \text{if } H(x) < 0, \end{cases} \tag{8.1}$$

where, for the time being we have suppressed any parameter dependence. In this chapter we shall assume that the degree of smoothness is uniformly 1. That is, we consider Filippov systems for which $F_2(x) \neq F_1(x)$ for all $x \in \Sigma$.

In Chapter 7, we considered grazing bifurcations in such systems, but only under the condition (7.7) that

$$\mathcal{L}_{F_1}(H(x))\mathcal{L}_{F_2}(H(x)) := (H_x F_1)(H_x F_2) \geq 0, \qquad \forall x \in \Sigma. \tag{8.2}$$

In particular, this would imply that at a grazing point *both* vector fields should be tangent to Σ; that is, $\mathcal{L}_{F_1}(H(x)) = \mathcal{L}_{F_2}(H(x)) = 0$. This is a non-generic assumption (but was motivated in Chapter 7 by a class of second-order oscillators with non-smooth forcing). In this chapter, we will relax this condition so that generically when $H_x F_1 = 0$ we have that $H_x F_2 \neq 0$ and so we are at the boundary of a region of sliding. Recall that sliding motion occurs when the product in (8.2) is negative; see Section 2.2.3 to which we refer for definitions and notation. Adopting the Utkin equivalent control method, we define the sliding flow as

$$F_s = \frac{F_1 + F_2}{2} + \frac{F_2 - F_1}{2}\beta(x), \tag{8.3}$$

where specifically the equivalent control is

$$\beta(x) = -\frac{\mathcal{L}_{F_1}(H(x)) - \mathcal{L}_{F_2}(H(x))}{\mathcal{L}_{F_1}(H(x)) - \mathcal{L}_{F_2}(H(x))}. \tag{8.4}$$

The sliding region is given by

$$\widehat{\Sigma} := \{x \in \Sigma : -1 \leq \beta \leq 1\}, \tag{8.5}$$

and its boundaries are

$$\partial\widehat{\Sigma}^{\pm} := \{x \in \Sigma : \beta = \pm 1\}. \tag{8.6}$$

Note that the boundary $\partial\widehat{\Sigma}^{+}$ corresponds to points where there is a tangency between the vector field F_1 and Σ (that is where $\mathcal{L}_{F_1}H(x) = 0$) and $\partial\widehat{\Sigma}^{+}$ to where there is a tangency between F_2 and Σ (so that $\mathcal{L}_{F_2}H(x) = 0$).

Except briefly in Section 8.6 below, we shall deal exclusively with the case of attracting sliding; that is where the sliding set $\widehat{\Sigma}$ is attracting from both sides.

8.1.1 The geometry of sliding bifurcations

sliding bifurcations are defined here as DIBs caused by the interaction between limit cycles of the Filippov system and the boundary of the sliding region $\widehat{\Sigma}$. Four different codimension-one DIBs involving interaction with the boundary $\partial\widehat{\Sigma}^{\pm}$ of a sliding region were originally identified by Feigin [98] and were subsequently analyzed by di Bernardo, Kowalczyk and coworkers in [84, 159, 85, 86] for general n-dimensional systems of the form (8.1). A three-dimensional schematic representation is given in Fig. 8.1, where we assume the local phase space topology introduced in Section 2.2.3 and depict only

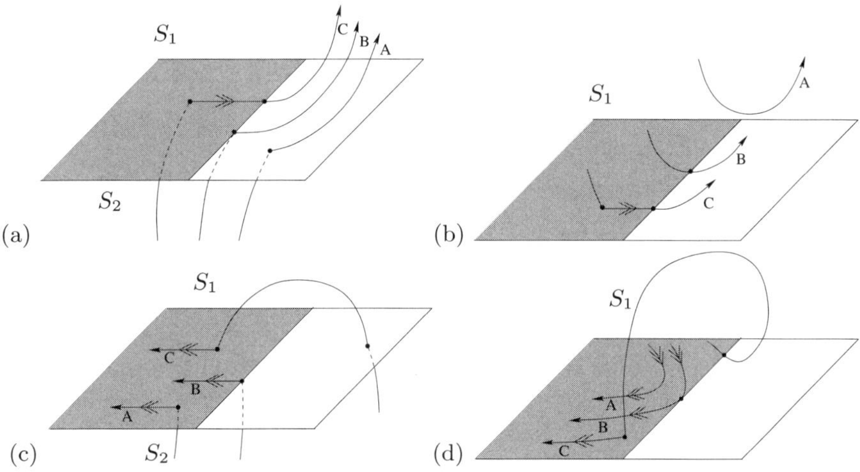

Fig. 8.1. Illustration in three dimensions of the four codimension-one bifurcation scenarios involving collision of a segment of trajectory with the boundary of the sliding region: (a) crossing-sliding, (b) grazing-sliding, (c) switching-sliding and (d) adding-sliding. In each case the discontinuity set Σ is a horizontal plane, with F_1 applying above Σ and F_2 below. The shaded portion represents the sliding region $\widehat{\Sigma}$, and the boundary in question is $\partial\widehat{\Sigma}^-$.

segments of trajectories ('A', 'B' and 'C') in a neighborhood of a boundary $\partial\widehat{\Sigma}^-$. In each case trajectory B is the critical one that defines the sliding bifurcation in question. Trajectories A and C would occur under perturbation of trajectory B, such that A → B → C would occur under a continuous small change in *initial conditions*. Later on, in Section 8.3, we will assume that what distinguishes the particular trajectories A and C is that they are small *parameter* perturbations from a limit cycle (trajectory B) that undergoes the DIB. (In fact, as we shall see in Section 8.3, in *grazing-sliding* — see Fig. 8.1b — limit cycles A and C might exist for the same parameter values. In the other three cases, close to the bifurcation, the cycles 'A', 'B' and 'C' must exist for distinct parameter values.) For definiteness, in Fig. 8.1 and throughout this chapter, we assume that the boundary in question is $\partial\widehat{\Sigma}^-$, but this is without loss of generality since we may relabel F_1 as F_2 and vice versa.

Let us now focus on Fig. 8.1. Panel (a) depicts the scenario we term a *crossing-sliding bifurcation*. Here, a trajectory crosses the switching manifold Σ transversally, precisely at the boundary of the sliding strip $\partial\widehat{\Sigma}^-$. This forms the boundary between two topologically distinct kinds of trajectory:

C that undergoes a small segment of sliding motion, and A that crosses Σ transversally, but outside the sliding region. Note that, by continuity, the sliding trajectory C evolves under the sliding flow F_s towards the boundary $\partial\widehat{\Sigma}^-$, at which point by definition [see (8.3), (8.6)] $F_s = F_1$, so the trajectory leaves the switching manifold tangentially there. The name 'crossing-sliding' refers to the fact that all trajectories under consideration approach Σ with non-zero speed, but as we move from A $\rightarrow$ B $\rightarrow$ C, we *cross* the sliding boundary $\partial\widehat{\Sigma}$

The second case depicted in Fig. 8.1(b) is the one we call a *grazing-sliding bifurcation*. Here, the trajectory A, lying locally within region S^+, is continuously perturbed into a trajectory B that forms a point of grazing with the switching manifold from above. Under further perturbation, we find a trajectory C, which has a section of sliding motion that evolves towards $\partial\widehat{\Sigma}^-$, at which point it leaves Σ tangentially. This scenario is the most directly analogous to the grazing bifurcation analyzed in Chapter 7, especially from the point of view of trajectory A; hence, the name *grazing*-sliding. Note that the conditions on the vector fields to define the geometry in this case are identical to those of the previous, crossing-sliding case (compare panels (a) and (b) of the figure). The difference comes in the trajectory segments we are interested in. In panel (a), we consider trajectories that arise from *below* the switching manifold, and in (b) trajectories from above. This nicely illustrates the non-uniqueness backwards in time associated with Filippov systems that undergo attracting sliding. The forward part of trajectories B is the same in both panels from the point of intersection with $\partial\Sigma$ onwards, but their prior histories are wildly different!

A third kind of bifurcation event, which we shall refer to as a *switching-sliding bifurcation*, is depicted in Fig.8.1(c). At first sight this scenario seems similar to the crossing-sliding case in panel (a). The distinction is in the sign of the curvature of trajectories under vector field F_1 that applies above Σ. Here trajectories curve downwards. Another way of stating this is that the sliding boundary $\partial\widehat{\Sigma}^-$ is repelling within the sliding region in this case, whereas in (a) it was attracting. Thus, if we perturb trajectory B to the right to form trajectory A, we will no longer cross Σ transversally as in panel (a), but now the downwards curvature makes us intersect Σ again, this time within the sliding region. Perturbing B to the left gives trajectory A, which evolves exclusively within the sliding region, away from the boundary $\partial\widehat{\Sigma}^-$. The name *switching*-sliding reflects the gaining of an extra switching transition (transversal crossing of Σ) under the sequence A $\rightarrow$ B $\rightarrow$ C.

The fourth and last case is the so-called *adding-sliding bifurcation*, shown in Fig. 8.1(d). It differs from the other three cases in that there is a segment of trajectory A that lies entirely within the sliding region $\widehat{\Sigma}$. The bifurcation event is that this trajectory is perturbed into B, which has a tangency with the boundary $\partial\widehat{\Sigma}^-$. Thus this is effectively a grazing bifurcation of the sliding motion itself. However, the effect is not as in a standard grazing bifurcation, because the perturbation into C causes the trajectory to leave Σ tangentially, to evolve above the Σ and then to land back within the sliding region. The name *adding-sliding* refers to the transition A $\to$ B $\to$ C, where a locally uninterrupted sliding motion is transformed into two separate pieces of sliding, separated by a region of free, non-sliding evolution. Thus, we have added one to the number of sliding segments in the trajectory. Note also, that the adding-sliding point itself, where the trajectory B grazes with the sliding boundary $\partial\widehat{\Sigma}$, is precisely the point at which the boundary switches between being attracting and repelling. Recall that the attraction or repulsion of this boundary was precisely what distinguished the cases (a) and (c) above. Hence one might easily anticipate a codimension-two DIB where any of the critical trajectories B in (a)–(c) happened additionally to interact with a point of $\partial\widehat{\Sigma}^-$ at which the sliding flow is tangent. We shall delay any treatment of codimension-two bifurcations until Chapter 9.

The above four cases in some sense represent the simplest ways that trajectories can interact with the boundary of a sliding region. The main aim of this chapter is to understand and classify the dynamical consequences of these bifurcations when the trajectory in question is a limit cycle, especially when this causes chaotic motion or a rapid change in attractor. We shall treat the general n-dimensional situation. Kuznetsov, Rinaldi & Gragnia [169] give a more complete classification of possible DIBs that involve sliding in two-dimensional Filippov systems with a single switching surface. This includes possibilities of global bifurcation, and also some equilibrium bifurcations that we covered in Chapter 5. In general n-dimensional systems, a complete classification remains unknown. In Section 8.6 below, we shall briefly treat two other cases of DIBs that involve repulsive sliding segments or multiple switching surfaces.

8.1.2 Normal form maps for sliding bifurcations

Having dealt with the geometry of sliding, let us now turn to an analytical description of the four kinds of sliding bifurcations we have just introduced. For each case, we can calculate a discontinuity mapping that accounts for the extra correction that must be added to account for the short extra passage of the more complex trajectory C in Fig. 8.1 when solving for a trajectory with the same event sequence as trajectory A. The results are summarized in Table 8.1, which gives the size of the leading-order term in the discontinuity mapping as a function of the size ε of a perturbation in initial conditions away from the critical trajectory B. The precise functional forms of these maps will be given shortly.

We start by giving analytical conditions that define each of the four bifurcation scenarios, along with appropriate non-degeneracy assumptions. In all four cases, the critical trajectory involved in the bifurcation event has a point of intersection with the boundary of the sliding region $\partial\widehat{\Sigma}^-$. Suppose this point of intersection occurs at $x = x^*$; then in all four cases, we have the following *defining* conditions:

Table 8.1. Summary of the singularities arising in each of the four sliding bifurcation scenarios.

Bifurcation type	DM leading-order term	Map singularity
crossing-sliding	$\varepsilon^2 + O(\varepsilon^3)$	2
grazing-sliding	$\varepsilon + O(\varepsilon^{3/2})$	1
switching-sliding	$\varepsilon^3 + O(\varepsilon^4)$	3
adding-sliding	$\varepsilon^2 + O(\varepsilon^{5/2})$	2

$$H(x^*) = 0, \qquad H_x(x^*) \neq 0, \tag{8.7}$$

$$\beta(x^*) = -1, \qquad \mathcal{L}_{F_1}(x^*) = 0. \tag{8.8}$$

(Note that $\beta(x^*) = -1$ implies $F_s(x^*) = F_1(x^*)$.) The first conditions (8.7) state that the point x^* belongs to the switching manifold, which is well defined; whereas the second ones (8.8) state that x^* is on the boundary of the sliding region, which without loss of generality we assume to $\partial\widehat{\Sigma}^-$.

Now let us turn to *non-degeneracy* conditions for each of the four sliding bifurcations. The first is that in a neighborhood of x^*, the vector field F_2 is not grazing and points towards Σ. That is

$$H_x F_2(x^*) > 0. \tag{8.9}$$

Other considerations involve the tangency of the sliding flow to $\partial\widehat{\Sigma}^-$. In order to define such a tangency, note that a convenient notation for the normal vector to $\partial\widehat{\Sigma}^- := \{x \in \Sigma : \beta(x) = -1\}$ is β_x, which given (8.8) implies

$$\beta_x = \frac{-2}{(\mathcal{L}_{F_2}H(x))^2}\frac{d}{dx}\mathcal{L}_{F_1}H(x); \tag{8.10}$$

see Fig. 8.2. Note that the denominator of (8.10) is positive, according to (8.9).

With reference to the geometry in Figs. 8.1 and 8.2, note that the crossing-sliding and grazing-sliding cases require the sliding flow to evolve locally *towards* $\partial\widehat{\Sigma}^-$. Hence we require

$$\left.\frac{\partial\beta(\Phi_s(x^*,0))}{\partial t}\right|_{t=0} < 0.$$

Where Φ_s is the flow operator corresponding to the sliding flow F_s. However, we have that $F_s = F_1$ at x^* by (8.8); hence, $\Phi_s(x^*,0) = \Phi_1(x^*,0)$. Moreover

$$\frac{\partial\beta(\Phi_1(x^*,0))}{\partial t} = \beta_x F_1(x^*) =: \mathcal{L}_{F_1}\beta(x^*).$$

Therefore the sign of $\mathcal{L}_{F_1}\beta(x^*)$ determines whether the boundary $\partial\widehat{\Sigma}^-$ is attracting or repelling with respect to the siding flow. crossing-sliding and grazing-sliding will therefore require the non-degeneracy condition

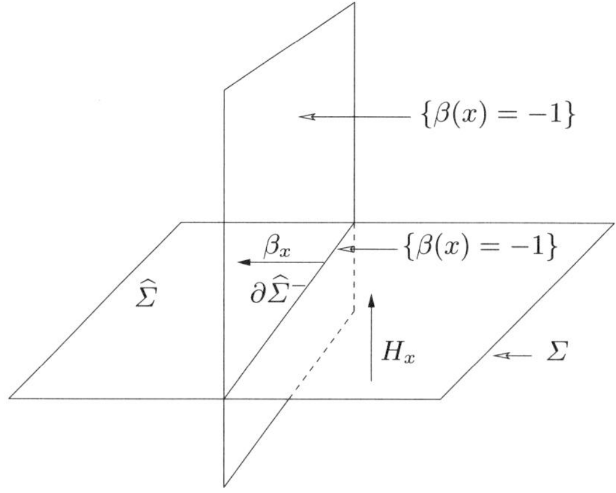

Fig. 8.2. Geometry of the boundary $\partial\widehat{\Sigma}^-$.

$$\mathcal{L}_{F_1}\beta(x^*) < 0, \tag{8.11}$$

whereas switching-sliding (case (c) of Fig. 8.1) requires

$$\mathcal{L}_{F_1}\beta(x^*) > 0, \tag{8.12}$$

so that the sliding flow points *away* from the boundary.

adding-sliding (case (d) of Fig. 8.1) is more subtle. Here we require an additional *defining condition* that there is a point of tangency of the sliding flow with $\partial\widehat{\Sigma}^-$ at the bifurcation point. That is

$$\mathcal{L}_{F_1}\beta(x^*) = 0. \tag{8.13}$$

Moreover, the geometry clearly implies that the sliding flow must reach a local minimum of β at the bifurcation point. Hence, we also require

$$\frac{\partial^2\beta(\Phi_s(x^*,0))}{\partial^2 t} > 0;$$

that is

$$\mathcal{L}^2_{F_1}\beta(x) := \beta_x F_{1x}F_1 + \beta_{xx}F_1^2 > 0. \tag{8.14}$$

We can now state the following Theorem on the form of the ZDM at each of the four sliding bifurcations. In each case, ZDM case describes the correction that must be made to trajectories of type A in Fig. 8.1 in order to obtain trajectories of type C.

Theorem 8.1 (ZDM for sliding bifurcations). *Suppose a piecewise-smooth system of the form (8.1) undergoes a sliding bifurcation at point x^*, defined by the conditions (8.7) and (8.8) under the non-degeneracy assumption (8.9). Then we have the following four cases:*

crossing-sliding; *under the additional non-degeneracy condition (8.11) the ZDM for trajectories starting in S_2 ($H(x) < 0$) to leading-order is given by*

$$x \mapsto \begin{cases} x, & \text{if } \mathcal{L}_{F_1} H(x) \le 0, \\ x + (\mathcal{L}_{F_1} H(x))^2 \dfrac{F_2(x) - F_1(x)}{2\mathcal{L}_{F_2} H(x) \mathcal{L}^2_{F_1} H(x)}, & \text{if } \mathcal{L}_{F_1} H(x) > 0, \end{cases} \tag{8.15}$$

where the error term for $\mathcal{L}_{F_1} H(x) > 0$ is $O(|x - x^*|^3)$.

grazing-sliding; *also under the additional non-degeneracy condition (8.11) the ZDM for trajectories starting in S_1 ($H(x) > 0$) to leading-order is given by*

$$x \mapsto \begin{cases} x, & \text{if } H(x) \ge 0, \\ x + \dfrac{H(x)(F_2(x) - F_1(x))}{\mathcal{L}_{F_2} H(x)}, & \text{if } H(x) < 0, \end{cases} \tag{8.16}$$

where the error term for $H(x) < 0$ is $O(|x - x^*|^{3/2})$.

switching-sliding; *under the additional non-degeneracy assumption (8.12), the ZDM may be written to leading-order in the form*

$$x \mapsto \begin{cases} x, & \text{if } \mathcal{L}_{F_1} H(x) \le 0, \\ x + \dfrac{2}{3} \dfrac{(\mathcal{L}_{F_1} H(x))^3}{(\mathcal{L}_{F_2} H(x))^2 (\mathcal{L}^2_{F_1} H(x))^2} Q, & \text{if } \mathcal{L}_{F_1} H(x) > 0, \end{cases} \tag{8.17}$$

where

$$Q = \mathcal{L}_{F_2} H(x)(F_{1,x} F_d - F_{d,x} F_1) - \mathcal{L}_{(F_{1,x} F_d - F_{d,x} F_1)} H(x) F_d$$

and $F_d = F_2 - F_1$. The error term for $\mathcal{L}_{F_1} H(x) > 0$ is $O(|x - x^*|^4)$.

adding-sliding; *under the additional defining condition (8.13) and non-degeneracy assumption (8.14), the ZDM to leading-order is*

$$x \mapsto \begin{cases} x, & \text{if } \mathcal{L}_{F_1} H(x) \ge 0, \\ x - \dfrac{9}{2} \dfrac{(\mathcal{L}_{F_1} H(x))^2}{(\mathcal{L}_{F_2} H(x))^2 \mathcal{L}^3_{F_1} H(x)} Q, & \text{if } \mathcal{L}_{F_1} H(x) < 0, \end{cases} \tag{8.18}$$

with Q defined as above. The error term for $\mathcal{L}_{F_1} H(x) > 0$ is $O(|x-x^*|^{5/2})$.

Remarks

1. The proof of this theorem follows from the explicit derivations in Section 8.3 below.

2. The corresponding PDMs can be derived by appropriate projection onto a chosen Poincaré section, which does not alter the leading-order discontinuity, as we will explain in Section 8.4. We also show how to compose this map with the natural Poincaré mapthat is constructed in the absence of sliding bifurcation. However, there is a subtlety because for most of these bifurcations, this normal form map can be shown to be singular on one side of the bifurcation. That is, the map has an eigenvalue precisely zero. This is an artifact of the loss of dimensionality of the flow associated with sliding. As we shall see, the subtlety is that, except in the case of the grazing-sliding bifurcation, this loss of dimensionality is not evident in the ZDM itself. It is only after projection with the smooth flow in order to construct the PDM that one can see it.
3. Note from the above forms of the maps, and indeed from Table 8.1, that it is only for grazing-sliding that the singularity of the map is 1. Hence, the dynamics can be analyzed using the analysis in Section 3.6. In all other cases, the induced map has a higher-order singularity so that it is both continuous and differentiable at the grazing bifurcation point and so will not lead to an immediate change in the attractor.
4. In the case that Σ is a locally flat manifold, $\Sigma := \{x \in \mathcal{D} : H_x(x - x^*) = 0\}$, then the various expressions in this theorem take on a particularly simple form. (Note that the conditions (8.7) mean that this flatness can always be obtained to sufficiently high order by near identity co-ordinate transformations, akin to those used to calculate a center manifold of a smooth flow, see e.g. [78].) Here, using (8.10), the condition that $\beta_x F_1 < 0$ reduces to $H_x F_{1x} F_1 > 0$. Moreover, if we suppose that $\partial\widehat{\Sigma}^-$ is flat also, then the condition (8.14) $\beta_x F_{1x} F_1 + \beta_{xx} F_1^2 > 0$ becomes simply $H_x F_{1x}^2 F_1 < 0$. Then simplified expressions for leading-order terms of the above mappings may be written, without Lie derivatives, in the forms (where combinations of F_1, F_2, H and their derivatives are evaluated at the grazing point $x = x^*$. In particular, we have:

   crossing-sliding; equation (8.15) may be rewritten as

$$x \mapsto \begin{cases} x, & \text{if } (H_x F_1)_x (x - x^*) \leq 0 \\ x + u, & \text{if } (H_x F_1)_x (x - x^*) > 0, \end{cases},$$

   where

$$u = \frac{1}{2} \frac{[(H_x F_1)_x (x - x^*)]^2}{(H_x F_2)\,[(H_x F_1)_x F_1]} F_d;$$

   grazing-sliding; equation

$$x \mapsto \begin{cases} x, & \text{if } H_x(x - x^*) \geq 0, \\ x - \dfrac{H_x(x - x^*)}{H_x F_2}(F_2 - F_1), & \text{if } H_x(x - x^*) < 0; \end{cases}$$

   switching-sliding;

$$x \mapsto \begin{cases} x, & \text{if } (H_x F_1)_x (x - x^*) \le 0, \\ x + w, & \text{if } (H_x F_1)_x (x - x^*) > 0, \end{cases}$$

where

$$w = \frac{2}{3} \frac{\left[(H_x F_1)_x (x - x^*)\right]^3}{(H_x F_2)^2 \left[(H_x F_1)_x F_1\right]^2} Q,$$

$$Q = \left[(H_x F_2)(F_{1,x} F_d - F_{d,x} F_1) - (H_x (F_{1,x} F_d - F_{d,x} F_1)) F_d\right];$$

adding-sliding; equation (8.18) may be rewritten as

$$x \mapsto \begin{cases} x, & \text{if } (H_x F_1)_x (x - x^*) \ge 0 \\ x + z, & \text{if } (H_x F_1)_x (x - x^*) < 0, \end{cases}$$

where

$$z = -\frac{9}{2} \frac{\left[(H_x F_1)_x (x - x^*)\right]^2}{(H_x F_2)^2 \left\{\left[(H_x F_1)_x F_1\right]_x F_1\right\}} Q.$$

Before proceeding to a derivation of such maps, we consider a motivating example of how these sliding bifurcations can organize the complicated dynamics.

8.2 Motivating example: a relay feedback system

We return to a more detailed description of the dynamics of case study III introduced in Chapter 1. As we shall see, this system undergoes both adding-sliding and grazing-sliding in different parts of its parameter space. Also, in Example 8.1 below, we shall show that a crossing-sliding bifurcation occurs in the same system at different parameter values. Examples of crossing-sliding and of switching-sliding will also be presented in Section 9.4.1 when discussing a friction oscillator example that has a codimension-two sliding bifurcation.

Let us focus on a particular three-dimensional relay system of type (1.22)–(1.24),

$$\begin{aligned} \dot{x} &= Ax + Bu, \\ y &= C^T x, \\ u &= -\text{sgn}(y), \end{aligned}$$

specifically taking

$$A = \begin{pmatrix} -(2\zeta\omega + \lambda) & 1 & 0 \\ -(2\zeta\omega\lambda + \omega^2) & 0 & 1 \\ -\lambda\omega^2 & 0 & 0 \end{pmatrix}, \quad B = \begin{pmatrix} k \\ 2k\sigma\rho \\ k\rho^2 \end{pmatrix}, \quad C = \begin{pmatrix} 1 \\ 0 \\ 0 \end{pmatrix}. \tag{8.19}$$

The parameters are chosen so that $\zeta \pm i\omega$ are a complex pair of poles of the transfer function

$$G(s) = k\frac{s^2 + 2\sigma\rho s + \rho^2}{(s^2 + 2\zeta\omega s + \omega^2)(s + \lambda)},$$

whereas $\sigma \pm i\rho$ are a complex pair of zeros of G. Moreover, $-\lambda$ is the single real pole and $\frac{k\rho^2}{\omega^2\lambda}$ is the steady-state gain.

We can write the three-dimensional relay system equivalently in the form (8.1) with

$$F_1 = Ax - B, \quad F_2 = Ax + B, \quad H(x) = C^T x.$$

Thus, we can define regions of smooth dynamics as

$$S_1 := \{x \in \mathbb{R}^3 \,:\, H(x) = x_1 > 0\},$$

$$S_2 := \{x \in \mathbb{R}^3 \,:\, H(x) = x_1 < 0\},$$

and the switching manifold Σ such that:

$$\Sigma := \{x \in \mathbb{R}^3 \,:\, H(x) = x_1 = 0\}.$$

A trajectory corresponding to typical symmetric self-oscillations for this system are presented in Fig. 8.3 with either no sliding or one sliding segment per half-period.

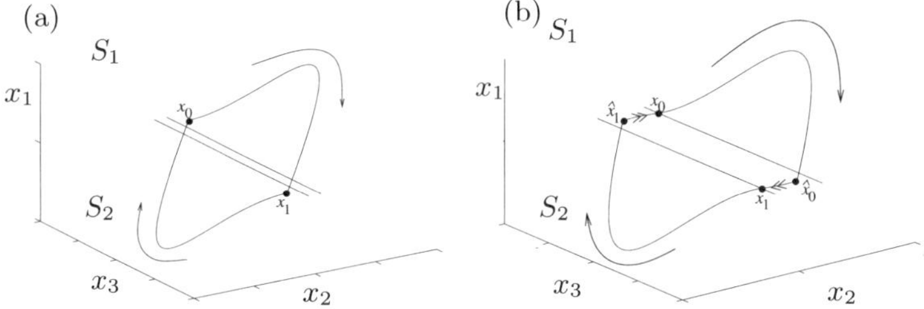

Fig. 8.3. (a) Typical non-sliding trajectory (corresponding to self-oscillations) of the relay feedback system (1.22)–(1.24) with $-\sigma = k = \lambda = \xi = \omega = 1$, $\rho = 3$. (b) A simple symmetric orbit with two sliding segments for $-\sigma = k = \lambda = \xi = \omega = 1$, $\rho = 1$.

More complex periodic orbits might contain many sliding segments per period. Solutions of this type are represented in Fig. 8.4. To avoid confusion, in what follows, we term a periodic solution characterized by N distinct sections of sliding motion per period as an *N-sliding orbit*; so that Fig. 8.3(b) shows a 2-sliding orbit and Fig. 8.4 two different 12-sliding orbits. Moreover, we refer to an orbit as being symmetric if one half-period may be mapped into the other under the transformation $x \mapsto -x$ (the simplest form of $\mathbb{Z}_2$ symmetry).

8.2.1 An adding-sliding route to chaos

We consider the bifurcation scenario that occurs for decreasing values of the parameter $\zeta \in [-0.08, -0.06]$ with all other parameters fixed at the values

$$k = 1, \quad \lambda = 0.05, \quad \omega = 10, \quad \rho = 1, \quad \sigma = -1. \tag{8.20}$$

This represents a region of parameter space where a seemingly chaotic solution was found [84]. We begin by considering the stable 12-sliding stable symmetric orbit shown in Fig. 8.4(a) and represented by the solid branch to the right of the point SB in the bifurcation diagram shown in Fig. 8.5. This Monte Carlo bifurcation diagram depicts the discrete values of the x_3 co-ordinate versus the bifurcation parameter ζ.

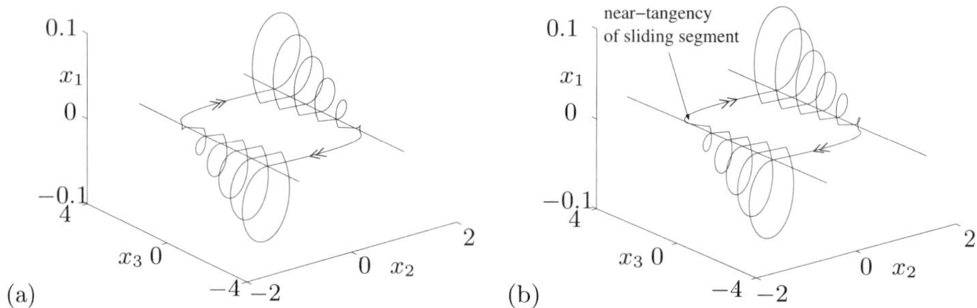

Fig. 8.4. (a) Symmetric 12-sliding orbit for parameter values (8.20) and $\zeta = -0.0628$. (b) Stable asymmetric 12-sliding orbit close to the adding-sliding bifurcation point. Note the near-tangency of one of its sliding sections.

This choice of co-ordinate follows from the fact that we can introduce a Poincaré section at the exiting portion of the boundary $\partial\widehat{\Sigma}^{\pm}$ of the sliding region. In our case, such a choice implies that x_1 and x_2 remain constant under variation of the bifurcation parameter ζ. Hence, all information is contained in the value of x_3 at these points, which immediately illustrates the loss of information inherent in sliding. After a sliding portion of trajectory, information on the point of entry to the sliding region $\widehat{\Sigma}$ has been lost by the evolution under the vector filed F_s, which lives in a lower-dimensional space. Thus we have effectively a one-dimensional Poincaré mapinstead of the usual two-dimensional map one would expect for a three-dimensional flow.

Consider Fig. 8.5 in detail, the isolated points of which were computed using a Monte Carlo approach augmented by dashed lines that represent the simplest unstable periodic orbits. The specific Poincaré section is given by the segment $x_1 = 0$, $x_2 = -1$ and $x_3 \in [2.3\,2.6]$.

Starting from the right-hand side of the figure we observe that the 12-sliding stable symmetric orbit undergoes a subcritical symmetry-breaking bifurcation at $\zeta \approx -0.0628$ (point SB in Fig. 8.5). This is an example of a smooth

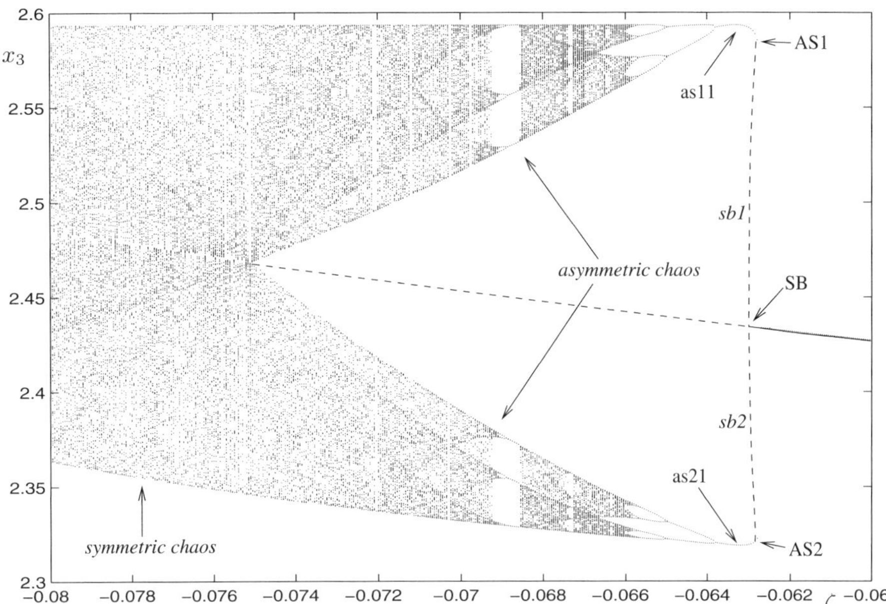

Fig. 8.5. The bifurcation diagram for the third-order relay feedback system under investigation for parameter values (8.20) with ζ varying in the interval $[-0.08, -0.06]$. See text for details.

bifurcation occurring in the Poincaré map associated with the flow around the periodic orbit. Upon tracing the two branches of asymmetric solutions, we find that they each restabilize in a saddle-node bifurcation at $\zeta \approx -0.0623$ [points AS1 and AS2 in Fig. 8.5, and solution depicted in Fig. 8.4(b)]. The new *stable* asymmetric periodic solution branches born there are labeled $as11$ and $as21$ in the figure. Despite the apparent similarity of this transition with a classical saddle-node bifurcation, we wish to emphasize that the bifurcation at AS1 (equivalently AS2) is strongly organized by the near-tangency of a sliding segment with $\partial\widehat{\Sigma}^{\pm}$. Such tangency of a limit cycle with the boundary of the sliding region corresponds to an *adding-sliding* in the notation we have just introduced. Note from Table 8.1 and from Theorem 8.1 that the adding-sliding discontinuity mapping leads to a map singularity that is quadratic. Therefore, according to the analysis in Chapter 4, there would not be an immediate jump in the slope of the map, and hence no immediate change in the attractor. Instead, we should expect a singularity in the second-derivative of the solution curve in the bifurcation diagram. This then explains the apparent sharpness in the change of slope near the saddle-node points in Fig. 8.5.

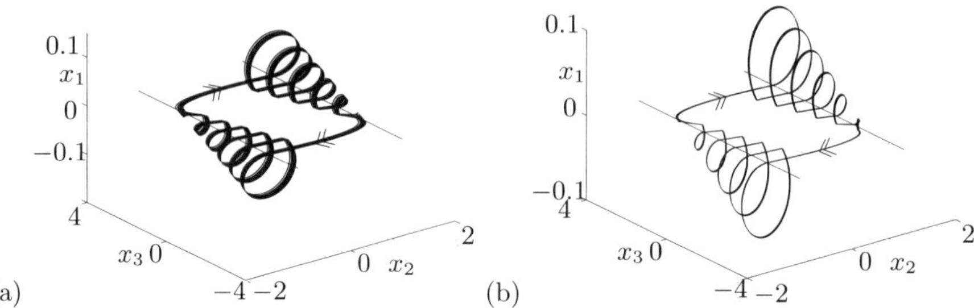

Fig. 8.6. (a) One of the two stable asymmetric chaotic attractors exhibited by the system for $\zeta = -0.07$. (b) Symmetric chaotic attractor exhibited by the system for $\zeta = -0.078$ formed by the merging of two bands of asymmetric chaos.

As we now reduce ζ from these AS points, the newly formed stable asymmetric sliding orbits undergo a sequence of period-doublings that accumulate into an asymmetric pair of fully developed chaotic attractors. As shown in Fig. 8.6(a), these chaotic attractors are organized by underlying asymmetric orbits with a multiple number of sliding segments. Decreasing ζ further, these two bands of asymmetric chaos then merge into one and the symmetric stable chaotic attractor depicted in Fig. 8.6(b) is created (eventually to be destroyed in a *boundary crisis* bifurcation for $\zeta < -0.08$; not depicted).

It might be worth mentioning that in the control theory literature it has often been assumed that self-oscillations of symmetric, unforced relay feedback systems such as (1.22)–(1.24) are also *symmetric*. Specifically, it is usually conjectured that asymmetric periodic solutions in these systems can only exist through *asymmetric* external forcing or relay characteristics (see for example the conjecture by Tsypkin [255, p. 179]). Here, we have shown that indeed a symmetric unforced relay feedback system may exhibit stable asymmetric self-oscillations through symmetry-breaking bifurcations of sliding periodic orbits like at the point SB in Fig. 8.5. Moreover, via an adding-sliding bifurcation and subsequent period-doubling cascade, there can be asymmetric chaotic solutions.

8.2.2 An adding-sliding bifurcation cascade

Let us next consider a different example of adding-sliding where an N-sliding orbit is created by a sequence of adding-sliding transitions in which it successively acquires extra sliding sections.

Taking the fixed parameter values

$$-\sigma = \rho = k = \lambda = 1 \quad \text{and} \quad \zeta = 0.05, \tag{8.21}$$

upon increase of ω through the value 10.24, it is found that a symmetric 4-sliding orbit transforms into a 6-sliding orbit through adding-sliding. See Fig. 8.7.

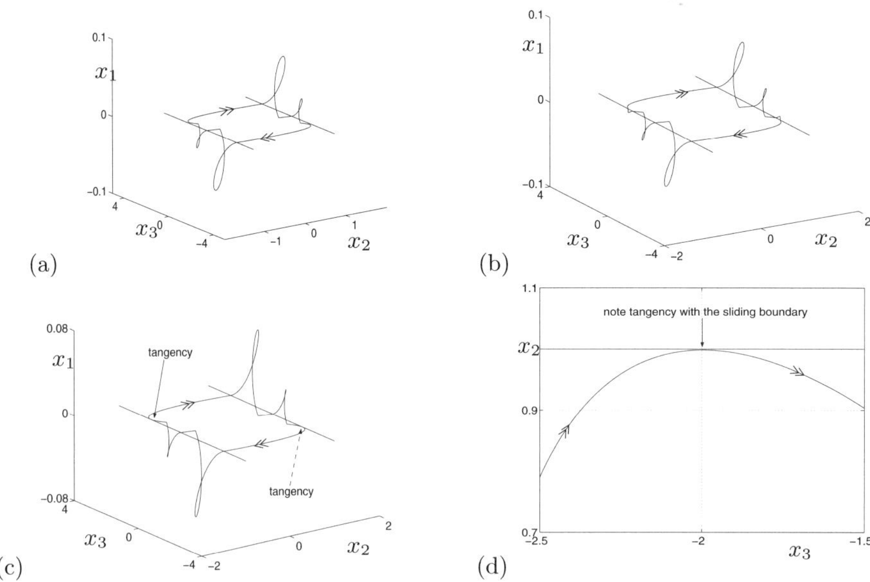

Fig. 8.7. The first adding-sliding bifurcation in the cascade for parameter values (8.21): (a) a symmetric 4-sliding orbit at $\omega = 10.17$; (b) a 6-sliding orbit at $\omega = 10.74$. Panels (c) and (d) depict the orbit at the bifurcation value $\omega = 10.24$.

Magnification of the bifurcating orbit in Fig. 8.7(d) clearly shows the tangency of the orbit with $\partial\widehat{\Sigma}$ at the adding-sliding point. It is easy to show that this is an adding-sliding DIB. The 4-sliding orbit instantaneously transforms into a 6-sliding orbit with no apparent change in slope of the path of solutions as it passes through the bifurcation. In the notation of Chapter 3, this is an $A \mapsto B$ transition (see the left-most transition in Fig. 8.8). This is an important point since the fold-like scenario we described in Fig. 8.5 actually occurred a short parameter distance away from the point AS1. In both of these manifestations of adding-sliding, the effect of the bifurcation is to cause a jump in the *second derivative* of the solution locus, which gives rise to a sharp, but smooth (C^1-differentiable) corner. In the case in Fig. 8.5 this corner gives rise to a saddle-node nearby, whereas in Fig. 8.8 the corner merely

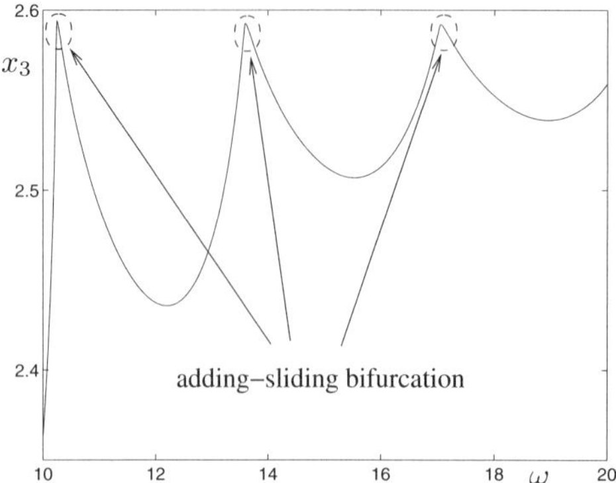

Fig. 8.8. Bifurcation diagram for the third-order relay feedback system under investigation, with ω varying in the interval $[10, \quad 20]$.

gives a rapid local rise in amplitude (as measured by the value of x_3 in the Poincaré section).

The bifurcation diagram in Fig. 8.8 shows how, with further increase of bifurcation parameter ω, the 6-sliding orbit acquires additional sliding segments through a cascade of qualitatively similar adding-sliding bifurcations. In this manner, a 2-sliding orbit at $\omega = 5$ eventually becomes a 14-sliding orbit at $\omega = 25$.

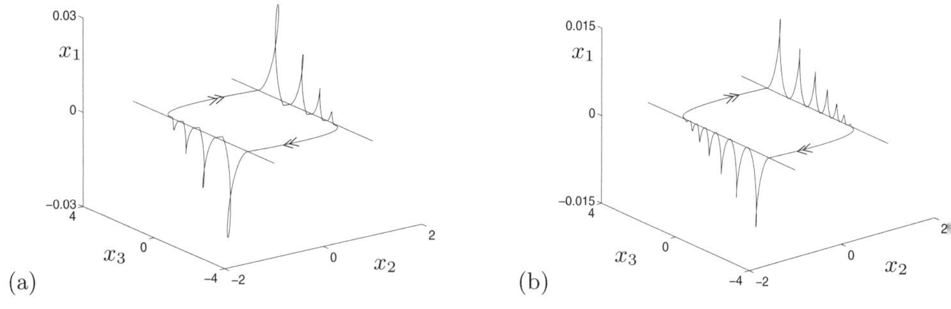

Fig. 8.9. (a) The 8-sliding at $\omega = 16$ and (b) the 14-sliding orbit at $\omega = 25$, from the adding-sliding cascade depicted in Fig. 8.8

.

The 8-sliding and 14-sliding orbits in the sequence of are depicted in Fig. 8.9.

8.2.3 A grazing-sliding cascade

Upon further exploration of the parameter space of the system under investigation, we find that the symmetric 12-sliding orbit that undergoes symmetry-breaking at point SB in Fig. 8.5 is born in a cascade of grazing-sliding bifurcations. To explain this cascade, let us return to parameter regime (8.20) and allow the bifurcation parameter ζ to vary between 0.02 and 0.05. The resulting bifurcation diagram is presented in Fig. 8.10.

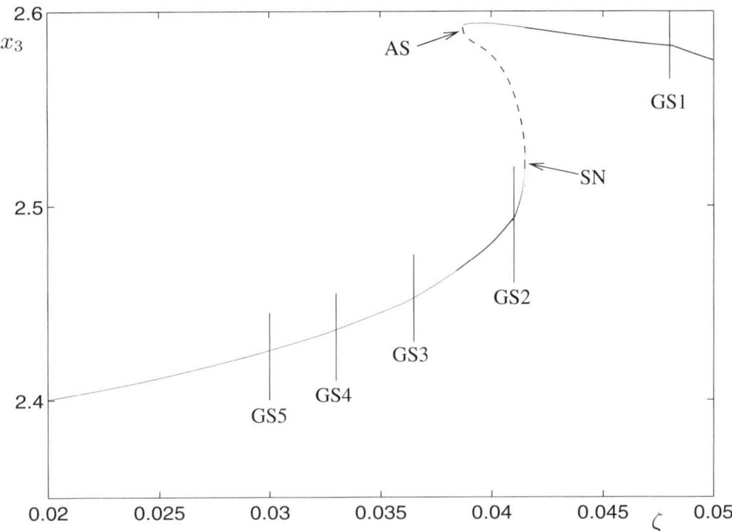

Fig. 8.10. Bifurcation diagram for the third-order relay feedback system under investigation for parameter values (8.20) with varying ζ, using the same Poincaré section as in Fig. 8.5.

Let us now explain the observed phenomena upon decreasing ζ. For $\zeta \approx 0.048$, there is a stable, symmetric 4-sliding orbit, which at label GS1 undergoes a grazing-sliding bifurcation and becomes a 6-sliding orbit. If we continue along this branch with decreasing values of the parameter ζ, the 6-sliding orbit disappears in a fold at $\zeta \approx 0.0385$ close to the adding-sliding point (labeled AS), akin to what we observed to occur for the asymmetric 12-sliding orbits at labels AS1 and AS2 in Fig. 8.5. The unstable branch from this bifurcation then bends back and restabilizes at $\zeta \approx 0.0415$ in a regular saddle-node bifurcation (label SN). This restabilized 4-sliding orbit then undergoes a sequence of grazing-sliding bifurcations labeled GS2-GS5 in Fig. 8.10. Each bifurcation adds a (symmetric pair of) sliding segments to the orbit, culminating in the existence for $\zeta < 0.03$ of a symmetric 12-sliding orbit.

Consider the grazing scenario labeled GS5 in more detail; see Fig. 8.11 As the parameter ζ is decreased, one of the loops making up the orbit changes its shape [Fig. 8.11(c)]. Further decrease of ζ causes the loop to touch the boundary of the sliding strip and causes an extra local piece of sliding to occur [Fig. 8.11(d)]. All the grazing-sliding bifurcations have the character

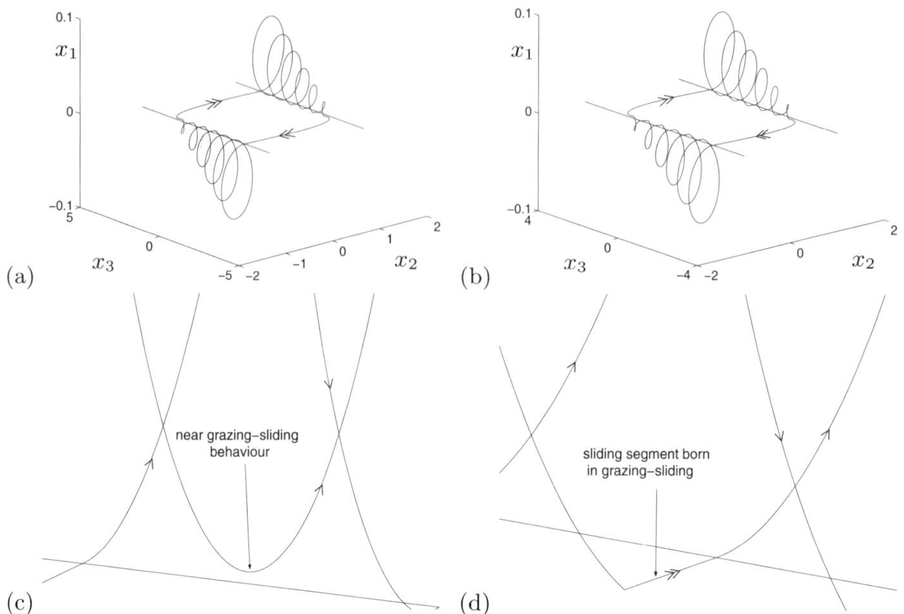

Fig. 8.11. Stable symmetric orbits with multiple sliding sections near point GS5 in Fig. 8.10; (a) 'before' ($\zeta = 0.032$) and (b) 'after' ($\zeta = 0.025$) the bifurcation. Panels (c) and (d) show zooms into the relevant part of the trajectories in (a) and (b), respectively.

that neither the stability nor the existence of the fundamental periodic orbit in question changes through the bifurcation; that is, we have a persistence scenario $A \mapsto B$ in the parlance of Chapter 3. It is worthwhile to note though that there is an appreciable change in the slope of the locus of periodic orbits in the bifurcation diagram as we pass through each grazing-sliding bifurcation. This is most pronounced for GS1 and GS2, but it is nevertheless there in the other manifestations of this bifurcation.

Fig. 8.12. The overall effect of the grazing-sliding cascade. The 4-sliding symmetric orbit depicted in (a) for $\zeta = 0.05$ becomes the 12-sliding orbit in (b) for $\zeta = 0.025$.

In case study IV, a friction oscillator, we also saw an example of a Filippov system that exhibits a grazing-sliding bifurcations, but whose effect is far less benign than in the above example, leading to an immediate jump to chaos. We shall return to that example in Section 8.4 below, where we will explicitly show the onset of robust chaos via an analysis of the discontinuity mapping.

Note that an analysis of the *cascading* of the adding-sliding and grazing-sliding bifurcations we have uncovered here is beyond the scope of this book. In particular, what kind of mechanism could lie at the accumulation of such cascades, and whether there could be some universal 'Feigenbaum' constant as for period-doubling cascades, remains an open question.

8.3 Derivation of the discontinuity mappings

Let us now return to a proof of Theorem 8.1 by explicit construction of the discontinuity mappings. The method will be the same as for the grazing bifurcations presented in Chapters 6 and 7 using Lie derivatives, but it will lead to somewhat more cumbersome expressions due to the presence of the sliding flow. We shall treat the grazing-sliding and crossing-sliding cases in some detail and merely sketch the steps necessary to derive the presented expressions for adding-sliding and switching-sliding.

For clarity, it is useful in what follows to write all expressions in terms of Lie derivatives with respect to F_1 and the difference vector field $F_d = F_2 - F_1$. Thus, using the definition of β from (8.4), we obtain

$$\beta(x) = -\frac{\mathcal{L}_{(F_1+F_2)}H(x)}{\mathcal{L}_{F_d}H(x)} = -1 - 2\frac{\mathcal{L}_{F_1}H(x)}{\mathcal{L}_{F_d}H(x)}, \tag{8.22}$$

and after writing the sliding vector field as

$$F_s = \frac{F_1 + F_2}{2} + \frac{F_d}{2}\beta(x),$$

we get the closed-form expression

$$F_s = F_1 - \frac{\mathcal{L}_{F_1}(H)}{\mathcal{L}_{F_d}(H)}F_d. \tag{8.23}$$

Also,

$$\mathcal{L}_{F_s} = \mathcal{L}_{F_1} - \frac{\mathcal{L}_{F_1}(H)}{\mathcal{L}_{F_d}(H)}\mathcal{L}_{F_d}. \tag{8.24}$$

Therefore,

$$\mathcal{L}_{F_s}H(x) = \mathcal{L}_{F_1}H(x) - \mathcal{L}_{F_1}H(x) = 0, \tag{8.25}$$

which accords with our geometrical intuition, since the sliding flow lies tangent to $\Sigma := \{x : H(x) = 0\}$ and $\mathcal{L}_{F_s}H(s)$, which is the time derivative of H along the sliding flow, should be zero.

Before proceeding further, let us establish a few useful results.

Lemma 8.1. *At a point* x^* *where* $\mathcal{L}_{F_1}H(x^*) = 0$*, we have* $\beta(x^*) = -1$ *and* $F_s(x^*) = F_1(x^*)$*. Also, for any smooth scalar function, say* $G(x)$*, we have* $\mathcal{L}_{F_s}G(x^*) = \mathcal{L}_{F_1}G(x^*)$*, Moreover,*

$$\mathcal{L}_{F_s}(\beta)(x^*) = -2\frac{\mathcal{L}^2_{F_1}(H)}{\mathcal{L}_{F_d}(H)}(x^*).$$

Proof. The first two results follow immediately from (8.22) and (8.23). The next results follow by direct calculation

$$\begin{aligned}\mathcal{L}_{F_s}(G)(x^*) &= G_x\left(F_1 - \frac{\mathcal{L}_{F_1}H}{\mathcal{L}_{F_d}H}F_d\right)(x^*)\\ &= G_x(x^*)F_1(x^*) - \frac{\mathcal{L}_{F_1}H(x^*)}{\mathcal{L}_{F_d}H(x^*)}G_xF_d(x^*)\\ &= \mathcal{L}_{F_1}(G)(x^*). \end{aligned} \tag{8.26}$$

Finally, by (8.26),

$$\mathcal{L}_{F_s}\beta(x^*) = \mathcal{L}_{F_1}\beta(x^*).$$

Applying $\mathcal{L}_{F_1}$ to (8.22), we then have

$$\begin{aligned}\mathcal{L}_{F_1}\left(-1 - 2\frac{\mathcal{L}_{F_1}H(x^*)}{\mathcal{L}_{F_d}H(x^*)}\right) &= -\mathcal{L}_{F_1}H(x^*)\mathcal{L}_{F_1}\left(\frac{2}{\mathcal{L}_{F_d}(H)(x^*)}\right) - 2\frac{\mathcal{L}^2_{F_1}H(x^*)}{\mathcal{L}_{F_d}H(x^*)}\\ &= -2\frac{\mathcal{L}^2_{F_1}H(x^*))}{\mathcal{L}_{F_d}H(x^*)}.\end{aligned}$$

Lemma 8.2. *At a point* x^* *for which* $\mathcal{L}_{F_1}H(x^*) = 0$*, where* H *satisfies (8.7), and* $\mathcal{L}^2_{F_1}H(x^*) \neq 0$*, then* $H_x(x^*)$ *and* $\frac{d}{dx}\mathcal{L}_{F_1}H(x^*)$ *are linearly independent. Hence, if two surfaces defined by* $H = 0$ *and* $\beta = -1$ *intersect at a point where* $\mathcal{L}^2_{F_1}H(x^*) \neq 0$*, they must intersect transversally and they define a codimension-two manifold as represented in Fig. 8.2.*

Proof. By (8.7), $H_x(x^*) \neq 0$. Suppose that $\frac{d}{dx}\mathcal{L}_{F_1}H(x^*) = \alpha H_x(x^*)$ for some non-zero scalar α; then

$$\mathcal{L}^2_{F_1}(H)(x^*) = \frac{d}{dx}\mathcal{L}_{F_1}H(x^*)F_1(x^*) = \alpha H_xF_1(x^*) = \alpha\mathcal{L}_{F_1}H(x^*) = 0,$$

which contradicts our assumption.

Using the above, we can rewrite certain of the defining and non-degeneracy assumptions purely in terms of F_1 and F_d. The defining condition (8.7) valid for all sliding bifurcations becomes

$$\mathcal{L}_{F_1}H(x^*) = 0. \tag{8.27}$$

The non-degeneracy assumption (8.9) valid for all sliding bifurcations becomes

$$\mathcal{L}_{F_d}(x^*) > 0, \tag{8.28}$$

and the extra non-degeneracy condition (8.11) for crossing-sliding and grazing-sliding becomes

$$\mathcal{L}^2_{F_1} H(x^*) > 0, \tag{8.29}$$

with the opposite inequality (8.12) applying at a switching-sliding now being written as

$$\mathcal{L}^2_{F_1} H(x^*) < 0. \tag{8.30}$$

Also, the defining condition for an adding sliding bifurcation (8.13) becomes

$$\mathcal{L}^2_{F_1} H(x^*) = 0, \tag{8.31}$$

and the non-degeneracy assumption (8.14) becomes

$$\mathcal{L}^3_{F_1} H(x^*) < 0. \tag{8.32}$$

Let us now treat each of the bifurcations in turn, treating the crossing-sliding and grazing-sliding cases in some detail, giving only the briefest details for the other two. Complete derivations may be found in [85].

8.3.1 Crossing-sliding bifurcation

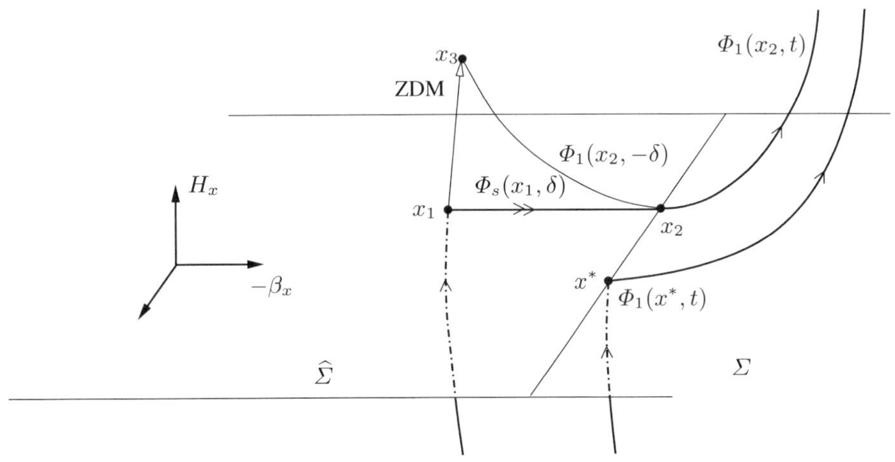

Fig. 8.13. A schematic representation of the ZDM derivation for the crossing-sliding bifurcation.

Suppose the defining conditions (8.7), (8.27) hold with non-degeneracy conditions (8.28) and (8.29) at a point x^* that is part of a critical trajectory that starts in region S_2, flows under Φ_2, intersects the point x^* on the boundary $\partial\widehat{\Sigma}^-$ and then continues under Φ_1. Figure 8.13 depicts a perturbation of this trajectory that flows under Φ_2 to hit the sliding region at some point, say $x_1 \in \widehat{\Sigma}$. This trajectory is then constrained to evolve within $\widehat{\Sigma}$ under the sliding flow $\Phi_s(x_1, t)$ until it hits the boundary $\partial\widehat{\Sigma}^-$ after some time, say δ, at the point $x_2 := \Phi_s(x_1, \delta)$. The trajectory then leaves Σ following the flow Φ_1.

The zero-time discontinuity mapping or ZDM in this context is the correction that needs to be applied to the flow at point x_1 in order to account for the presence of the sliding region. Specifically, this correction must be such that the evolution from the point at which Σ is first hit onwards may be described entirely by applying the discontinuity map and using flow Φ_1. As depicted in Fig. 8.13, the ZDM maps x_1 to the point x_3, which is the image of the point x_2 under the flow $\Phi_1(x_2, -\delta)$. Hence the total elapsed time from x_1 to x_3 is zero.

To get an analytical expression for the ZDM, we therefore need to consider the following combination of flows:

$$x_3 = \Phi_1(\Phi_s(x_1, \delta), -\delta). \tag{8.33}$$

We start by expanding the operator (8.33) as a Taylor series in time using Lie derivatives. Specifically, for x close to x^*, let $P(x)$ be any smooth scalar function; then we have

$$\begin{aligned} P(\Phi_1(\Phi_s(x,t),-t)) &= (I + t\mathcal{L}_{F_s} + \frac{t^2}{2}\mathcal{L}_{F_s}^2 + O(t^3))(I - t\mathcal{L}_{F_1} + \frac{1}{t^2}\mathcal{L}_{F_1}^2 + \\ &\quad +O(t^3))P(x_1), \\ &= (I + (\mathcal{L}_{F_s} - \mathcal{L}_{F_1})t + \frac{t^2}{2}(\mathcal{L}_{F_s}^2 - 2\mathcal{L}_{F_s}\mathcal{L}_{F_1} + \mathcal{L}_{F_1}^2) + \\ &\quad +O(t^3))P(x). \end{aligned} \tag{8.34}$$

Substituting for $\mathcal{L}_{F_s}$ using (8.24), we find the coefficient of the $O(t)$ term to be

$$\frac{v(x)}{\mathcal{L}_{F_d}H(x)}\mathcal{L}_{F_d}, \qquad \text{where} \quad v(x) = \mathcal{L}_{F_1}H(x). \tag{8.35}$$

Note that $v(x)$ is a small quantity since by (8.25) $v(x^*) = 0$ and x is close to x^*. Hence the first term of the expansion (8.34) is of $O(vt)$. After operator multiplication and carrying out various simplifications on the $O(t^2)$ term, we then obtain

$$P(\Phi_1(\Phi_s(x_1,t),-t)) = (I - \frac{v(x)t}{\mathcal{L}_{F_d}H(x)}\mathcal{L}_{F_d} - \frac{t^2}{2}\frac{\mathcal{L}_{F_1}^2H(x)}{\mathcal{L}_{F_d}H(x)}\mathcal{L}_{F_d} + O((v,t)^3))P(x).$$

Taking $P(x)$ to be each of the components of x in turn, we thus obtain the general expression

$$\Phi_1(\Phi_s(x,t),-t) = x - \frac{v(x)t}{\mathcal{L}_{F_d}H(x)}F_d(x) - \frac{t^2}{2}\frac{\mathcal{L}^2_{F_1}H(x)}{\mathcal{L}_{F_d}H(x)}F_d(x) + O((v,t)^3). \tag{8.36}$$

We next need to find the time δ. To do this we must solve

$$\mathcal{L}_{F_1}H(\Phi_s(x_1,\delta(x_1))) = 0, \tag{8.37}$$

for $\delta(x_1)$. Expansion of (8.37) in δ gives

$$\mathcal{L}_{F_1}H(\Phi_s(x_1,\delta)) = v(x_1) + \mathcal{L}^2_{F_1}H(x_1)\delta + O((v,\delta)^2) = 0. \tag{8.38}$$

Now, since $\mathcal{L}^2_{F_1}(H)(x_1)$ is non-zero by the non-degeneracy hypothesis (8.29), the Implicit Function Theorem guarantees that there is a locally unique solution of the form

$$\delta = -\frac{v(x_1)}{\mathcal{L}^2_{F_1}H(x_1)} + O(v^2), \tag{8.39}$$

where $v(x_1)$ is given defined in (8.35). Note that δ is positive, since within the sliding set, $v(x_1) = \mathcal{L}_{F_1}H(x_1) < 0$ and $\mathcal{L}^2_{F_1}H(x_1)$ is positive by hypothesis (8.29).

Substitution of (8.39) for $t = \delta$ into the expansion (8.36) evaluated at a point x_1 gives that the leading-order expression for the ZDM is

$$x_3(x_1,v) = \begin{cases} x_1, & v \geq 0, \\ x_1 + v(x_1)^2\dfrac{F_d(x_1)}{2\mathcal{L}_{F_d}H(x_1)\mathcal{L}^2_{F_1}H(x)} + O(v^3), & v < 0. \end{cases} \tag{8.40}$$

Finally, substituting for v from (8.35) into (8.40) and expanding around $x_1 = x = x^*$ gives the leading-order expression for $ZDM(x_1) = x_3(x_1, v(x_1))$:

$$\text{ZDM}(x) = \begin{cases} x, & \mathcal{L}_{F_1}H(x) \geq 0, \\ x + \mathcal{L}_{F_1}H(x^*)^2\dfrac{F_d(x^*)}{2\mathcal{L}_{F_d}H(x^*)\mathcal{L}^2_{F_1}H(x^*)}, & \mathcal{L}_{F_1}H(x) \leq 0, \end{cases} \tag{8.41}$$

which is equivalent to the final form of the ZDM stated in Theorem 8.1.

Notice that (8.40) is expressed as a function of two independent variables. However, when the ZDM map is formulated, the initial point x_1 determines the variable v.

8.3.2 Grazing-sliding bifurcation

We now take exactly the same defining and non-degeneracy conditions (8.7), (8.27), (8.28) and (8.29), but this time suppose that the critical trajectory evolves entirely in the region S_1 apart from the point x^* at which it hits $\partial\widehat{\Sigma}^-$ tangentially. See Fig. 8.14.

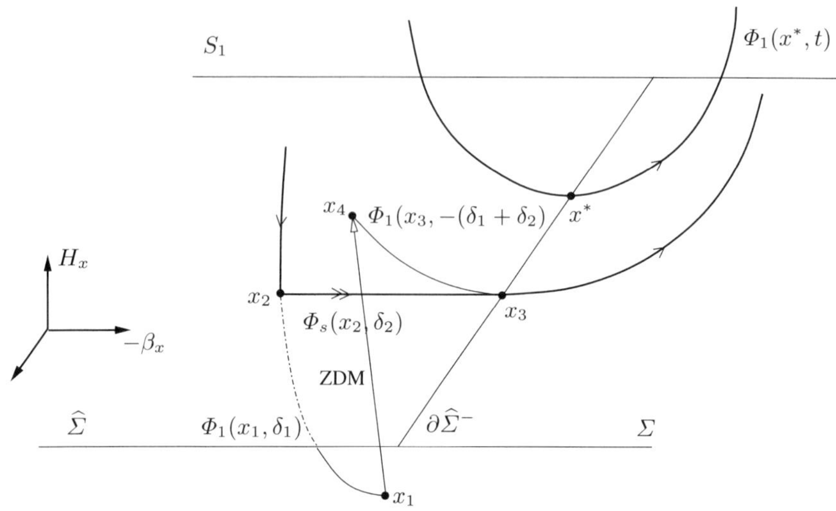

Fig. 8.14. Schematic representation of the construction of the ZDM for the grazing-sliding bifurcation.

We consider the motion of a perturbed trajectory close to the critical one, starting from a general point x_1 that after evolution under the flow Φ_1 through a time δ_1 would hit the switching manifold, $\widehat{\Sigma}$, at some point x_2. Note that δ_1 may be positive or negative. If $\delta_1 > 0$, then $x_1 \in S_1$ is a point on a physical trajectory of the piecewise-smooth system. If $\delta_1 < 0$, then we are in the situation depicted in Fig. 8.14 where $x_1 \in S_2$ is a virtual point that would be reached if the flow Φ_1 were followed through Σ as if the discontinuity boundary were not there. In order to construct the ZDM, we follow the evolution from the point x_2 within $\widehat{\Sigma}$ using the sliding flow, for a time $\delta_2 > 0$ until we reach $\partial\widehat{\Sigma}^-$ at the point x_3. The ZDM correction is then given by the flow $\Phi_1(x_3, -(\delta_1 + \delta_2))$ to the point x_4. Hence the total time spent in going from x_1 to x_4 is zero. The construction of the analytical form of the ZDM follows in a similar fashion to the equivalent derivations for grazing bifurcations in Chapters 6 and 7. Consider the following combination of flows:

$$x_4 = \Phi_1(\Phi_s(\Phi_1(x_1, \delta_1), \delta_2), -(\delta_1 + \delta_2)), \tag{8.42}$$

and perform a Taylor expansion for small times $\delta_{1,2}$ using Lie derivatives. Thus, for any scalar function $P(x)$ that is smooth near $x = x^*$ we have

$$\begin{aligned}
&P(\Phi_1(\Phi_s(\Phi_1(x,t), s), -(t+s))) = \\
&(I + t\mathcal{L}_{F_1} + \frac{t^2}{2}\mathcal{L}_{F_1}^2 + O(t^3))(I + s\mathcal{L}_{F_s} + \frac{s^2}{2}\mathcal{L}_{F_s}^2 + O(s)^3) \\
&(I + (-t-s)\mathcal{L}_{F_1} + \frac{(-t-s)^2}{2}\mathcal{L}_{F_1}^2 + O(t+s)^3)P(x).
\end{aligned}$$

After operator multiplication and collection of terms at successive orders of t and s, we get

$$P(\Phi_1(\Phi_s(\Phi_1(x,t),s),-(t+s))) = \\ (I + s(\mathcal{L}_{F_s} - \mathcal{L}_{F_1}) + st(\mathcal{L}_{F_1}\mathcal{L}_{F_s} - \mathcal{L}_{F_s}\mathcal{L}_{F_1}) + \\ \frac{s^2}{2}(\mathcal{L}_{F_s}^2\mathcal{L}_{F_1}^2 - 2\mathcal{L}_{F_s}\mathcal{L}_{F_1}) + sO(s,t)^2)P(x). \tag{8.43}$$

We now substitute for $\mathcal{L}_{F_s}$ using (8.24), and to lowest order, we find

$$P(\Phi_1(\Phi_s(\Phi_1(x,t),s),-(t+s))) = \\ (I + (-st - \frac{s^2}{2})\frac{\mathcal{L}_{F_1}^2 H(x)}{\mathcal{L}_{F_d} H(x)}\mathcal{L}_{F_d} + \mathcal{L}_{F_1} H(x) Q(x) + \\ sO(s,t)^2)P(x),$$

where $Q(x)$ represents a somewhat lengthy expression that we do not spell out here. Taking $P(x)$ to be each of the components of x in turn, we obtain

$$\Phi_1(\Phi_s(\Phi_1(x,t),s),-(t+s)) - \mathcal{L}_{F_1} H(x) Q(x) = \\ x + (-st - \frac{s^2}{2})\frac{\mathcal{L}_{F_1}^2 H(x)}{\mathcal{L}_{F_d} H(x)} F_d + sO(s,t)^2. \tag{8.44}$$

Now, as we did in Chapters 6 and 7, we first derive the ZDM under the assumption that the starting point x_1 is chosen such to lie on the *normal* Poincaré section

$$\mathcal{L}_{F_1} H(x_1) = 0. \tag{8.45}$$

In this case, the term proportional to $Q(x)$ in (8.44) vanishes, which leads to considerably simpler expressions. At the end of the calculation, we take a general point x_1 and show that leading-order expression for the ZDM does not change. For the time being then, let us assume (8.45) and calculate leading-order expressions for the times $t = \delta_1(x_1)$ and $s = \delta_2(x_1)$.

The time δ_1 is defined as the time to get from x_1 to the point of first intersection with Σ. That is,

$$H(\Phi_1(x_1, \delta_1)) = 0.$$

This step follows exactly the same methodology as for the derivation of the equivalent time δ in the ZDM for grazing bifurcations derived in Chapters 6 and 7. Thus, to leading-order, we obtain

$$\delta\sqrt{\frac{H(\Phi_1(x_1,\delta_1)) - H(x) - \mathcal{L}_{F_1} H(x)\delta_1}{\delta_1^2}} + y = 0,$$

where $y = \sqrt{-H(x_1)}$. For small y, the Implicit Function Theorem shows the existence of a unique, smooth function $\delta_1(x_1, y)$

$$\delta_1(x_1, y) = y\left(-\sqrt{\frac{2}{\mathcal{L}^2_{F_1} H(x_1)}} + O(y)\right) \tag{8.46}$$

satisfying $\delta_1(x^*, 0) = 0$.

We next need to similarly solve for the time δ_2. The equation determining δ_2 can be written to leading-order as

$$\mathcal{L}_{F_1} H(\Phi_s(\Phi_1(x_1, \delta_1)), \delta_2) = 0, \tag{8.47}$$

where δ_1 is the time at which sliding begins. Expanding the operator on the left-hand side of (8.47) in δ_1 and δ_2 we get

$$\left[I + \mathcal{L}_{F_1}\delta_1 + O(\delta_1^2)\right]\left[I + \mathcal{L}_{F_s}\delta_2 + O(\delta_2^2)\right]\left[\mathcal{L}_{F_1} H(x_1)\right]$$

and hence

$$\mathcal{L}_{F_1} H(\Phi_s(\Phi_1(x_1, \delta_1)), \delta_2) - \mathcal{L}_{F_1} H(x_1)\left(1 - \delta_2 \frac{\mathcal{L}_{F_d}\mathcal{L}_{F_1} H(x_1)}{\mathcal{L}_{F_d} H(x_1)}\right) =$$
$$(\delta_1 + \delta_2)\mathcal{L}^2_{F_1} H(x_1) + O(\delta_1, \delta_2)^2.$$

Using (8.47) and $\mathcal{L}_{F_1} H(x_1) = 0$, we see that

$$\delta_2(x, \delta_1) = \delta_1(-1 + O(\delta_1)). \tag{8.48}$$

Substituting (8.46) into (8.48) gives us

$$\delta_2(x_1, \delta_1(x_1, y)) = y\left(\sqrt{\frac{2}{\mathcal{L}^2_{F_1} H(x_1)}} + O(y)\right). \tag{8.49}$$

After substituting for δ_2 (8.49), and for δ_1 (8.46) into (8.44), we finally obtain the ZDM for the trajectories that contain a sliding segment ($H(x_1) \leq 0$)

$$\widehat{\mathrm{ZDM}}(x_1) = x_1 + y^2\left(\frac{1}{\mathcal{L}_{F_d} H(x_1)} F_d(x_1) + O(y)\right) := x_1 + \Gamma(x_1, y)y^2, \tag{8.50}$$

where $y(x_1) = \sqrt{-H(x_1)}$ and Γ is a smooth function of its two arguments.

However, we obtained (8.50) under the assumption that $\mathcal{L}_{F_1} H(x_1) = 0$. If we consider any point x_1 such that $\mathcal{L}_{F_1} H(x_1) = v$, then to get the leading-order term, we need to project the ZDM by a smooth flow mapping Φ_1 and obtain our ZDM by additionally flowing for some time, say δ_0 to the point where $\mathcal{L}_{F_1}(H)(\Phi_1(x_1, \delta_0)) = 0$. Thus, we consider the following flow combination

$$\Phi_1(\widehat{\mathrm{ZDM}}(\Phi_1(x, \delta_0), y), -\delta_0) = x + y^2(\widehat{\beta}(x, y) + O(t_1)) \tag{8.51}$$

and

$$y = \sqrt{-H(\Phi_1(x_1, \delta_0))} = \sqrt{-H_{\min}(x_1)} \tag{8.52}$$

since $\Phi_1(x_1, \delta_0)$ is the point where H has its local minimum along a trajectory through x_1.

The algebra is identical to that for the grazing bifurcation in Chapters 6 and 7, so we merely state the result that

$$H_{\min}(x_1, v) = H(x) + O(v^2)$$

and that the ZDM is unchanged by this additional projection. That is, we can finally write the ZDM for the grazing-sliding as

$$\mathrm{ZDM}(x) = x + \begin{cases} 0, & H_{\min} \geq 0, \\ \Gamma(x, y, v) y^2, & H_{\min} < 0, \end{cases} \tag{8.53}$$

with $v = \mathcal{L}_{F_1} H(x)$, $y = \sqrt{-H_{\min}(x, v)}$, $H_{min}(x, 0) = H(x)$ and

$$\Gamma(x, 0, 0) = \frac{F_d(x)}{\mathcal{L}_{F_d} H(x)}.$$

Note that to leading-order all the Lie derivatives can be evaluated at $x = x^*$, which gives the form of the ZDM map for grazing-sliding given in Theorem 8.1.

8.3.3 Switching-sliding bifurcation

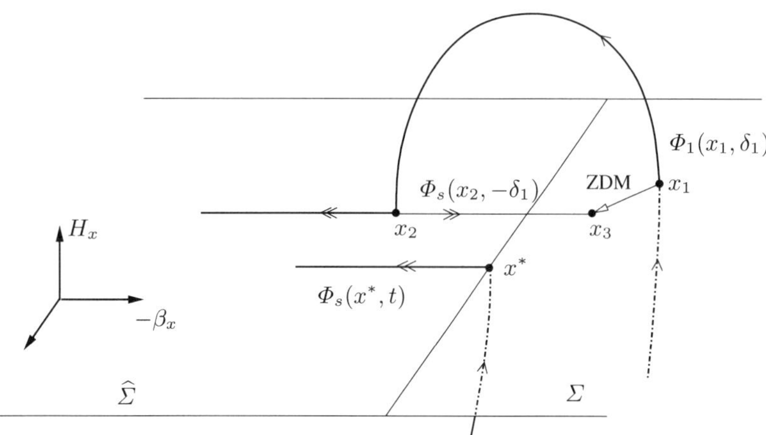

Fig. 8.15. A schematic representation of the construction of the ZDM for the switching-sliding bifurcation.

Suppose now the same conditions (8.7), (8.27) and (8.28) hold but the opposite inequality (8.30) applies. Then the critical trajectory, which starts in region S_2 before hitting the sliding region at a point x^* in the boundary $\partial\widehat{\Sigma}^-$, subsequently slides away from $\partial\widehat{\Sigma}^-$ within the sliding region $\widehat{\Sigma}$. See Fig. 8.15. Let us consider a perturbed trajectory that hits the switching manifold at a point x_1 outside the sliding region in a neighborhood of the point x^*. The ZDM in this case represents the correction that must be applied to the point x_1 in order that the subsequent evolution can be locally described using just the sliding flow Φ_s. The analytical construction of the ZDM proceeds in two steps: First, we consider the flow from x_1 using flow Φ_1 until it reaches the sliding region at a point x_2 after some time δ; then we flow backwards in time for the same amount of time from the point x_2 using the sliding flow Φ_s, reaching the point x_3. As usual, the ZDM is then the mapping from x_1 to x_3.

Thus the ZDM can be written as

$$x_3 = \Phi_s(\Phi_1(x_1, \delta), -\delta), \tag{8.54}$$

where the time δ is found by solving

$$\frac{H(\Phi_1(x_1, \delta)) - H(x_1)}{\delta} = 0.$$

Note that such an expression is uniformly valid for x_1 in a sufficiently small neighborhood of x^* since $H(\Phi_1(x, \delta))$ is locally quadratic in δ at $x = x^*$. The form of the ZDM stated in the Theorem is then calculated by Taylor expansion of the flow combination (8.54) using appropriate Lie derivatives.

8.3.4 Adding-sliding bifurcation

We have come now to the last of the four sliding bifurcation scenarios. Here, the same conditions (8.7), (8.27) hold but the non-degeneracy condition is replaced with the equality (8.31) and the degeneracy condition (8.32).

As shown in Fig. 8.16, in this case a segment of the critical trajectory lies locally entirely in $\widehat{\Sigma}$ and has a quadratic tangency with the boundary of the sliding region $\partial\widehat{\Sigma}^-$ at the point x^*. Consider a perturbed trajectory, with initial point x_1, that crosses the boundary $\partial\widehat{\Sigma}^-$ at some point, say x_2. The system then switches to flow Φ_1, until it reaches Σ again at x_3. The ZDM in this case represents the correction that needs to be applied at point x_1 so that the local description of the trajectory is governed entirely by the sliding flow Φ_s. To derive such a mapping we need to consider three different steps (see Fig. 8.16): We first need to find the time δ_1 at which the trajectory crosses the boundary of the sliding region at x_2; then we must find the time of flight δ_2 under Φ_1 in order to reach x_3 in Σ; finally we flow the system using Φ_s through time $-(\delta_1 + \delta_2)$ so that the total time spent is zero.

The ZDM is thus encapsulated in the flow composition

$$x_4 = \Phi_s(\Phi_1(\Phi_s(x_1, \delta_1), \delta_2), -(\delta_1 + \delta_2)), \tag{8.55}$$

where the times δ_1 and δ_2 are found by solving

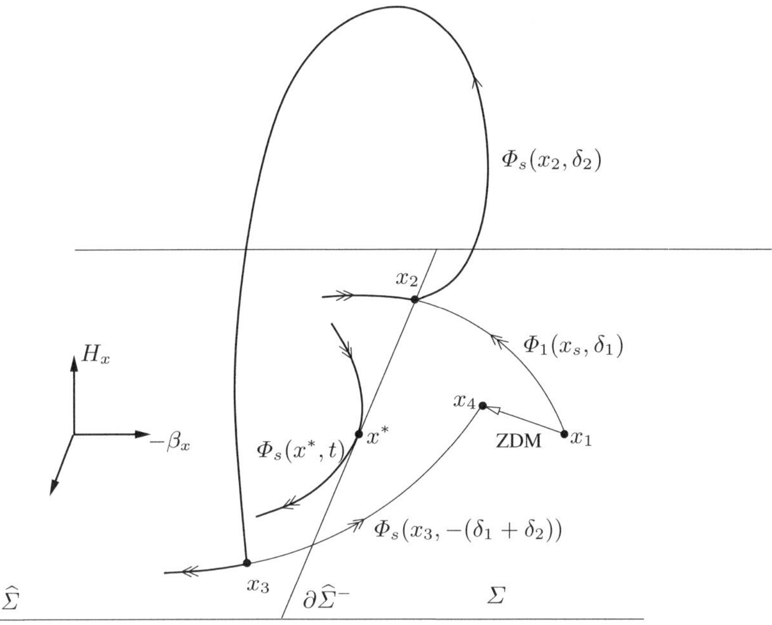

Fig. 8.16. A schematic representation of constructing ZDM for adding-sliding bifurcation case.

$$\mathcal{L}_{F_1} H(\Phi_s(x_1, \delta_1)) = 0, \quad \text{for} \quad \delta_1$$

and

$$H(\Phi_1(\Phi_s(x_1, \delta_1), \delta_2)) = 0, \quad \text{for} \quad \delta_2.$$

The form of the ZDM stated in the Theorem is then obtained by Taylor expansion of the flow combination (8.55) using appropriate Lie derivatives.

8.4 Mapping for a whole period: normal form maps

Let us now consider a parameterized version of (8.1)

$$\dot{x} = \begin{cases} F_1(x, \mu), & \text{if } H(x, \mu) > 0, \\ F_2(x, \mu), & \text{if } H(x, \mu) < 0, \end{cases} \tag{8.56}$$

where $x \in \mathcal{D} \subset \mathbb{R}^n$ and $\mu \in \mathbb{R}$. We suppose that one of the bifurcations analyzed above occurs with respect to the critical trajectory that passes through the point x^* when $\mu = \mu^*$. We suppose that the critical trajectory is a hyperbolic limit cycle.

We shall be interested in the bifurcation scenarios caused when the critical trajectory undergoing the sliding bifurcation is part of a limit cycle. Specifically we shall see how discontinuities in the ZDMs translate into discontinuities in the Poincaré mapping around the periodic orbit. We then try to understand the dynamics of this non-smooth 'normal form map' using the classification strategies introduced in Chapters 3 and 4. The idea behind the use of discontinuity mappings is that we can construct a Poincaré map for a given orbit as a composition of mappings. These are all smooth except for the discontinuity mapping itself. Since the type of discontinuity is generically preserved under composition with a smooth mapping, we can thus state that the type of discontinuity found in the discontinuity mapping will be the one characterizing the full Poincaré map. However, as we shall see, there are subtleties when sliding is involved, because the creation of a new portion of sliding motion leads to a loss of system dimension that results in maps that are singular in one region of definition (as analyzed in Section 3.6). This loss of dimension can already be seen in the Poincaré discontinuity mapping (PDM), which also contains the discontinuity inherent in the ZDM. Composing this map with the regular Poincaré map defined around the periodic orbit in the absence of the extra degeneracy associated with the sliding bifurcation leads to the normal form map. However, due to the presence of sliding in some of the cases, the choice of Poincaré section used to define the PDM needs to be treated with some delicacy.

We shall take each of the four cases in turn in each subsection below. Section 8.5 that follows considers the grazing sliding case in more detail, because that is the only case that can lead to an abrupt change in attractor precisely at the DIB point.

8.4.1 Crossing-sliding bifurcation

For simplicity, we begin by considering a Filippov system with a single discontinuity boundary Σ, which at some parameter value $\mu = \mu^*$ has a hyperbolic limit cycle that undergoes a crossing-sliding bifurcation that without loss of generality we assume to take place at $\partial\widehat{\Sigma}^-$, as depicted in Fig. 8.17. Such an orbit starts at a point A, where it satisfies the conditions for crossing-sliding (8.7), (8.8), (8.11). From this point the trajectory leaves Σ tangentially and moves into region S_1 following the flow Φ_1. Eventually, it crosses the switching manifold at the point B (outside $\widehat{\Sigma}$), moves into region S_2 following the flow Φ_2, before closing itself at the point A.

Suppose we choose as a (remote) Poincaré section, the segment $\Pi_1 \subset \Sigma$ that includes the point B (see Fig. 8.17). We then consider two auxiliary sections Π_2 and Π_3, which are entry and exit sections local to the crossing-sliding point A.

Then, the compound Poincaré map $P_N := \Pi_1 \mapsto \Pi_1$ describing the orbit can be obtained as the composition of the following maps:

$$P_{12} : \Pi_1 \mapsto \Pi_2, \quad \mathrm{PDM} : \Pi_2 \mapsto \Pi_3, \quad P_{31} : \Pi_3 \mapsto \Pi_1.$$

Notice that maps P_{12} and P_{31} are obtained by considering the flows Φ_2 and Φ_1, respectively. Hence, they are smooth and invertible since the flow is locally

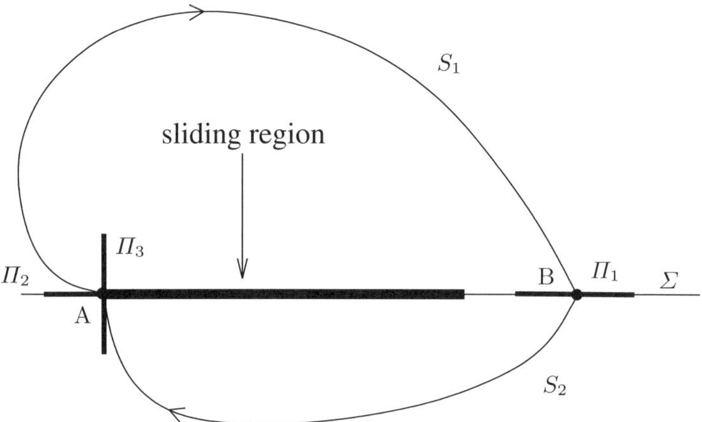

Fig. 8.17. A periodic orbit undergoing the crossing-sliding scenario

transversal to the starting and ending sections. Therefore, the discontinuity in P_N must be introduced by the PDM.

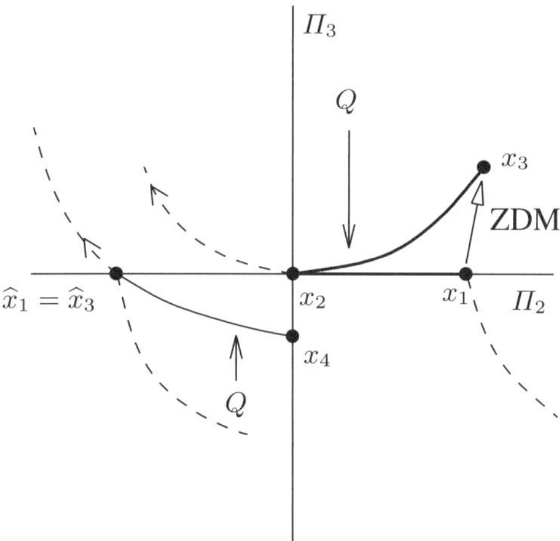

Fig. 8.18. Action of the PDM applied to a point $\widehat{x}_0$ that lies outside of the sliding region and to a point x_0 that lies within the sliding set.

Let us first discuss the geometric construction of the PDM in this case, as it is more subtle than in all other examples. Here the entry and exit Poincaré sections Π_2 and Π_3 are allowed to be different. This is because we need to choose an entry Poincaré section that is transverse to all incoming trajectories, irrespective of whether they slide. Similarly, the exit section should be transverse to all outgoing trajectories. For this reason we choose the incoming section to be

$$\Pi_2 = \Sigma = \{x \in \mathcal{D} : H(x) = 0\},$$

and the outgoing section

$$\Pi_3 = \{x \in \mathcal{D} : \mathcal{L}_{F_1} H(x) = 0\}.$$

Note that the grazing point $A = x^*$ for $\mu = \mu^*$ is in both Poincaré sections by definition. Now, from this geometry it is intuitively obvious where the loss of dimensionality comes from.

Consider the two points $\widehat{x}_1$ and x_1 in Fig. 8.18. Consider first points $\widehat{x}_1$ in Π_2 outside the sliding region. The action of the PDM for such points is reduced to that of the projection mapping, since the ZDM for such points is the identity. That is $\widehat{x}_1 = \widehat{x}_3$, where $\widehat{x}_3$ is the image of $\widehat{x}_1$ under the action of the ZDM. The flow Φ_1 starting from these points is almost tangential to Σ. Thus, we will see a near loss of rank by 1 in the PDM mapping (i.e., this segment of Π_2 is squeezed into a small segment of Π_3). This is depicted schematically in Fig. 8.18 where the projection mapping, labeled Q and indicated by a solid line, maps $\widehat{x}_1 = \widehat{x}_3$ to x_4 lying on Π_3.

Now consider points x_1 in $\widehat{\Sigma} \cap \Pi_2$. Here the ZDM contains a non-zero quadratic term and, by construction, maps points onto the surface, G say, formed by computing trajectories backwards in time via flow Φ_1 from the boundary of the sliding region $\partial\widehat{\Sigma}^-$ (see Fig. 8.13). This surface is everywhere tangential to the flow Φ_1, and therefore an exact loss of rank by 1 in G will be observed. In Fig. 8.18 we can see that the point x_1 is mapped by the ZDM to $x_3 \in G$, which is then mapped by the projection map Q (following flow Φ_1) to x_2, which lies on the boundary $\partial\widehat{\Sigma}^-$.

Therefore, the ZDM causes a discontinuity in the second-derivative terms, and the projection map introduces the rank loss such that the mapping is singular and has co-rank 1 on the sliding side of this discontinuity. Since P_{12} and P_{31} are smooth, the full Poincaré mapping $P_N = P_{31} \circ PDM \circ P_{12}$ also has these properties. In particular, let us suppose that at $\mu = \mu^*$ $P_{31}(x^*) = x_*$ and $P_{12}(x_*) = x^*$, and that the linearization of the maps can be written in the form

$$P_{31}(x^*, \mu) = x^* + N_1(x - x^*) + M_1(\mu - \mu^*) + O\left(|x - x^*|^2, (\mu - \mu^*)^2\right) \quad (8.57)$$

$$P_{12}(x_*, \mu) = x_* + N_2(x - x_*) + M_2(\mu - \mu^*) + O\left(|x - x_*|^2, (\mu - \mu^*)^2\right) \quad (8.58)$$

for $n \times n$ matrices $N_{1,2}$ and $1 \times n$ matrix $M_{1,2}$ satisfying

$$N_1 := \frac{\partial}{\partial x} P_{31}\bigg|_{x=x^*,\mu=\mu*} \quad \text{and} \quad M_1 := \frac{\partial}{\partial \mu} P_{31}\bigg|_{x=x^*,\mu=\mu*},$$

$$N_2 := \frac{\partial}{\partial x} P_{12}\bigg|_{x=x_*,\mu=\mu*} \quad \text{and} \quad M_2 := \frac{\partial}{\partial \mu} P_{21}\bigg|_{x=x_*,\mu=\mu*}.$$

Moreover, without loss of generality, we suppose that the local PDM map is independent of parameters. Then we can state the following:

Theorem 8.2 (normal form map for a crossing-sliding bifurcation). *Suppose a hyperbolic periodic orbit $p(t;\mu)$ with event sequence as in Fig. 8.17 of a piecewise-smooth system written in local co-ordinates in the form (8.56) has a regular crossing-sliding at $(x,\mu) = (x^*,\mu^*)$. Then the Poincaré map P_N from Π_1 to itself can be written as*

$$P_N(x,\mu) = P_{31}(\mathrm{PDM}(P_{12}(x,\mu)),\mu), \tag{8.59}$$

where P_{31} and P_{12} are given by (8.57) and (8.58), respectively, and

$$\mathrm{PDM}(x) = \begin{cases} \mathrm{PDM}_L(x, \mathcal{L}_{F_1}H(x)) & \text{when } \mathcal{L}_{F_1}H(x) \le 0, \\ \mathrm{PDM}_R(x, \mathcal{L}_{F_1}H(x)), & \text{when } \mathcal{L}_{F_1}H(x) \ge 0, \end{cases} \tag{8.60}$$

where PDM_L and PDM_R are given by

$$\begin{aligned}\mathrm{PDM}_L(x,v) = {} & x - v\left(\frac{1}{\mathcal{L}^2_{F_1}H(x)} + v^2\left(\frac{\mathcal{L}_{F_d}\mathcal{L}_{F_1}H(x)}{2\mathcal{L}_{F_d}H(x)\mathcal{L}^2_{F_1}H(x)} + \right.\right. \\ & \left.\left. + \frac{\mathcal{L}^3_{F_1}H(x)}{2(\mathcal{L}^2_{F_1}H(x))^2}\right)\frac{1}{\mathcal{L}^2_{F_1}H(x)}\right)\mathcal{L}_{F_1} + v^2\frac{\mathcal{L}_{F_d}}{2\mathcal{L}_{F_d}H(x)\mathcal{L}^2_{F_1}H(x)} \\ & + v^2\frac{\mathcal{L}^2_{F_1}}{2(\mathcal{L}^2_{F_1}H(x))^2} + O(v^3),\end{aligned} \tag{8.61}$$

$$\begin{aligned}\mathrm{PDM}_R(x,v) = {} & x - \left(\frac{1}{\mathcal{L}^2_{F_1}H(x)}v + \frac{1}{2}\frac{\mathcal{L}^3_{F_1}H(x)}{(\mathcal{L}^2_{F_1}H(x))^3}v^2\right)\mathcal{L}_{F_1} + \\ & + \frac{1}{2}\frac{1}{(\mathcal{L}^2_{F_1}H(x))^2}v^2\mathcal{L}^2_{F_1} + O(v^3).\end{aligned} \tag{8.62}$$

Remarks

1. If we compare (8.61) and (8.62) we can clearly see that the two equations differ at $O(v^2)$. Hence, the projection mapping indeed does not cancel the leading-order correction that needs to be applied to the system due to the crossing-sliding scenario.
2. More complex event sequences than in Fig. 8.17 can also be captured by this normal form. If the critical periodic orbit crosses other remote discontinuity boundaries and potentially has other regions of sliding, providing all crossings and entry points of sliding happen transversely with respect to both μ and x. Then the Poincaré maps P_{31} and P_{12} will be replaced by the smooth Poincaré mappings that result from the compositions of the relevant flow maps and transverse discontinuity mappings.

Proof. To prove the theorem we just need to find an explicit expression for the PDM by projecting $x \in \Sigma$ from the ZDM given by (8.40) onto Π_2 using flow Φ_1. Thus, we have

$$\Phi_1(x,t) = x + \mathcal{L}_{F_1} t + \frac{1}{2}\mathcal{L}_{F_1}^2 t^2 + O(t^3). \tag{8.63}$$

We can find the time δ, required to reach Σ, by solving equation

$$\mathcal{L}_{F_1} H(\Phi_1(x,\delta)) = 0. \tag{8.64}$$

Expanding (8.64) in δ and noting that $\mathcal{L}_{F_1} H(x) := v = O(\delta)$ we can solve (8.64) for δ as a power series in v. The existence of the solution for sufficiently small v is guaranteed by the Implicit Function Theorem. We have

$$\delta(x,v) = -\frac{1}{\mathcal{L}_{F_1}^2 H(x)} v - \frac{1}{2}\frac{\mathcal{L}_{F_1}^3 H(x)}{(\mathcal{L}_{F_1}^2 H(x))^2} v^2 + O(v^3). \tag{8.65}$$

We can now substitute (8.65) into (8.63).

Thus, the projection mapping for $x \in \{x \in \Sigma : v \geq 0\}$ to $O(v^2)$ becomes

$$\begin{aligned}\mathrm{PDM}_R(x,v) = x &- \left(\frac{1}{\mathcal{L}_{F_1}^2 H(x)} v + \frac{1}{2}\frac{\mathcal{L}_{F_1}^3 H(x)}{(\mathcal{L}_{F_1}^2 H(x))^3} v^2\right)\mathcal{L}_{F_1} + \\ &+ \frac{1}{2}\frac{1}{(\mathcal{L}_{F_1}^2 H(x))^2} v^2 \mathcal{L}_{F_1}^2 + O(v^3).\end{aligned} \tag{8.66}$$

We next need to obtain the PDM for points $x \in \Sigma : v \leq 0$. To do so we project points mapped by the ZDM using flow Φ_1 onto Π_2. Thus, expanding $\Phi_1(x_3,t)$ in t, we get

$$\Phi_1(x_3,t) = x_3 + \widehat{\mathcal{L}_{F_1}} t + \widehat{\mathcal{L}_{F_1}}^2 \frac{t^2}{2} + O(t^3), \tag{8.67}$$

where the hat symbol denotes that the quantities are evaluated at x_3 which is the image of a point $x \in \Sigma$ for $v \leq 0$ under the action of the ZDM. The correction of $\mathcal{L}_{F_1}$ and to $\mathcal{L}_{F_1}^2$ due to the fact that the starting points do not lie on Σ is of $O(v^2)$. Hence, we can rewrite (8.67) as

$$\Phi_1(x_3,t) = x_3 + \mathcal{L}_{F_1} t + \mathcal{L}_{F_1}^2 \frac{t^2}{2} + O((v,t)^3),$$

or equivalently as

$$\Phi_1(x_3,t) = x + v^2 \frac{\mathcal{L}_{F_d}}{2\mathcal{L}_{F_d} H(x) \mathcal{L}_{F_1}^2 H(x)} + \mathcal{L}_{F_1} t + \frac{1}{2}\mathcal{L}_{F_1}^2 t^2 + O((v,t)^3). \tag{8.68}$$

where we have used the explicit form of the ZDM (8.41). We can find the time δ as a function of v by solving

$$\mathcal{L}_{F_1}(H)(\Phi_1(x_3,\delta)) = 0, \tag{8.69}$$

for $\delta(x_3,v)$. Expanding (8.68) in t, yields

$$\mathcal{L}_{F_1}H(\Phi_1(x_3,\delta)) = v + v^2\frac{\mathcal{L}_{F_d}\mathcal{L}_{F_1}H(x)}{2\mathcal{L}_{F_d}H(x)\mathcal{L}^2_{F_1}H(x)} + \mathcal{L}^2_{F_1}H(x)\delta +$$
$$+\mathcal{L}^3_{F_1}H(x)\frac{\delta^2}{2} + O((\delta,v)^3) = 0. \tag{8.70}$$

Solving (8.70) for δ as a power series in v, to quadratic order in v, gives

$$\delta(x,v) = -\frac{1}{\mathcal{L}^2_{F_1}H(x)}v - \left(\frac{\mathcal{L}_{F_d}\mathcal{L}_{F_1}H(x)}{2\mathcal{L}_{F_d}H(x)\mathcal{L}^2_{F_1}H(x)} + \frac{\mathcal{L}^3_{F_1}H(x)}{2(\mathcal{L}^2_{F_1}H(x))^2}\right)\frac{1}{\mathcal{L}^2_{F_1}H(x)}v^2.$$

Substituting (8.4.1) into (8.68) will give us an expression for the PDM for x that belongs to the sliding region. Hence, we have

$$\mathrm{PDM}_L(x,v) = x - v\left(\frac{1}{\mathcal{L}^2_{F_1}H(x)} + v^2\left(\frac{\mathcal{L}_{F_d}\mathcal{L}_{F_1}H(x)}{2\mathcal{L}_{F_d}H(x)\mathcal{L}^2_{F_1}H(x)} +\right.\right.$$
$$\left.\left.+\frac{\mathcal{L}^3_{F_1}H(x)}{2(\mathcal{L}^2_{F_1}H(x))^2}\right)\frac{1}{\mathcal{L}^2_{F_1}H(x)}\right)\mathcal{L}_{F_1} + v^2\frac{\mathcal{L}_{F_d}}{2\mathcal{L}_{F_d}H(x)\mathcal{L}^2_{F_1}H(x)}$$
$$+v^2\frac{\mathcal{L}^2_{F_1}}{2(\mathcal{L}^2_{F_1}H(x))^2} + O(v^3). \tag{8.71}$$

Finally, the PDM (8.60) is written by combining (8.71) and (8.66).

Now let us consider the implications of the structure of this normal form map on the dynamical unfolding of a stable hyperbolic periodic orbit undergoing a crossing-sliding bifurcation at a critical parameter value μ^*. Since the first derivative of the Poincaré mapping is continuous, we expect the orbit to persist under small parameter variations. The main effect of the DIB is that the orbit will acquire a section of sliding motion. However, the Poincaré mapping around the limit cycle will not be topological equivalent. In particular, although the map local to the fixed point corresponding to a non-sliding solution is invertible, that built around a sliding solution is not. So, this DIB corresponds to a bifurcation in the classical sense of Definition 2.16, even though there will not be a change in the local attractor.

In fact, we can say what happens to the multipliers of the periodic orbit undergoing the bifurcation. Since the Poincaré mapping is piecewise-smooth, the multipliers should vary smoothly on either side of the discontinuity. They should be continuous across the boundary, but their derivative with respect to parameters is expected to have a jump over the discontinuity. Moreover, on the sliding side, at least one multiplier of the Poincaré mapping must be zero, since the Poincaré mapping here has co-rank 1, and by continuity, that multiplier must be approach zero as the bifurcation point is reached from the other side.

Example 8.1. Consider the third-order relay feedback system (8.19) (8.19), at the parameter values $\rho \approx 2.09$ and $-\sigma = k = \lambda = k = \zeta = \omega = 1$. Here it is

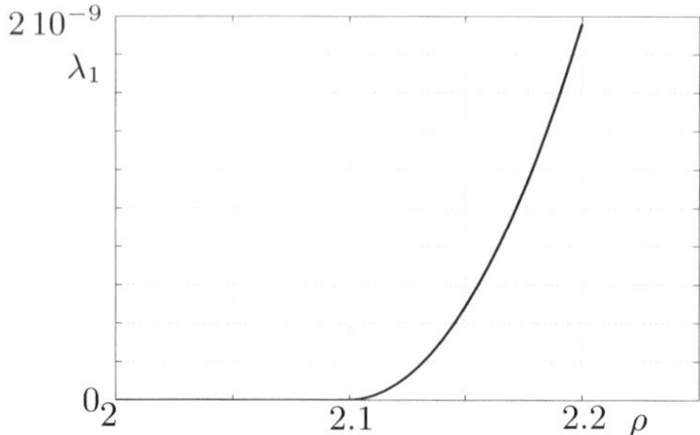

Fig. 8.19. One of the multipliers of the Poincaré map calculated for a periodic point of an orbit with (left) and without (right) sliding segments (see Example 8.1).

found that a simple symmetric periodic orbit (that is an orbit formed by two segments — one generated by vector field F_1 and the other by vector field F_2) undergoes a crossing-sliding bifurcation. Figure 8.19 shows the behavior of one multiplier of the associated Poincaré map as the parameter ρ is varied through the bifurcation point (note that we have two multipliers as the system is three dimensional).

We can see from the figure that the multiplier is identically 0 for periodic orbits that feature a sliding segment, and that it continuously increases from this value as we move away from the bifurcation along the branch of non-sliding orbits. Notice that the multipliers change in a continuous and smooth way. This follows from the fact that the bifurcating orbit is symmetric with real multipliers. If we consider a map which captures the dynamics of half of the orbit then we would observe that the multipliers change continuously but in a non-smooth way under the variation of the bifurcation parameter.

8.4.2 Grazing-sliding bifurcation

Consider a periodic orbit, shown in Fig. 8.20, that grazes Σ at a point A on the boundary of the sliding region $\partial\widehat{\Sigma}^-$ satisfying conditions (8.7), (8.8), (8.11) for a grazing-sliding, but otherwise lying entirely in region S_1.

That is, without loss of generality, we assume that the grazing orbit does not contain any sliding segments apart from the zero-length sliding segment. (For more general situations, we only need assume that the flow that takes the grazing sliding point to itself around the limit cycle crosses all other discontinuity boundaries — or enters sliding regions — transversally, so that the Poincaré mapthat takes the grazing sliding point to itself is smooth in both μ and x.)

To study the bifurcation upon varying a parameter we consider some section Π_1 transversal to the combination of the system flows Φ_1 and Φ_s in region S^+ and the section $\Pi_2 := \{x \in \mathbb{R}^n : \mu(x) = -1\}$ going through point

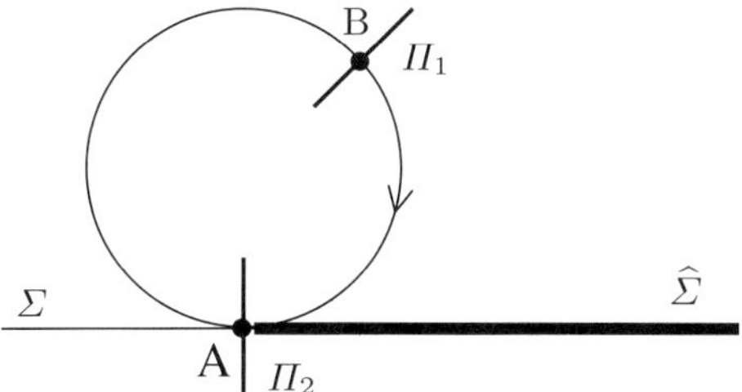

Fig. 8.20. Simplest orbit undergoing grazing-sliding bifurcation

A transversal to flow Φ_1. The full Poincaré map P then maps Π_1 back to itself and is a combination of flows Φ_1 and Φ_s obtained by composition of the following mappings:

$$P_{12} : \Pi_1 \mapsto \Pi_2, \quad PDM : \Pi_2 \mapsto \Pi_2, \quad P_{21} : \Pi_2 \mapsto \Pi_1.$$

A similar argument to the previous cases shows that maps P_{12} and P_{21} are smooth and of full rank. The discontinuity therefore must be due to the PDM, which is formed from a composition of the zero-time discontinuity map for grazing-sliding (8.16) and a projection mapping Q taking points back to Π_2 by flow Φ_1.

Consider now the PDM. The ZDM for this case is the identity for trajectories that remain in region S_1, and contains a linear leading-order correction term otherwise. Moreover, from (8.74) we see that this correction cannot be parallel to the vector field F_1, because it has the direction of $(F_2 - F_1)$ (if $(F_2 - F_1)$ were parallel to F_1, then sliding would not be possible). Hence, the composition of the ZDM with the projection map M does not cancel out the leading-order term and the discontinuity is still of linear leading-order.

We can easily obtain an expression for the PDM by choosing the Poincaré section Π_2 to be the zero level set of a function $\mathcal{L}_{F_1} H(x)$. Note that this surface necessarily crosses $\partial\widehat{\Sigma}^-$ transversally by the non-degeneracy assumptions of grazing-sliding (see Fig. 8.20).

Then, we simply project the ZDM onto the Poincaré section. We follow the same steps that were used to express the ZDM for a general point. That is we can expand

$$\Phi_1(\mathrm{ZDM}(x,y), -\delta) = x + y^2 \Gamma(x,y) - \delta \mathcal{L}_{F_1} + \text{ h.o.t}, \tag{8.72}$$

where $\mathrm{ZDM}(x,y)$ is given by (8.50) with $y(x) = \sqrt{-H(x)}$ and

$$\Gamma(x,y) = \frac{1}{\mathcal{L}_{F_d} H(x_1)} F_d(x_1) + O(y).$$

The value of the vector field F_1 at some point x that lies on the chosen Poincaré section, and at some x_3 that is the image of x under the action of the ZDM, differ by $O(y^2)$. That is, the remainder term contains also terms of $O(ty^2)$.

We now need to determine a leading-order expression for the time t. We know that $\mathcal{L}_{F_1}(H)(\Phi_1(x_3, -\delta)) = 0$. Hence, expanding this expression in t we get

$$\mathcal{L}_{F_1}H(\Phi_1(x_3, -\delta)) = \mathcal{L}_{F_1}H(x_3) - \delta\mathcal{L}^2_{F_1}H(x_3) + \text{ h.o.t} = 0. \tag{8.73}$$

Note that the first term of (8.73) is small and of $O(t)$. Equation (8.73) can now be solved for δ as a power series in $v_3 := \mathcal{L}_{F_1}(H)(x_3)$ (for sufficiently small v_3), which yields

$$\delta = \frac{v_3}{\mathcal{L}^2_{F_1}H(x_3)} + \text{ h.o.t.}$$

We can further express v_3 as a function of $y = \sqrt{-H(x)}$. To leading-order in y, we have

$$v_3 = \mathcal{L}_{F_1}(H)(x + y^2\Gamma(x, y)) = \mathcal{L}_{F_1}H(x) + \mathcal{L}_{F_1}(H)(\Gamma(x, y))y^2.$$

Hence,

$$\frac{\mathcal{L}_{F_1}(H)(\Gamma(x, y))y^2}{\mathcal{L}^2_{F_1}H(x)}$$

in the direction of the vector field F_1 is the sought correction to the ZDM, which gives us the PDM to the leading-order for the grazing-sliding bifurcation.

Expanding (8.72) around x^* to leading-order in x, we get the PDM:

$$\text{PDM}(x) = x + \begin{cases} 0, & H_{\min} \geq 0, \\ H_{\min}\left[\dfrac{F_2(x^*)}{\mathcal{L}_{F_d}H(x^*)} - \dfrac{\mathcal{L}_{F_2}\mathcal{L}_{F_1}(H)(x^*)}{\mathcal{L}^2_{F_1}H(x^*)}F_1(x^*)\right], & H_{\min} < 0, \end{cases} \tag{8.74}$$

with H_{min} defined as for the ZDM.

Finally, consider the surface G which is the forward image under the flow Φ_1 of the sliding region. By construction, the grazing-sliding ZDM maps points below this surface back to G and therefore is rank-deficient for these points. Therefore, the PDM is also rank-deficient. We can summarize the construction as follows:

Theorem 8.3 (normal form map for a grazing-sliding bifurcation). *Suppose a hyperbolic periodic orbit $p(t;\mu)$ with event sequence as in Fig. 8.20 of a piecewise-smooth system that is written in local co-ordinates in the form (8.56) has a regular grazing-sliding at $(x, \mu) = (x^*, \mu^*)$. Let $T(x, \mu)$ be the time of flight upon following trajectories of flow Φ_1 from Π_2 to Π_2 by composing the Poincaré maps P_{12} and P_{21}. Then we can express $P_{21} \circ P_{12}(x)$ as $\Phi_1(x, T)$.*

The normal form map $P_N : \Pi_2 \to \Pi_2$ can then be expressed as

$$P_N(x,\mu) = \begin{cases} \Phi_1(x,T(x,\mu)), & \text{when } H(\Phi_1(x,T(x,\mu)),\mu) \geq 0, \\ \mathrm{PDM}(\Phi_1(x,T(x,\mu)), & \text{when } H(\Phi_1(x,T(x,\mu)),\mu) < 0. \end{cases} \quad (8.75)$$

where PDM(x) *is given by (8.74).*

Remarks

1. From the above considerations, we see that the derivative of the Poincaré mapP_N is discontinuous at the bifurcation point x^*. Thus, for such mappings we cannot therefore conclude that the periodic orbit will persist under parameter variation through the bifurcation point. To unfold the bifurcation, we need instead to use the classification technique developed in Chapter 3 for border-collision bifurcations in locally piecewise-linear maps that are noninvertible in one region. This unfolding forms the subject of Section 8.5 below.
2. Finally, if the orbit survives the bifurcation, we can expect a jump in multipliers as the periodic orbit acquires a sliding portion. The jump in multipliers is nicely illustrated by the fact that a sliding periodic orbit must have at least one multiplier 0, whereas no such restriction exists for an orbit entirely in region S_1.
3. Recall the case of grazing bifurcations in systems without sliding but which nevertheless have degree of smoothness 1 (see Chapter 7): the normal form map is characterized by a square-root singularity. When grazing in the presence of sliding is considered the normal form map instead has a linear leading-order singularity. Thus, the occurrence of sliding makes a significant change to the nature of the dynamics.
4. This Theorem also applies for grazing-sliding of more general periodic orbits, which may have other transverse interactions with discontinuity boundaries away from x^*. In that case $\Phi_1(x,T(x,\mu))$ should be replaced with the appropriate smooth combination of flows and transverse discontinuity mappings that describe the event sequence of the critical periodic orbit.

8.4.3 Switching-sliding bifurcation

Consider a periodic orbit starting at a point A on the boundary of the sliding region $\partial\widehat{\Sigma}^-$ that slides up to a point $B \in \partial\widehat{\Sigma}^+$, crosses the boundary transversally and then follows flow Φ_2 until closing itself (see Fig. 8.21). To construct the full map we introduce three sections. The first is the Poincaré section Π_1, which we can simply assume to be a segment of the sliding boundary $\partial\widehat{\Sigma}^+$ containing point B. Section Π_2 is taken to be a segment of Σ containing A and Π_3 to be a portion of $\partial\widehat{\Sigma}^-$ containing A as well. The full Poincaré map $P_N : \Pi_1 \mapsto \Pi_1$ is obtained by composing appropriate submappings. Although the two maps $P_{21} : \Pi_1 \mapsto \Pi_2$ and $P_{31} : \Pi_3 \mapsto \Pi_1$ generated by flows Φ_2 and

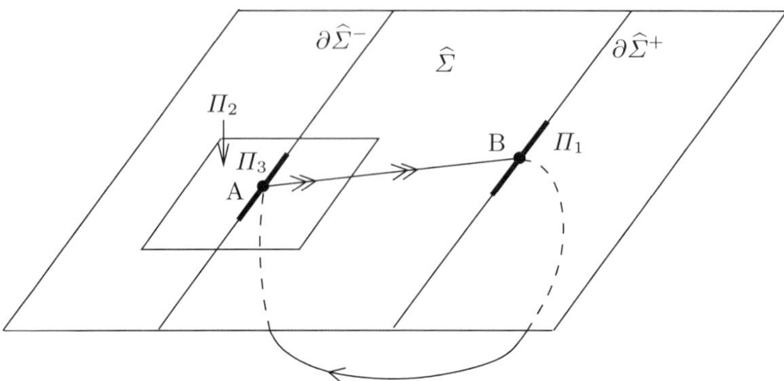

Fig. 8.21. Simplest orbit undergoing a switching-sliding bifurcation

Φ_s, respectively, are smooth and of full rank, the discontinuity is contained in the map $PDM : \Pi_2 \mapsto \Pi_3$.

The PDM is itself formed from a composition of the ZDM with appropriate projection mapping Q that takes points from the switching manifold Σ to Π_3 using flow Φ_s. Summarizing, we have:

Theorem 8.4 (normal form map for a switching-sliding bifurcation). *Suppose a hyperbolic periodic orbit $p(t;\mu)$ with event sequence as in Fig. 8.17 of a piecewise-smooth system that is written in local co-ordinates in the form (8.56) has a regular switching-sliding at $(x,\mu) = (x^*,\mu^*)$ Then the Poincaré mapPi$_N$ from Π_1 to itself can be written as*

$$P_N(x,\mu) = P_{31}(\mathrm{PDM}(P_{12}(x,\mu)),\mu),$$

where P_{31} and P_{12} are given by (8.57) and (8.58), respectively, and $\mathrm{PDM}(x) = Q(\mathrm{ZDM}(x))$ *where Q is the smooth projection mapping onto Π_3 and $ZDM(x)$ is given by (8.17).*

Remarks

1. In this case, Q could cancel the leading- and higher-order discontinuity introduced by the ZDM. For example, in planar systems, Π_3 is a point and therefore the PDM cannot be discontinuous. Assuming this is not the case (which would be a generic assumption in three or more dimensions), then the map has the same properties as the ZDM; see Table 8.1 and Theorem 8.1.
2. Thus, in the case of switching-sliding, we have continuous derivatives at least up to order 2 and discontinuous derivatives of higher order. Hence, as for crossing-sliding, a hyperbolic trajectory will persist under parameter variation since the mapping has continuous first derivative. Moreover, the multipliers are continuous with continuous first derivative with respect to

the parameter, but their second derivative with respect to the bifurcation parameter will in general be discontinuous.

3. As in Theorem 8.2, the above result is also valid for more complex event sequences of the critical periodic orbit, provided all other discontinuity boundaries are encountered transversally, with P_{31} and P_{12} representing the appropriate compositions of flow and transverse discontinuity mappings.

8.4.4 Adding-sliding bifurcation

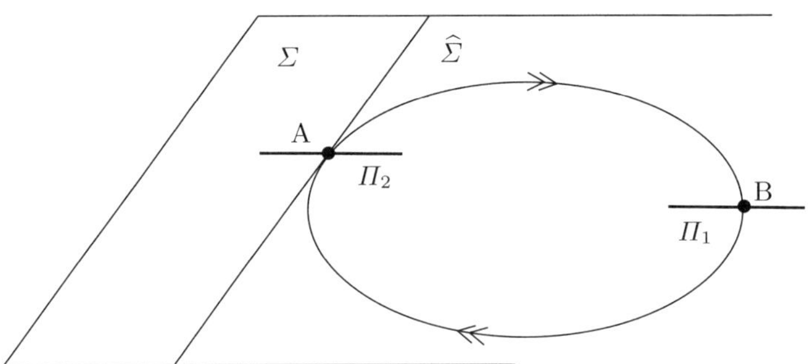

Fig. 8.22. Simplest orbit undergoing adding-sliding bifurcation.

Consider now a periodic orbit entirely contained in the sliding region $\widehat{\Sigma}$ that grazes $\partial\widehat{\Sigma}^-$ at one point, say A (see Fig. 8.25). We can express the normal form map in a similar fashion as in the grazing-sliding case:

Theorem 8.5 (normal form map for adding-sliding bifurcation). *Suppose a hyperbolic periodic orbit $p(t;\mu)$ with event sequence as in Fig. 8.25 of a piecewise-smooth system that is written in local co-ordinates in the form (8.56) has a regular adding-sliding at $(x,\mu) = (x^*,\mu^*)$. Let $T(x,\mu)$ be the time of flight upon following trajectories of flow Φ_1 from Π_2 to Π_2 by composing the Poincaré maps P_{12} and P_{21}. Then we can express $P_{21} \circ P_{12}(x)$ as $\Phi_1(x,T)$.*

The normal form map $P_N : \Pi_2 \to \Pi_2$ can then be expressed as

$$P_N(x,\mu) = \begin{cases} \Phi_1(x,T(x,\mu)), & \text{when } \mathcal{L}_{F_1}H(\Phi_1(x,T(x,\mu)),\mu) \geq 0, \\ \mathrm{PDM}((\Phi_1(x,T(x,\mu)))), & \text{when } \mathcal{L}_{F_1}H(\Phi_1(x,T(x,\mu)),\mu) < 0. \end{cases}$$

where PDM(x) *is given by* $Q(\mathrm{ZDM}(x))$ *where Q is the smooth projection onto Π_2 using flow Φ_2 and* ZDM(x) *is given by (8.18).*

Remarks

1. By arguments similar to the ones presented for the grazing-sliding case, the Poincaré map P_N for an orbit undergoing a adding-sliding bifurcation will be smooth on the sliding side but will have a jump in second derivative across the boundary.
2. Since the first derivative is continuous, a hyperbolic orbit will persist under parameter variation, but the first derivative with respect to the parameter of its multipliers will have a jump as we cross the boundary. Similarly, the second derivative of the multipliers approaches infinity on one side, due to the presence of a $\mathcal{O}(5/2)$ singularity in the Poincaré mapping.
3. As in Theorem 8.3, this theorem is also valid for general adding-sliding bifurcations of periodic orbits, provided the adding-sliding point is the only structurally unstable event. Then $\Phi_1(x, T(x, \mu))$ should be replaced by the relevant composition of flow maps and transverse discontinuity mappings.

8.5 Unfolding the grazing-sliding bifurcation

We now extend the analysis of the previous section for the case of the grazing-sliding in order to understand the bifurcation structure of the simplest periodic orbits. In order to do so, we use the explicit form of the normal form (8.75) to extend the classification presented in Section 3.6. To fix notation we suppose that the hyperbolic periodic orbit $p(t; \mu^*)$ undergoing the bifurcation has period T^*. That is, in the notation of Theorem 8.2, $T(x^*, \mu^*) = T^*$.

8.5.1 Non-sliding period-one orbits

Consider first the conditions for a branch of non-sliding period-one (i.e., with period close to T^*) limit cycles to emanate from the grazing-sliding orbit. Suppose the period of the orbit we seek is $\bar{T}$ and let $\bar{x}$ be its intersection with the Poincaré section $\Pi_2 = \{\mathcal{L}_{F_1} H = 0\}$ local to x^*. Note from the geometry that at such a point $\bar{x}$, H must attain a local minimum $H(\bar{x}) = H_{\min}$. For such an orbit to be admissible [that is, to correspond to an actual orbit of the piecewise-smooth system (8.56)] we require $H_{\min} \leq 0$.

The equations defining such a fixed point of (8.75) are

$$\mathcal{L}_{F_1}(H)(\bar{x}, \mu) = 0, \qquad \bar{x} - \Phi_1(\bar{x}, \bar{T}, \mu) = 0. \tag{8.76}$$

At the bifurcation point, when $\mu = \mu^*$ we know that $x = x^*$, $T = T^*$ is a solution. Therefore we can seek a solution to (8.76) in the form $\bar{x}(\mu)$, $\bar{T}(\mu)$ for μ close to μ^*. Such a solution branch is guaranteed by the Implicit Function Theorem provided the Jacobian derivative of the pair of equations (8.76) with respect to $\bar{x}$ and μ is non-singular at $\mu = \mu^*$. That is, provided the matrix

$$L^* = \begin{pmatrix} V^* & 0 \\ I - J^* & -F^* \end{pmatrix},$$

where

$$J^* = \Phi_{1x}(x^*, T^*; \mu^*), \qquad V^* = \mathcal{L}_{F_1}(H)_x(x^*, \mu^*),$$
$$F_1^* = \Phi_{1t}(x^*, T^*; \mu^*) = F_1(x^*, \mu^*)$$

is non-singular The determinant of this matrix may be obtained by using the theory of bordered matrices [121]. We obtain

$$\det(L^*) = V^* F^* \det(I - N^*),$$

where

$$N^* = (I - \frac{F^* V^*}{V^* F^*}) J^*, \qquad V^* F^* = \mathcal{L}^2_{F_1}(H)(x^*, \mu^*) > 0, \tag{8.77}$$

with the last inequality holding by virtue of non-degeneracy assumption (8.11) written in the form (8.29).

Now, the assumption that the critical periodic orbit is hyperbolic means that when viewed as a non-sliding orbit its linearization cannot have a multiplier equal to unity other than the trivial Floquet multiplier 1 corresponding to the flow direction. In fact, if we recall the construction of Poincaré maps from Chapter 2, the linearization (8.77) is nothing else but the construction of the linearization of the Poincaré map$N$ from the monodromy matrix J^* [cf. (2.17)]. There we showed that the trivial eigenvalue 1 of J^* is changed to a trivial eigenvalue 0 of N^*. Therefore the matrix $I - N^*$ has no eigenvalues equal to 1, which means that $I - N^*$ is non-singular, and so the Implicit Function Theorem applies. This gives us a unique smooth branch of solutions $\bar{x}(\mu)$, $\bar{T}(\mu)$ for μ close to μ^*.

Moreover, we define

$$\nu_0(\mu) = H(\bar{x}(\mu), \mu) \tag{8.78}$$

it is clear that $\bar{x}(\mu)$ corresponds to a unique admissible non-sliding orbit of period one, if and only if $\nu_0 > 0$. Nevertheless, the function $\bar{x}(\mu)$ is well defined and smooth for all μ close to μ^*.

8.5.2 Sliding orbit of period-one

Now consider the possibility of a branch of period-one sliding orbits emanating from the grazing-sliding bifurcation. Here we shall use the explicit from of the ZDM map (8.53) for $H_{\min} < 0$, namely

$$\mathrm{ZDM}(x, y, v) = x + \Gamma(x, y, v) y^2$$

with $v = \mathcal{L}_{F_1} H(x)$, $y = \sqrt{-H_{\min}(x, v)}$ and

$$\Gamma(x, 0, 0) = \frac{F_d(x)}{\mathcal{L}_{F_d} H(x)}.$$

Analogous to the definition conditions (8.75), the equations that define a period-one orbit that has a small portion of sliding near x^* can be written

$$\begin{aligned} \mathcal{L}_{F_1}(H)(x',\mu) &= 0, & x' - \Phi_1(x'',T'';\mu) &= 0, \\ x'' - x' - \Gamma(x',y,0;\mu)y^2 &= 0, & y^2 + H(x',\mu) &= 0. \end{aligned} \tag{8.79}$$

Disregarding the final condition (8.79) for a moment, and viewing y as an independent variable, we find that the non-sliding period-one orbit provides a solution to these equations with $y = 0$ with $x' = x'' = \bar{x}$, $T'' = \bar{T}$, for all μ close to μ^*. Now, for (small, fixed) non-zero y we can eliminate x'' using the first equation in (8.79) and similarly appeal to the Implicit Function Theorem to obtain a smooth solution branch $x'(\mu; y)$ with

$$x'(\mu; y) = \bar{x}(\mu) + \left[\bar{N}(I - \bar{N})^{-1}\bar{E} + O(y)\right] y^2,$$

where

$$\begin{aligned} \bar{N}(\mu) &= \left(I - \frac{\bar{F}_1\bar{V}}{\bar{V}\bar{F}_1}\right)\bar{J}, & \bar{J}(\mu) &= \Phi_{1x}(\bar{x}(\mu),\bar{T}(\mu);\mu), \\ \bar{F}_1(\mu) &= F_1(\bar{x}(\mu),\mu), & \bar{V}(\mu) &= \mathcal{L}_{F_1}(H)_x(\bar{x}(\mu),\mu), \\ \bar{E}(\mu) &= \Gamma(\bar{x}(\mu),0,0;\mu), & \bar{C}^T(\mu) &= H_x(\bar{x}(\mu),\mu). \end{aligned}$$

Substitution of this expression into the final condition (8.79) and expansion in y gives

$$\nu_0(\mu) + y^2(\nu_2(\mu) + O(y)) = 0, \tag{8.80}$$

$$y \geq 0, \tag{8.81}$$

where we have introduced

$$\nu_2(\mu) := 1 + \bar{C}^T\bar{N}(I - \bar{N})^{-1}\bar{E}. \tag{8.82}$$

Note that at the grazing point μ^*, $\nu_0(\mu^*) = 0$, and let us think of ν_2 as a second, independent unfolding parameter. As we shall see the sign of ν_2 determines whether this sliding orbit coexists with the non-sliding periodic orbit, or exists for the other sign of the parameter perturbation $\mu - \mu^*$. Before doing so, it is worth pointing out that it is implicit in the above construction that the stability of the sliding orbit is governed by the non-trivial eigenvalues $(I - \bar{C}^T\bar{E})\bar{N}$; note that the linearization of the ZDM with respect to the x variable gives $I - C^TE$ and the linearization around the periodic point of a cycle, disregarding the presence of switching, is simply N. Hence, the composition of these two linearizations around the periodic point $\bar{x}$ gives $(I - \bar{C}^T\bar{E})\bar{N}$.

8.5.3 Conditions for persistence or a non-smooth fold

Now we are in a position to derive analytic conditions that classify whether a persistence or non-smooth fold transition occurs for the primary periodic orbit undergoing the grazing-sliding bifurcation.

Grazing for the non-sliding orbit happens when $\nu_0 \to 0$, Comparing equations (8.78) and (8.79), we see that it makes sense to compare ν_0 for a non-sliding orbit with $-y^2$ for a sliding orbit, as they both measure the local minimum value of H along the trajectory. Hence from (8.79) we find that

$$\nu_2 = \frac{\nu_0}{-y_{\text{sliding}}^2} = \frac{H_{\min}(\text{non-sliding})}{H_{\min}(\text{sliding})}$$

in the limit as $\nu_0 \to 0$ whenever $\nu_2 \neq 0$. Since $H_{\min}$ must be positive for a non-sliding orbit and negative for a sliding one, a positive value of ν_2 means that the non-sliding orbit exists for $\nu_0 > 0$ and the sliding for $\nu_0 < 0$, whereas a negative value of ν_2 means that both orbits exist for $\nu_0 > 0$ a single sign of ν_0. It then follows that the sign of the quantity ν_2 given by (8.82) determines the simplest possible outcome of the grazing-sliding bifurcation, that is, whether persistence of the primary cycle occurs ($\nu_2 > 0$) or a non-smooth fold ($\nu_2 < 0$).

It is possible to derive further analytic conditions for the existence of period-two, period-three orbits, and so on, using the same techniques. Instead, in the following example we shall appeal to the fact that the leading-order normal form map one can derive in a neighborhood of a grazing-sliding bifurcation will have the leading-order form given by the maps studied in Section 3.6. Thus we can appeal directly the classification of the dynamics that was developed there.

8.5.4 A dry-friction example

Example 8.2 (dry-friction oscillator). Recall the dry-friction oscillator model in case study IV, presented in Chapter 1, which was found to exhibit a grazing-sliding bifurcation causing the sudden onset of chaos. Using the affine approximation of normal form P_N for the grazing-sliding we will show how the above theory can explain this behavior.

Let us first reintroduce the system of interest, which can be written as

$$\ddot{u} + u = c(1 - \dot{u}) + A\cos(\nu t), \tag{8.83}$$

where

$$c(v) = \alpha_0 \text{sgn}(v) - \alpha_1 v + \alpha_2 v^3 \tag{8.84}$$

is a kinematic friction characteristic and for $v = 0$ is set valued, i.e., $-\alpha_0 < v < \alpha_0$. The grazing-sliding of a simple limit cycle (without any sliding segments) can be examined numerically and is found to occur to occur for the parameter values

$$\alpha_0 = 1.5, \quad \alpha_1 = 1.5, \quad \alpha_2 = 0.45 \quad A = 0.1 \quad \nu = 1.7077997,$$

and has period $8\pi/\nu$, i.e., $4T$ where $T = 2\pi/\nu$ is the fundamental driving frequency.

We start by putting the system (8.83) into the general form (8.1). Setting $\nu t = \tau$, $x_1 = u$, $x_2 = \dot{u}$, we can express (8.83) as a set of first-order ODEs with discontinuous right-hand side of the form

$$\begin{aligned} \dot{x}_1 &= x_2, \\ \dot{x}_2 &= -x_1 + \alpha_0 \mathrm{sgn}(1 - x_2) - \alpha_1(1 - x_2) + \alpha_2(1 - x_2)^3 + A\cos(\tau), \\ \dot{\tau} &= \nu. \end{aligned}$$

The switching surface Σ in this case can be defined as

$$\Sigma := \{x \in \mathbb{R}^3 : H(x) := 1 - x_2 = 0\}. \tag{8.85}$$

Thus, the normal to Σ is the vector

$$H_x = \begin{bmatrix} 0 & -1 & 0 \end{bmatrix}. \tag{8.86}$$

The dynamics of the system is smooth and continuous when $H(x)$ is non-zero and is governed by the vector fields

$$F_1 = \begin{pmatrix} x_2 \\ -x_1 + \alpha_0 - \alpha_1(1 - x_2) + \alpha_2(1 - x_2)^3 + F\cos(\tau) \\ \nu \end{pmatrix}, \quad \text{if } H(x) > 0 \tag{8.87}$$

and

$$F_2 = \begin{pmatrix} x_2 \\ -x_1 - \alpha_0 - \alpha_1(1 - x_2) + \alpha_2(1 - x_2)^3 + F\cos(\tau) \\ \nu \end{pmatrix}, \quad \text{if } H(x) < 0. \tag{8.88}$$

According to our analysis, sliding motion (sticking in the parlance of the friction systems) is possible if condition (8.5) is satisfied, that is, if

$$\alpha_0 > 0,$$

which is true since it is assumed that the coefficients of the kinematic friction characteristic (8.84) are positive.

Using Utkin's method we can define the vector field F_s which governs the flow on the switching manifold. Substituting (8.87) and (8.88) into (8.3), we then get the following expression for the sliding flow F_s:

$$F_s = \begin{pmatrix} x_2 \\ -x_1 - \alpha_1(1 - x_2) + \alpha_2(1 - x_2)^3 + F\cos(\tau) - \beta(x)\alpha_0 \\ \nu \end{pmatrix}, \tag{8.89}$$

where $-1 \leq \beta(x) \leq 1$. Since the vector field F_s must lie on the switching manifold Σ, we have

$$H_x F_s = 0, \tag{8.90}$$

and using (8.4), we can express $\beta(x)$ as

$$\beta(x) = -\frac{x_1 + \alpha_1(1 - x_2) - \alpha_2(1 - x_2)^3 - F\cos(\tau)}{\alpha_0}. \tag{8.91}$$

The sliding region $\widehat{\Sigma}$ can then be defined as

$$\widehat{\Sigma} = \{x \in \Sigma : -1 \leq \frac{-x_1 + A\cos(\tau)}{\alpha_0} \leq 1\}. \tag{8.92}$$

In our case, the bifurcation point is at

$$(x_1^*, x_2^*, \tau^*) = (\alpha_0 + A\cos(\tau^*), 1, \tau^*) = (1.4198, 1, 3.7828).$$

We first check that the set of analytical conditions (8.7), (8.8) and (8.11), which define a grazing-sliding bifurcation are indeed satisfied at the bifurcation point under investigation. In fact, we get:

1. $H(x^*) = 0$,
2. $\beta(x^*) = -1 \Rightarrow \mathcal{L}_{F_1}(H)(x^*) = 0$,
3. $\mathcal{L}^2_{F_1}(H)(x^*) = H_x F_{1x} F_1 = 1 + \nu F \sin(\tau^*) = 0.8971 > 0$.

Thus, at the calculated value of x^*, the system satisfies all three conditions and therefore the bifurcation event described in Fig. 1.23 of Chapter 1 is indeed due to a grazing-sliding bifurcation. We now show how knowledge of this can be used to classify analytically the observed bifurcation scenario and hence explain the sudden appearance of a chaotic attractor using the theory of border-collisions.

We first calculate the quantity ν_2 given by (8.82) above, which we expect to be positive. We shall also calculate the eigenvalues of N above that give us information on the stability of the $4T$-grazing orbits, with and without the zero-length sliding segment. A numerical integration of the system along the grazing orbit of period $4T$ (with the calculated parameter values given above) allows the monodromy matrix [the linearization of $\Phi_1(x^*, 4T)$] to be found, and it has the form

$$J^* = \begin{pmatrix} 0.1885 & -0.4852 & -0.1104 \\ -0.9743 & 2.5043 & 0.5705 \\ 3.1507 & -8.1102 & -1.8448 \end{pmatrix}.$$

Using the above theory the stability the non-sliding orbit is given simply by the multipliers of the Poincaré map, which are the non-trivial eigenvalues of J^*, which are, $\lambda_{1N} = 0.8407$ and $\lambda_{2N} = 0.0072$. Also, the stability of the sliding orbits that bifurcate is governed by the eigenvalues of $(I - E^* C^{*T}) N^*$, using the notation introduced above. Now, for this example $C^T = [0, \; -1, \; 0]$, and the ZDM is simple to calculate, giving $E = [0, -1, 0]^T$. The relevant eigenvalues are then found to be $\lambda_{1Ns} = -1.6564$ and $\lambda_{2Ns} = 0$. For stability of each of these orbits, we require these multipliers to lie inside the unit circle. Finally, we can use (8.82) to compute $\nu_2(\mu^*) = 16.8006$. Putting this information together, we find that the effect of the grazing-sliding bifurcation is a persistence of the *stable* non-sliding orbit of period $4T$ to an *unstable* orbit of the same period with a single sliding segment. This agrees with what is observed numerically.

The dynamics of the friction oscillator can be further classified by obtaining the piecewise-affine approximation of the normal form map P_N and by studying the resulting expression *a posteriori* for the existence of other attractors. The map has co-rank one or greater, and thus for the friction system it suffices to use two co-ordinates to describe it. Further we can let one of the co-ordinates be the value of H, and we can use the other one to put the affine approximation of P_N in the canonical form used in Chapter 3 where border-collision bifurcations in noninvertible piecewise-smooth planar maps were studied.

That is, we can now easily obtain a mapping of the form

$$\Sigma(\tilde{x}_1, \tilde{x}_2, \tilde{\mu}) = \underline{\tilde{x}} = \begin{cases} \begin{pmatrix} \tau_1 & 1 \\ 0 & 0 \end{pmatrix} \begin{pmatrix} \tilde{x}_1 \\ \tilde{x}_2 \end{pmatrix} + \tilde{\mu} \begin{pmatrix} 1 \\ 0 \end{pmatrix}, & \text{if } \tilde{x}_1 \leq 0, \\ \begin{pmatrix} \tau_2 & 1 \\ -\delta_2 & 0 \end{pmatrix} \begin{pmatrix} \tilde{x}_1 \\ \tilde{x}_2 \end{pmatrix} + \tilde{\mu} \begin{pmatrix} 1 \\ 0 \end{pmatrix}, & \text{if } \tilde{x}_1 \geq 0, \end{cases} \tag{8.93}$$

where τ_1, τ_2 are traces of $(I - E^* C^*)N^*$ and N^*, respectively, δ_2 is the product of the nontrivial eigenvalues of N^*; and $\tilde{\mu} = (1 - \tau_2 + \delta_2)\nu_0$ (for a co-ordinate transformations that give the canonical form (8.93), see the first section of Chapter 3).

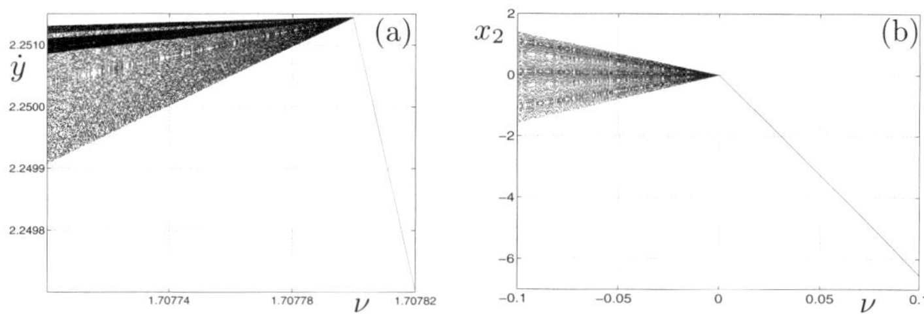

Fig. 8.23. Bifurcation diagrams obtained from the (a) numerical integration of the system (8.83) at the parameter values given in the text, and (b) from the map (8.93).

After some algebra (see [158] for the details), we find $\tau_1 = 0.8479$, $\tau_2 = -1.6564$ and $\delta_2 = 0.0061$, which satisfy the following inequality relations:

$$\tau_1 < -\frac{1}{1 + \tau_2}, \tag{8.94}$$

$$\tau_1(\tau_2+1) - \delta_2(1+\frac{1}{\tau_2}) < 0, \tag{8.95}$$

$$\tau_1(\tau_2+1) - \delta_2(1+\frac{1}{\tau_2}) + 1 > 0. \tag{8.96}$$

According to the theory of Section 3.6, these are precisely the conditions required for a border-collision bifurcation from a fixed point attractor to a chaotic attractor, which is indeed what we observe numerically. See Fig. 8.23 for the comparison between the bifurcation diagram computed numerically and that obtained from application of the discontinuity mapping.

8.6 Other cases

We have clearly not exhausted all possible DIBs that involve sliding in Filippov systems. For example, we could look at boundary-intersection crossings as identified in Chapter 7, but in the presence of sliding. Also, we could look at special trajectories in a neighborhood of a point where $\partial\widehat{\Sigma}^+$ and $\partial\widehat{\Sigma}^-$ intersect (see, e.g., the work by Teixera in [248]). In the next two subsections, we consider yet more possibilities.

8.6.1 Grazing-sliding with a repelling sliding region — catastrophe

We look at catastrophic discontinuity-induced bifurcations that lead to destruction of an attractor. Specifically this can come about due to an attractor that grazes with a switching manifold Σ within its repelling sliding region.

Consider the following simple, constructed example, where such a scenario occurs.

Example 8.3. Consider a planar Filippov system that can be written as

$$F_1 = \begin{pmatrix} (1+\mu-(x_1^2+x_2^2))x_1 - (x_1^2+x_2^2)x_2 \\ (x_1^2+x_2^2)x_1 + (1+\mu-(x_1^2+x_2^2))x_2 \end{pmatrix} \quad \text{if} \quad x_1 < 1,$$

$$F_2 = \begin{pmatrix} 1 \\ 0 \end{pmatrix} \quad \text{if} \quad x_1 > 1,$$

with μ being a bifurcation parameter. Note that the vector field F_1 is like the normal form for the Hopf bifurcations [168]. The switching manifold Σ is defined as

$$\Sigma := \{x \in \mathbb{R}^2 : x_1 = 1\}.$$

For $\mu < 0$ the system has a stable limit cycle that does not intersect Σ. An example of such an orbit is depicted for $\mu = -0.1$ in Fig. 8.3(a). For $\mu = 0$ the limit cycle undergoes a grazing-sliding bifurcation but with a repelling sliding set [see Fig. 8.3(b)]. For $\mu > 0$ the limit cycle is destroyed, because the invariant set forming the cycle crosses Σ (see Fig. 8.25). Note that there is now no attracting motion in the vicinity of the limit cycle. This might therefore be described as a kind of 'blue sky catastrophe' for piecewise-smooth systems.

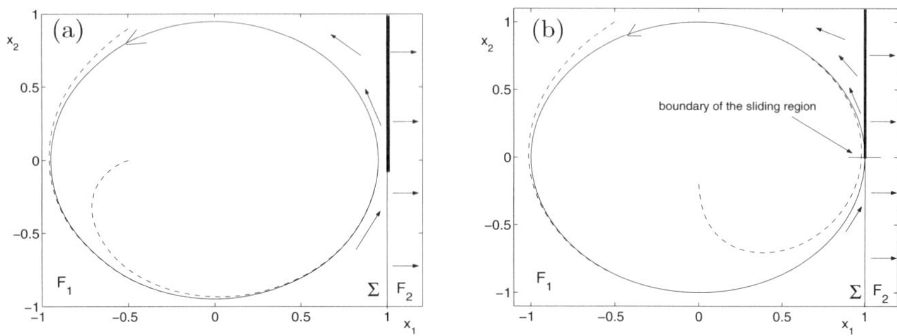

Fig. 8.24. Limit cycle undergoing a 'catastrophe' through a grazing-sliding bifurcations with a boundary of repelling sliding region, (a) before the bifurcation and (b) at the bifurcation

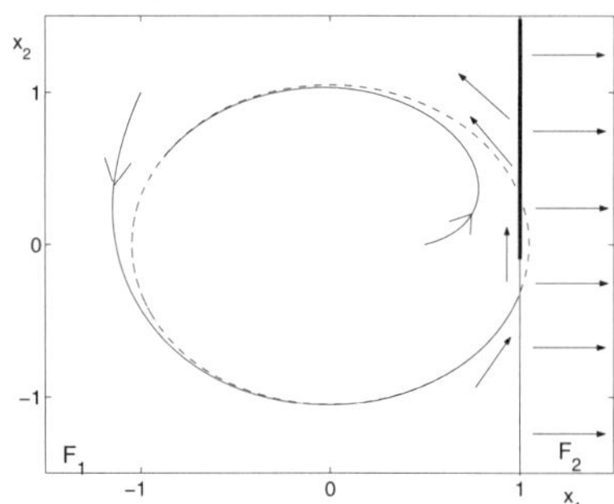

Fig. 8.25. Phase space representation after the grazing-sliding leading to a catastrophe. The dashed circle depicts the flow lines along the limit cycle ignoring the existence of the switching boundary.

8.6.2 Higher-order sliding

Other possible instances of DIBs involve *higher-order sliding* motion. higher-order sliding might occur when two switching manifolds characterized by sliding regions intersect transversally (see Chapter 2).

Let us consider Filippov type system of the form

$$\dot{x} = \begin{cases} F_1(x), & \text{if} \quad H_1(x) > 0,\ H_2(x) > 0, \\ F_2(x), & \text{if} \quad H_1(x) < 0,\ H_2(x) > 0, \\ F_3(x), & \text{if} \quad H_1(x) < 0,\ H_2(x) > 0, \\ F_4(x), & \text{if} \quad H_1(x) < 0,\ H_2(x) < 0, \end{cases} \tag{8.97}$$

with $x \in \mathbb{R}^n$ and $F_i : \mathbb{R}^n \mapsto \mathbb{R}^n$ for $i = 1, \dots, 4$ and $H_1, H_2 : \mathbb{R}^n \mapsto \mathbb{R}$. We define the switching manifolds Σ_1 and Σ_2 as

$$\Sigma_1 := \{x \in \mathbb{R}^n : H_1(x) = 0\} \tag{8.98}$$

and

$$\Sigma_2 := \{x \in \mathbb{R}^n : H_2(x) = 0\}. \tag{8.99}$$

We assume that sliding regions, say $\widehat{\Sigma}_1$ and $\widehat{\Sigma}_2$, are formed within the discontinuity sets Σ_1 and Σ_2, respectively. Moreover, we assume that $\widehat{\Sigma}_1$ and $\widehat{\Sigma}_2$ intersect along a codimension-2 submanifold, say $\widehat{\Sigma}_{12}$, which is attracting from both $\widehat{\Sigma}_1$ and $\widehat{\Sigma}_2$. The system evolution within $\widehat{\Sigma}_{12}$ will be termed higher-order sliding. In the three-dimensional case depicted in Fig. 8.26 $\widehat{\Sigma}_{12}$ is a one-dimensional manifold. In the figure we depict a schematic trajec-

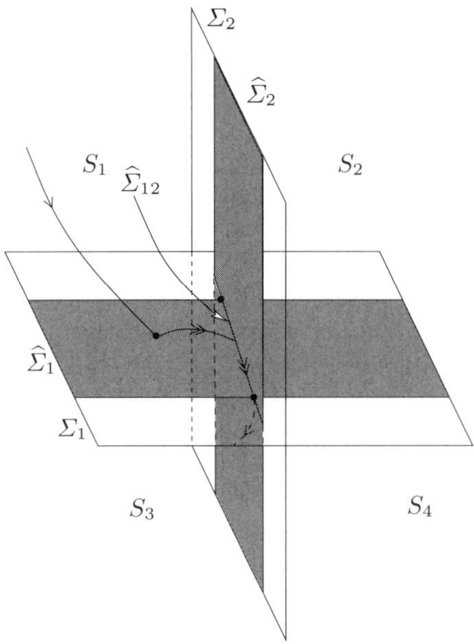

Fig. 8.26. Sliding within a set formed by the union of two sliding regions.

tory containing a segment of higher-order sliding. This trajectory consists of segments generated by different vector fields. Namely, the first segment is generated by the vector field F_1 until reaching Σ_1 within its sliding region $\widehat{\Sigma}_1$. From here the trajectory slides within $\widehat{\Sigma}_1$ until the point of intersection with the second switching manifold Σ_2 within its sliding regions $\widehat{\Sigma_2}$. From this point higher-order sliding ensues within $\widehat{\Sigma}_{12}$. The evolution continues until the boundary of $\widehat{\Sigma}_1$ is reached. Then, the trajectory leaves Σ_1 and evolves within $\widehat{\Sigma}_2$.

Let us now suppose that the described trajectory forms part of a limit cycle that depends on some parameter, say μ. Continuous parameter variation will move the point of intersection of the trajectory with Σ_1 and the point at which the higher-order sliding region $\widehat{\Sigma}_{12}$ is reached. Parameter variation might then lead to the situation depicted in Fig. 8.28 where the trajectory after evolution within $\widehat{\Sigma}_1$ reaches the boundary of the sliding region before getting to $\widehat{\Sigma}_{12}$. Thus, the segment of higher-order sliding is 'lost'. Such a transition implies a DIB. At the instant of the bifurcation, a trajectory reaches the boundary of the sliding region exactly at the same time as when $\widehat{\Sigma}_{12}$ is reached — see Fig. 8.27 where we depict a trajectory reaching a point labeled as C that belongs to the boundary of $\widehat{\Sigma}_1$ and to the switching manifold Σ_2. We claim

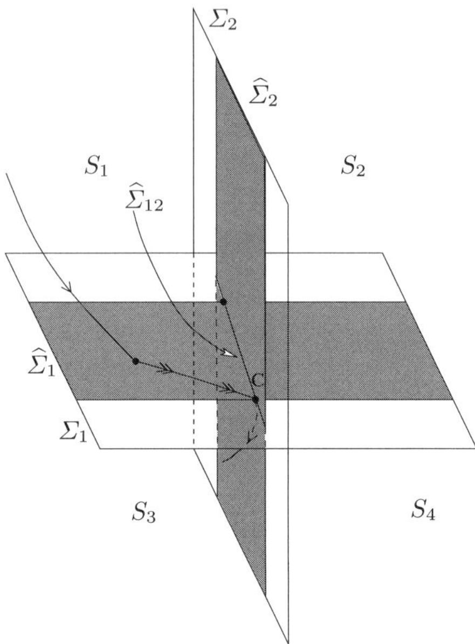

Fig. 8.27. A trajectory at a DIB reaching the boundary of $\widehat{\Sigma}_1$ and the switching manifold Σ_2 at the same time.

that this scenario is of codimension-1 since generically a continuous variation of one parameter only is sufficient to observe a transition from a trajectory depicted in Fig. 8.26 to a trajectory presented in Fig. 8.28.

A detailed analysis of such higher-order sliding bifurcations is beyond the scope of this book. Although we should point out that other codimension-one DIB phenomena involving higher-order sliding are possible. Consider, for example, a trajectory evolving within $\widehat{\Sigma}_2$ such that the sliding flow crosses Σ_1. That is, the vector field generating the sliding motion is discontinuous across Σ_1, but in such a way that $\widehat{\Sigma}_2$ persists on both sides of Σ_1 and, while on $\widehat{\Sigma}_2$, the discontinuity across Σ_1 is such that the trajectory in question hits Σ_1 transversally and crosses it. Now, suppose that under parameter variation the sliding flow hits tangentially the boundary of $\widehat{\Sigma}_1$. Under appropriate non-degeneracy conditions, further parameter variation would then lead the trajectory to acquire a segment of higher-order sliding. Such a scenario would correspond to the codimension-1 crossing-sliding bifurcation, but for the

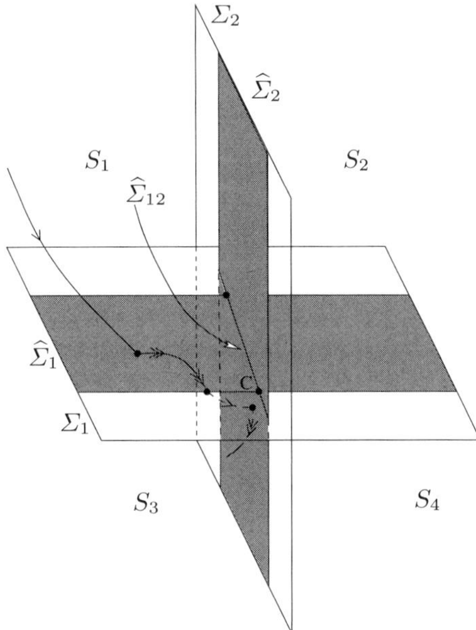

Fig. 8.28. A trajectory after a higher-order sliding segment 'was lost' due to a DIB of codimension-1.

already sliding flow. Similarly, switching-sliding and grazing-sliding involving a segment of higher-order sliding can also occur. If the phase space is four-dimensional and the manifold $\widehat{\Sigma}_{12}$ is two-dimensional, adding-sliding is possible where the higher-order sliding interacts with the boundary of the sliding region.

The following model system of relevance to earthquake engineering provides a potential example that may well exhibit several of these higher-order sliding bifurcations within its parameter space:

Example 8.4 (A two-mass stick-slip system). Galvanetto [110, 111, 112, 113] considers the idealized two mass stick-slip dry-friction oscillator model represented in Fig. 8.4. It can be expressed in dimensionless form as

$$m_1\ddot{u}_1 + u_1 + k_{12}(u_1 - u_2) = C_{k1}(\dot{u}_1 - V_{dr}), \tag{8.100}$$

$$m_2\ddot{u}_2 + u_2 + k_{12}(u_2 - u_1) = C_{k2}(\dot{u}_2 - V_{dr}), \tag{8.101}$$

where m_1, m_2 denote masses of the blocks; k_1, k_2, and k_{12} denote stiffness of springs attaching masses to rigid walls and stiffness of the spring between the masses, V_{dr} denotes velocity of the driving belt; and C_{k1} and C_{k2} are functions describing kinematic friction. The function C_{k1} is defined as

$$C_{k1}(\dot{u}_1 - V_{dr}) = \begin{cases} \dfrac{1-\delta}{1-\gamma v} + \delta + \eta v^2, & \text{if} \quad v < 0, \\ -\dfrac{1-\delta}{1+\gamma v} - \delta - \eta v^2, & \text{if} \quad v > 0, \end{cases} \tag{8.102}$$

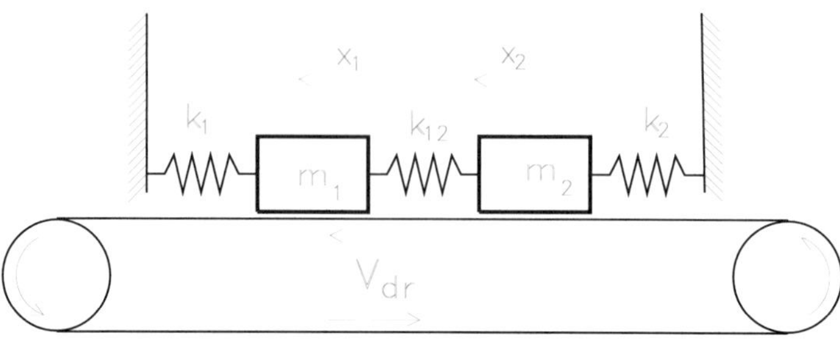

Fig. 8.29. Two block stick-slip system. (Reprinted from [113] with permission from Elsevier.)

where $v = \dot{u}_1 - V_{dr}$ and $C_{k2} = \beta C_{k1}$ with β being a positive parameter.

As shown in [110, 113] the system can exhibit a plethora of different types of stick-slip periodic and aperiodic motions. However, the transitions between the different types of such oscillations are not well understood as yet. Many of the periodic motions observed involve higher-order sliding, which in this case refers to the situation when both masses stick to the driving belt.

9

Further applications and extensions

The aim of this chapter is to present some further examples drawn from a number of applications and experiments, which illustrate the preceding analysis, and also briefly introduce further extensions to the theory that go beyond the detailed scope of this book.

We begin in Section 9.1 by presenting the study of an experimental impact oscillator and by comparing the results with analytical calculations, building on the treatment of case study I from the Chapter 1. The results serve to illustrate the delicate effects of noise and parameter uncertainty on the dynamics of experimental non-smooth systems. This leads, in Secs. 9.2 and 9.3, to a discussion of two models, motivated by applications, for the rattling of gears and for a force-limited hydraulic damper. Through the former example, we discuss how the global behavior of a compliant system (for which grazing bifurcations lead to a Poincaré mapwith a (3/2)-type singularity) can give very similar results to an impacting model of the same system. Also, we illustrate how to derive analytically necessary conditions of existence of the periodic solutions and their bifurcations. The latter example is used to motivate a preliminary treatment of the dynamics associated with the interaction between an invariant torus of a smooth flow and a discontinuity surface, extending the analysis for limit cycles presented in Chapters 6 and 7. Finally, we consider the global dynamics of friction oscillators, where we can develop an understanding of the whole of parameter space by unfolding various possible *codimension-two discontinuity-induced bifurcations.*

9.1 Experimental impact oscillators: noise and parameter sensitivity

In case study I on the impact oscillator in Chapter 1, we showed data taken from laser Doppler measurements of a carefully constructed experimental impact oscillator comprising a massive, stiff and lightly damped beam striking a rigid obstacle, together with numerical simulation results at corresponding parameter values of a simple model, from the work of Popp *et. al.* [208]. There is generally good agreement between theory and experiment both in the phase space representation of (periodic and chaotic) attractors (cf. Figs. 1.5 and 1.6) and also in the one-parameter bifurcation diagrams [cf. Fig. 1.12(a) and (b)]. Nevertheless, we notice some important distinctions between the dynamics of the single-degree-of-freedom impact oscillator model (1.1). Most notably, immediately after each impact there is apparent high-frequency noise in the experimental plots, which quickly damps away.

9.1.1 Noise

If we consider again the experimental results from the impact oscillator in Chapter 1, reproduced here as Fig. 9.1, we see that the noise in the exper-

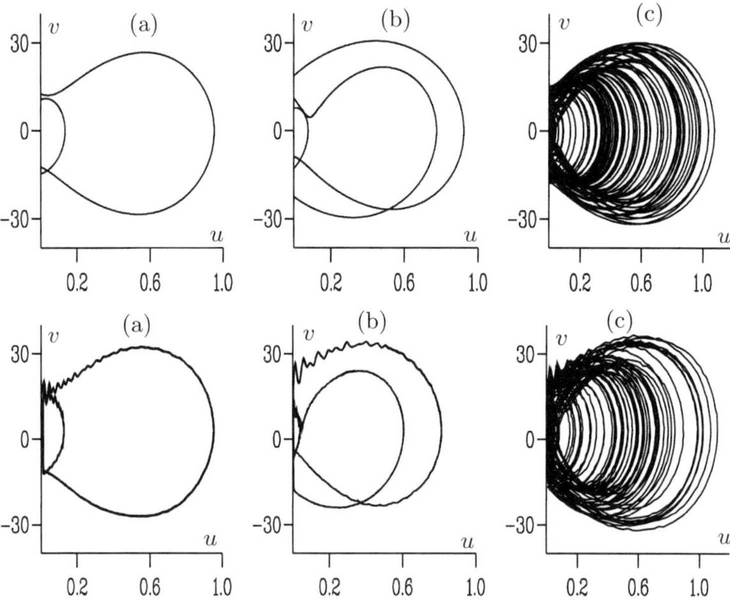

Fig. 9.1. Top: Phase space projection of an impact oscillator (see Chapter 1 for details) for three different input frequencies. Bottom: The corresponding results from the experimental system.

imental phase potraits comprises a fairly regular transient higher-frequency oscillation superimposed on the main attractor. This can be understood from the fact that the stiff beam in the physical impacting system can oscillate in many different modes. Whilst these are in general highly damped (apart from the principal mode), the impact itself excites higher-order modes. Thus, rather than modeling the rigid beam system by a single-degree-of-freedom model, we should more realistically take a higher-dimensional model from the discretization of a cantilevered beam (as in Example 6.3). This in turn leads to a higher-dimensional impact law where energy is transferred at impact from the fundamental mode to higher modes. However, the experiment was specifically constructed so that these higher-order modes were highly damped. Thus when the trajectory next returns to the impacting surface, the effect on the overall dynamics due to the higher-order modes is negligible (see Fig. 1.5 where the higher-frequency motion created for $\dot{u} > 0$ is virtually undetectable for $\dot{u} < 0$). Thus, these high-dimensional effects can be well represented by using a single degree-of-freedom model by adjusting the effective coefficient of restitution to account for the energy loss from the fundamental model. This approximation is a good one provided that impacts are well separated so that transient high frequency motion has a chance to decay. It is likely to be less accurate when impacts are close together.

These observations serve to illustrate two points. First, that 'coefficients of restitution' are typically not just a function of the local properties of the two materials in contact at the moment of impact. Rather they are a property of the ability of the material to dissipate the shock waves created at impact. We know this from everyday life. A drumstick hitting a free symbol has a completely different dynamics from the same stick hitting the same symbol that is being held. In the former case, the energy is dissipated through largely undamped sound waves in the symbol. In the latter, these same waves are felt as rapidly damped vibrations in the hand and arm of the person holding the symbol and we hear almost nothing. In the case of a more flexible cantilever beam (see for example the discussion in [31]), significant energy is imparted to the higher-frequency modes and these modes decay much more slowly than the case above. This leads to very significant energy loss from the primary mode and to model the dynamics, in terms of a single-degree-of-freedom impact oscillator, a low value of the coefficient of restitution, r, must be taken. For example, in [31] r was set equal to 0.2. In this case, though, the single-degree-of-freedom model did not compare particularly well with the experimental results. This simple discussion immediately shows us that any consideration of the nature of impacting for a general system is going to be difficult. We have in this book deliberately kept away from the delicate issues associated with *tribology* or the precise mechanism of interaction between two impacting bodies. Cases involving either multiple contacting bodies (such as the Newton's cradle toy) or impacts in the presence of friction (as in the so-called Painlevé paradox [213]) are beyond the scope of what we cover here. More information is given in the books [219, 38, 39], see also the paper [46] for the first steps in bifurcation analysis in such more complex hybrid systems.

The second point is that noise clearly plays a key role in non-smooth systems. The theory of bifurcations in the presence of noise even for smooth systems is not complete [6]. What is clear is that noise can delay bifurcations, unfold bifurcations or smear out bifurcation diagrams [11]. A few results have been published recently on the effect of noise on discontinuity-induced bifurcations. For example [123] consider the effect of noise on the bifurcation diagram in DC–DC converters (case study V in Chapter 1). It seems that noise can have a far more fundamental role in non-smooth systems than in smooth ones. A possible explanation for this phenomenon is that a small amount of noise added to a trajectory close to a discontinuity surface could cause it to intersect the surface, with a consequent major effect on its resulting dynamics. Indeed, there is evidence to suggest that noise can effectively smooth out the bifurcation diagram and that even small amounts of noise can cause large qualitative details in the fine structure of bifurcation diagrams. A careful *stochastic analysis* of the effects of noise on DIBs is a non-trivial task and is certainly beyond the scope of this book. However, the general issue of (deterministic) smoothing of non-smooth systems is a ripe topic to study (see for example [136] which considers how a period-adding cascade can arise from a period-doubling cascade in a smoothed-out version of the square-root map). Such analysis suggests that the fine detail of whether the local discontinuity mapping has an $O(3/2)$ or a square-root singularity is not as important as the global dynamics caused by the grazing. We already saw examples of this in Chapters 4, 7 and 8 where grazing bifurcations whose discontinuity mappings show a jump that is higher order than linear and hence do not destroy or change the stability of the bifurcating orbit instantly, nevertheless cause a catastrophic change at a nearby parameter value. We shall return to this issue in Sec. 9.2 below.

First though let us consider a direct experimental illustration of grazing in an impacting system.

9.1.2 An impacting pendulum: experimental grazing bifurcations

The experimental work of Popp *et. al* [208] described above amply demonstrates many of the predicted features of impact oscillator dynamics experimentally, but does not contain a simple *primary* grazing bifurcation where a non-impacting orbit first starts to impact. In order to engineer just such an event, Piiroinen *et. al.* [220] considered experimental results for a forced pendulum system run in a reduced gravity environment, based on the earlier design presented in [26].

A simple rigid-arm pendulum that strikes a vertical impact surface is an easily realized single-degree-of-freedom impact oscillator. By horizontally shaking the supporting pivot of the pendulum a variety of dynamic behavior can be observed. However, with the impact barrier located at the static equilibrium position (see Fig. 9.2), the velocity of impact tends to be relatively high and thus grazing bifurcations of the fundamental *period-one* (i.e. with the same period as the forcing) do not typically occur at achievable frequencies. In [220, 26] an effective lowering of gravity was achieved by inclining the angle of the plane in which the pendulum is free to move. Then, if the impact surface is placed sufficiently far from static equilibrium and the forcing amplitude is relatively low, impacts will not take place. For intermediate angles of impact, as the forcing frequency is gradually increased towards the first resonant frequency, a critical value is reached at which the steady periodic response first makes contact with the obstacle.

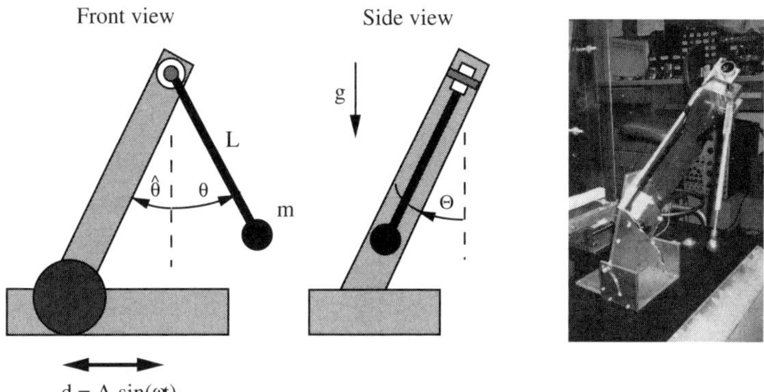

Fig. 9.2. The pendulum/impact barrier assembly; see [220] for dimensions and parameter details. (Reprinted from [220] with permission from Springer-Verlag.)

The experimental configuration used is shown in Fig. 9.2. The pendulum is constructed using a relatively light aluminum arm of length 305 mm and a steel mass of diameter 25.4 mm attached at the end. The pivot of the pendulum consists of low-friction bearings, and a rotational potentiometer measures the angle $\theta(t)$ to a relatively high degree of accuracy. The assembly is mounted on a shaking table that imparts a harmonic base displacement and is inclined at $\Theta = 76.2°$ from the vertical (giving an effective gravitational constant $g_e = 0.24g$). The angle of contact with the impact barrier $\widehat{\theta}$ and the forcing frequency ω are used as the primary control parameters: The former was fixed at discrete values at intervals of $\widehat{\theta} = 10°$ and the latter was varied statically by small amounts.

A crucial issue in modeling many mechanical systems is the estimation of the amount of damping. Damping in this system arises through energy loss at impact and also between impacts through Coulomb friction in the bearing and viscous air drag. In order to estimate the overall damping both between impacts and at impact, a simple logarithmic decrement method was used, taking $\widehat{\theta} = 0$ and with no external forcing. Successive peak amplitudes A_k, $k = 1, 2, \ldots$, were recorded, and an overall damping factor D was calculated by the formula

$$D = \frac{1}{4\pi} \ln \left(\frac{A_k}{A_{k+1}} \right) \tag{9.1}$$

giving a measured value for $D = 0.07$, that is, 7% of critical damping, which includes both an effect from impacts and from other sources.

Figures 9.3 and 9.4 summarize, for different values of $\widehat{\theta}$, the response of the pendulum when the forcing frequency is gradually increased through the range of primary resonance. The forcing frequency is expressed in terms of the ratio η of the forcing frequency to that of the natural frequency of an impacting pendulum with impact at $\theta = 0$. In this case the primary resonance (in the absence of impact) occurs when $\eta = 1/2$.

For each value of $\widehat{\theta} \leq 40°$, a critical forcing frequency $\eta_c(\widehat{\theta})$ slightly less than $\eta = 1/2$ was found such that for $\eta < \eta_c$ the attracting motion is a period-one limit cycle that does not impact. At $\eta = \eta_c$ this limit cycle just grazes with the obstacle, and impacting motion is found for $\eta > \eta_c$. We therefore might expect to see behavior resembling that presented in Chapter 6 (see also Chapter 4) in the unfolding of the grazing bifurcation. For η close to, and greater than η_c, we observe a complex sequence of chaotic and/or periodic motions until eventually, for large enough η, the motion settles back to period-one non-impacting motion (except for $\widehat{\theta} = 10°$ in which the impacting motion persists up to large η). This final bifurcation is also a grazing event. For $\widehat{\theta} \geq 50°$ it was found that the pendulum amplitude is always less than the barrier angle $\theta(t) < \widehat{\theta}$ and hence no impacting motion takes place.

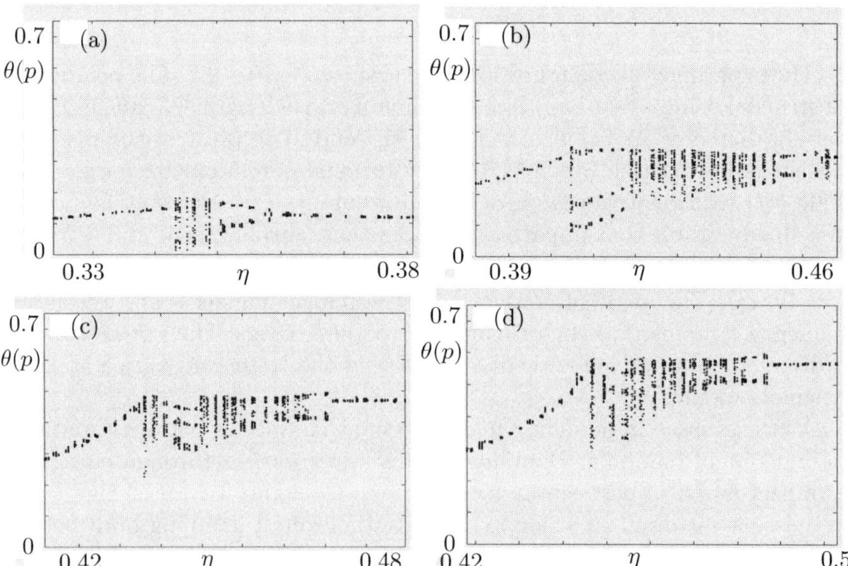

Fig. 9.3. Experimental bifurcation diagrams in which the response is sampled once during a forcing cycle (at an arbitrary but consistent phase). (a) $\widehat{\theta} = 10°$, (b) $\widehat{\theta} = 20°$, (c) $\widehat{\theta} = 30°$, (d) $\widehat{\theta} = 40°$. (Reprinted from [220] with permission from Springer-Verlag.)

Note that η_c increases with $\widehat{\theta}$ and also the size of the window in which impacting motion occurs shrinks. We conjecture that there is a point for some $\widehat{\theta}$ value between 40° and 50° at which the two grazing bifurcations come together, thus destroying entirely the window of complex dynamics.

Now consider some of the features of the bifurcation diagrams in Fig. 9.3. For each value of $\widehat{\theta}$ we find that the initial grazing bifurcation causes a significant change of the behavior of the system. In Fig. 9.3(b), when $\widehat{\theta} = 20°$ the stable period-one orbit jumps immediately to a stable period-three orbit that eventually becomes chaotic as η increases. This is similar to Fig. 4.15 of Chapter 4, in which we see the creation of a period-three maximal orbit. In contrast, if we look at Fig. 9.3(d) when $\widehat{\theta} = 40°$ there appears to be more complex behavior at the bifurcation with some evidence for chaos and period-adding. Time-series data extracted from the ensuing dynamics are shown in Fig. 9.4(a). However for each value of $\widehat{\theta}$ there is also at least one significant window of periodic motion, the first of which is: *period-two* for $\widehat{\theta} = 10°$ (for $0.35 < \eta < 0.36$); *period-three* for $\widehat{\theta} = 20°$ ($0.406 < \eta < 0.42$); *period-four* for $\widehat{\theta} = 30°$ ($0.44 < \eta < 0.45$); and *period-five* for $\widehat{\theta} = 40°$ [around $\eta = 0.45$, see Fig. 9.3(c)].

An explanation of this observed period-adding-like behavior, is that we see in each case a stable maximal periodic orbit created as an unstable orbit at the grazing bifurcation, in the manner described in Sec. 4.3 in Chapter 4, which has then restabilized. The increase of $\widehat{\theta}$ is equivalent to moving through the diagram presented in Figure 4.15 so that we see a progressive increase in the period of the observed maximal orbit.

A dimensionless model for the motion for the pendulum is given by

$$\ddot{\theta} + \frac{2\beta}{\eta}\dot{\theta} + \frac{1}{4\eta^2}\sin(\theta) = \alpha\cos(\theta)\sin(t) \quad \theta > \widehat{\theta}. \tag{9.2}$$

Here $\alpha = 0.2258$ is the constant forcing amplitude, the forcing frequency is scaled to unity and the natural frequency is $1/2\eta$ of the undamped non-impacting pendulum, and β is the damping ratio in the absence of energy loss at impact. At $\theta = \widehat{\theta}$ we assume that there is simple impact law given by

$$\dot{\theta} \mapsto -r\dot{\theta}, \tag{9.3}$$

where $0 \le r \le 1$ is the coefficient of restitution. A key issue is finding values of the parameters β and r consistent with the measured value of D estimated according to (9.1).

A numerical measure for D was found by computing the decrease in amplitude of the model system over 10 peaks of the unforced system, see Fig. 9.5(a). This experiment was repeated for different values of r and β to find a curve in the (r, β)-plane that corresponds to $D = 0.07$ [Fig. 9.5(b)]. A realistic choice of r for steel on steel contact lies somewhere between 0.5 and 0.95 (see for example [208]), depending on the geometry, and the consequent possibilities for the spread of acoustic waves. A value of $r = 0.8$ corresponds to a linear

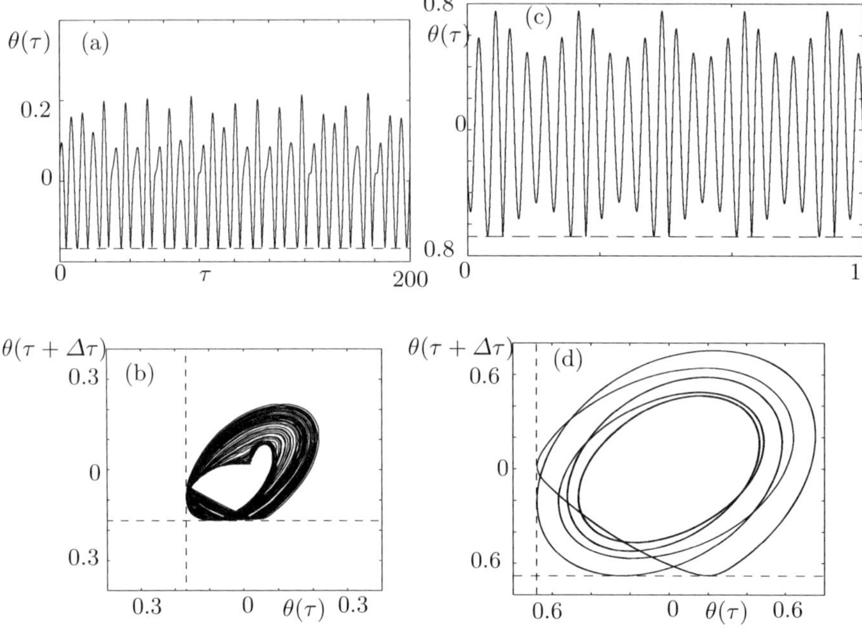

Fig. 9.4. Sample experimental time series and phase potrait projections. (a) and (b) $\widehat{\theta} = 10°, \eta = 0.35$; (c) and (d) $\widehat{\theta} = 40°, \eta = 0.45$. (Reprinted from [220] with permission from Springer-Verlag.)

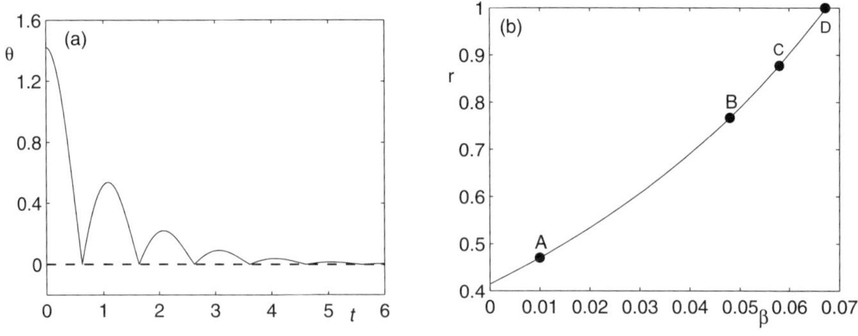

Fig. 9.5. (a) Free decay for the impacting pendulum with no forcing. (b) The relation between r and β to get a damping ratio that is 7% of critical. See text for significance of labeled points. (Reprinted from [220] with permission from Springer-Verlag.)

damping coefficient β of 0.015, and $r = 0.5$ to $\beta = 0.05$. One way of finally fitting was to use the data in Fig. 9.3(d) for $\widehat{\theta} = 40°$, to match the value of η_c for the first grazing bifurcation. This results in the values of $r = 0.8$ and $\beta = 0.05$ marked as point B on the curve in Fig. 9.5(b), which we shall now use.

With these values we may now look at some numerical simulations of the system and compare these with the experimental results. A series of these for a range of values of $\widehat{\theta}$ is presented in Fig. 9.6.

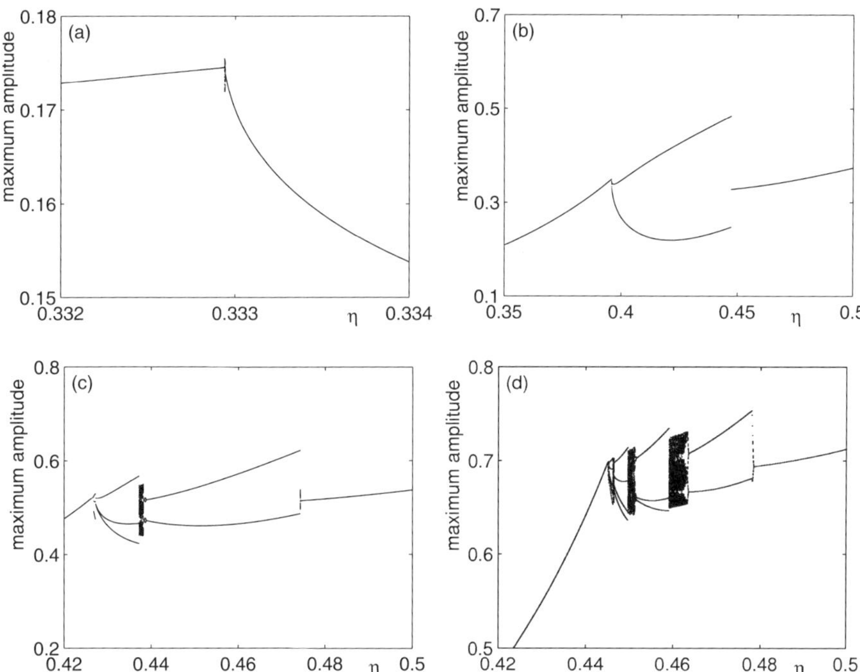

Fig. 9.6. Bifurcation diagrams under variation of η and where (a) $\widehat{\theta} = 10°$, (b) $\widehat{\theta} = 20°$, (c) $\widehat{\theta} = 30°$, and (d) $\widehat{\theta} = 40°$ obtained using direct numerical simulation. (Reprinted from [220] with permission from Springer-Verlag.)

A close-up of one of these figures is given in Fig. 9.7(a), which depicts a numerically simulated bifurcation diagram of (9.2) using a Poincaré section at $\dot{\theta} = 0$ ($\ddot{\theta} < 0$). We can see that at $\eta \approx 0.44$ the behavior changes at a grazing bifurcation followed by a reversed period-adding cascade interspersed with chaos. The correspondence with the experimental results for the same value of $\widehat{\theta}$ given in Figure 9.3(d) is good. In the numerical calculations we see the expected onset of the grazing bifurcation followed by chaos with a reversed period-adding cascade (the period of the periodic windows decreases by one for successive windows with increasing η). Eventually as η increases there is a second grazing bifurcation that takes the motion back to single-impact-per-period motion.

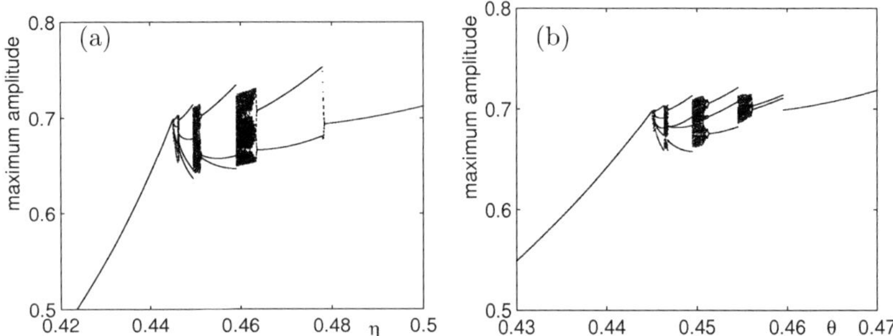

Fig. 9.7. Bifurcation diagram of the forced impacting pendulum under variation of η and where $\widehat{\theta} = 40°$ using (a) the direct numerical simulation and (b) the discontinuity mapping approach (notice the difference in the axis scaling). (Reprinted from [220] with permission from Springer-Verlag.)

When $\widehat{\theta} < 40°$ the numerical results indicate that there is a transition from the non-impacting orbit to a (maximal) periodic orbit. This agrees qualitatively with the experimental results, although the quantitative correspondence between the experimental and numerical results is less good for smaller values of $\widehat{\theta}$. For example, whilst we do see the creation of a maximal periodic orbit when $\widehat{\theta} = 20°$, the period differs (it is period-two in the numerical simulation and period-three in the experiments). Nevertheless, we see the same broad features; there is a first grazing bifurcation from non-impacting period-one periodic motion to impacting chaotic and/or periodic motion. As in the experimental results, the period of the first appreciable periodic window increases with increasing $\widehat{\theta}$; in this case period-one for 10°, two for 20°, three for 30° and four (actually, briefly, five) for 40°. Also, for all cases other than $\widehat{\theta} = 10°$, there is a second grazing bifurcation for higher η that returns the system to non-impacting motion.

To analyze the observed bifurcation we apply the discontinuity mapping approach to calculate the ZDM for this system using the methods from Chapter 6. Specifically, writing the system in first-order form for a variable $x(t) = (x_1(t), x_2(t), x_3(t))$ with $x_1 = \theta$, $x_2 = \dot{\theta}$ and $x_3 = t$, we get

$$\begin{aligned} \dot{x}_1 &= x_2, \\ \dot{x}_2 &= \alpha \cos(x_1) \sin(x_3) - \frac{2\beta}{\eta} x_2 - \frac{1}{4\eta^2} \sin(x_1), \\ \dot{x}_3 &= 1, \end{aligned} \tag{9.4}$$

with impact occurring when $H(x) = \widehat{\theta} - x_1 = 0$. Hence we find, in the notation of Theorem 6.2, that $H_x = (1,0,0)^T$, $W = (0, -(1+r), 0)^T$, $y = x_2$ and a is $\dot{x}_2$ evaluated at the impact time $x_3 = t_i$, $x_1 = \widehat{\theta}$ and $x_2 = 0$. Thus we get a local discontinuity mapping

$$x \mapsto \begin{cases} x, & \text{if} \quad x_1 \le \widehat{\theta}, \\ x + \begin{pmatrix} 0 \\ 1+r \\ 0 \end{pmatrix} \sqrt{2(x_1 - \widehat{\theta})a(x)} + O(x_1), & \text{if} \quad x_1 > \widehat{\theta}, \end{cases} \tag{9.5}$$

where the acceleration at impact is given by

$$a(x) = \alpha \cos(\widehat{\theta}) \sin(t_i) - \frac{2\beta}{\eta}\dot{\theta} - \frac{1}{4\eta^2}\sin(\widehat{\theta}).$$

Fig. 9.7(b) shows the result of using the discontinuity map when $\widehat{\theta} = 40°$ combined with a numerical solution of the ordinary differential equations (9.4) to obtain the Poincaré map from the Poincaré section $x_2 = 0$, $\dot{x}_2 > 0$ to itself, close to impact, at which point the restitution law is replaced by the mapping 9.5. The similarity between the direct numerical calculations and those given by applying the discontinuity mapping is clear, at least for η in the vicinity of the grazing impact.

For a more complete analysis we must calculate the matrix A corresponding to the linearization of the Poincaré map about the periodic grazing orbit, and its associated eigenvalues $\lambda_{1,2}$. For $\widehat{\theta} = 20°$, $\lambda_{1,2} = -0.0066 \pm 0.4668i$. For $\widehat{\theta} = 40°$, $\lambda_{1,2} = 0.4646 \pm 0.2137i$, and for slightly larger $\widehat{\theta}$ the eigenvalues become real. The analysis presented in Chapter 4 then implies that for $\widehat{\theta} < 40°$ we expect to see a jump to a (maximal) periodic orbit at grazing, whereas for slightly larger $\widehat{\theta}$ we expect to see chaotic behavior interspersed with period-adding as we have a real leading eigenvalue with $1/4 < \lambda_1 < 2/3$.

In Fig. 9.8(b) a *delay plot* θ_{n+1} against θ_n is depicted using the global discontinuity map evaluated at $\eta = 0.4458$. This gives a representation of the chaotic attractor. The square-root term in the local map (9.5) is clearly visible as the almost vertical lines together with the other 'one-dimensional' sets comprising the attractor. Similar mappings were also obtained both numerically and experimentally by Fedriksson *et. al.* [105] for an impacting pipe conveying fluid.

9.1.3 Parameter uncertainty

Clearly although each of the experiments, the numerical calculations and the theory show a period-adding cascade for the impacting pendulum when $\widehat{\theta} = 40°$, the qualitative agreement between experiments and theory is nowhere near as good as that obtained for the impacting beam experiment we reviewed in Sec. 9.1.1, especially for smaller values of $\widehat{\theta}$. One reason for this could be the

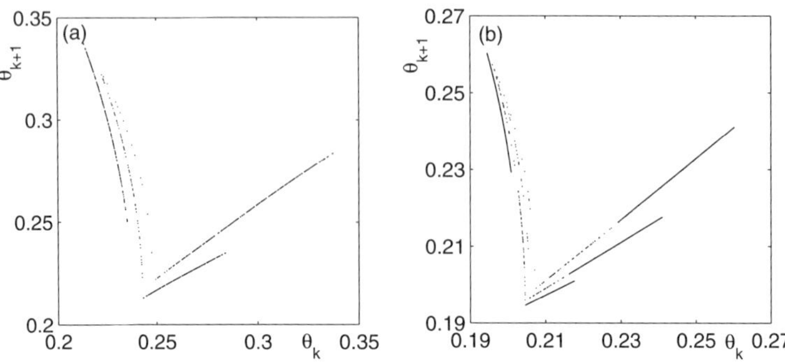

Fig. 9.8. Delay plots for the impacting pendulum near grazing using (a) the discontinuity map at section $h(x) = x_3 = 0$ and (b) the direct numerical simulation. (Reprinted from [220] with permission from Springer-Verlag.)

sensitivity to noise of the fundamental grazing bifurcation, as we have already intimated. However, we also show here that uncertainty in the parameters can also cause huge variability in what is observed. Numerical results in [220] show that both the location of η_c and the eigenvalues of the linearization around the corresponding grazing periodic orbits are highly sensitive to damping. In Fig. 4.14 of Chapter 4, it is clear that small changes in the values of the eigenvalues, and hence of a_1 and a_2 (in the notation of that chapter), can lead to large changes in the period of the observed maximal orbit. Thus it is not surprising that the results when $\widehat{\theta}$ is small (in particular the period of the observed orbit) are very sensitive to small changes in the parameters of the system. When $\widehat{\theta}$ takes larger values, and the eigenvalues of the linearization are real, we expect to see less sensitivity in the results.

To examine the sensitivity, the authors in [220] varied the parameters β and r along the curve in Fig. 9.5(b) and found no significant difference in the qualitative picture of the bifurcation diagrams, but huge quantitative differences. Generally speaking, cases with smaller β (and hence r closer to unity) have significantly wider intervals of chaos and also wider, more appreciable windows of the higher-period orbits, which seems closer to the experimental data. However, such values tend to greatly overestimate the maximum value $\widehat{\theta}$ for grazing. There might be a number of reasons for this other than uncertainty in the damping and restitution parameters, such as vibrations in the experimental setup, errors in the parameter measurements, or wrong assumptions in the mathematical modeling (for example damping may enter through a nonlinear velocity-dependent term due to Coulomb friction in the pivot).

To highlight this uncertainty further, we consider a lower value of forcing amplitude 0.71α, taking the coefficient of restitution $r = 0.4707$ and the damping coefficient $\beta = 0.01$ fixed [point B in Fig. 9.5(b)]. One can argue that this is effectively similar to increasing the overall damping in the structure, since a decrease of either r or β has the effect of lowering the amplitude of the resonance peak. The resulting bifurcation diagrams under η variation for $\widehat{\theta} = 10°, 20°, 30°, 40°$ are shown in Figs. 9.9(a)–(d), respectively. If we now compare these results with the experimental bifurcation diagrams in Figs. 9.3(a)–(d) the match is better than that presented in Figs. 9.6(a)–(d) with this perturbed amplitude especially for $\widehat{\theta} = 10°$ and $40°$.

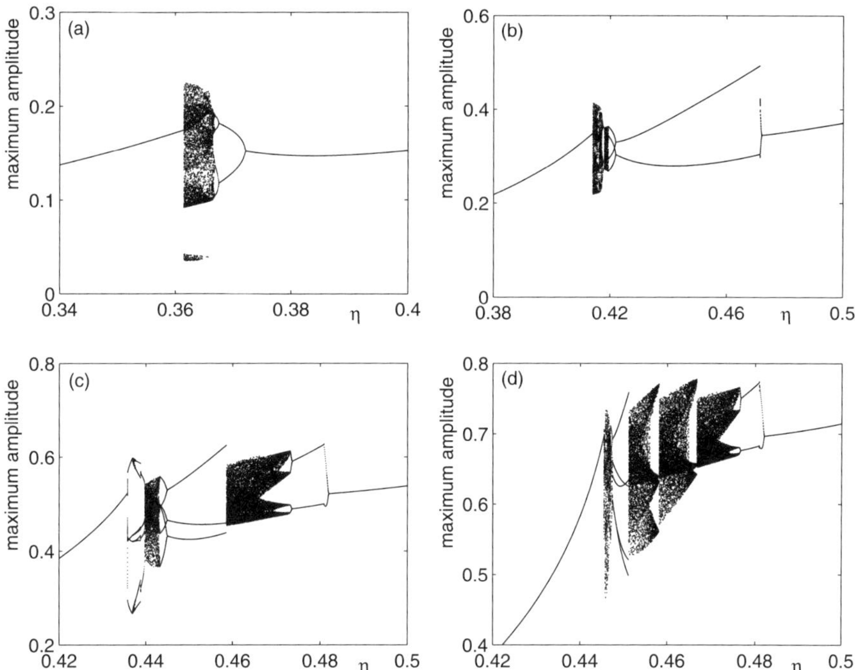

Fig. 9.9. Changes in existence and stability of the forced impacting pendulum under variation of η and where $\widehat{\theta} = 10°$ (a), $\widehat{\theta} = 20°$ (b), $\widehat{\theta} = 30°$ (c) and $\widehat{\theta} = 40°$ (d). The driving frequency is 0.71α, the coefficient of restitution $r = 0.4707$ and the damping coefficient $\beta = 0.01$. (Reprinted from [220] with permission from Springer-Verlag.)

Clearly much more work needs to be done to understand the effects of parameter uncertainty and noise on the nature of the bifurcations close to grazing.

9.2 Rattling gear teeth: the similarity of impacting and piecewise-smooth systems

This section is concerned with backlash, another piecewise-smooth nonlinearity common to a wide range of applications. In particular, we investigate the dynamics of lightly loaded meshing gears in the large stiffness limit. Such systems have a tendency to rattle unpredictably with different levels and frequencies of noise [154, 274, 61] because of gears moving through their backlash. The key example is that of the unloaded gears in an automotive manual transmission system [61].

In the literature, systems with backlash have been mostly studied by considering a piecewise-linear model of the nonlinearity (see Fig. 9.11). Natsiavas [195] showed how to construct solutions by matching trajectory sections together and solving a single transcendental equation. It has also been shown that backlash systems can exhibit complicated dynamics. The occurrence of chaotic behavior in a backlash system was discussed by Mahfouz & Badrakhan [186] mainly through numerical simulations. The possible bifurcations were examined, again numerically, in more detail in by Kleczka *et. al.* [156] and by Wiercigroch [266], and bifurcation scenarios leading to chaotic behavior were further investigated numerically by Luo & Menon in [182].

Other research has also analyzed the dynamics of gear models that contain a time-varying parametric stiffness term to model the changes in the number of teeth in contact. These models are typically solved using harmonic balance methods or other approximate strategies [34, 147, 149, 249, 250, 262].

An alternative modeling approach is that of considering the limit of infinite backlash stiffness. In this case, for example, solutions that impact only one side of their backlash are equivalent to the dynamics of a ball bouncing on a massive oscillating table (see [236] and, for example, [181] and references therein). More generally, the dynamics of a system with backlash, under this modeling assumption, becomes directly related to the motion of a rigid block that can impact walls placed symmetrically about its rest position [133]. This system is affected by impacting events and is characterized by the presence of three discontinuity boundaries in phase space. For an exhaustive review of these results as well as those concerning impact oscillators and impact dampers we refer to the book by Brogliato [38].

Experiments demonstrating the validity of simple backlash models are described by Wiercigroch & Sin in [270]. However, there is currently no full analytical explanation of the mechanisms leading to the onset of bifurcations and chaos in backlash systems. Here, we focus on a simple model of a gear pair. Specifically, the system we study is a single-degree-of-freedom oscillator with a backlash nonlinearity. The aim is to derive analytical conditions for the existence and stability of several families of periodic rattling solutions and to study their bifurcations. In so doing, we wish to compare the results obtained by using (i) an impacting model of backlash with (ii) those derived for a piecewise-linear backlash model. We show that the results obtained using these two different modeling approaches are the same to leading order (further illustrating the calculation made in Chapter 1 for the bi-linear oscillator with large stiffness ratio).

An interesting feature of the system we investigate is the coexistence of several different attractors so that, even when the parameters are set for silent operation, different rattling periodic solutions might coexist. To further investigate this issue, two-parameter bifurcation diagrams can be derived analytically and validated through numerical simulations. It is shown that systems with backlash can exhibit bifurcations such as cyclic-folds or period-doubling (also observed in smooth systems) in addition to discontinuity-induced bifurcations such as grazing events. More details of the analysis can be found in the paper by Halse *et. al.* [126].

9.2.1 Equations of motion

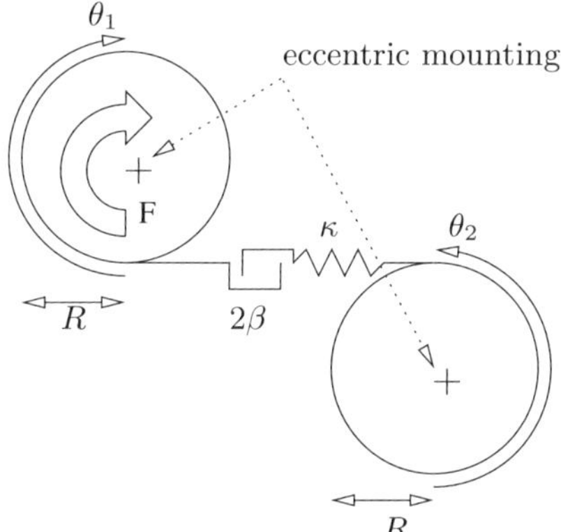

Fig. 9.10. Basic gear model with backlash.

We consider a simple model of two meshing gears, one driven by an oscillating torque F and the other not driven. We assume that the gears have the same moment of intertia I, radius R and linear viscous friction coefficient C. The rotational displacement of the driven gear is described by $\theta_1(\tau)$ and the rotational displacement of the free gear by $\theta_2(\tau)$ (see Fig. 9.10). Resolving moments, the equations of motion of the gears are thus

$$I\ddot{\theta}_1 + C\dot{\theta}_1 + R\Psi(\theta_1, \theta_2) = F(\tau), \tag{9.6}$$

$$I\ddot{\theta}_2 + C\dot{\theta}_2 - R\Psi(\theta_2, \theta_2) = 0. \tag{9.7}$$

Here Ψ is the interaction force between the gears given by

$$\Psi(\theta_1, \theta_2) = \kappa B(\theta_1 - \theta_2, \beta),$$

where B is the backlash function

$$B(u,\beta) = \begin{cases} u-\beta, & u > +\beta, \\ 0, & |u| < \beta, \\ u+\beta, & u < -\beta. \end{cases} \tag{9.8}$$

We note that if the gears are turning at an approximately constant speed, then the mean forcing scales with the dissipation. To see this, let $\bar{\dot{\theta}}_1$, $\bar{\dot{\theta}}_2$, $\bar{\Psi}$ and $\bar{F}$ denote the average values of $\dot{\theta}_1$, $\dot{\theta}_2$, Ψ and $F(t)$, and observe that

$$\begin{aligned} \frac{C}{I}\bar{\dot{\theta}}_1 + \frac{R}{I}\bar{\Psi} &= \frac{\bar{F}}{I}, \\ \frac{C}{I}\bar{\dot{\theta}}_2 - \frac{R}{I}\bar{\Psi} &= 0. \end{aligned}$$

If we add these two equations and assume that the average rotational speed is $\bar{\dot{\theta}}_1 = \bar{\dot{\theta}}_2 = 2\,\pi\Omega$, we have

$$\bar{F} = 4\pi C\Omega.$$

Now, let $u(\tau) = \theta_1(\tau) - \theta_2(\tau)$, then from (9.6) and (9.7), we obtain

$$\ddot{u} + \frac{C}{I}\dot{u} + \frac{2R\kappa}{I}B(u,\beta) = \frac{F(\tau)}{I}.$$

We assume that the periodic forcing can be represented by a small amplitude sinusoidal offset from $\bar{F}$ and with a frequency of once per rotation, i.e.

$$F(\tau) = 4\pi C\Omega + \gamma\cos(2\pi\Omega\tau).$$

We define a dimensionless time t by $t = \Omega\tau$. Then, if we denote differentiation with respect to t by a prime, we have

$$u'' + \frac{C}{I\Omega}u' + \frac{2R\kappa}{I\Omega^2}B(u,\beta) = \frac{4\pi C}{I\Omega} + \frac{\gamma}{I\Omega^2}\cos(2\pi t).$$

The rotational co-ordinates (angles) are already dimensionless, so therefore the coefficients are dimensionless. We define the new non-dimensional parameters

$$\delta = \frac{C}{I\Omega}, \qquad K = \frac{R\kappa}{\Omega^2 I}, \qquad \epsilon = \frac{\gamma}{I\Omega^2},$$

and hence

$$u'' + \delta u' + 2KB(u,\beta) = 4\pi\delta + \epsilon\cos(2\pi t). \tag{9.9}$$

Note that here we have represented the forcing term as an external source, as might come from an automotive engine. A similar equation can be derived if we assume an alternative source of periodic excitation; for example, eccentricity in the mountings of the two gears (see [274]).

The nonlinearity arises because the motion $u(t)$ will typically not be confined to any one of the regions for all time, but will swap whenever u passes through the values $\pm\beta$.

For typical machines, the stiffness K is very large, and so we can also think of taking an impacting model of backlash $[\lim_{K\to\infty} B(u,\beta,K)]$ in which the system states are reset by an instantaneous impact condition applied whenever $u = \pm\beta$. In what follows we will model the contact in both ways, as shown in Fig. 9.11.

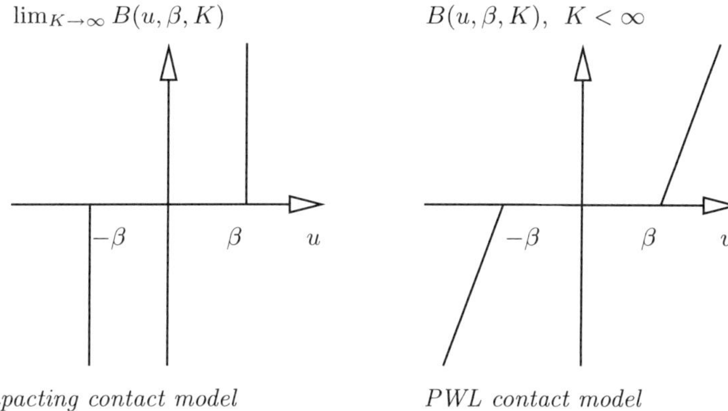

Fig. 9.11. Backlash models.

First, note that geared systems should be designed to operate so that the gear teeth remain in contact for all time, with no rattle. It is straightforward to derive a condition for such a solution to exist [126]. That is, we must have

$$\epsilon < \frac{2\pi\delta}{K}\sqrt{(2K - 4\pi^2)^2 + 4\pi^2\delta^2}. \tag{9.10}$$

In the limit of large stiffness ($K \to \infty$), the condition for silent operation reduces to

$$\epsilon < 4\pi\delta. \tag{9.11}$$

There is a range of anecdotal evidence to suggest that satisfying this bound in real geared systems is a major challenge. However, choosing parameters such that (9.10) is satisfied is not enough to guarantee the desired silent operation; stable rattling solutions may also coexist. Throughout we suppose that (9.11) holds as this simplifies the subsequent analysis considerably.

9.2.2 An illustrative case

To illustrate the similarity between results obtained using the impacting model and those obtained using a piecewise-linear approximation of backlash, we investigate the existence of a particular type of solution using both models. Specifically, we choose to study those solutions that contact only the boundary

$u = \beta$. Such solutions will be labeled as $P(m, 1, 0)$ solutions to indicate that they are m-periodic, with one crossing (or impact) on one side of the backlash and no crossing on the other. (For more details and the analysis of other types of solutions, see [126].)

We shall show that the solutions for both kinds of model have the same leading-order behavior, and present two-parameter (δ, ϵ) bifurcation diagrams. These results are shown in Figs. 9.14 and 9.15. In fact, the assumption (9.11) limits the study to a triangular region of parameter space $\epsilon < 4\pi\delta$, so throughout we have changed our x-axis co-ordinate as shown in Fig. 9.13.

It is of particular interest that we appear to have codimension-two bifurcation points, i.e. points where fold bifurcations and grazing bifurcations occur simultaneously, i.e. a smooth and a discontinuity-induced bifurcation occur together. We shall return to a discussion of possible kinds of codimension-two DIBs in Sec. 9.4 below.

9.2.3 Using an impacting contact model

We consider the existence and stability of the orbits of interest with the impacting contact model assumed to be perfectly elastic ($r = 1$). We then consider the classical bifurcations that may occur.

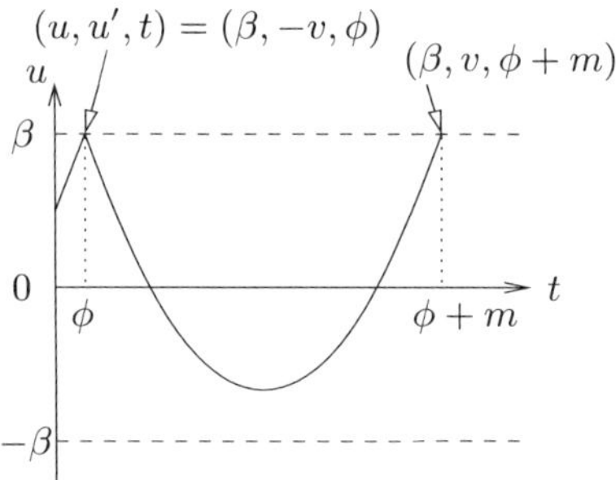

Fig. 9.12. Notation for the $P(m, 1, 0)$ solution with the impacting contact model.

Here we should point out that we can only have $P(m, 1, 0)$ solutions that contact the boundary at $u = \beta$. We cannot have solutions that only impact the boundary $u(t) = -\beta$ as this would violate assumption (9.11). Therefore we consider the construction of solutions where the period $m \in \mathbb{N}$.

Our first step is to construct the unknown part of our solutions by solving (9.9). With reference to Fig. 9.12, we construct the solution such that, immediately after impact, we have

$$\begin{aligned} u(\phi) &= \beta, \\ u'(\phi) &= -V, \end{aligned}$$

for some unknown initial time ϕ and velocity $-V$. The periodicity conditions than imply that immediately before impact

$$u(m+\phi) = \beta,$$
$$u'(m+\phi) = V.$$

Then, solving these equations, we obtain

$$\begin{aligned} \beta &= 4\pi\phi - \frac{\epsilon}{2\pi\sqrt{\delta^2+4\pi^2}} \cos\left[2\pi\phi + \arctan\left(\frac{\delta}{2\pi}\right)\right] + c_1 + c_2 e^{-\delta\phi}, \\ -V &= 4\pi + \frac{\epsilon}{\sqrt{\delta^2+4\pi^2}} \cos\left[2\pi\phi - \arctan\left(\frac{2\pi}{\delta}\right)\right] - c_2\delta e^{-\delta\phi}, \end{aligned} \tag{9.12}$$

$$\begin{aligned} \beta &= 4\pi(\phi+m) - \frac{\epsilon}{2\pi\sqrt{\delta^2+4\pi^2}} \cos\left[2\pi\phi + \arctan\left(\frac{\delta}{2\pi}\right)\right] + c_1 + c_2 e^{-\delta(\phi+m)}, \\ V &= 4\pi + \frac{\epsilon}{\sqrt{\delta^2+4\pi^2}} \cos\left[2\pi\phi - \arctan\left(\frac{2\pi}{\delta}\right)\right] - c_2\delta e^{-\delta(\phi+m)}, \end{aligned} \tag{9.13}$$

where c_1 and c_2 are the undetermined integration constants. Equating the two equations for β we have

$$0 = -4\pi m + c_2 e^{-\delta\phi}(1 - e^{-\delta m}),$$

and using (9.13) and (9.12), we obtain

$$-2V = -c_2\delta e^{-\delta\phi}(1 - e^{-\delta m}).$$

Hence, we find that, for the orbit to exist, we must have

$$V = 2\pi\delta m, \tag{9.14}$$

$$c_2 = \frac{4\pi m e^{\delta\phi}}{1 - e^{-\delta m}}. \tag{9.15}$$

Substituting these values for v and c_2 into (9.13) and solving for ϕ, we have

$$2\pi\phi = \arcsin\left[\frac{\sqrt{\delta^2+4\pi^2}}{\epsilon}\left(\frac{4\pi m\delta e^{-\delta m}}{1-e^{-\delta m}} + 2\pi\delta m - 4\pi\right)\right) - \arctan\left(\frac{\delta}{2\pi}\right]. \tag{9.16}$$

There are two admissible solutions to (9.16). We can expand these solutions in terms of our small parameters,

$$\phi = \begin{cases} \frac{m^2\pi\delta^2}{3\epsilon} - \frac{\delta}{4\pi^2} + O(\delta^3), & \textit{in-phase} \text{ solution}, \\ \frac{1}{2} - \frac{m^2\pi\delta^2}{3\epsilon} + \frac{\delta}{4\pi^2} + O(\delta^3), & \textit{out-of-phase} \text{ solution}. \end{cases} \tag{9.17}$$

Note that these solutions cannot exist if the argument of the arcsin function in (9.16) is greater than unity. Thus, expanding the argument of the arcsin in terms of the small parameters δ and ϵ, we find

$$\frac{\sqrt{\delta^2+4\pi^2}}{\epsilon}\left(\frac{4m\pi\delta e^{-m\delta}}{1-e^{-m\delta}}+2m\pi\delta-4\pi\right)=\frac{2m^2\pi^2\delta^2}{3\epsilon}+O(\delta^2).$$

For this quantity to be less than one, we require to leading order,

$$\epsilon>\frac{2\pi^2m^2\delta^2}{3}+O(\delta^3). \tag{9.18}$$

We also must consider whether the trajectory has the correct itinerary, so that it does not hit the boundary at $u=-\beta$. To check this we must find the minimum displacement. Again, we can only find this minimum point approximately; we first find the point at which the velocity is zero, i.e. $\widehat{t}$ such that $u'(\widehat{t})=0$. We try a power series approximation of the form

$$\widehat{t}=\phi+\frac{m}{2}+\widehat{t}_0+\widehat{t}_1\delta+\widehat{t}_2\delta^2+\ldots,$$

and we solve for the coefficients $\widehat{t}_i$ by comparing terms of $O(\delta^k)$ in turn. We then substitute this series expression for $\widehat{t}$ into our ODE solution $x(t)$, and expand this as a series as well to find the minimum displacement of the candidate periodic orbit $\widehat{x}=x(\widehat{t})$.

After some algebraic manipulation, we find that, for the in-phase solution $[\phi=O(\delta)]$ not to contact the lower boundary, we require

$$\beta>\begin{cases}\frac{m^2\pi\delta}{4}+\frac{\epsilon}{4\pi^2}-O(\delta^2), & \text{if } m \text{ odd},\\ \frac{m^2\pi\delta}{4}+O(\delta^2), & \text{if } m \text{ even},\end{cases} \tag{9.19}$$

and for the out-of-phase solution $[\phi=\frac{1}{2}+O(\delta)]$ not to contact the lower boundary, we require

$$\beta>\begin{cases}\frac{m^2\pi\delta}{4}+\frac{\epsilon}{4\pi^2}+O(\delta^2), & \text{if } m \text{ odd},\\ \frac{m^2\pi\delta}{4}+O(\delta^2), & \text{if } m \text{ even}.\end{cases} \tag{9.20}$$

To assess stability, we need to consider the eigenvalues of the matrix $A_1:=Q(\phi+m,2\pi m\delta)\Phi_1(m)$ obtained by composing the system flow $\Phi_1(m)$ with an appropriately derived, transverse, discontinuity map Q (for the sake of brevity we omit such derivation here),

$$A_1=\begin{bmatrix}-1 & \frac{1}{\delta}(e^{-\delta m}-1)\\ \frac{4}{m}+\frac{\epsilon}{\pi m\delta}\cos(2\pi\phi) & \frac{-1}{\delta}(e^{-\delta m}-1)\left(\frac{4}{m}+\frac{\epsilon}{\pi m\delta}\cos(2\pi\phi)\right)-e^{-\delta m}\end{bmatrix}, \tag{9.21}$$

where ϕ is given by (9.16).

The eigenvalues of A_1 are the Floquet multipliers of the orbit. For stability these multipliers must be within the unit circle. The bifurcations that may occur as these eigenvalues cross the unit circle are considered in the next section. For the in-phase solution $[\phi=O(\delta)]$ found above, we find the leading-order term of the Floquet multipliers to be

$$\lambda_{1,2} = 1 + \frac{\epsilon}{2\pi\delta} \pm \frac{1}{2\pi}\sqrt{\frac{\epsilon}{\delta^2}(\epsilon + 4\pi\delta)} + O(\delta),$$

of which one is outside the unit circle for $4\pi\delta > \epsilon$. For the out-of-phase $[\phi = \frac{1}{2} + O(\delta)]$ solution the Floquet multipliers are

$$\lambda_{1,2} = \left(1 - \frac{m\delta}{2}\right) f_{1,2} + O(\delta^2),$$

where

$$f_{1,2} = 1 - \frac{\epsilon}{2\pi\delta} \pm \frac{i}{2\pi}\sqrt{\frac{\epsilon}{\delta^2}(4\pi\delta - \epsilon)},$$
$$|f_{1,2}| = 1.$$

Therefore, $|\lambda_{1,2}| \approx 1 - \frac{m\delta}{2}$, and hence this solution is stable for $4\pi\delta > \epsilon$ and $\epsilon \sim \delta$.

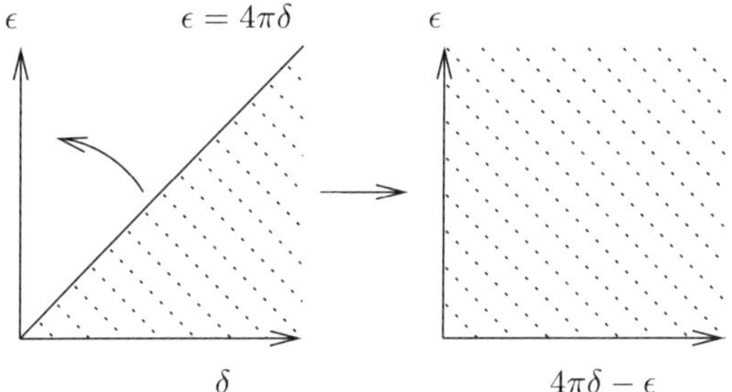

Fig. 9.13. Transformation of parameter space for visualization.

Now let us turn to possible bifurcations in this model. At a smooth local bifurcation point we have at least one Floquet multiplier on the unit circle. We denote the two Floquet multipliers as λ_1 and λ_2. There are three possibilities for a bifurcation.

1. Complex conjugate Floquet multipliers on the unit circle: $\lambda_1 = e^{i\theta}, \lambda_2 = e^{-i\theta}$. This cannot occur for $\delta > 0$ as we require $\lambda_1\lambda_2 = 1 = e^{-\delta m}$ [the product of the eigenvalues of A_1 is $\det(A_1)$]. This implies that there are no secondary Hopf or Neimark–Sacker bifurcations in the region of interest.
2. A Floquet multiplier on the unit circle on the negative real axis: $\lambda_1 = -1$ (and therefore $\lambda_2 = -e^{-\delta m}$). We can then use the fact that $\mathrm{tr}A_1 = \lambda_1 + \lambda_2$, giving

$$-1 - \frac{1}{\delta}(e^{-\delta m} - 1)\left(\frac{4}{m} + \frac{\epsilon}{\pi m\delta}\cos(2\pi\phi)\right) + e^{-\delta m} = -1 - e^{-\delta m},$$

which reduces to the condition

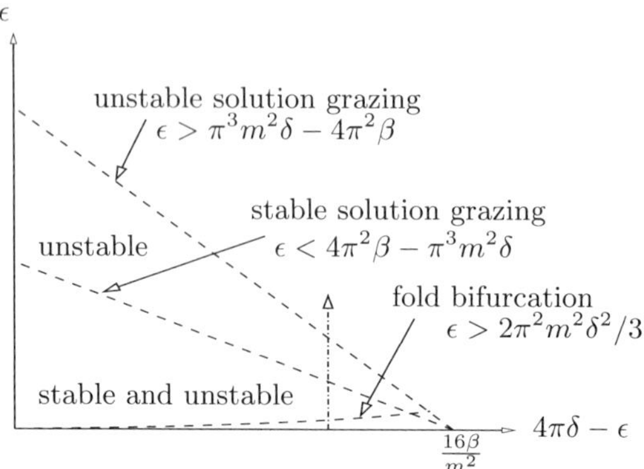

Fig. 9.14. Sketch of the bifurcations of the $P(m, 1, 0)$ solutions, with m odd. If we follow the dash-dot line we have a sequence of bifurcations: a fold bifurcation where the unstable and stable solutions are created, a grazing bifurcation where the stable solution impacts the boundary at $x = -\beta$, and finally a grazing bifurcation where the unstable solution impacts the boundary at $x = -\beta$.

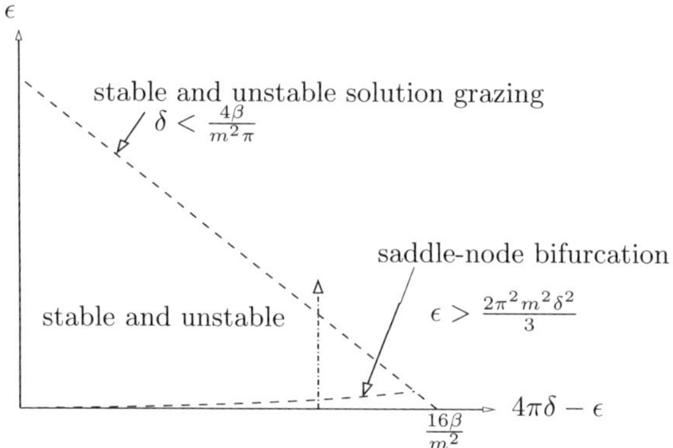

Fig. 9.15. Sketch of bifurcations of $P(m, 1, 0)$ solutions, m even. As we follow the dash-dotted line we have a saddle-node bifurcation where the unstable and stable solutions are born, and then a simultaneous (to $O(\delta^2)$) grazing of the opposing boundary $x = -\beta$.

$$\cos(2\pi\phi) = \frac{-4\pi\delta}{\epsilon}.$$

This equation can only have solutions for $\epsilon > 4\pi\delta$, which is outside the region of interest, and consequently period-doubling does not occur.

3. A Floquet multiplier on the unit circle on the positive real axis: $\lambda_1 = +1$ (and therefore $\lambda_2 = e^{-\delta m}$), corresponding to period-doubling bifurcations. Again we use the $\mathrm{tr}(A_1) = \lambda_1 + \lambda_2$ condition to find

$$\cos(2\pi\phi) = \frac{-2\pi\delta(\delta m - 2 + e^{-\delta m}(\delta m + 2))}{\epsilon(e^{-\delta m} - 1)}. \tag{9.22}$$

We can substitute the expression (9.16) for ϕ into (9.22) to find the exact location of the bifurcation in the $\epsilon - \delta$ plane. To leading order this location is given by

$$\epsilon = \frac{2\pi^2 m^2 \delta^2}{3} + O(\delta^4). \tag{9.23}$$

An eigenvalue at +1 implies that we are at a transcritical, cyclic fold or symmetry-breaking bifurcation. The solution is not symmetric so we can eliminate the possibility of symmetry-breaking. The solutions (in this form) cannot exist past the bifurcation point [see (9.18)] so we cannot have a transcritical bifurcation. Therefore we have a cyclic-fold bifurcation, where unstable and stable solutions meet and disappear.

In addition to the smooth bifurcations highlighted above, backlash oscillators, can also undergo discontinuity-induced bifurcations associated with tangential intersections of the system trajectory with the backlash boundaries defined by $|u| = \beta$. The orbits under investigation can only graze the boundary at $u = -\beta$. It is clear that the locus of grazing bifurcations is the same as the locus of the conditions for the existence of these solutions derived above.

We present these findings on two-parameter bifurcation diagrams in Figs. 9.14 and 9.15 obtained by using the conditions of existence that correspond to grazing orbits and also the changes in stability that correspond to the smooth bifurcations described above. We see many coexisting stable impacting solutions. Provided that $4\pi\delta - \epsilon < 16\beta$, we increase ϵ from zero while keeping $4\pi\delta - \epsilon$ constant (following the dash/dotted lines in Figs. 9.14 and 9.15), we first see a cyclic fold bifurcation that simultaneously gives birth to a pair of solutions that hits $y = \beta$ once per period, one stable and one unstable. Increasing ϵ by an order of magnitude then destroys the solutions through grazing bifurcations. We now turn our attention to the effects of the large finite (rather than infinite) torsional stiffness K.

9.2.4 Using a piecewise-linear contact model

One of the aims of this section is to compare the results obtained by using an impacting model of backlash against those derived using a piecewise-linear contact model. To address this issue, we now consider again solutions of type $P(m, 1, 0)$ but analyze their existence using the piecewise-linear contact model for the backlash nonlinearity.

When the piecewise-linear model is used, the orbit of interest has the form sketched in Fig. 9.16. We see that, now, the solution spends some finite time σ in the region $u > \beta$ where $B(u, \beta) = u - \beta$. To give conditions of existence of such a solution, we proceed as follows.

We write our parameters in terms of the single small parameter δ,

$$\mathbf{p} = \begin{bmatrix} \delta \\ \epsilon \\ \beta \\ K \end{bmatrix} = \begin{bmatrix} \delta \\ \epsilon_1 \delta \\ \beta_1 \delta \\ K_{-2}/\delta^2 \end{bmatrix}. \tag{9.24}$$

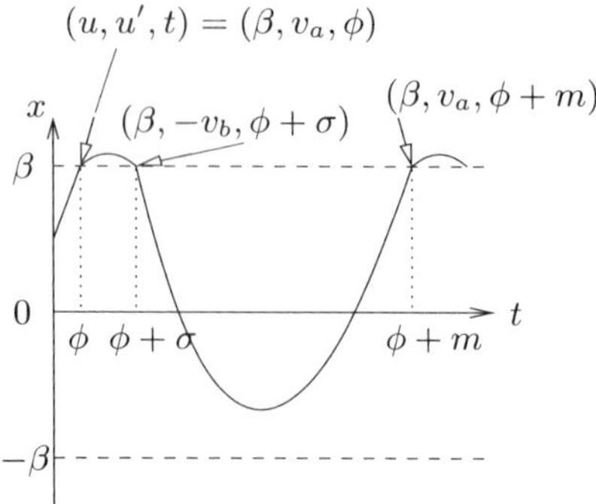

Fig. 9.16. Notation for the $P(m, 1, 0)$ solution with the piecewise-linear contact model.

We then have to find the unknowns $\mathbf{y} = [v_a, v_b, \phi, \sigma]^T$, which characterize the solution of interest. To this aim, we write such unknown variables as a series in the small parameter δ,

$$\mathbf{y} = \begin{bmatrix} v_{a0} \\ v_{b0} \\ \phi_0 \\ \sigma_0 \end{bmatrix} + \begin{bmatrix} v_{a1} \\ v_{b1} \\ \phi_1 \\ \sigma_1 \end{bmatrix} \delta + \begin{bmatrix} v_{a2} \\ v_{b2} \\ \phi_2 \\ \sigma_2 \end{bmatrix} \delta^2 + \ldots. \tag{9.25}$$

We can then find $\mathbf{y}$ by solving the set of four equations obtained by considering the matching conditions on each section of the trajectory, 'gluing' each section to the next to form the solution under investigation. Each of these equations is derived from the solutions of our linear ODEs for displacement and velocity. In particular, considering the sketch diagram in Fig. 9.16, the first equations are from the solution of the ODE for $u > \beta$,

$$u'' + \delta u' + 2K(u - \beta) = 4\pi\delta + \epsilon \cos(2\pi t),$$

with initial conditions $u(\phi) = \beta$, $u'(\phi) = v_a$. We then have two matching conditions to glue the trajectory segments together at time $t = \phi + \sigma$, namely

$$u(\phi + \sigma) = \beta, \tag{9.26}$$
$$u'(\phi + \sigma) = -v_b. \tag{9.27}$$

A similar technique generates the second pair of equations. First the ODE in the freeplay region is solved with initial conditions $u(\phi+\sigma) = \beta$, $u'(\phi+\sigma) = -v_b$, and then we apply the matching conditions

$$u(\phi + m) = \beta, \tag{9.28}$$
$$u'(\phi + m) = v_a. \tag{9.29}$$

The four equations (9.26), (9.27), (9.28) and (9.29) now form a system of equations for the unknowns $\mathbf{y} = (\phi, \sigma, v_a, v_b)^T$. We substitute (9.24) and (9.25) (with the additional assumption that, to leading order, the trajectory in the region $x > \beta$ is a sinusoid, which implies that $\sigma = \pi/\sqrt{2K} + \sigma_2\delta^2$) and expand in terms of our remaining small parameter δ. This enables us to find solutions for each set of coefficients in turn.

In summary we have two sets of solutions (dependent on the original parameters δ, ϵ, β and K). Namely, to $O(\delta^3)$ we have, for the in-phase solution $\phi_0 = 0$,

$$v_a = 2\pi m\delta - \frac{\pi(4\pi\delta + \epsilon)}{2\sqrt{2K}} + O(\delta^3), \tag{9.30}$$

$$v_b = 2\pi m\delta - \frac{\pi(4\pi\delta + \epsilon)}{2\sqrt{2K}} + O(\delta^3), \tag{9.31}$$

$$\phi = \left(-\frac{\pi}{2\sqrt{2K}} + \frac{\pi m^2\delta^2}{3\epsilon} - \frac{\delta}{4\pi^2}\right) + \left(-\frac{\pi^2 m\delta^2}{2\epsilon\sqrt{2K}} - \frac{1}{mK} - \frac{\epsilon}{4\pi mK\delta}\right) + O(\delta^3),$$

$$\sigma = \frac{\pi}{\sqrt{2K}} + \frac{4\pi\delta + \epsilon}{2\pi mK\delta} + O(\delta^3), \tag{9.32}$$

and for the out-of-phase solution $\phi_0 = 1/2$,

$$v_a = 2\pi m\delta - \frac{\pi(4\pi\delta - \epsilon)}{2\sqrt{2K}} + O(\delta^3), \tag{9.33}$$

$$v_b = 2\pi m\delta - \frac{\pi(4\pi\delta - \epsilon)}{2\sqrt{2K}} + O(\delta^3), \tag{9.34}$$

$$\begin{aligned}\phi &= \frac{1}{2} + \left(-\frac{\pi}{2\sqrt{2K}} - \frac{\pi m^2\delta^2}{3\epsilon} - \frac{\delta}{4\pi^2}\right) \\ &\quad + \left(\frac{\pi^2 m\delta^2}{2\epsilon\sqrt{2K}} - \frac{1}{mK} + \frac{\epsilon}{4\pi mK\delta}\right) + O(\delta^3),\end{aligned}$$

$$\sigma = \frac{\pi}{\sqrt{2K}} + \frac{4\pi\delta - \epsilon}{2\pi mK\delta} + O(\delta^3). \tag{9.35}$$

Examining these expressions we see that, at first, they might appear very different to the corresponding conditions of existence of the same orbit, given by (9.14) and (9.17), computed when an impacting model of backlash is considered. As we will not show, the two set of conditions are actually identical to leading order.

First, note that we have

$$v_a - v_b = O(\delta^3),$$

so to $O(\delta^3)$ the impacting model with coefficient of restitution equal to one is appropriate. Furthermore, as $K \to \infty$, we find

$$v_a, v_b \to 2\pi\delta m = \widehat{v},$$

where $\widehat{v}$ is the impact velocity predicted by the impacting contact model (9.16). Moreover, the impact times predicted by the two models also match. Indeed, the 'mid-impact' time $\phi + \sigma/2$ is equal, to order δ^2, to that given by the impacting model (9.16) for both the in- and out-of-phase solutions. Additionally, as $K \to \infty$, $\sigma \to 0$, as we might expect.

Thus, the necessary conditions for the existence of these solutions in the impacting and piecewise-linear contact models are identical, to leading order, as are the solutions themselves. We now have to check *a posteriori* whether the solution has the correct itinerary, that is, whether each of the trajectory sections remain in the assumed region of phase-space: namely $u(t) > -\beta$. It may be shown that the condition for this to hold is again the same to leading order as that found with the impacting contact model.

When the piecewise-linear contact model is used, stability of solutions is given by the eigenvalues of the matrix $A_2 := \Phi_1(m - \sigma)\Phi_2(\sigma)$. Finding expansions for the eigenvalues is cumbersome; upon numerical investigation of the formula for the eigenvalues, it is possible to show that, as in the impacting case, we have a stable out-of-phase solution and an unstable in-phase solution (see [126] for more details). Turning now to possible DIBs in the piecewise-linear contact mode, analytical calculations given in [126] show that grazing bifurcations occur on exactly the same parameter set (to leading order) as for the impacting model.

Thus, to leading order in the small parameter δ, the piecewise-linear finite stiffness model of backlash produces the same bifurcations, both smooth and discontinuity-induced, as the impacting model, in the case of solutions impacting only on one boundary. The same can be proved for more complicated family of orbits as, for instance, those with impacts on both sides of the lash [126].

Using the derivation presented so far, it is possible to derive complete bifurcation diagrams for the orbits of interest. These diagrams can then be validated numerically using appropriate continuation methods. It can be shown that, in general, several stable solutions can coexist for the same parameter values. For example, Fig. 9.17 shows some of the coexisting solutions obtained by integrating (9.9) with the piecewise-linear contact model of the backlash nonlinearity.

9.3 A hydraulic damper: non-smooth invariant tori

We now turn to an application of the analysis of DIBs to the industrially motivated problem of explaining the complex dynamics of a fully parametric model of a hydraulic damper. Such devices are used in many diverse engineering applications such as in earthquake-resistant buildings [140] and car shock-absorbers [261]. Figure 9.18(a) illustrates a particular design that incorporates a symmetric pair of 'blow-off' relief valves that are intended to allow high-velocity motion but to provide high damping at low velocities [Fig. 9.18(b)]. A model for such a situation was derived in Eyres *et. al.* [92] based on the earlier work by Surace *et. al.* [244, 245]. Here we describe the results of embedding such a model in a simple closed-loop spring-mass system,

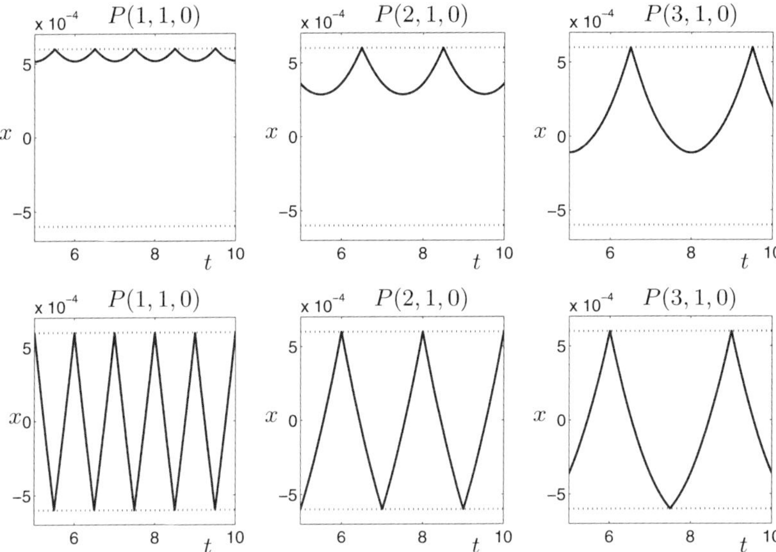

Fig. 9.17. Numerical integration of the differential equation showing coexisting stable $P(m, 1, 0)$ and $P(m, 1, 1)$ solutions for $1 \leq m \leq 3$ with the piecewise-linear contact model, $\delta = 0.5 \times 10^{-4}$, $\epsilon = 10^{-4}$, $\beta = 6 \times 10^{-4}$, $K = 10^8$. These solutions all coexist with the permanent contact solution, a sinusoid of amplitude $\epsilon/(2K)$ centered at $x = \beta + 2\pi\delta/K$.

as in Fig. 9.18(c), subject to harmonic forcing. Further details can be found in the bifurcation study given in [93].

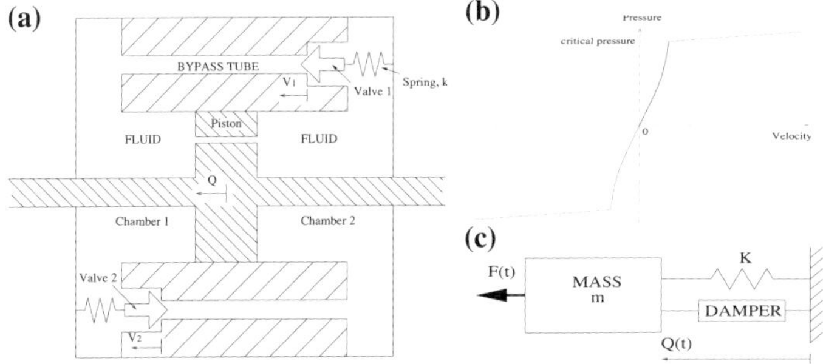

Fig. 9.18. (a) Sketch of the hydraulic damper with relief valves. (b) The desired pressure-velocity characteristic. (c) The damper embedded in a simple closed-loop system. (Reprinted from [92] with permission from Springer-Verlag.)

9.3.1 The model

The positions $V_i(t)$ of the two relief valves and the pressure difference $P(t)$ between the two main chambers can be described by the following second-order system [92]:

$$\ddot{V}_1 = \begin{cases} K_1 - (\delta_1 \dot{V}_1 + k_1 V_1)/m_{v1}, & \text{if } \quad V_1 < 0, \\ K_1, & \text{if } V_1 = 0 \text{ and } A_{v1} \Delta K_1 < 0, \\ 0, & \text{otherwise}, \end{cases} \tag{9.36}$$

$$\ddot{V}_2 = \begin{cases} K_2 - (\delta_2 \dot{V} + k_2 V_2)/m_{v2}, & \text{if } V_2 > 0, \\ K_2, & \text{if } V_2 = 0 \text{ and } K_2 > 0, \\ 0, & \text{otherwise}, \end{cases} \tag{9.37}$$

$$\dot{P} = \frac{1+\zeta}{\zeta\beta\overline{V}} \left[A\dot{Q} - \text{sign}(P) h \left(\frac{-D_1 + \sqrt{D_1^2 + 4D_2|P|}}{2D_2} + R\sqrt{|P|} \right) \right], \tag{9.38}$$

where

$$K_i(P) = A_{vi} P - k_i V_{ci}, \quad i = 1, 2,$$

$$R(V_1, V_2, t) = C_{po} \left(\frac{\gamma(\Xi_1 + \Xi_2)^2}{1 + \gamma(\Xi_1 + \Xi_2)} \right) \pi d_v \sin(\alpha) \sqrt{\frac{2}{\rho}},$$

and

$$\Xi_1 = \begin{cases} -V_1, & \text{if } V_1 < 0, \\ 0, & \text{otherwise}, \end{cases} \qquad \Xi_2 = \begin{cases} V_2, & \text{if } V_2 > 0, \\ 0, & \text{otherwise}. \end{cases} \tag{9.39}$$

Here, A is the cross-sectional area of the main valve, A_{vi} is the area of blow-off valve i, ζ is the proportional volume of chamber 1 compared with chamber 2, β is the compressibility constant of the fluid in the damper and $\overline{V}$ is the average volume of chamber 1. Taking right to left as the positive displacement direction, the sign convention is that valve 1 will be open when the displacement $V_1 < 0$ is negative (9.39), whereas $V_2 > 0$ indicates that valve 2 is open. The losses due to a given valve deflection are characterized by an initial slope proportional to γ and maximum loss for the valve fully open equal to C_{po}. The equation of motion for valve i is defined by the mass of the valve m_{vi}, spring stiffness k_i and damping term δ_i. The parameter R is a measure of the magnitude of flow going through the valve and is strictly positive for values of valve half angle $\alpha < \frac{\pi}{2}$. The direction of flow is considered in (9.38) through the term sign(P). The constants D_1 and D_2 represent the linear and quadratic head losses due to the fluid flow through the orifice and are a function of the piston and orifice geometry, fluid viscosity and density ρ.

The constant V_{c_i} represents the precompression of the springs in relief valve i given by

$$V_{c_i} = \frac{P_{\text{crit}} A_v}{k}, \quad i = 1, 2, \tag{9.40}$$

where P_{crit} is the critical pressure for blow off, see Fig. 9.18(b). At impact $V_i = 0$ we assume a highly dissipative impact law with *zero coefficient of restitution* so that

$$\dot{V}_i(t^+) = 0. \tag{9.41}$$

This law is a good approximation for systems where a light structure impacts with a spatially extended object that can easily dissipate sound, e.g. in the clanging of church bells [33]. Here, we additionally have the impact occurring within a viscous fluid, which makes significant rebound unlikely.

The simple equation for the displacement $Q(t)$ of the closed-loop system can be written as

$$M\ddot{Q} + KQ + AP = \sigma \sin(\omega t) \tag{9.42}$$

where σ and ω are the amplitude and frequency of the harmonic forcing. Assuming the chambers are of equal volume so $\zeta = 1$ and we have symmetric springs so $A_{v1} = A_{v2}$, $\delta_1 = \delta_2$, $k_1 = k_2$ and $m_{v1} = m_{v2}$ we can non-dimensionalize by setting

$$\begin{aligned} &X_1 = V_1,\ X_3 = V_2,\ Q_1 = Q,\ p = \frac{P}{P_{\text{crit}}}, \\ &X_2 = \dot{V},\ X_4 = \dot{V},\ Q_2 = \dot{Q}, \end{aligned} \tag{9.43}$$

and writing (9.38)–(9.42) as a system of first-order equations:

$$\begin{aligned}
\dot{Q}_1 &= Q_2, \\
\dot{Q}_2 &= C_1 \sin(\omega t) - C_2 Q_1 - C_3 p, \\
\dot{X}_1 &= X_2, \\
\dot{X}_2 &= \begin{cases} C_4 p - C_5 X_2 - C_6 X_1 - C_7, & \text{if } X_1 < 0, \\ C_4 p - C_7, & \text{if } X_1 = 0 \text{ and } C_4 p - C_7 < 0, \\ 0, & \text{otherwise,} \end{cases} \\
\dot{X}_3 &= X_4, \\
\dot{X}_4 &= \begin{cases} C_4 p - C_5 X_4 - C_6 X_3 + C_7, & \text{if } X_3 > 0, \\ C_4 p + C_7, & \text{if } X_3 = 0 \text{ and } C_4 p + C_7 > 0, \\ 0, & \text{otherwise,} \end{cases} \\
\dot{p} &= \left\{ C_8 Q_2 - \operatorname{sign}(p) \left(C_9 + \sqrt{C_9^2 + C_{10}|p|} + r\sqrt{|p|} \right) \right. ,
\end{aligned} \tag{9.44}$$

where

$$r(X_1, X_3, t) = C_{11} \left(\frac{(-X_1 \Theta(-X_1) + X_2 \Theta(X_2))^2}{1 + C_{12}(-X_1 \Theta(-X_1) + X_2 \Theta(X_2))} \right) \tag{9.45}$$

and Θ is the Heaviside step function. The constants $C_1 - C_{12}$ are defined in Table 9.1, which are indicative of a macro-scale device, and the impact law (9.41) is used for the valve/valve seat impact. The impact surfaces are $\Sigma_1 := \{X_1 = 0\}$ and $\Sigma_2 := \{X_3 = 0\}$.

Table 9.1. Parameter values used for simulations of the damper model (9.44)

Parameter	Physical meaning	Value
C_1	$\frac{\sigma}{M}$	4×10^4
C_2	$\frac{K}{M}$	4×10^6
C_3	$\frac{AP_{\text{crit}}}{M}$	2×10^2
C_4	$\frac{A_v P_{\text{crit}}}{m_v}$	1.98×10^4
C_5	$\frac{\delta}{m_v}$	5×10^3
C_6	$\frac{k}{m_v}$	2×10^7
C_7	$\frac{V_c}{m_v}$	1.96×10^4
C_8	$\frac{(1+\zeta)A}{\zeta\beta\overline{V}P_{\text{crit}}}$	4.21×10^5
C_9	$-\frac{(1+\zeta)D_1}{\zeta\beta\overline{V}P_{\text{crit}}2D_2}$	-1.66×10^4
C_{10}	$\frac{(1+\zeta)^2\sqrt{P_{crit}}}{(\zeta\beta\overline{V}P_{\text{crit}})^2 D_2}$	2.89×10^3
C_{11}	$\frac{(1+\zeta)C_{po}\gamma\pi d_v \sin(\alpha)\sqrt{2}}{\zeta\beta\overline{V}\sqrt{P_{\text{crit}}}\sqrt{\rho}}$	1.31×10^{14}
C_{12}	γ	4×10^5

Impacting systems with zero coefficient of restitution have the property that they can stick immediately without chatter. An analysis of a single-degree-of-freedom system with a zero coefficient of restitution law was given in the work of Shaw and Holmes [238]. It was shown there that such systems can have complex dynamics, arising as we might expect, from grazing bifurcations. The bifurcation diagram in Fig. 9.19 for the hydraulic damper shows that extremely rich chaotic dynamics can also arise in this system for sufficiently high forcing frequencies (which we should stress are well beyond the range likely to be encountered in most everyday applications of hydraulic dampers).

We will now attempt to explain some features of the observed dynamics through the analysis of the associated grazing bifurcations.

9.3.2 Grazing bifurcations

We first note that there are several kinds of discontinuity in the model. Using the notation from Chapter 2, the degree of smoothness of the flow is equal to 2, whenever the pressure reaches the critical value for the values to open (i.e. when the Heaviside function in r is switched on or off). In contrast, there is an impact, corresponding to a degree of smoothness equal to 0, whenever the valve impacts with the valve seat. We shall focus on the dynamics caused by grazing bifurcations happening with respect to the latter discontinuity surface. Figure 9.20 depicts the geometry of phase space close to such an event. For definiteness we suppose the second valve undergoes the grazing. Owing to the symmetry between the second and third set of equations in (9.44), similar considerations apply to the first valve.

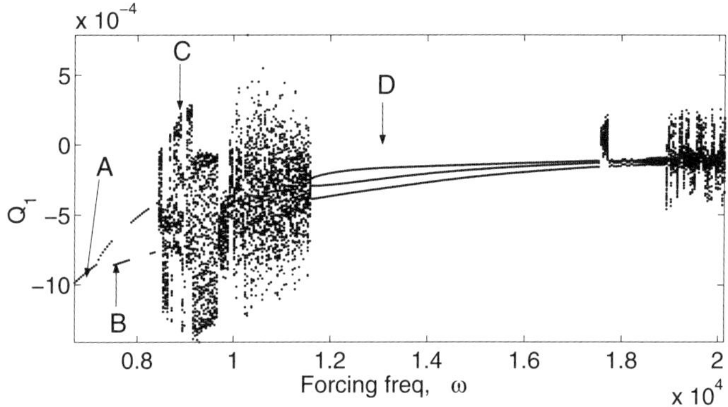

Fig. 9.19. Numerically computed bifurcation diagram showing mass displacement versus forcing frequency at the Poincaré section $Q_2 = 0$. Labels A–D illustrate different dynamical regimes. At A (and all frequency values less than this) stable, symmetric period-one motion is observed for which both valves open. Region B corresponds to symmetry-broken period-one motion, where one valve opens for longer than the other. In region C there is chaotic dynamics and in D stable period-three asymmetric motion. (Reprinted from [93] with permission from SIAM. Copyright (c)2005 Society for Industrial and Applied Mathematics.)

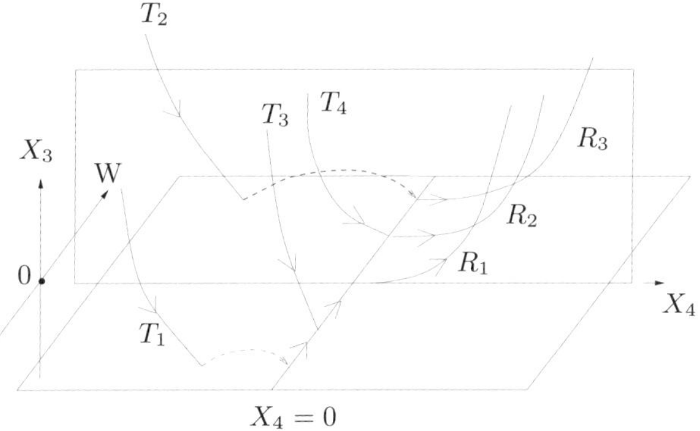

Fig. 9.20. Sketch of the possible dynamics close to a grazing of valve 2, projected onto the directions associated with X_3, X_4 and $W = p - C_7/C_4$. Solid lines correspond to trajectory segments, and dashed lines to the impact map. Here trajectory segments T_1 and T_4 correspond to impacting and sticking behavior, and both end up rebounding along segment R_1. Trajectory T_2 impacts in the non-sticking region and rebounds along R_3. Trajectory $T_4 - -R_2$ represents a grazing trajectory of the kind we analyze here. (Reprinted from [93] with permission from SIAM. Copyright (c)2005 Society for Industrial and Applied Mathematics.)

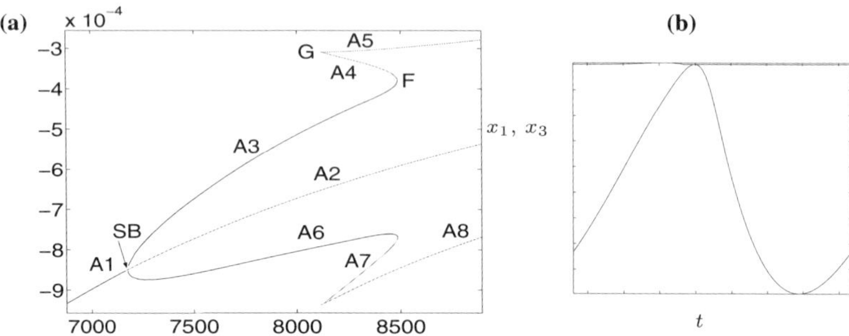

Fig. 9.21. (a) Numerical continuation results unfolding the first few bifurcations of the symmetry-broken solutions that emerge from the main stable period branch A1. Here SB refers to a symmetry-breaking pitchfork, F to a fold and G to a grazing bifurcation point. Only branches A1, A3 and A6 are stable. (b) Trajectory corresponding to grazing of valve 1 for branch A5 at the point G. Note that x_3 is zero throughout this solution as valve 2 never opens. (Reprinted from [93] with permission from SIAM. Copyright (c)2005 Society for Industrial and Applied Mathematics.)

A bifurcation diagram close to the onset of observed chaotic motion is presented in Fig. 9.21, where the path following technique described in Chapter 2 is used to perform the computations. In fact, this bifurcation diagram is part of an increasingly complex bifurcation diagram of the simplest few periodic orbits for higher ω-values, which involves period-doubling bifurcations, further grazings, Neimark–Sacker bifurcations, quasi-periodic motion and a wide parameter interval of period-three motion. Further details of the dynamics are presented in [93].

Let us focus on explaining the onset of chaos near to the grazing bifurcation of the period-one limit cycle at $\omega = 8116.1$. In order to construct a PDM associated with the flow, the Poincaré section Π containing the grazing point (for valve 2) is taken to be

$$\Pi := \{(Q_1, Q_2, X_1, X_2, X_3, X_4, p) \; : \; X_4 = 0\} .$$

The numerical continuation algorithm used in [93] enabled the authors to compute the linearized Poincaré mapof the orbits on branches A4 and A5. They found that branch A4 has one real positive eigenvalue larger than unity which increases exponentially in magnitude as the solution approaches the grazing point. On the other hand, branch A5 has two complex conjugate eigenvalues outside the unit circle that remain bounded as the solution approaches the grazing point.

Moreover, this can be predicted by applying the bifurcation analysis of Chapter 4, Sec. 4.3. In particular, we find, in the notation of that theory,

$$N_1 = \begin{bmatrix} 0.0467 & 4.4042 \times 10^{-4} & 0.3208 & 8.889 \times 10^{-6} & 0.0015 & -9.315 \times 10^{-8} & -0.0155 \\ -1861 & 0.0931 & -377.25 & -0.0059 & -76.58 & -0.0124 & 8.807 \\ -0.9529 & 3.677 \times 10^{-4} & -1.2709 & -2.180 \times 10^{-5} & -0.0155 & 1.558 \times 10^{-5} & 0.0220 \\ -7050 & -11.6379 & 56164 & 0.9760 & 7897 & 1.1580 & -1233 \\ 0 & 0 & 0 & 0 & 0 & 0 & 0 \\ 0 & 0 & 0 & 0 & 0 & 0 & 0 \\ 0.6160 & -0.0027 & 11.15 & 1.95 \times 10^{-4} & 1.2554 & 1.27 \times 10^{-4} & -0.2081 \end{bmatrix},$$

$$E = \begin{bmatrix} 0 \; 0 \; 1278.1 \; 0 \; 0 \; 0 \; 0 \end{bmatrix}^T,$$

$$C^T = \begin{bmatrix} 0 \; 0 \; 1 \; 0 \; 0 \; 0 \; 0 \end{bmatrix},$$

$$M = \begin{bmatrix} 39392 \; 5.3954 \times 10^7 \; 36804 \; 1.4788 \times 10^8 \; 0 \; 0 \; -56776 \end{bmatrix}^T.$$

From this we can calculate directly that the only physically possible bifurcation scenario is $a, ab \mapsto \varnothing$ according to the notation introduced in Chapter 3. Specifically, there is an unstable period-two solution ab that exists on the same side of the bifurcation together with the non-impacting unstable orbit a.

In this example then, grazing is not directly responsible for the onset of stable chaotic motion. However, clearly we can construct a chaotic repeller close to the grazing bifurcation point, and it seems that this set becomes stable due to other features present in the dynamics. In particular the onset of chaos in the Monte Carlo bifurcation diagrams appears to be close to the fold F in Fig. 9.21 where branch A_3 becomes unstable. Presumably at or near such a point the chaotic repeller becomes stable by some form of boundary crisis.

Also, clearly many features of the chaotic attractor vary as we pass the region C in Fig. 9.19. As argued in [93] grazing bifurcations of higher-period limit cycles play a role in organizing these qualitative changes of attractor. For example, we are about to see the chaotic attractor associated with a period-nine orbit that is close to grazing.

Fig. 9.22 depicts the observed attractor as ω varies, depicted via a plot of one variable against its first iterate under a simple Poincaré map. An interesting feature is that for higher forcing frequencies within this parameter interval of complex motion, there appears to be two-frequency motion associated with motion on an invariant torus. For example, the attractor in panels (k) and (l) of Fig. 9.22 appears to be topological equivalent to a circle. This would correspond to an invariant torus in the flow.

9.3.3 A grazing bifurcation analysis for invariant tori

In Chapter 2 we stated that motion on an invariant torus in smooth dynamical systems can either be phase locked onto a periodic orbit, or can be genuinely quasi-periodic. However, for non-smooth systems it seems that things are more complicated. For example, panel (i) of Fig. 9.22 is a good example. Here there appears to be a chaotic attractor that is close to a phase-locked period-nine

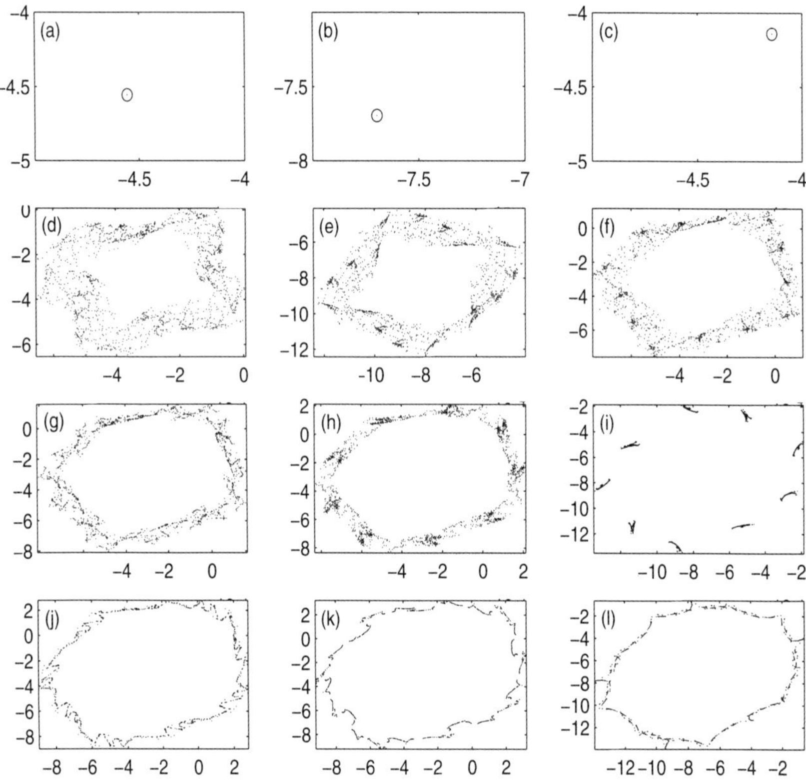

Fig. 9.22. Representations of observed attractor for a series of forcing frequencies ω as (a) 8228.7, (b) 8318.2, (c) 8407.6, (d) 8497.1, (e) 8586.5, (f) 8675.9, (g) 8764, (h) 8854.8, (i) 8944.3, (j) 9033.7 (k) 9123.2, (l) 9212.6. See [93] for details of Poincaré section used. In the first three plots, a circle has been drawn around the single fixed point of the map for ease of viewing. The values on the x- and y-axes are multiplied by the factor 10^{-4}, which is not indicated in the figure for ease of viewing. (Reprinted from [93] with permission from SIAM. Copyright (c)2005 Society for Industrial and Applied Mathematics.)

solution. Fig. 9.23 shows a delay plot of the ninth iterate of the Poincaré map, where we see that the attractor approximately aligns along the $X_{n+9} = X_n$ line. However the shape of the attractor approaches that of a square-root mapping associated with the grazing bifurcation analyzed in Chapter 6. Also, the dynamics of panels (d)–(h) also appear to lie on 'fat tori'. This is suggestive that, as ω decreases, an invariant torus has become unstable in some kind of non-smooth bifurcation, creating a nearby chaotic attractor.

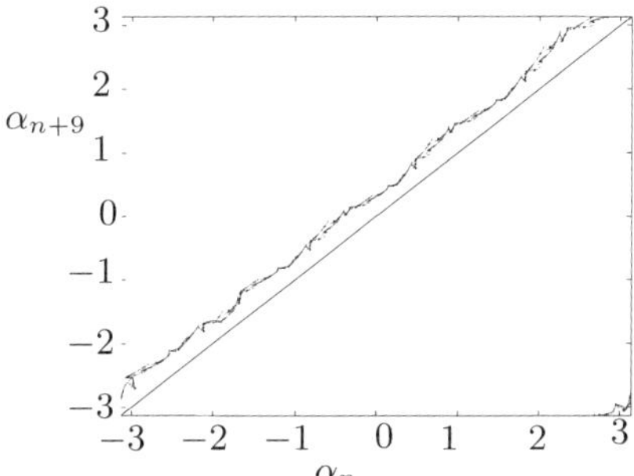

Fig. 9.23. Delay plots of the ninth iterate Poincaré mapcorresponding to section Σ for ω =9212. The co-ordinate α used is the angle associated with the vector X_1, X_3. The line $y = x$ is drawn for reference. (Reprinted from [93] with permission from SIAM. Copyright (c)2005 Society for Industrial and Applied Mathematics.)

A complete analysis of grazing bifurcations of invariant tori in impacting systems is not known and certainly goes beyond the scope of this book. A preliminary analysis has however been carried out by Dankowicz *et. al.* in [65], which we shall summarize briefly now.

Consider a smooth vector field

$$\dot{x} = F(x)$$

whose dynamics contains a smooth invariant torus $\mathcal{T}$, which is strongly attracting and normally hyperbolic, that is the speed of attraction onto the torus is greater than the time constants associated with the dynamics on the torus (see, e.g. [272]). We suppose that the torus can be parameterized by two angle co-ordinates $\alpha, \beta \in [0, 2\pi)$ and that at a parameter value $\mu = 0$ the torus grazes with a discontinuity surface $\Sigma : \{H(x) = 0\}$ at which an impact map of the form

$$x \to Rx = x + w(x)H_x(x)$$

applies. Let us suppose without loss of generality that the point of tangency is given in the local co-ordinates on the torus by $(\alpha, \beta) = (0, 0)$, and suppose further that for $\mu < 0$ there is no intersection between $\mathcal{T}$ and Σ locally. For $\mu > 0$, however, we assume that, if the effect of impact with Σ were ignored, the invariant torus penetrates Σ by a distance (measured in the orthogonal direction to the torus) proportional to μ. Now, we will make the strong assumption that the dynamics away from a neighborhood of this intersection region is sufficiently attracting in the directions *off* the torus that the flow around the torus will bring any points mapped under R back into an infinitesimal neighborhood of $\mathcal{T}$ when they return to a neighborhood of Σ. Hence, we shall approximate the entire dynamics using the torus co-ordinates (α, β). See Fig. 9.24

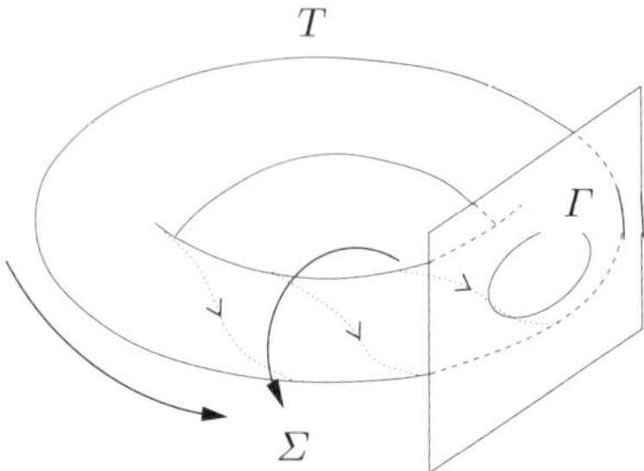

Fig. 9.24. The geometry of the intersection Γ between the discontinuity surface Σ and the invariant torus $\mathcal{T}$.

For μ small and positive, the intersection between Σ and $\mathcal{T}$ will in general be elliptical. Let Γ denote this set of points, and let us suppose that the co-ordinate directions α and β have been chosen such that they align with the principle axes of this ellipse:

$$\Gamma := \{\alpha, \beta : \frac{\alpha^2}{a^2} + \frac{\beta^2}{b^2} = \mu^2\} \tag{9.46}$$

for some positive constants a and b. Now, in general there will be two points

$$\gamma_0^\pm = (\alpha_0^\pm, \beta_0^\pm)$$

at which the vector field on the torus for $\mu > 0$ is tangent to Σ. These two points divide the ellipse into two line segments

$$\Gamma_+ = \Gamma \cup \Sigma^+, \quad \Gamma_- = \Gamma \cup \Sigma^-,$$

where $\Sigma^\pm$ are as defined in Chapter 6. Now, $\gamma_0^\pm$ lie on the grazing set G within Σ defined by $H_x F = 0$. Locally, we can approximate G by a straight line on Σ. Under variation of μ, such a straight line will intersect Γ in a straight line through the origin (plus cubic correction) in the torus co-ordinates (α, β), see Fig. 9.25. In what follows we wish only to derive the leading-order expression for the discontinuity map associated with the torus grazing, so we shall approximate this set by a straight line that passes through the origin

$$\Gamma_0 := \{\alpha, \beta : \alpha = k\beta\}, \quad \text{for some } k \in \mathbb{R}. \tag{9.47}$$

Here, we make the generic assumption that k is finite.

We now wish to study what happens to points in T that intersect Σ in forward time along Γ_+. The analysis is purely geometric and is carried out in the torus co-ordinates. The two additional pieces of information we need are an approximation to the flow F and the impact map R near Γ.

Consider first the direction of F at the points $\gamma_0^\pm$. Here, the flow is tangent both to the torus and to Σ. But, by definition the set Γ is everywhere tangent to both T and Σ. Hence if we do not assume that Γ and Σ have two directions in common at $\gamma_0^{\pm}$ (which would be highly non-generic) we have that the flow at $\gamma_0^{\pm}$ must be in the direction tangent to Γ. That is, in torus co-ordinates,

$$\frac{d\dot{\alpha}}{d\dot{\beta}} = \frac{d\alpha}{d\beta} = -k\frac{b^2}{a^2},$$

which can be reasoned by implicit differentiation of the definition of the ellipse Γ. Locally, we shall assume that the flow can be approximated by its leading-order term, so that in (α, β) co-ordinates we have

$$F \propto (1, kb^2/a^2).$$

For definiteness, let us assume that this constant of proportionality is positive (see Fig. 9.25). That is, flow occurs along straight lines

$$(\alpha(t), \beta(t)) = (\alpha_0 + t, \beta_0 + tkb^2/a^2), \tag{9.48}$$

where t is a scaling of the true time.

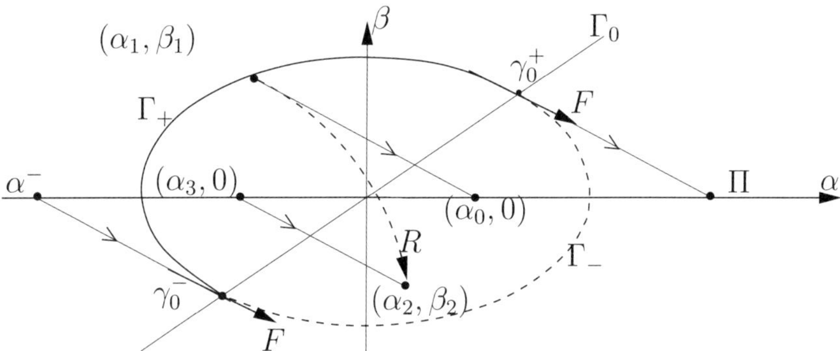

Fig. 9.25. Derivation of the PDM close to a grazing bifurcation of an invariant torus. Under the mapping, the point $\alpha_0 \in \Pi$ is mapped to α_3.

Let us now consider the reset map R. Now, by definition, Γ_0 corresponds to the grazing line $H_x F = 0$. Therefore, the leading-order form of the discontinuity mapping R must take the form

$$R : \begin{pmatrix} \alpha \\ \beta \end{pmatrix} \mapsto \begin{pmatrix} \alpha \\ \beta \end{pmatrix} + \begin{pmatrix} w_1 \\ w_2 \end{pmatrix} [\alpha - k\beta],$$

for constants w_1 and w_2.

We shall construct a PDM by taking a Poincaré section

$$\Pi : \beta = 0.$$

The construction is illustrated in Fig. 9.25. We consider a general point $(\alpha_0, 0) \in \Pi$, which is the forward image under F of a point in Γ_+. For this to be the case we must have that $\alpha^- < \alpha < \alpha^+$, where $\alpha^{\pm}$ are the images under the flow of the points $\gamma_0^{\pm}$. Specifically, solving (9.46) and (9.47) for the co-ordinates of $\gamma_0^{\pm}$, and evolving using (9.48) until $\beta = 0$, we obtain

$$\alpha^{\pm} = \pm \frac{a}{kb} \sqrt{\mu(1 + k^2 b^2)}.$$

To calculate the PDM we *first* evolve such a point $(\alpha_0, 0)$ (back) under the flow through $t = \delta$ to a point $(\alpha_1, \beta_1) \in \Gamma_+$

$$(\alpha_1, \beta_1) = \left(\alpha_0 \delta, -k \frac{a^2}{b^2} \delta \right), \tag{9.49}$$

where

$$\frac{\alpha^2}{a^2} + \frac{\beta^2}{b^2} = \mu. \tag{9.50}$$

Second, we apply the discontinuity map to obtain

$$\begin{pmatrix} \alpha_2 \\ \beta_2 \end{pmatrix} \mapsto \begin{pmatrix} 1 + w_1 & -kw_1 \\ w_2 & 1 - kw_2 \end{pmatrix}. \tag{9.51}$$

Finally we evolve this through time $t = \Delta$ under the flow to a point on Π again

$$(\alpha_3, 0) = \left(\alpha_2 + \Delta, \beta_2 - k \frac{a^2}{b^2} \right). \tag{9.52}$$

Now, in order to solve for α_3 as a function of α_0, we shall substitute everything into the only nonlinear equation, (9.50). Using (9.49) we can express α_1 and β_1 in terms of α_0 and δ. It remains to find an expression for δ. From (9.52) we can eliminate Δ and express β_2 in terms of α_3 and α_2. From (9.51) we then get two equations for α_2, which we can use to express δ in terms of α_0 and α_3. The final result is

$$c_1(\alpha_3 - \alpha_0)^2 + c_2 \alpha_0^2 = \mu,$$

where c_1 and c_2 are positive constants given by

$$c_1 = \frac{b^4 k^2}{(b^2 k w_1 + a^2 w_2)^2 (a^2 + k^2 b^2)}, \quad c_2 = \frac{b^2 k^2}{a^2 (a^2 + k^2 b^2)}.$$

From our assumption about the direction of the flow, we get that $\alpha_3 < \alpha_0$ and hence the leading-order expression for the PDM is

$$P_{\mathrm{PDM}} : \alpha \mapsto \begin{cases} \alpha, & c_2 \alpha^2 > \mu, \\ \alpha - \sqrt{\frac{\mu - c_2 \alpha^2}{c_1}} & c_2 \alpha^2 < \mu, \end{cases} \tag{9.53}$$

where c_1 and c_2 are given above. Note the characteristic square-root form of this mapping, as we saw for grazing bifurcations of periodic orbits in impacting systems.

In order to understand the dynamical implications of such a map, we need to compose it with another Poincaré mapthat describes the dynamics of the non-impacting torus. A suitable way to do that is via a *circle map* P_π that evolves the global Poincaré section $\Pi : \{\alpha, \beta : 0 \leq \alpha < 2\pi, \beta = 0\}$ to itself.

The simplest possible case is that P_π describes quasi-periodic motion, which we approximate by a rigid rotation

$$P_\pi : \alpha \mapsto \alpha + 2\pi\eta \mod 2\pi,$$

where η is the *rotation number* on the torus (see remark 1 below). The composition of this map with the above discontinuity mapping will then approximately describe the complete dynamics for $\mu > 0$. Thus we get

$$P_N = P_{PDM} \circ P_\pi : \alpha \mapsto \alpha - \sqrt{\max\left\{\frac{\mu - c_2\alpha^2}{c_1}, 0\right\}} + 2\pi\eta \mod 2\pi. \quad (9.54)$$

We suppose that η is close to a rational number p/q and consider the qth iterate of the map. For simplicity consider the case where q is small (mod 2π). If the penetration μ is large enough, the small semi-elliptical 'boil' on the unperturbed map crosses the 45° fixed-point line and creates a stable and an unstable fixed point. Because of the shape of the boil, under increasing penetration, the stable fixed point rapidly becomes unstable via a period-doubling bifurcation and eventually chaotic dynamics occurs. Also, small perturbations to the map can case the square-root part of the map to cross the 45° line which creates a grazing bifurcation.

There is much more that can be said about possible dynamics of circle maps composed with discontinuity mappings of the form (9.53). More details and also a example model system is given in [65]. We confine ourselves to a few remarks

Remarks

1. Note that the map (9.54) is continuous, despite having a square root singularity, but it is not invertible. So Theorem 2.1 presented in Chapter 2 for continuous circle maps, does not apply. In fact, the existence of chaotic dynamics leads to an interval of different rotation numbers for the map.
2. In contrast to the case periodic orbits for which grazing is a codimension-one event and nearby periodic attractors intersect the transversally, we find here that under variation of μ, points of grazing between $\mathcal{T}$ and Σ persist. That is, after the bifurcation, there remain trajectories on the torus that graze with the discontinuity surface. Hence since we might expect to find mode locking to attracting periodic orbits on the torus as we vary μ, we might find many separate grazing bifurcations of periodic orbits occurring in the neighborhood of the original grazing of the torus.

3. A similar general analysis for the grazing of invariant tori for piecewise-smooth flows with degree of smoothness 1 or greater is currently unknown. However, we note that results exist for torus bifurcations in more complex systems of DC–DC converters than the one analyzed in case study V in Chapter 1. See for example the intricate numerically constructed bifurcation diagrams in the work of Mosekilde *et. al.* [193].

9.4 Two-parameter sliding bifurcations in friction oscillators

Throughout this book we have focused on discontinuity-induced bifurcations that are unique to non-smooth systems. We have shown how these underlie sudden transitions between attractors, in particular the onset of chaotic dynamics. Through discontinuity mappings we have established a technique for analyzing DIBs of simple invariant sets such as equilibria, limit cycles and tori. However, so far our focus has been mainly on codimension-one events. Yet we know from smooth systems that codimension-two events play an important role as organizing centers for the dynamics in a parameter plane. The same must also be true of codimension-two DIBs. However, even if we restrict just to limit cycle bifurcations and only allow local bifurcations in the sense that they can be unfolded purely in terms of Poincaré maps defined in a neighborhood of grazing points, there remain many possible cases to consider. We cannot hope to be exhaustive here. In the review paper [161] an attempt is made to provide a framework for the classification of such codimension-two bifurcations (in each case giving examples for either hybrid, Filippov or continuous systems):

Type I: **Degenerate grazing point**. This is a point for which there is a degeneracy of one of the analytical conditions determining the properties of the vector fields local to the grazing event. A simple example of this would be a grazing impact with zero acceleration $a(x)$ leading to a breakdown in the local formulae for the ZDM and PDM. This is analogous to degenerate normal form coefficients for smooth bifurcations. Geometrically, this may often be regarded as the non-transverse intersection or non-quadratic tangency of the limit cycle with an appropriate set such as Σ, or $\widehat{\Sigma^{-}}$, or the failure of an assumption about the lack of sliding or chattering. This is likely to influence the leading order term of the normal form map derived via the discontinuity mapping. In the analysis of chatter presented in Chapter 6 we saw many examples of such degenerate grazing points. These were manifested as singular points (such as end points) on the projection of the discontinuity set onto the Poincaré surface and, as we saw, played an important role in the understanding of chattering and related behavior.

Type II: **Grazing of degenerate cycles**. These are bifurcations where the linear part of the Poincaré map around the trajectory, not taking account of impact, contains a degeneracy. The most obvious case is that the critical cycle is non-hyperbolic. Thus this scenario can be seen as a combination of a smooth and a non-smooth bifurcation occurring at the same parameter values. Another, more subtle possibilities is that the coefficients of this Poincaré maparound the cycle are such that, when composed with the local discontinuity map, they lead to a degeneracy in one of the dynamical consequences. An example would be a transition between a period-adding and a birth of chaos scenario in the unfolding of the square-root map associated with grazing in hybrid systems that was analyzed in Chapter 6.

Type III: **Simultaneous occurrence of two grazings** at two different points along the critical orbit. The possibilities here are large. Each of the codimension-one bifurcations outlined in the previous Chapters could occur along lines in a parameter plane. Independently, at another point along the critical periodic orbit, a second grazing event could occur. This would then form the intersection point in two-parameter space between these two lines of independent codimension-two bifurcations. However, in an unfolding one might well find that other bifurcation curves necessarily emerge from such a codimension-two point.

In order to illustrate this classification, we provide here an example of each occurring in the context of sliding bifurcations in the simple dry friction oscillator models similar to that in case study IV.

9.4.1 A degenerate crossing-sliding bifurcation

In [160], Kowalczyk & di Bernardo consider a version of the dry friction oscillator given by

$$\ddot{u} + u = \sin(\omega t) - A\,\mathrm{sign}(\dot{u}). \tag{9.55}$$

Here u is the position of an oscillating mass while ω and A represent the frequency of an external forcing and the amplitude of the friction characteristic respectively. Consider a region of parameter space where a stable, symmetric periodic orbit of period $\frac{2\pi}{\omega}$ exists, as depicted in Fig. 9.26(a). Under variation of A, such an orbit is found to undergo a crossing-sliding bifurcation at $A = 0.6656$ for $\omega = 2/3$. It can be checked [160] that at this point all the conditions introduced in Chapter 8 for the crossing-sliding bifurcation to occur are satisfied. For A-values beyond the bifurcation point, the orbit acquires a segment of sliding motion giving rise to the orbit depicted in Fig. 9.26(b).

The above bifurcation is just one point on the analytical curve given by

$$A = \omega^2 \sin\left[\arctan\left(\frac{\omega\sin(\pi\omega^{-1})}{1+\cos(\pi\omega^{-1})}\right) + \pi\right]\frac{1}{\omega^2 - 1} \quad \text{for} \quad \omega \in (0.5, \infty),\ \omega \neq 1.$$

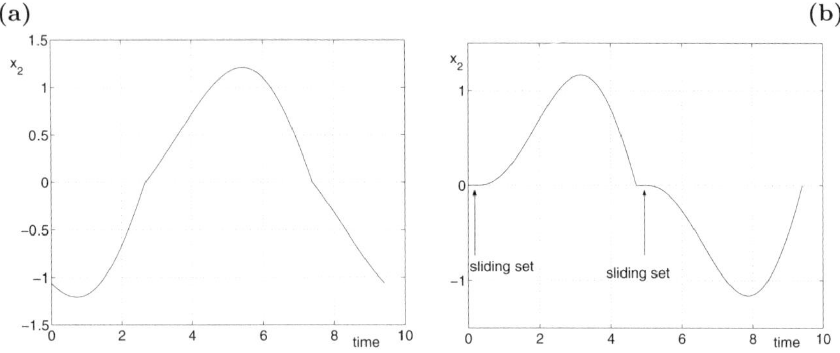

Fig. 9.26. (a) A simple symmetric periodic orbit of (9.55) for $\omega = 2/3$ with two transversal crossings of the switching surface per period just 'before' the crossing-sliding bifurcation, and (b) the same orbit just after the bifurcation. The crossing-sliding for this ω-value occurs for $F = 0.6656$. (Reprinted from [161] with permission from Elsevier.)

at which a branch of crossing-slidings occur for this symmetric limit cycle. In [162], the limit cycles on this curve in the parameter plane were followed using numerical continuation. In particular, it was found that the branch does indeed terminate as $\omega \to 0.5^+$, $(A \to 1/3)$. It was then shown in [160] that the termination is caused by a failure in the non-degeneracy condition of the crossing-sliding, hence a type-I codimension-two DIB. Moreover, inevitably at such a bifurcation point (provided a further non-degeneracy condition is satisfied, which it is for this example), two additional curves of codimension-one sliding bifurcations must emerge, as we shall now explain.

Recall from Chapter 8 the defining conditions (8.7) and (8.8) for a crossing-sliding bifurcation. The additional condition defining the codimension-two point is the failure of the non-degeneracy condition (8.11), which yields

$$H_x F_{1x} F_1 = 0. \tag{9.56}$$

From a geometric viewpoint, this means that now the outgoing flow at the bifurcation point is parallel to the boundary of the sliding region. Hence, in order for the trajectory to leave the switching manifold, the vector field F_1 should exhibit a local maximum with respect to $\partial\widehat{\Sigma}^-$. Thus, a non-degeneracy condition for this codimension-two event can be written as

$$\left.\frac{d^3(H(\Phi_1(x,t)))}{dt^3}\right|_{t=0} = H_x(F_{1x})^2 F_1 > 0. \tag{9.57}$$

The key to understanding the dynamics in a neighborhood of such a codimension-two sliding bifurcation is to understand the topology of the grazing set $\partial\Sigma$ close to the point where the degenerate limit cycle intersects it. The situation is depicted in Fig. 9.27. Here, there is a degenerate point along the grazing line that bounds the region of sliding. Three different qualitative behaviors can then be identified corresponding to different regions of initial conditions on the switching manifold Σ:

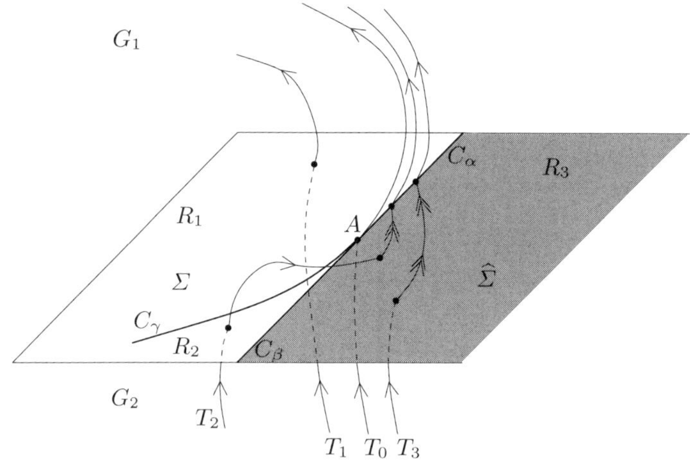

Fig. 9.27. Phase space topology around the codimension-two degenerate crossing-sliding point. Trajectory labeled as T_0 denotes part of the critical limit cycle interacting with the boundary of the sliding set in the codimension-two scenario. Perturbations applied to the critical cycle result in possible limit cycles rooted in regions R_1, R_2 or R_3. This is schematically depicted by trajectories T_1, T_2 and T_3 (rooted correspondingly in R_1, R_2 and R_3 regions), which differ from T_0 by the number of segments which form part of a limit cycle locally to the codimension-two point. (Reprinted from [160] with permission from Elsevier.)

1. Trajectories starting from region R_1 leave the switching manifold towards region G_1.
2. Trajectories starting from region R_2 still leave the switching surface towards G_1 but, after some finite time, hit the switching manifold again within the sliding region $\widehat{\Sigma}$. Then, they evolve according to the sliding flow until crossing the boundary $\partial\widehat{\Sigma}^-$, where they finally leave the switching manifold (see trajectory T_2 in Fig. 9.27).
3. Trajectories rooted in region R_3 will evolve according to the sliding flow until $\partial\widehat{\Sigma}^-$ is reached (see trajectory T_3 in Fig. 9.27). Then, they will leave the switching manifold towards G_1.

Fig. 9.28 shows the corresponding partitioning of the switching manifold for the example (9.55). The corresponding trajectories are shown in Fig. 9.29.

It is possible in principle to use the theory of discontinuity mappings as introduced in Chapter 8 to the situation where a limit cycle passes through the degenerate point on the grazing line. We do not present such an analysis here but present the results schematically in Fig. 9.30 with more details in

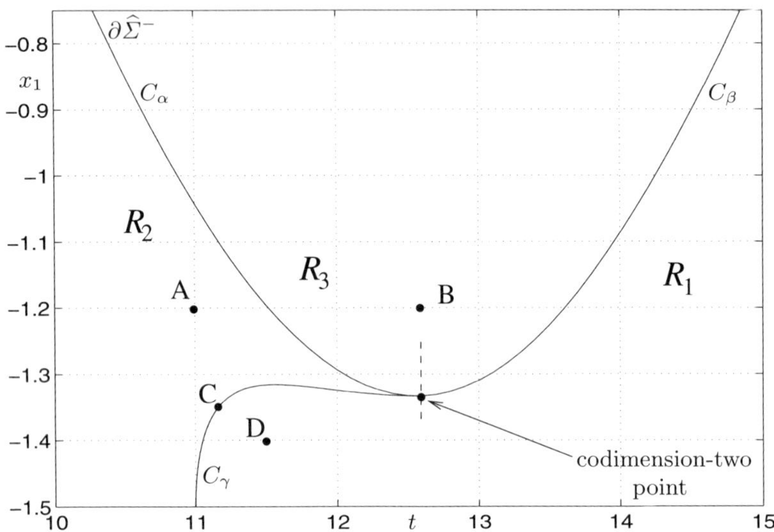

Fig. 9.28. Numerically evaluated boundaries corresponding to the theoretical curves in Fig. 9.27. (Reprinted from [160] with permission from Elsevier.)

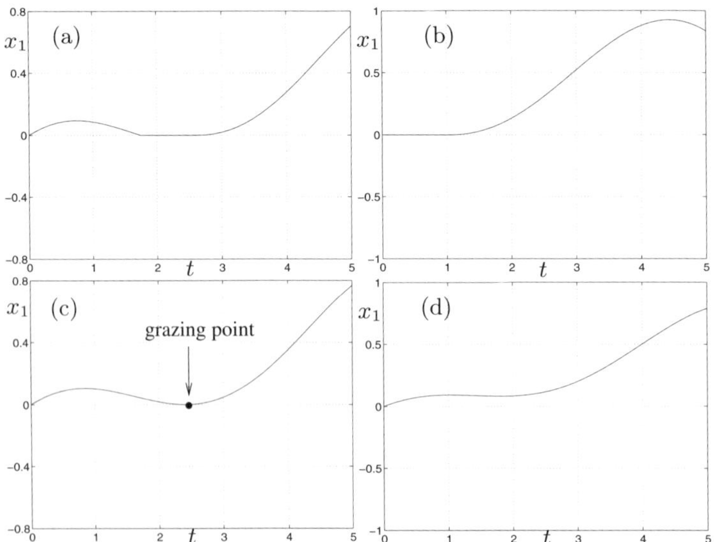

Fig. 9.29. Qualitatively different trajectories around the codimension-two node starting from regions R_1, R_2 and R_3. Time series representing the velocity coordinate of the trajectories starting in every of the three regions are shown in each of the panels. Each panel is labeled with the letter indicating the starting point in phase space as shown in Fig. 9.28. (Reprinted from [160] with permission from Elsevier.)

[160]. In particular note that curves of switching-sliding and grazing-sliding bifurcations are an inevitable consequence of the codimension-two crossing-sliding bifurcation. The corresponding curves for the friction oscillator are

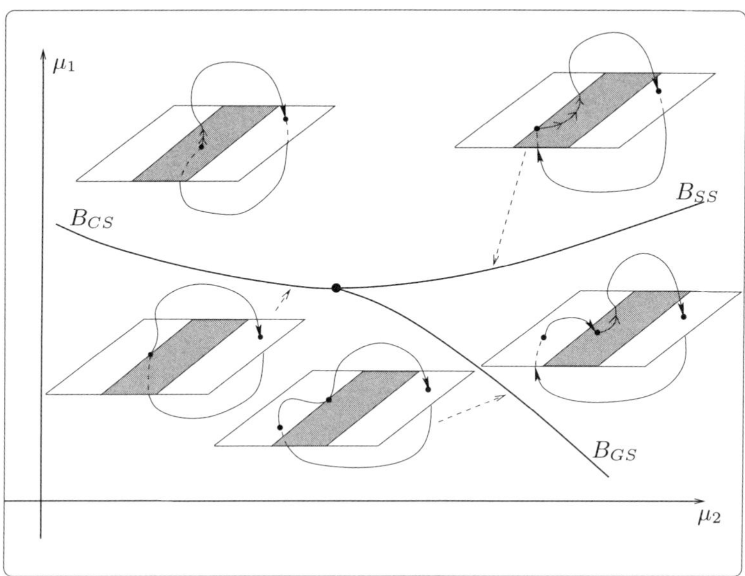

Fig. 9.30. Qualitative representation of the theoretical unfolding of the codimension-two degenerate crossing-sliding bifurcation. (Reprinted from [160] with permission from Elsevier.)

presented in Fig. 9.31.

Note that this whole bifurcation scenario, in the specific example of interest, does not cause any change in the existence of limit cycles, nor their stability. In fact all the orbits computed are stable.

9.4.2 Fold bifurcations of grazing-sliding limit cycles

Let us now turn to an example of a Type-II codimension-two bifurcation point in a friction oscillator that *does* lead to a qualitative change in the system dynamics.

We consider the dry friction oscillator with slightly more complex friction law studied as case study IV in Chapter 1:

$$\ddot{u} + u = f(1 - \dot{u}) + A\cos(\nu t), \tag{9.58}$$

where

$$f(z) = \alpha_0 \mathrm{sgn} z - \alpha_1 z + \alpha_2 z^3. \tag{9.59}$$

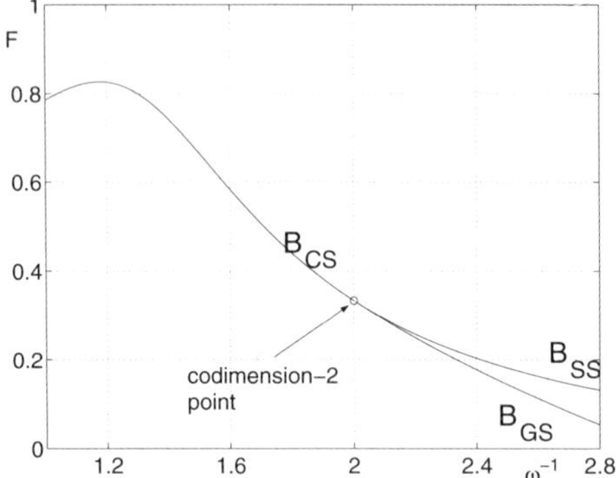

Fig. 9.31. Numerically computed two-parameter bifurcation diagram of the friction oscillator near the degenerate crossing-sliding bifurcation. (Reprinted from [161] with permission from Elsevier.)

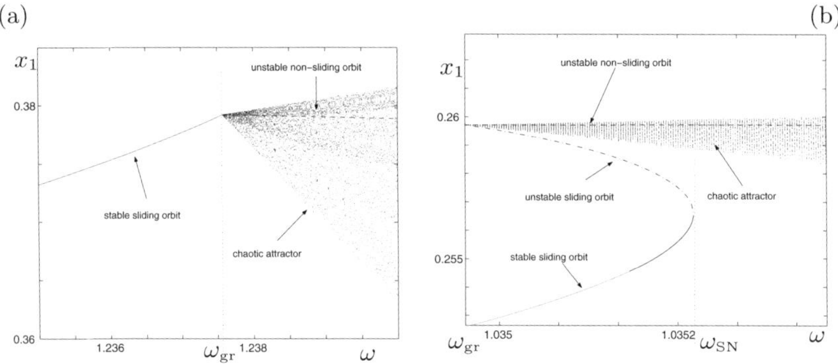

Fig. 9.32. One-parameter bifurcation diagrams for (9.58), (9.59) with (a) $\alpha_4 = 0.7$ and (b) $\alpha_4 = 0.4$ obtained by direct numerical simulation (a dashed line denotes an unstable orbit found using numerical continuation).

A branch of period-one orbits (i.e. having the same period $T = \frac{2\pi}{\omega}$ as the drive) undergoing grazing-sliding bifurcations has been detected for $\alpha_1 = \alpha_2 = 1.5$, $\alpha_3 = 0.45$ and for a range of ω and α_4 values. Figure 9.32 illustrates two different one-parameter scenarios with varying ω. In panel (a), for $\alpha_4 = 0.7$, we see that upon increasing ω, a stable period-one orbit that incorporates a sliding segment is destroyed at the grazing-sliding point. On the other side of the bifurcation there exists an unstable orbit with the same period that does not slide. This coexists with a chaotic attractor. In contrast for $\alpha_4 = 0.4$ [panel (b)] two unstable period-one orbits, one with a sliding segment and the other without, are created as ω is increased, accompanied by the birth of a chaotic attractor. At a higher ω-value there is a fold bifurcation where the unstable sliding orbit coalesces with a stable sliding orbit. This stable orbit can then be traced with decreasing ω. Thus, for ω-values between the grazing-sliding and the fold we have bi-stability between a chaotic attractor and a period-one orbit.

From these numerical calculations we conclude that there should exist a curve of fold bifurcations connecting to the curve of grazings at a codimension-two point. A numerically computed two-parameter bifurcation diagram shows that the branch of fold bifurcations joins the curve of grazing bifurcations tangentially, at the codimension-two point $(\alpha_4, \omega) = (0.5199, 1.1136)$. The grazing point on the limit cycle is $(y, \dot{y}, \omega t) = (1.3309, 1, 1.9020)^T$. From which values, together with evaluation of various derivatives of the vector field, it can be verified that the defining and non-degeneracy conditions for the grazing-sliding bifurcation (8.7), (8.8) and (8.11) from Chapter 8 are satisfied at the grazing point. However, what defines the codimension-two point is a degeneracy in the linearization around the critical limit cycle. We cannot directly use linear theory to determine the stability of the grazing cycle, as any Poincaré mapping will be non-differentiable at the corresponding fixed point, but if we consider the grazing cycle as having no sliding segment, the non-trivial eigenvalues of the cycle do not lie on the unit circle and their numerical values are -5.2652 and -0.0151. However, if we consider the cycle as having one zero-length sliding segment the non-trivial eigenvalues are found to be 1 and 0. Thus one could say the sliding version of the grazing cycle is non-hyperbolic, and this opens up the possibility of having several different cycles with a short sliding segment as parameters are varied.

In Nordmark & Kowalczyk [202] analyze such non-hyperbolic grazing-sliding events using discontinuity mappings. The inevitability of the fold emanating from the codimension-two point is predicted, as are the asymptotics of the fold curve. Note that the bifurcation has an immediate effect in that the creation of a region of bi-stability between the chaotic attractor and a stable sliding limit cycle [202]. It was also found that the codimension-two point also has an indirect consequence for yet lower values of α_4. Specifically, Fig. 9.33 shows the one-parameter bifurcation diagram for $\alpha = 0.3$. Here the chaotic attractor is not created immediately at the grazing bifurcation point upon increasing ω. Instead the chaotic attractor existing for large ω is destroyed upon decreasing ω by collision with the unstable sliding orbit. This *boundary-crisis* bifurcation destroys the chaotic attractor and leads to a short ω-interval where the stable sliding limit cycle is the only attractor.

9.4.3 Two simultaneous grazings

We return to the simple friction oscillator example (9.55) to show an example of a type-III codimension-two discontinuity-induced bifurcation. Specifically, a codimension-two situation was identified by Feigin [98] where adding-sliding and grazing-sliding occur at two distinct points along a limit cycle. In later work, Kowalczyk *et. al.* [160], [161] analyzed such a codimension-two bifurcation point has been located at $\omega^{-1} = 7.76990$, $A = 0.299984$, with the

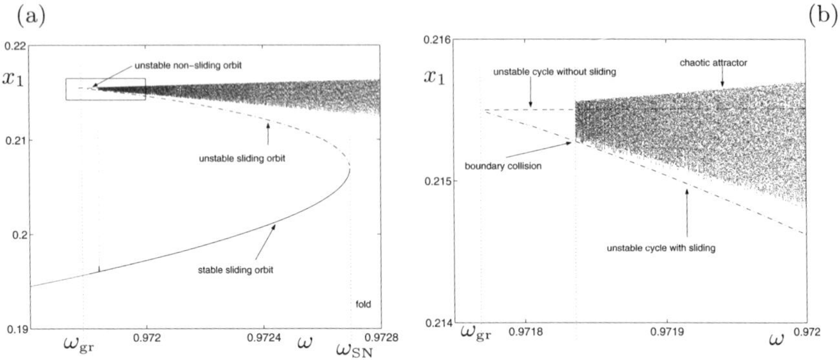

Fig. 9.33. bifurcation diagram via direct numerical simulation with varying ω for $\alpha_4 = 0.3$. Panel (b) shows a zoom of the boxed area of panel (a)

grazing-sliding occurring at $u = 0.100044137$, $\dot{u} = 0$, $\omega t = 0.411547546$ and the adding-sliding at $u = 0.700016$, $\dot{u} = 0$, $\omega t = \frac{\pi}{2}$. The period of the bifurcating orbit is equal to $T = \frac{2\pi}{\omega} = 15.53980\pi$.

The time series of the components of the orbit exhibiting the above adding-sliding and grazing-sliding bifurcation scenarios, are depicted in Fig. 9.34. Projection of the position component onto the switching manifold Σ, in panel (a) allows us to easily spot the instant of adding-sliding; whereas observing the velocity component, panel (b), captures the grazing-sliding interaction.

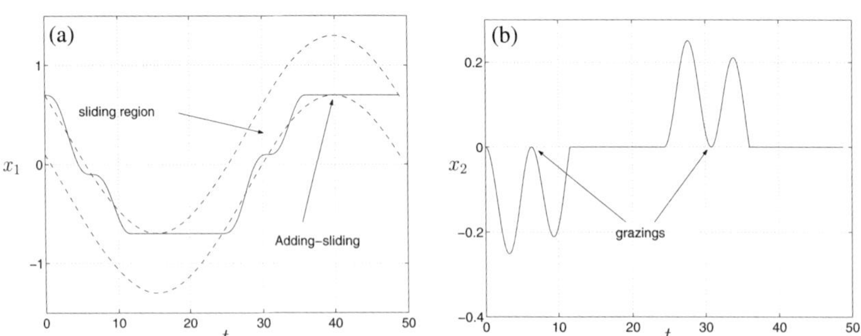

Fig. 9.34. Time series of (a) the position component of the system (9.55) (b) the velocity component for $\omega^{-1} = 7.76990$, $F = 0.299984$. (Reprinted from [161] with permission from Elsevier.)

Variation of the bifurcation parameters A and ω will cause the trajectory to cross the grazing-sliding and the adding-sliding boundaries in the two-parameter space. Depending on the character of the ZDM map that captures the dynamics of the system, a limit cycle undergoing a grazing-sliding scenario might be destroyed. Therefore, generically at the codimension-two point under consideration a branch of adding-sliding bifurcations should terminate at the codimension-two point. However, in our dry-friction oscillator example such a situation does not occur and a stable orbit exists in all regions around the codimension-two point. bifurcation diagrams depicting branches of grazing and adding-sliding which cross at the codimension-two point are presented in Fig. 9.35.

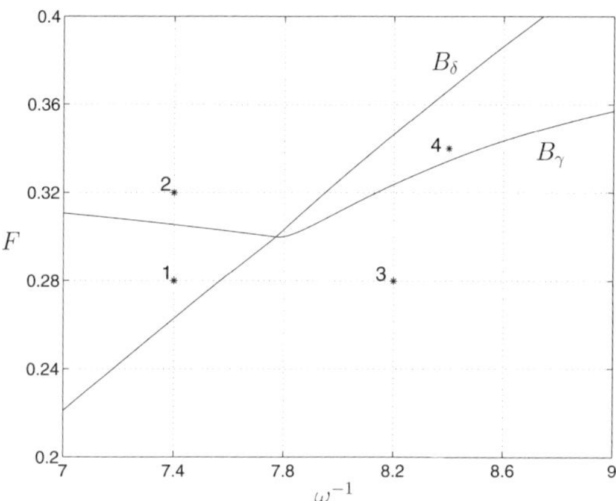

Fig. 9.35. Two-parameter bifurcation diagram around the codimension-two node. B_γ denotes a branch of grazing-sliding and B_δ a branch of adding-sliding bifurcations that cross at the codimension-two point. (Reprinted from [161] with permission from Elsevier.)

A limit cycle near the codimension-two point is depicted in Fig. 9.1. Variation of any of the two bifurcation parameters might lead the orbit to cross the bifurcation boundaries B_δ or B_γ (which denote respectively adding-sliding and grazing-sliding bifurcations) and hence become topologically distinct limit cycles from the point of view of discontinuity-induced bifurcation introduced in Chapter 2. For example, Fig. 9.36 depicts two orbits to the right of the curve B_δ, i.e. 'after' the adding-sliding bifurcation takes place. Additional non-sliding segments, which form small lobes in the phase plots of the limit cycles, are clearly visible. Although the topology of all of these limit cycles changes, in this example it is found that each of these orbits is an attractor of the system, so as in Sec. 9.4.1 this codimension-two point does not represent a bifurcation in the sense of a change in system attractor.

Clearly, there are many possibilities for two-parameter discontinuity-induced bifurcations in piecewise-smooth systems, and even the enumeration of all possibilities remains a considerable research challenge. It is then

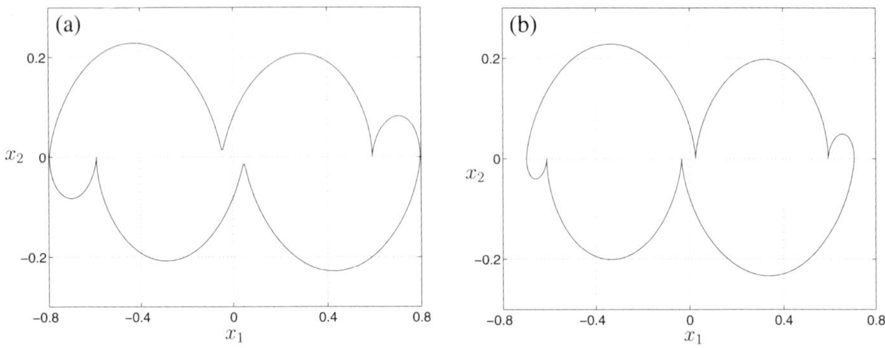

Fig. 9.36. (a) A limit cycle for $\omega^{-1} = 8.2$ and $F = 0.34$ corresponding to asterisk '3' in Fig. 9.35 and (b) a limit cycle for $\omega^{-1} = 8.4$ and $F = 0.34$ corresponding to asterisk '4' in Fig. 9.35. (Reprinted from [161] with permission from Elsevier.)

necessary to provide an unfolding of the dynamics close to such points, including identifying which curves of codimension-one bifurcations necessarily emanate from the codimension-two point. Then, uniquely for piecewise-smooth systems, there may be dynamical events that do not occur directly at the critical parameter value, but occur at nearby parameter values due to a rapid change in higher-derivatives of the Poincaré map(as we saw in Example 7.5). Clearly such phenomena are likely to be affected by the presence of a codimension-two singularity. An evaluation of which of these codimension-two events are 'important' is likely to be influenced by what is observed in applications.

References

1. J. Adolfsson, H. Dankowicz, and A. Nordmark. 3D passive walkers: Finding periodic gaits in the presence of discontinuities. *Nonlinear Dynamics*, 24:205–229, 2001.
2. M.A. Aizerman and F.R Gantmakher. On the stability of periodic motions. *J. Appl. Math. Mech.*, pages 1065–1078, 1958.
3. R. Alzate, M. di Bernardo, U. Montanaro, and S. Santini. Experimental and numerical verification of bifurcations and chaos in cam-follower impacting systems. *Nonlinear Dynamics*, 2007. to appear (DOI 10.1007/s11071-006-9188-8) 2007.
4. D. Amrani and D.P. Atherthon. Designing autonomous relay systems with chaotic motion. In *Proc. 28th IEEE Conf. on Decision and Control*, volume 1, pages 932–936, Tampa, FL, 1989.
5. A. A. Andronov, S. E. Khaikin, and A. A Vitt. *Theory of Oscillators*. Pergamon Press, Oxford, 1965.
6. L. Arnold. *Random Dynamical Systems*. Springer-Verlag, 1998.
7. V.I. Arnold. *Geometrical Methods in the Theory of Ordinary Differential Equations*. Springer-Verlag, 1988.
8. D. G. Aronson, M. A. Chory, G. R. Hall, and R. P. McGehee. Bifurcations from an invariant circle for two-parameter families of maps of the plane: A computer-assisted study. *Communications in Mathematical Physics*, 83:303–354, 1982.
9. D. K. Arrowsmith and C. M. Place. *An Introduction to Dynamical Systems*. CUP, Cambridge, UK, 1990.
10. U.M. Ascher and L.R. Petzold. *Computer methods for ordinary differential equations and differential-algebraic equations*. SIAM, Philadelphia, 1998.
11. P. Ashwin, E. Covas, and R. Tavakol. Influence of noise on scalings for in-out intermittency. *Physical Review E*, 64(6):066204, December 2001.
12. M. Aslam. Earthquake rocking response of rigid bodies. *Journal of the structural division - ASCE*, 106:377–392, 1980.
13. K. J. Åström. Oscillations in Systems with Relay Feedback. In K. J. Åström, G. C. Goodwin, and P. R. Kumar, editors, *Adaptive Control, Filtering, and Signal Processing*, volume 74 of *IMA Volumes in Mathematics and its Applications*, pages 1–25. Springer-Verlag, 1995.

14. V. Avrutin and M. Schanz. On the scaling properties of the period-increment scenario in dynamical systems. *Chaos, Solitons & Fractals*, 11:1949–1955, 2000.
15. V. Avrutin and M. Schanz. Border-collision period doubling scenario. *Physical Review E*, 70:020101, 2004.
16. V. Avrutin and M. Schanz. Period-doubling scenario without flip bifurcations in a one-dimensional map. *International Journal of Bifurcations and Chaos*, 15:1267–1284, 2005.
17. V. Avrutin and M. Schanz. On multi-parametric bifurcations in a scalar piecewise-linear map. *Nonlinearity*, 19:1875–1906, 2006.
18. B. Azejczyk, T. Kapitaniak, J. Wojewoda, and R. Barron. Experimental-observation of intermittent chaos in a mechanical system with impacts. *J. Sound Vib.*, 178:272–275, 1994.
19. V. I. Babitskii. *Theory of Vibroimpact Systems. Approximate methods.* Nauka, Moscow, 1978.
20. V.I. Babitsky and A.M. Veprik. Universal bumpered vibration isolator for severe environment. *J. Sound Vib.*, 218:269–292, 1998.
21. S. Banerjee and C. Grebogi. Border collision bifurcations in two-dimensional piecewise smooth maps. *Physical Review E*, 59:4052–4061, 1999.
22. S. Banerjee, P. Ranjan, and C. Grebogi. Bifurcations in two-dimensional piecewise smooth mapsTheory and applications in switching circuits. *IEEE Transactions on Circuits and Systems I*, 47(5):633–643, May 2000.
23. S. Banerjee, J. York, and C. Grebogi. Robust Chaos. *Physical Review E*, 59:4052–4061, 1999.
24. S. Banerjee, J. A Yorke, and C. Grebogi. Robust Chaos. *Physical Review Letters*, 80:3049–3052, 1998.
25. N.N. Bautin and E.A. Leontovich. *Methods and Techniques for Qualitative Analysis of Dynamical Systems on the Plane.* Nauka, Moscow, 1976. [in Russian].
26. P V Bayly and L N Virgin. An experimental study of an impacting pendulum. *Journal of Sound and Vibration*, 164(2):364–374, 1993.
27. L. Benadero, A. El Aroudi, G. Olivar, E. Toribio, and E. Gomez. Two-dimensional bifurcation diagrams. Background pattern of fundamental DC-DC converters with PWM control. *International Journal of Bifurcation and Chaos*, 13(2):427–451, February 2003.
28. S. Bennett. *A history of control engineering 1930 -1955.* IEE London, U.K, 1993.
29. A. Berghuvud and A. Stensson. Consequences of Nonlinear Characteristics of a Secondary Suspension in a Three-Piece Freight Car Bogie. *Vehicle System Dynamics*, 36:37–55, 2001.
30. S.R. Bishop. Impact oscillators. *Phil. Trans. Roy. Soc. A*, 347:347–351, 1994.
31. S.R. Bishop, M.G. Thompson, and S. Foale. Prediction of period-1 impacts in a driven beam. *Proc. Roy. Soc. Lond. A*, 452:2579–2592, 1996.
32. S. Blackmore, K. Salmon, J. Vogwell, I. G. Turner, and S. R. Brown. Development of a test to measure the impact properties of coated tablets. *Journal of Pharmacy and Pharmacology*, 56:133, 2004.
33. A. Blakeborough. An analytical response of church bells to earthquake excitation. *Journal of Earthquake Engineering*, 5(1):69–92, January 2001.
34. G. Blankenship and A. Kahraman. Steady State Forced Response of a Mechanical Oscillator with Combined Parameter Excitation and Clearance Nonlinearity. *Journal of Sound and Vibration*, 185(5):743–765, 1995.

35. B. Blazejczyk-Okolewska, K. Czolczynski, and J. Kapitaniak, T. Wojewoda. *Chaotic Mechanics in Systems with Impacts and Friction.* World Scientific, Singapore, 1999.
36. V.D. Blondel and J.N. Tsitsiklis. Complexity of stability and controllability of elementary hybrid systems. *Automatica*, 35(3):479–490, 1999.
37. P. C. Bressloff and J. Stark. Neuronal dynamics based on discontinuous circle maps. *Physics Letters A*, 150:87–195, 1990.
38. B. Brogliato. *Nonsmooth Mechanics – Models, Dynamics and Control.* Springer–Verlag, New York, 1999.
39. B. Brogliato. *Impacts in Mechanical Systems – Analysis and Modelling.* Springer–Verlag, New York, 2000. Lecture Notes in Physics, Volume 551.
40. B. Brogliato. Some perspectives on the analysis and control of complimentarity systems. *IEEE Transactions Automatic Control*, 48:918–935, 2003.
41. B. Brogliato, A.A. ten Dam, L. Paoli, F. Génot, and M. Abadie. Numerical simulation of finite dimensional multibody nonsmooth mechanical systems. *Applied Mechanics Reviews*, 5:107–150, 2002.
42. C. Budd and F. Dux. Chattering and related behaviour in impact oscillators. *Philosophical Transactions of the Royal Society London A*, 347:365–389, 1994.
43. C. Budd and F. Dux. Intermittency in impact oscillators close to resonance. *Nonlinearity*, 7:1191–1224, 1994.
44. C. Budd, F. Dux, and K. A. Cliffe. The effect of frequency and clearance variations on a single degree of freedom impact oscillators. *Journal of sound and vibration*, 184:475–502, 1995.
45. C.J. Budd and A.G. Lee. Double impact orbits of periodically forced impact oscillators. *Proc. Roy. Soc. Lond. A*, 452, 1996.
46. C.J. Budd and P. Piiroinen. Corner bifurcations in non-smoothly forced impact oscillators. *Physica D*, 220:127–145, 2006.
47. B. Buffoni and J. Toland. *Analytic theory of global bifurcation, an introduction.* Princeton University Press, 2003.
48. R. S. Burns. *Advanced Control Engineering.* Butterworth Heinemann Oxford, 2001.
49. M.K. Camlibel, W.P.M.H. Heemels, and J.M. Schumacher. Stability and controllability of planar bimodal complementarity systems. In *Proceedings IEEE Conference on Decision and Control*, volume 2, pages 1651– 1656, 2003.
50. V. Carmona, E. Freire, E. Ponce, and F. Torres. On simplifying and classifying piecewise-linear systems. *IEEE Transactions on Circuits and Systems I*, 49(5):609–620, May 2002.
51. V. Carmona, E. Freire, E. Ponce, and F. Torres. Invariant manifolds of periodic orbits for piecewise linear three-dimensional systems. *IMA Journal of Applied Mathematics*, 69(1):71–91, February 2004.
52. V. Carmona, E. Freire, E. Ponce, and F. Torres. Bifurcation of invariant cones in piecewise linear homogeneous systems. *International Journal of Bifurcation And Chaos*, 15(8):2469–2484, August 2005.
53. D.R.J. Chillingworth. Discontinuity geometry for an impact oscillator. *Dynamical Systems*, 17:380–420, 2002.
54. W. Chin, E. Ott, H. E. Nusse, and C. Grebogi. Grazing bifurcations in impact oscillators. *Physical Review E*, 50:4427–4444, 1994.
55. S.N. Chow and J.K. Hale. *Methods of bifurcation theory.* Springer-Verlag, 1982.
56. P. Collet and J.P. Eckmann. *Iterated maps on the interval as dynamical systems.* Birkhauser, 1980.

57. H.J. Connors. Flow-induced vibration and wear of steam generator tubes. *Nucl. Technology*, 55:311–331, 1982.
58. P. A. Cook. Simple feedback systems with chaotic behaviour. *Systems & Control Letters*, 6:223–227, 1985.
59. P. A. Cook. *Nonlinear Dynamical Systems.* Englewood Cliffs, NJ, 1986.
60. S. Coombes and S. H. Doole. Neuronal population dynamics with post inhibitory rebound: A reduction to piecewise linear discontinuous circle maps. *Dynamics and Stability of Systems*, 11:193–217, 1996.
61. P. Couderc, J. Callenaere, J. DerHagopian, G. Ferraris Besso, A. Kassai, Y. Borjesson, L. Verdillon, and S. Gaimard. Vehicle Driveline Dynamic Behaviour: Experimentation and Simulation. *Journal of Sound and Vibration*, 219(1):133–157, 1998.
62. P. Cvitanovic. *Universality in chaos.* Bristol, Adam Hilger, Ltd., 1984, 519 p., 1984.
63. H. Dankowicz. On the modeling of dynamic friction phenomena. *Zeitschrift fuer Angewandte Mathematik und Mechanik*, 79(6):399–409, 1999.
64. H. Dankowicz and A. B. Nordmark. On the origin and bifurcations of stick-slip oscillations. *Physica D*, 136(3-4):280–302, February 2000.
65. H. Dankowicz, P. Piiroinen, and A. B. Nordmark. Low-velocity impacts of quasiperiodic oscillations. *Chaos Solitons & Fractals*, 14(2):241–255, August 2002.
66. H. Dankowicz and P.T. Piiroinen. Exploiting discontinuities for stabilization of recurrent motions. *Dynamical Systems*, 17:317–342, 2002.
67. J. de Weger, D. Binks, Molenaar. J., and W. van de Water. Generic behavior of grazing impact oscillators. *Phys. Rev. Lett.*, 76:3951–3954, 1996.
68. J. H. B. Deane and D. C. Hamill. Analysis, simulation and experimental study of chaos in the buck concerter. In *Proceedings of the Power Electronics Specialists Conf. (PESC 1990)*, pages 491–8, New York, 1990. IEEE Press.
69. K. Deimling. *Multivalued Differential Equations.* W de Gruyter, New York, 1992.
70. J. P. den Hartog. Forced Vibrations with combined Columb and viscous friction. *Transaction of the American Society of Mechanical Engineering*, 53:107–115, 1931.
71. A. J. Van der Schaft and J. M. Schumacher. *An introduction to Hybrid Dynamical Systems.* Springer - Verlag, 2000.
72. F. Dercole and Yu.A. Kuznetsov. User Guide to SlideCont 1.0. Technical report, Utrecht University, The Netherlands, 2003. `http://www.math.uu.nl/people/kuznet/cm/slidecont.pdf`.
73. R.L. Devaney. *An introduction to chaotic dynamical systems.* Addison-Wesley, Reading, Mass., second edition, 1989.
74. A. Dhooge, W. Govaerts, and Yu. A. Kuznetsov. MATCONT: A MATLAB package for numerical bifurcation analysis of ODEs. *ACM Trans. Math. Softw.*, 29(2):141–164, 2003.
75. M. di Bernardo, C. J. Budd, and A. R. Champneys. Grazing, skipping and sliding: analysis of the nonsmooth dynamics of the DC/DC buck converter. *Nonlinearity*, 11:858–890, 1998.
76. M. di Bernardo, C. J. Budd, and A. R. Champneys. Corner Collision implies Border–Collision Bifurcation. *Physica D*, 154:171–194, 2001.
77. M. di Bernardo, C. J. Budd, and A. R. Champneys. Corner-Collision implies Border-Collision Bifurcation. *Physica D*, 154:171–194, 2001.

78. M. di Bernardo, C. J. Budd, and A. R. Champneys. Grazing bifurcations in n-dimensional piecewise-smooth dynamical systems. *Physica D*, 160:222–254, 2001.
79. M. di Bernardo, C.J. Budd, A.R. Champneys, P.S. Kowalczyk, A. Nordmark, G. Olivar, and P.T. Piiroinen. Bifurcations in nonsmooth dynamical systems. *SIAM Review*, 2006. SIAM Review, to appear.
80. M. di Bernardo, M. I. Feigin, S. J. Hogan, and M. E. Homer. Local Analysis of C-bifurcations in n-dimensional piecewise smooth dynamical systems. *Chaos, Solitons and Fractals*, 10:1881–1908, 1999.
81. M. di Bernardo, E. Fossas, G. Olivar, and F. Vasca. Secondary Bifurcations and High Periodic Orbits in Voltage Controlled Buck Converter. *International Journal of Bifurcations and Chaos*, 7:2755–2771, 1997.
82. M. di Bernardo, F. Garofalo, L. Glielmo, and F. Vasca. Switchings, Bifurcations and Chaos in DC/DC converters. *IEEE Transactions on Circuits and Systems, Part I*, 45:133–141, 1998.
83. M. di Bernardo, F. Garofalo, L. Ianelli, and F. Vasca. Bifurcations in piecewise-smooth feedback systems. *International Journal of Control*, 75(16):1243–1259, 2002.
84. M. di Bernardo, K. H. Johansson, and F. Vasca. Self-Oscillations and Sliding in Relay Feedback Systems: Symmetry and Bifurcations. *International Journal of Bifurcations and Chaos*, 11(4):1121–1140, 2001.
85. M. di Bernardo, P. Kowalczyk, and A. Nordmark. Bifurcations of dynamical systems with sliding: derivation of normal-form mappings. *Physica D*, 170:175–205, 2002.
86. M. di Bernardo, P. Kowalczyk, and A. Nordmark. Sliding Bifurcations: A novel mechanism for the sudden onset of chaos in friction oscillators. *International Journal of Bifurcations and Chaos*, 13:2935–2948, 2003.
87. M. di Bernardo, A. Nordmark, and G. Olivar. Structural stability and bifurcations of equilibria in piecewise-smooth dynamical systems. submitted to *Physica D*, 2007.
88. E.J. Doedel, A.R. Champneys, T.F. Fairgrieve, Yu.A. Kuznetsov, B. Sandstede, and X. Wang. AUTO97: Continuation and bifurcation software for ordinary differential equations (with HomCont). Technical report, Concordia University, 1997.
89. P. G. Drazin. *Nonlinear systems*. Cambridge University Press, 1992.
90. M. Dutta, H. E. Nusse, E. Ott, J. A. Yorke, and G. H. Yuan. Multiple attractor bifurcations: A source of unpredictability in piecewise smooth systems. *Physical Review Letters*, 83(21):4281–4284, November 1999.
91. J. Emans, M. Wiercigroch, and A. M. Krivtsov. Cumulative effect of structural nonlinearities: chaotic dynamics of cantilever beam system with impacts. *Chaos Solitons & Fractals*, 23(5):1661–1670, March 2005.
92. R. D. Eyres, A. R. Champneys, and N. A. J. Lieven. Modelling and dynamic response of a damper with relief valve. *Nonlinear Dynamics*, 40(2):119–147, April 2005.
93. R. D. Eyres, P. T. Piiroinen, A. R. Champneys, and N. A. J. Lieven. Grazing bifurcations and chaos in the dynamics of a hydraulic damper with relief valves. *SIAM Journal On Applied Dynamical Systems*, 4(4):1076–1106, 2005.
94. J. Farey. On a curious property of vulgar fractions. *Philos. Mag. & Journal, London*, 47:385–386, 1816.

95. M. I. Feigin. Doubling of the oscillation period with C-bifurcations in piecewise continuous systems. *Journal of Applied Mathematics and Mechanics (Prikladnaya Matematika i Mechanika)*, 34:861–869, 1970.
96. M. I. Feigin. On the generation of sets of subharmonic modes in a piecewise continuous system. *Journal of Applied Mathematics and Mechanics (Prikladnaya Matematika i Mechanika)*, 38:810–818, 1974.
97. M. I. Feigin. On the structure of C-bifurcation boundaries of piecewise continuous systems. *Journal of Applied Mathematics and Mechanics (Prikladnaya Matematika i Mechanika)*, 42:820–829, 1978.
98. M. I. Feigin. *Forced Oscillations in systems with discontinuous nonlinearities.* Nauka, Moscow, 1994. [In Russian].
99. M. I. Feigin. The increasingly complex structure of the bifurcation tree of a piecewise-smooth system. *Journal of Applied Mathematics and Mechanics*, 59:853–863, 1995.
100. A. F. Filippov. *Differential Equations with Discontinuous Righthand Sides.* Kluwer Academic Publishers, Dortrecht, 1988.
101. I. Flugge-Lotz. *Discontinuous automatic control.* Princenton University Press, 1953.
102. S. Foale and S. R. Bishop. Dynamical complexities of forced impacting systems. *Phil. Trans. Roy. Soc. London A*, 338:547–556, 1992.
103. S. Foale and S. R. Bishop. Bifurcations in Impact Oscillators. *Nonlinear Dynamics*, 6:285–299, 1994.
104. E. Fossas and G. Olivar. Study of Chaos in the Buck Converter. *IEEE Transactions on Circuits and Systems - I: Fundamental Theory and Applications*, 43:13–25, 1996.
105. M H Fredriksson, D Borglund, and A B Nordmark. Experiments on the Onset of Impacting Motion Using a Pipe Conveying Fluid. *Nonlinear Dynamics*, 19:261–271, 1999.
106. M. H. Fredriksson and A. B. Nordmark. Bifuractions caused by grazing incidence in many degrees of freedom impact oscillators. *Proc. Royal Soc. London A*, 453:1261–1276, 1997.
107. M. H. Fredriksson and A. B. Nordmark. On normal form calculations in impact oscillators. *Proceedings of the Royal Society London A*, 456:315–329, 2000.
108. E. Freire, E. Ponce, F. Rodrigo, and F. Torres. Bifurcation sets of continuous piecewise linear systems with two zones. *International Journal of Bifurcation and Chaos*, 8(11):2073–2097, 1998.
109. L. M. Fridman and A. Levant. Higher order sliding modes as a natural phenomenon in control theory. In *Robust Control via Variable Structure and Lyapunov Techniques. Lecture notes in control and information science*, volume 217, pages 107–133. Springer-Verlag, 1996.
110. U. Galvanetto. Bifurcations and chaos in a four-dimensional mechanical system with dry-friction. *Journal of Sound and Vibration*, 204(4):690–695, 1997.
111. U. Galvanetto. Computational techniques for nonlinear dynamics in multiple friction oscillators. *Computer Methods in Applied Mechanics and Engineering*, 163:373–382, 1998.
112. U. Galvanetto. Nonlinear dynamics of multiple friction oscillators. *Computer methods in applied mechanics and engineering*, 178:291–306, 1999.
113. U. Galvanetto. Some Discontinuous Bifurcations in a two block stick-slip system. *Journal of Sound and Vibration*, 284(4):653–669, 2001.

114. U. Galvanetto and S. R. Bishop. Computational techniques for nonlinear dynamics in multiple friction oscillators. *Computer methods in applied mechanics and engineering*, 163:373–382, 1998.
115. A. Ganguli and S. Banerjee. Dangerous bifurcation at border collision: When does it occur? *Physical Review E*, 71(5):057202, May 2005.
116. R. Genesio and A. Tesi. Chaos Prediction in a Third order relay system. Internal Report RT 29/90. Dipartimento di Sistemi ed Informatica, University of Florence, Italy., 1990.
117. F. Giannakopoulos and K. Pliete. Qualitative analysis of a planar system of piece-wise linear differential equations with line of discontinuity. *Nonlinearity*, 14:1611–1632, 2001.
118. L. Glass and M.C. Mackey. A simple model for phase locking of biological oscillators. *Journal of Mathematical Biology*, 7:339–352, 1979.
119. M. Golubitsky and D. G. Schaeffer. *Singularities and Groups in Bifurcation Theory*, volume vol. 1. Springer-Verlag: New York, Heidelberg, Berlin, 1985.
120. M. Golubitsky, D.G. Schaeffer, and I.A. Stewart. *Singularities and Groups in Bifurcation Theory: Volumes 1 and 2*. Springer-Verlag, 1988.
121. W.J.F. Govaerts. *Numerical Methods for Bifurcations of Dynamical Equilibria*. SIAM, Philidelphia, 1987.
122. H. Goyder and C. Teh. A study of the impact dynamics of loosely supported heat exchanger tubes. *J. Pressure Vessel Technology*, 111:394–401, 1989.
123. T. Griffin, S.J. Hogan, G. Olivar, and V. Moreno. Dynamics of stochastic nonsmooth systems. In *Proceedings of the European Conference on Nonlinear Oscillations (ENOC)*, 2005.
124. J. Guckenheimer and P. Holmes. *Nonlinear Oscillations, Dynamical Systems, and Bifurcations of Vector Fields*. Springer–Verlag, New York, 1983. Applied Mathematical Sciences, Volume 42.
125. C. Halse, M. Homer, and M. di Bernardo. C-bifurcations and period-adding in one-dimensional piecewise-smooth maps. *Chaos, Solitons and Fractals*, 18:953–976, 2003.
126. C.K. Halse, R.E. Wilson, M. di Bernardo, and M. E. Homer. Coexisting solutions and bifurcations in mechanical oscillators with backlash. *International Journal of Sound and Vibration*, 2007. doi:10.1016/j.jsv.2007.05.010.
127. W. P. M. H. Heemels and B. Brogliato. The complementarity class of hybrid dynamical systems. *European Journal of Control*, 9(2-3):322–360, 2003.
128. F. Hendricks. Bounce and chaotic motion in impact print hammers. *IBM J.*, 27:273–280, 1983.
129. M. Henon. A two-dimensional mapping with a strange attractor. *Communications in Mathematical Physics*, 50:69–77, 1976.
130. N. Hinrichs, M. Oestreich, and K. Popp. Dynamics of oscillators with impact and friction. *Chaos Solitons & Fractals*, 8(4):535–558, April 1997.
131. N. Hinrichs, M. Oestreich, and K. Popp. On the modelling of friction oscillators. *Journal of Sound and Vibration*, 216(3):435–459, 1998.
132. N. Hinrichs, M. Oestreich, and K. Popp. Experiments, modelling and analysis of friction and impact oscillators. *Zeitschrift fuer Angewandte Mathematik und Mechanik*, 79:S95–S96, 1999.
133. S. J. Hogan. On the dynamics of rigid-block motion under harmonic forcing. *Proc. Roy. Soc. London A*, 425:441–476, 1989.
134. S. J. Hogan. Damping in rigid block dynamics contained between side walls. *Chaos, Solitons and Fractals*, 11:495–506, 2000.

135. S. J. Hogan and M. E. Homer. Graph theory and piecewise smooth dynamical systems of arbitrary dimension. *Chaos Solitons & Fractals*, 10:1869–1880, 1999.
136. S.J. Hogan. Effects of smoothing on bifurcation scenarios of piecewise-smooth continuous maps. Personal communication, 2005.
137. S.J. Hogan, L. Higham, and T.C.L. Griffin. Dynamics of a piecewise linear map with a gap. *Proceedings of the Royal Society A-Mathematical Physical and Engineering Sciences*, 463:49–65, 2007.
138. P. Holmes, R.J. Full, D. Koditschek, and J. Guckenheimer. Dynamics of legged locomotion: Models, analyses and challenges. *SIAM Review*, 48(2):207–304, 2006.
139. M.E. Homer, S. J. Hogan, M. di Bernardo, and C. Williams. The importance of choosing attractors for optimizing chaotic communications. *IEEE Trans. on Circuits and Systems II*, 51:511–516, 2004.
140. G.W. Housner, L.A. Bergman, T.K. Caughey, A.G. Chassiakos, R.O. Claus, S.F. Masri, R.E. Skelton, T.T. Soong, B.F. Spencer, and J.T.P. Yau. Structural engineering: past, present and future. *Journal of Engineering Mechanics*, 123:117–134, 1997.
141. G. Iooss and D.D. Joseph. *Elementary Stability and Bifurcation Theory.* Springer-Verlag, 1980.
142. A.P. Ivanov. Impact oscillations: linear theory of stability and bifurcations. *J. Sound Vib.*, 178:361–378, 1994.
143. P. Jain and S. Banerjee. Border-collision bifurcations in one-dimensional discontinuos maps. *International Journal of Bifurcations and Chaos*, 13:3341–3351, 2003.
144. K. H. Johansson. *Relay Feedback and Multivariable Control.* PhD thesis, Lund Institute of Technology, 1997.
145. K. H. Johansson, M. Egerstedt, J. Lygeros, and S. Sastry. On the regularization of Zeno hybrid automata. *Systems and Control Letters*, 38(3):141–150, October 1999.
146. A. Kaharaman and R. Singh. Nonlinear dynamics of a spur gear pair. *J. Sound Vib.*, 142:49–75, 1990.
147. A. Kahraman and G. Blankenship. Effect of Involute Contact Ratio on Spur Gear Dynamics. *Journal of Mechanical Design*, 121:112–118, 1999.
148. T. Kapitaniak and Y. Maistrenko. Riddling bifurcations in coupled piecewise linear maps. *Physica D-Nonlinear Phenomena*, 126(1-2):18–26, February 1999.
149. K. Karagiannis and F. Pfeiffer. Theoretical and Experimental Investigations of Gear-Rattling. *Nonlinear Dynamics*, 2:367–387, 1991.
150. J. G. Kassakian, M. F. Schlecht, and G. C. Verghese. *Principles of Power Electronics.* Addison-Wesley, 1991.
151. J. Keener and J. Sneyd. *Mathematical Physiology.* Springer-Verlag, 1998.
152. J.P. Keener. Chaotic behaviour in piecewise continuous difference equations. *Trans. American Mathematical Society*, 261:589–604, 1980.
153. H B Keller. *Lectures on Numerical Methods In Bifurcation Problems.* Springer-Verlag, 1987.
154. U. Kiencke and L. Nielsen. *Automotive Control Systems for Engine, Driveline and Vehicle.* Springer-Verlag, 2000.
155. A.A. Kilbas, H.M. Srivastava, and J.J. Trujillo. *Theory and Applications of Fractional Differential Equations.* Elsevier, 2006.
156. M. Kleczka, E. Kreuzer, and W. Schiehlen. Local and global stability of a piecewise linear oscillator. *Phil. Trans. Roy. Soc. Lond. A*, 338:533–546, 1992.

157. L. E. Kollar, G. Stepan, and J. Turi. Dynamics of piecewise linear discontinuous maps. *International Journal of Bifurcation and Chaos*, 14(7):2341–2351, July 2004.
158. P. Kowalczyk. Robust chaos and border-collision bifurcations in non-invertible piecewise-linear maps. *Nonlinearity*, 18:485–504, 2005. Submitted for publications to Nonlinearity.
159. P. Kowalczyk and M. di Bernardo. On a novel class of bifurcations in hybrid dynamical systems - the case of relay feedback systems. In *Proceedings of Hybrid Systems Computation and Control*, pages 361–374. Springer–Verlag, 2001.
160. P. Kowalczyk and M. di Bernardo. Two-parameter degenerate sliding bifurcations in Filippov systems. *Physica D-Nonlinear Phenomena*, 204(3-4):204–229, May 2005.
161. P. Kowalczyk, M. di Bernardo, A.R. Champneys, S.J. Hogan, M. Homer, Yu.A. Kuznetsovz, A. Nordmarkx, and P.T. Piiroinen. Two-parameter nonsmooth bifurcations of limit cycles: classification and open problems. *International Journal of Bifurcations and Chaos*, 6:601–629, 2006.
162. P. Kowalczyk and P. Piiroinen. Numerical continuation of sliding bifurcations in a friction oscillator. Personal Communication, 2006.
163. A. M. Krivtsov and M. Wiercigroch. Dry friction model of percussive drilling. *Meccanica*, 34(6):425–435, December 1999.
164. A. Kumar, S. Banerjee, and D. P. Lathrop. Dynamics of a piecewise smooth map with singularity. *Physics Letters A*, 337(1-2):87–92, March 2005.
165. M. Kunze. *Non-smooth Dynamical Systems*. Springer–Verlag, Berlin, 2000. Lecture Notes in Mathematics, Volume 1744.
166. M. Kunze and M.M. Marques. *Impacts in Mechanical Systems*. Springer-Verlag, 2000.
167. T. Küpper and S. Moritz. Generalized Hopf bifurcation for non-smooth planar systems. *Phil. Trans. Royal Society A*, 359:2483–2496, 2001.
168. Yu A. Kuznetsov. *Elements of Applied Bifurcation Theory (third edition)*. Springer–Verlag, New York, second edition, 2004. Applied Mathematical Sciences, Volume 112.
169. Yu. A. Kuznetsov, S. Rinaldi, and A. Gragnani. One-parameter bifurcations in planar Filippov systems. *International Journal of Bifurcations and Chaos*, 13(8):2157–2188, 2003.
170. H. Lamba. Chaotic, regular and unbounded behavior in the elastic impact oscillator. *Physica D*, 82(1-2):117–135, April 1995.
171. H. Lamba and C.J. Budd. Scaling of Lyapunov exponents at nonsmooth bifurcations. *Phys. Rev. E*, 20:84–91, 1994.
172. G.H. Lean. Subharmonic motions of moored ships due to wave action. *Trans. R. Naval Architects*, 113:387–399, 1971.
173. R Leine. *Bifurcations in Discontinuous Mechanical Systems of Filippov-Type*. PhD thesis, Teknische Universiteit Eindhoven, The Netherlands, 2000.
174. R. I. Leine, B. Brogliato, and H. Nijmeijer. Periodic motion and bifurcations induced by the Painleve paradox. *European Journal of Mechanics A-Solids*, 21(5):869–896, 2002.
175. R. I. Leine and D. H. van Campen. Discontinuous fold bifurcations in mechanical systems. *Archive of Applied Mechanics*, 72(2-3):138–146, April 2002.

176. R. I. Leine, D. H. van Campen, and B. L. van de Vrande. Bifurcations in nonlinear discontinuous systems. *Nonlinear Dynamics*, 23(2):105–164, October 2000.
177. R.I. Leine and H. Nijmeijer. *Dynamics and Bifurcations in Non-Smooth Mechanical Systems.* Springer-Verlag, 2004.
178. T.Y. Li and J.A. Yorke. Period three implies chaos. *Amer. Math. Monthly*, 82:985–992, 1975.
179. D. Liberzon. *Switchings in Systems and Control.* Birkhauser, 2003.
180. T. LoFaro. Period-adding bifurcations in a one parameter family of interval maps. *Mathematical and Computer Modelling*, 24:27–41, 1996.
181. A. C. J. Luo and R. P. S. Han. The dynamics of a bouncing ball with a sinusoidally vibrating table revisited. *Nonlinear Dynamics*, 10(1):1–18, 1996.
182. A. C. J. Luo and S. Menon. Global Chaos in a periodically forced, linear system with a dead-zone restoring force. *Chaos, solitons and fractals*, 19(5):1189–1199, 2004.
183. J. Lygeros, K. H. Johansson, S. N. Simic, J. Zhang, and S. S. Sastry. Dynamical properties of hybrid automata. *IEEE Transactions on Automatic Control*, 48(1):2–17, January 2003.
184. R.S. MacKay and J.D. Miess. *Hamiltonian Dynamical Systems.* Hilger, Bristol, 1987.
185. R. N. Madan, editor. *Chua's Circuit: A Paradigm for Chaos.* World Scientific, 1993.
186. I. A. Mahfouz and F. Badrakhan. Chaotic behaviour of some piecewise-linear systems. Part 2: systems with clearance. *International Journal of Sound and Vibration*, 143:289–328, 1990.
187. V. Maistrenko, Y. Maistrenko, and E. Mosekilde. Torus breakdown in noninvertible maps. *Physical Review E*, 67(4):046215, April 2003.
188. V. Maistrenko, Y. Maistrenko, and I. Sushko. Noninvertible 2-dimensional maps arising in radiophysics. *International Journal of Bifurcation And Chaos*, 4(2):383–400, April 1994.
189. Y. L. Maistrenko, V. L. Maistrenko, and S. I. Vikul. On period-adding sequences of attracting cycles in piecewise linear maps. *Chaos Solitons and Fractals*, 9(1-2):67–75, 1998.
190. C. Mira. About two-dimensional piecewise continuous non-invertible maps. *Int. J. Bifurc. Chaos*, 6:893–918, 1996.
191. C. Mira, C. Rauzy, Y. Maistrenko, and I.S. Sushko. Some properties of a two-dimensional piecewise linear non-invertible maps. *International Journal of Bifurcations and Chaos*, 6:2299–2320, 1996.
192. J. J. Moreau. Sur les mesures differentielles de fonctions vectorielles et certain problemes d'evolution. *C. R. Acad. Sci. Paris A*, 282:837–840, 1976.
193. E. Mosekilde and Zh. Zhusubalyev. *Bifurcations and Chaos in Piecewise-smooth dynamical systems.* World Scientific, 2003.
194. P.C. Mueller. Calculation of Lyapunov exponents for dynamical systems with discontinuities. *Chaos, Solitons & Fractals*, 5:1671–1681, 1995.
195. S. Natsiavas. Periodic Response and Stability of Oscillators with Symmetric Trilinear Restoring Force. *Journal of Sound and Vibration*, 134(2):315–331, 1989.
196. A. Nordmark. *Grazing conditions and chaos in impacting systems.* PhD thesis, Royal Institute of Technology, Stockholm, Sweden, 1992.

197. A. B. Nordmark. Non-periodic motion caused by grazing incidence in impact oscillators. *Journal of Sound and Vibration*, 2:279–297, 1991.
198. A. B. Nordmark. Universal Limit Mapping in Grazing Bifurcations. *Physical Review E*, 55:266–270, 1997.
199. A B Nordmark. Existence of priodic orbits in grazing bifurcations of impacting mechanical oscillators. *Nonlinearity*, 14:1517–1542, 2001.
200. A. B. Nordmark. Discontinuity mappings for vector fields with higher-order continuity, 2002. Personal communication.
201. A. B. Nordmark. Transitions from complete to incomplete chattering in impacting systems. personal communication, 2007.
202. A. B. Nordmark and P. Kowalczyk. A codimension-two scenario of sliding solutions in grazing-sliding bifurcations. *Nonlinearity*, 19(1):1–26, January 2006.
203. A.B. Nordmark and P.T. Piiroinen. Simulation and stability analysis of impacting systems with complete chattering. in preparation.
204. H. Nusse, E. Ott, and J. Yorke. Border collision bifurcations: an explanation for observed bifurcation phenomena. *Phys. Rev. E*, 49:1073–1076, 1994.
205. H. E. Nusse and J. A. Yorke. Border-Collision bifurcations including period 2 to period 3 for piecewise smooth systems. *Physica D*, 57(1-2):39–57, June 1992.
206. H. E. Nusse and J. A. Yorke. Border collision bifurcation: an explanation for observed bifurcation phenomena. *Physical Review E*, 49:1073–1076, 1994.
207. H. E. Nusse and J. A. Yorke. Border-collision bifurcations for piecewise-smooth one-dimensional maps. *International Journal of Bifurcations and Chaos*, 5(1):189–207, February 1995.
208. M. Oestreich, N. Hinrichs, and K. Popp. Bifurcation and stability analysis for a non-smooth friction oscillator. *Archive of Applied Mechanics*, 66(5):301–314, April 1996.
209. M. Oestreich, N. Hinrichs, K. Popp, and C. J. Budd. Analytical and Experimental Investigation of an Impact Oscillator. *Proceedings of the ASME 16th Biennal Conf. on Mech. Vibrations and Noise*, DETC97VIB-3907:1–11, 1997.
210. G. Osorio, M. Di Bernardo, and S. Santini. Chattering and complex behavior of a Cam-follower system. In *Proceedings of the European Nonlinear Dynamics Conference (ENOC)*, 2005. Eindhoven, The Netherlands.
211. G. Osorio, M. Di Bernardo, and S. Santini. Corner impact bifurcations: a novel class of discontinuity-induced bifurcations in cam-follower systems. *SIAM Journal on Applied Dynamical Systems*, 2007. to appear.
212. M.P. Paidoussis and G.X. Li. Cross-flow-induced chaotic vibrations of heat exchanger tubes impacting on loose supports. *J. Sound Vib.*, 152:305–326, 1992.
213. P. Painlevé. Sur les lois du frottement de glissement. *C.R Acad. Sci. Paris*, 121:112–115, 1905.
214. E. Pavlovskaia and M. Wiercigroch. Modelling of vibro-impact system driven by beat frequency. *International Journal of Mechanical Sciences*, 45(4):623–641, April 2003.
215. E. Pavlovskaia, M. Wiercigroch, and C. Grebogi. Modeling of an impact system with a drift. *Physical Review E*, 64:056224, 2001.
216. F. Peterka. Part 1: Theoretical analysis of n-multiple $(1/n)$-impact solutions. *CSAV Acta Technica*, 26(2):462–473, 1974.
217. F. Peterka. Results of analogue computer modelling of the motion. Part 2. *CSAV Acta Technica*, 19:569–580, 1974.

218. L. R. Petzold. Subroutine DDASSL. Technical report, Lawrence Livermore National Laboratory, 1989.
219. F. Pfeiffer and C. Glocker. *Multibody dynamics with unilateral constraints.* Wiley-VCH, Berlin, 1996.
220. P. T. Piiroinen, L. N. Virgin, and A. R. Champneys. Chaos and Period-Adding; Experimental and Numerical Verification of the Grazing Bifurcation. *Journal of Nonlinear Science*, 14:383–404, 2004.
221. P.T. Piiroinen, H.J. Dankowicz, and A.B. Nordmark. Breaking Symmetries and Constraints: Transitions from 2D to 3D in Passive Walkers. *Multibody System Dynamics*, 10:147–176, 2003.
222. P.T. Piiroinen and Yu. A. Kuznetsov. An event-driven method to simulate Filippov systems with accurate computing of sliding motions. *ACM Transactions on Mathematical Software*, 34(3):to appear, 2008.
223. K. Popp, N. Hinrichs, and M. Oestreich. Dynamical behaviour of friction oscillators with simultaneous self and external excitation. *Sadhana (Indian Academy of Sciences)*, 20:627–654, 1995.
224. K. Popp and P. Shelter. Stick–slip vibrations and chaos. *Philosophical Transactions of the Royal Society A*, 332(1624):89–105, 1990.
225. T. K. Pratt and R. Williams. Non-linear analysis for stick-slip motion. *Journal of Sound and Vibration*, 74:531–542, 1981.
226. S. X. Qu, S. Wu, and D. R. He. A multiple devils staircase in a discontinuous map. *Physics Letters A*, 231:152–158, 1997.
227. S. X. Qu, S. Wu, and D.R. He. Multiple devils staircase and type-V intermittency. *Physical Review E*, 57:402–411, 1998.
228. M. Renouf, F. Dubois, and P. Alart. A parallel version of the Non Smooth Contact Dynamics Algorithm applied to the simulation of granular media. *Journal of Computational and Applied Mathematics*, 168:375–382, 2004.
229. K. Sato, Sumio Yamamoto, Kazutaka Yokota, Katsutoshi Yoshida, Takanori Ohmori, Toshihiro Aoki, and Shu Karube. Chaotic Behaviour in Gear System. *JSME International Journal Series C*, 41(3):577–581, 1998.
230. M. Schatzman. A class of nonlinear differential equations of second order in time. *Nonlinear Analysis: Theory, Methods and Applications*, 2:355–373, 1978.
231. S. Scott. *Chemical Chaos.* Oxford University Press, 1994.
232. R. Seydel. *From equilibrium to chaos: practical stability analysis.* Springer–Verlag, New York, 1988.
233. R. Seydel. *Practical bifurcation and stability analysis: from equilibrium to chaos.* Springer, 1994.
234. M. Sharif-Bakhtiar and S. W. Shaw. The dynamic response of a centrifugal pendulum vibration absorber with motion-limiting stops. *Journal of Sound Vibration*, 126:221–235, 1988.
235. A.N. Sharkovskii. Coexistence of cycles of a continuous map of a line into itself. *Ukrainian Mathematical Journal*, 16:61–71, 1964.
236. S. W. Shaw and P. J. Holmes. Periodically forced linear oscillator with impacts: Chaos and long-period motions. *Phys. Rev. Lett.*, 51:623–626, 1983.
237. S. W. Shaw and P. J. Holmes. A periodically forced piecewise linear oscillator. *J. Sound and Vibration*, 90:129–144, 1983.
238. S. W. Shaw and P. J. Holmes. A periodically forced piecewise linear oscillator with large dissipation. *ASME J. Appl. Math.*, 50:849–857, 1983.

239. S. N. Simic, K. H. Johansson, J. Lygeros, and S. Sastry. Towards a geometric theory of hybrid systems. *Dynamics of continuous discrete and impulsive systems – B*, 12(5-6):649–687, October 2005.
240. E. Sontag. *Mathematical Control Theory: Deterministic Finite Dimensional Systems*. Springer-Verlag, 1998.
241. C.T. Sparrow. Chaos in A 3-Dimensional Single Loop Feedback-System with a Piecewise Linear Feedback Function. *Journal of Mathematical Analysis And Applications*, 83:275–291, 1981.
242. J.C. Sprott. *Chaos and Time-series Analysis*. Oxford University Press, 2003.
243. D.E. Stewart. Rigid-body dynamics with friction and impacts. *SIAM Review*, 42:3–39, 2000.
244. C. Surace, K. Worden, and G.R. Tomlinson. An improved nonllinear model for an automotive shock absorber. *Nonlinear Dynamics*, 3:413–429, 1992.
245. C. Surace, K. Worden, and G.R. Tomlinson. On the nonlinear characteristics of automotive shock absorbers. *Proceedings of the Institute of Mechanical Engineers Part D: Journal of Automobile Engineering*, 206:1–16, 1992.
246. I. Sushko, A. Agliari, and L. Gardini. Bistability and border-collision bifurcations for a family of unimodal piecewise smooth maps. *Discrete and Continuous Dynamical Systems-Series B*, 5(3):881–897, August 2005.
247. I. Sushko, A. Agliari, and L. Gardini. Bifurcation structure of parameter plane for a family of unimodal piecewise smooth maps: Border-collision bifurcation curves. *Chaos Solitons & Fractals*, 29(3):756–770, August 2006.
248. M.A. Teixera. Generic bifurcation of sliding vector fields. *Journal of Mathematical Analysis and Applications*, 176:436–457, 1993.
249. S. Theodossiades and S. Natsiavas. Non-linear Dynamics of Gear-Pair Systems with Periodic Stiffness and Backlash. *Journal of Sound and Vibration*, 229(2):287–310, 2000.
250. S. Theodossiades and S. Natsiavas. Periodic and Chaotic Dynamics of Motor-driven Gear-pair Systems with Backlash. *Chaos, Solitons and Fractals*, 12, 2427-2440 2001.
251. J. M. T. Thompson and R. Ghaffari. Chaotic dynamics of an impact oscillator. *Phys. Rev. A*, 27:1741–1743, 1983.
252. J.M.T. Thompson, A.R. Bokaian, and R. Ghaffari. Subharmonic responses and chaotic motions of a bilinear oscillator. *IMA J. Appl. Math.*, 31:207–234, 1983.
253. J.M.T. Thompson, A.R. Bokaian, and R. Ghaffari. Subharmonic and chaotic motions of compliant offshore structures and articulated mooring towers. *ASME J. Energy Resources Tech.*, 106:191–198, 1984.
254. M. Toda. Wave Propagation in Anharmonic Lattices. *Journal of the Physical Society of Japan*, 23:501–506, 1967.
255. Ya. Z. Tsypkin. *Relay Control Systems*. Cambridge University Press, Cambridge, U.K., 1984.
256. P.C. Tung and S.W. Shaw. A method for the improvement of impact printer performence. *J. Vib. Acoustics Stress Reliability Des.*, 110:528–532, 1988.
257. V. I. Utkin. *Sliding Modes in Control Optimization*. Springer-Verlag, Berlin, 1992.
258. S. Varigonda and T. Georgiou. Dynamics of relay relaxation oscillators. *IEEE Transactions on Automatic Control*, 46:65–77, 2001.
259. D. J. Wagg. Multiple nonsmooth events in multi-degree-of-freedom vibro-impact systems. *Nonlinear Dynamics*, 43:137–148, 2006.

260. D.J. Wagg and S.R. Bishop. Application of non-smooth modelling techniques to the dynamics of a flexible impacting beam. *J. Sound Vib.*, 256:803–820, 2002.
261. J. Wallaschek. Dynamics of nonlinear automobile shock absorbers. *International Journal of Non-Linear Mechanics*, 25:299–308, 1990.
262. J. Warminski, G. Litak, and K. Szabelski. Dynamic phenomena in gear boxes. In *Applied nonlinear dynamics and chaos of mechanical systems with discontinuities.* World Scientific, 2000.
263. G. S. Whiston. Global dynamics of a vibro impacting linear oscillator. *Journal of Sound and Vibrations*, 118:395–429, 1987.
264. G. S. Whiston. The vibro-impact response of a harmonically excited and preloaded one-dimensional linear oscillator. *J. Sound Vib.*, 115:303–324, 1987.
265. G.S. Whiston. Singularities in vibro-impact dynamics. *J. Sound Vib.*, 152:427–460, 1992.
266. M. Wiercigroch. Bifurcation-analysis of harmonically excited linear-oscillator with clearance. *Chaos solitons & fractals*, 4(2):297–303, February 1994.
267. M. Wiercigroch and E. Budak. Sources of nonlinearities, chatter generation and suppression in metal cutting. *Philosophical Transactions of The Royal Society Of London Series A*, 359(1781):663–693, April 2001.
268. M. Wiercigroch and B. de Kraker. *Applied Nonlinear Dynamics and Chaos of Mechanical Systems with Discontinuities.* World Scientific Publishing, 2000.
269. M. Wiercigroch, R.D. Neilson, and M.A. Player. Material removal rate prediction for ultrasonic drilling of hard materials using an impact oscillator approach. *Phys. Lett. A*, 259:91–96, 1999.
270. M. Wiercigroch and V. W. T. Sin. Experimental study of a symmetrical piecewise base-excited oscillator. *Journal of Applied Mechanics-Transactions of the ASME*, 65(3):657–663, September 1998.
271. S. Wiggins. *Global bifurcations and Chaos: analytical methods.* Springer-Verlag, 1988.
272. S. Wiggins. *Normally Hyperbolic Invariant Manifolds in Dynamical Systems.* Springer-Verlag, 1994.
273. S. Wiggins. *Introduction to Applied Nonlinear Dynamical Systems and Chaos (2nd edition).* Springer-Verlag, 2003.
274. R. E. Wilson and A. R. Champneys. Mathematical modelling of vibrations of the HMB1800 Booster. Technical report, BOC Edwards, 2003.
275. A.T. Winfree. Electrical turbulence in three-dimensional heart muscle. *Science*, 266:1003–1006, 1994.
276. L.A. Wood and K.P. Byrne. Experimental investigations of a random repeated impact process. *J. Sound Vib.*, 85:53–69, 1981.
277. Y. Yoshitake and A. Sueoka. Forced Self-Excited Vibration with Dry Friction. In Marian Wiercigroch and Bram de Kraker, editors, *Applied nonlinear dynamics and chaos of mechanical systems with discontinuities*, pages 237–259. World Scientific, 2000.
278. G. Yuan, S. Banerjee, E. Ott, and J.A. Yorke. Border-collision bifurcations in the buck converter. *IEEE Transactions on Circuits and Systems-I*, 45:707–716, 1998.
279. G. Zhou and S.R. Reid, editors. *Impact on Composites.* Woodhead Publishers, 2000.

280. Z. T. Zhusubaliyev and E. Mosekilde. Birth of bilayered torus and torus breakdown in a piecewise-smooth dynamical system. *Physics Letters A*, 351(3):167–174, February 2006.
281. Z. T. Zhusubaliyev, E. A. Soukhoterin, and E. Mosekilde. Border-collision bifurcations on a two-dimensional torus. *Chaos Solitons & Fractals*, 13(9):1889–1915, July 2002.
282. Z. T. Zhusubaliyev, E. A. Soukhoterin, and E. Mosekilde. Quasi-periodicity and border-collision bifurcations in a DC-DC converter with pulsewidth modulation. *IEEE Transactions On Circuits And Systems I*, 50(8):1047–1057, August 2003.
283. Y. Zou and T. Küpper. Generalized Hopf bifurcation for non-smooth planar dynamical systems: the corner case. *Northeastern Mathematical Journal*, 17:383–386, 2001.
284. Y. Zou and T. Küpper. Hopf bifurcation for non-smooth planar dynamical systems. *Northeast. Math. Journal*, 17:261–264, 2001.
285. Y. Zou, T. Küpper, and W. J. Beyn. Generalized hopf bifurcation for planar filippov systems continuous at the origin. *Journal of Nonlinear Science*, 16(2):159–177, April 2006.
286. Y. K. Zou and T. Küpper. Generalized hopf bifurcation emanated from a corner for piecewise smooth planar systems. *Nonlinear Analysis-Theory Methods & Applications*, 62(1):1–17, July 2005.

Index

Applied Mathematical Sciences

(continued from page ii)

61. *Sattinger/Weaver:* Lie Groups and Algebras with Applications to Physics, Geometry, and Mechanics.
62. *LaSalle:* The Stability and Control of Discrete Processes.
63. *Grasman:* Asymptotic Methods of Relaxation Oscillations and Applications.
64. *Hsu:* Cell-to-Cell Mapping: A Method of Global Analysis for Nonlinear Systems.
65. *Rand/Armbruster:* Perturbation Methods, Bifurcation Theory and Computer Algebra.
66. *Hlavácek/Haslinger/Necasl/Lovísek:* Solution of Variational Inequalities in Mechanics.
67. *Cercignani:* The Boltzmann Equation and Its Application.
68. *Temam:* Infinite Dimensional Dynamical Systems in Mechanics and Physics, 2nd ed.
69. *Golubitsky/Stewart/Schaeffer:* Singularities and Groups in Bifurcation Theory, Vol. II.
70. *Constantin/Foias/Nicolaenko/Temam:* Integral Manifolds and Inertial Manifolds for Dissipative Partial Differential Equations.
71. *Catlin:* Estimation, Control, and The Discrete Kalman Filter.
72. *Lochak/Meunier:* Multiphase Averaging for Classical Systems.
73. *Wiggins:* Global Bifurcations and Chaos.
74. Mawhin/Willem: Critical Point Theory and Hamiltonian Systems.
75. *Abraham/Marsden/Ratiu:* Manifolds, Tensor Analysis, and Applications, 2nd ed.
76. *Lagerstrom:* Matched Asymptotic Expansions: Ideas and Techniques.
77. *Aldous:* Probability Approximations via the Poisson Clumping Heuristic.
78. *Dacorogna:* Direct Methods in the Calculus of Variations.
79. *Hernández-Lerma:* Adaptive Markov Processes.
80. *Lawden:* Elliptic Functions and Applications.
81. *Bluman/Kumei:* Symmetries and Differential Equations.
82. *Kress:* Linear Integral Equations, 2nd ed.
83. *Bebernes/Eberly:* Mathematical Problems from Combustion Theory.
84. *Joseph:* Fluid Dynamics of Viscoelastic Fluids.
85. *Yang:* Wave Packets and Their Bifurcations in Geophysical Fluid Dynamics.
86. *Dendrinos/Sonis:* Chaos and Socio-Spatial Dynamics.
87. *Weder:* Spectral and Scattering Theory for wave Propagation in Perturbed Stratified Media.
88. *Bogaevski/Povzner:* Algebraic Methods in Nonlinear Perturbation Theory.
89. *O'Malley:* Singular Perturbation Methods for Ordinary Differential Equations.
90. *Meyer/Hall:* Introduction to Hamiltonian Dynamical Systems and the N-body Problem.
91. *Straughan:* The Energy Method, Stability, and Nonlinear Convection, 2nd ed.
92. *Naber:* The Geometry of Minkowski Spacetime.
93. *Colton/Kress:* Inverse Acoustic and Electromagnetic Scattering Theory, 2nd ed.
94. *Hoppensteadt:* Analysis and Simulation of Chaotic Systems.
95. *Hackbusch:* Iterative Solution of Large Sparse Systems of Equations.
96. *Marchioro/Pulvirenti:* Mathematical Theory of Incompressible Nonviscous Fluids.
97. *Lasota/Mackey:* Chaos, Fractals, and Noise: Stochastic Aspects of Dynamics, 2nd ed.
98. *de Boor/Höllig/Riemenschneider:* Box Splines.
99. *Hale/Lunel:* Introduction to Functional Differential Equations.
100. *Sirovich (ed):* Trends and Perspectives in Applied Mathematics.
101. *Nusse/Yorke:* Dynamics: Numerical Explorations, 2nd ed.
102. *Chossat/Iooss:* The Couette-Taylor Problem.
103. *Chorin:* Vorticity and Turbulence.
104. *Farkas:* Periodic Motions.
105. *Wiggins:* Normally Hyperbolic Invariant Manifolds in Dynamical Systems.
106. *Cercignani/Ilner/Pulvirenti:* The Mathematical Theory of Dilute Gases.
107. *Antman:* Nonlinear Problems of Elasticity, 2nd ed.
108. *Zeidler:* Applied Functional Analysis: Applications to Mathematical Physics.
109. *Zeidler:* Applied Functional Analysis: Main Principles and Their Applications.
110. *Diekman/van Gils/Verduyn Lunel/Walther:* Delay Equations: Functional-, Complex-, and Nonlinear Analysis.
111. *Visintin:* Differential Models of Hysteresis.
112. *Kuznetsov:* Elements of Applied Bifurcation Theory, 3rd ed.
113. *Hislop/Sigal:* Introduction to Spectral Theory.
114. *Kevorkian/Cole:* Multiple Scale and Singular Perturbation Methods.
115. *Taylor:* Partial Differential Equations I, Basic Theory.
116. *Taylor:* Partial Differential Equations II, Qualitative Studies of Linear Equations.
117. *Taylor:* Partial Differential Equations III, Nonlinear Equations.
118. *Godlewski/Raviart:* Numerical Approximation of Hyperbolic Systems of Conservation Laws.
119. *Wu:* Theory and Applications of Partial Functional Differential Equations.
120. *Kirsch:* An Introduction to the Mathematical Theory of Inverse Problems.
121. *Brokate/Sprekels:* Hysteresis and Phase Transitions.
122. *Gliklikh:* Global Analysis in Mathematical Physics: Geometric and Stochastic Methods.

(continued on next page)

Applied Mathematical Sciences

(continued from previous page)

123. *Khoi Le/Schmitt:* Global Bifurcation in Variational Inequalities: Applications to Obstacle and Unilateral Problems.
124. *Polak:* Optimization: Algorithms and Consistent Approximations.
125. *Arnold/Khesin:* Topological Methods in Hydrodynamics.
126. *Hoppensteadt/Izhikevich:* Weakly Connected Neural Networks.
127. *Isakov:* Inverse Problems for Partial Differential Equations, 2nd ed.
128. *Li/Wiggins:* Invariant Manifolds and Fibrations for Perturbed Nonlinear Schrödinger Equations.
129. *Müller:* Analysis of Spherical Symmetries in Euclidean Spaces.
130. *Feintuch:* Robust Control Theory in Hilbert Space.
131. *Ericksen:* Introduction to the Thermodynamics of Solids, Revised Edition.
132. *Ihlenburg:* Finite Element Analysis of Acoustic Scattering.
133. *Vorovich:* Nonlinear Theory of Shallow Shells.
134. *Vein/Dale:* Determinants and Their Applications in Mathematical Physics.
135. *Drew/Passman:* Theory of Multicomponent Fluids.
136. *Cioranescu/Saint Jean Paulin:* Homogenization of Reticulated Structures.
137. *Gurtin:* Configurational Forces as Basic Concepts of Continuum Physics.
138. *Haller:* Chaos Near Resonance.
139. *Sulem/Sulem:* The Nonlinear Schrödinger Equation: Self-Focusing and Wave Collapse.
140. *Cherkaev:* Variational Methods for Structural Optimization.
141. *Naber:* Topology, Geometry, and Gauge Fields: Interactions.
142. *Schmid/Henningson:* Stability and Transition in Shear Flows.
143. *Sell/You:* Dynamics of Evolutionary Equations.
144. *Nédélec:* Acoustic and Electromagnetic Equations: Integral Representations for Harmonic Problems.
145. *Newton:* The N-Vortex Problem: Analytical Techniques.
146. *Allaire:* Shape Optimization by the Homogenization Method.
147. *Aubert/Kornprobst:* Mathematical Problems in Image Processing: Partial Differential Equations and the Calculus of Variations.
148. *Peyret:* Spectral Methods for Incompressible Viscous Flow.
149. *Ikeda/Murota:* Imperfect Bifurcation in Structures and Materials.
150. *Skorokhod/Hoppensteadt/Salehi:* Random Perturbation Methods with Applications in Science and Engineering.
151. *Bensoussan/Frehse:* Regularity Results for Nonlinear Elliptic Systems and Applications.
152. *Holden/Risebro:* Front Tracking for Hyperbolic Conservation Laws.
153. *Osher/Fedkiw:* Level Set Methods and Dynamic Implicit Surfaces.
154. *Bluman/Anco:* Symmetries and Integration Methods for Differential Equations.
155. *Chalmond:* Modeling and Inverse Problems in Image Analysis.
156. *Kielhöfer:* Bifurcation Theory: An Introduction with Applications to PDEs.
157. *Kaczynski/Mischaikow/Mrozek:* Computational Homology.
158. *Oertel:* Prandtl–Essentials of Fluid Mechanics, 10th Revised Edition.
159. *Ern:* Theory and Practice of Finite Elements.
160. *Kaipio:* Statistical and Computational Inverse Problems.
161. *Ting:* Viscous Vortical Flows II.
162. *Ammari/Kang:* Polarization and Moment Tensors: With Applications to Inverse Problems and Effective Medium Theory.
163. *Bernardo/Budd/Champneys/Kowalczyk:* Piecewise-smooth Dynamical Systems: Theory and Applications.

Printed in the United States of America